Alberto Guzman

Derivatives and Integrals of Multivariable Functions

Birkhäuser
Boston • Basel • Berlin

Alberto Guzman
The City College of New York, CUNY
Department of Mathematics
New York, NY 10031
U.S.A.

Library of Congress Cataloging-in-Publication Data

A CIP record for this book is available from the Library of Congress, Washington D.C., U.S.A.

AMS Subject Classifications: 26-01, 26B10, 26B12, 26B15, 26B20

Printed on acid-free paper

Birkhäuser

ISBN 0-8176-4274-9 SPIN 10851356
ISBN 3-7643-4274-9

Typeset by TEXniques, Inc., Cambridge, MA.
Printed in the United States of America.

9 8 7 6 5 4 3 2 1

A member of BertelsmannSpringer Science+Business Media GmbH

To the fathers of my intellect:

Richard Beals

Arnold Darrow

Gerald Freilich

Alvin Hausner

Preface

This text is appropriate for a one-semester course in what is usually called advanced calculus of several variables. The approach taken here extends elementary results about derivatives and integrals of single-variable functions to functions in several-variable Euclidean space. The elementary material in the single- and several-variable case leads naturally to significant advanced theorems about functions of multiple variables.

In the first three chapters, differentiability and derivatives are defined; properties of derivatives reducible to the scalar, real-valued case are discussed; and two results from the vector case, important to the theoretical development of curves and surfaces, are presented. The next three chapters proceed analogously through the development of integration theory. Integrals and integrability are defined; properties of integrals of scalar functions are discussed; and results about scalar integrals of vector functions are presented. The development of these latter theorems, the vector-field theorems, brings together a number of results from other chapters and emphasizes the physical applications of the theory.

We presuppose that the reader is familiar with the properties of continuous functions and with the topology of Euclidean space. Some references will be made to the author's previous book, *Continuous Functions of Vector Variables*, in which continuity and topology are examined in the context of normed linear spaces. Ideally, the two texts can be used sequentially for a year-long introduction to multivariable advanced calculus. To understand the present book, however, it suffices to have familiarity with the basic theory of the real line. Thus, the reader with a good understanding of the properties of continuous single-variable functions and knowledge of the topology of real numbers should be able to extrapolate to higher dimensions.

Considerable background in linear algebra is needed, particularly for the study of derivatives. We assume that the reader is conversant with vector spaces, subspaces, linear combinations, and the notions of linear span, linear independence, basis, and dimension. Regarding linear maps, we only use the definition, a few basic properties, and the concept of the matrix representation of a linear map. As for matrices themselves, we use a great deal more: definitions and characteristics of inverses, determinants, adjoints, and rank, and the applications of these ideas to the solution of systems of linear equations.

The text constitutes a sequence of tightly interrelated topics. Nevertheless, for a course shorter than a full semester, it is possible to modify what is presented, yet still maintain logical flow. Sections 1.1–1.3 and 5.1–5.3 and Chapters 2, 4, and 6 should be covered in their entirety. These are essential for two reasons: first, the relevant results and proofs reveal the nature of multivariable calculus and demonstrate why the subject cannot simply be reduced to the study of one variable at a time; and second, the geometric or physical interpretations of the theory are straightforward and have a wide variety of applications. The latter part of Section 1.4, on third and higher-order derivatives, may be skipped. Chapter 3 and Sections 5.4–5.6 contain several important results, but painstaking examination of the proofs is extreme effort for modest reward. For example, the reader should know the statement of the implicit function theorem, especially the significance of Jacobians, but he or she need not track the full chain of logic in the proof. Similarly, the characterizations of integrals in Sections 5.4–5.6 are indispensable for Chapter 6, but a complete understanding of all of their proofs is not necessary.

I am grateful to the people at Birkhäuser for turning the class notes I started in 1995 into both this work and its predecessor. I express special appreciation to Elizabeth Loew for her long and hard work in producing the text, and to Ann Kostant for believing that my mathematical and pedagogical viewpoint would contribute to a worthwhile book.

Alberto Guzman
July, 2003

Contents

Derivatives and Integrals of Multivariable Functions

1
Differentiability of Multivariable Functions

In this chapter we use ideas based on the elementary definition of derivative to define derivatives of a vector function of a Euclidean vector variable. We relate these new derivatives to the elementary ones, and generalize an important result from the real-variable case.

1.1 Differentiability

To define differentiability with respect to a vector variable, it helps first to review the definition of derivative of a function of a scalar variable.

Suppose $f(t)$ is defined at $t = a$. If the limit $\lim_{t\to a} \frac{f(t)-f(a)}{(t-a)}$ exists, then we say that f is differentiable at $t = a$ and refer to the limit as the derivative of f. Thus, the derivative comes from calculating the changes in t and f, forming their quotient, and taking the limit of the quotient.

Suppose now that $\mathbf{g}$ is a function mapping a domain D from $\mathbf{R}^n$ to $\mathbf{R}^m$, and $\mathbf{b} \in D$. The changes from $\mathbf{b}$ to $\mathbf{x}$ and from $\mathbf{g}(\mathbf{b})$ to $\mathbf{g}(\mathbf{x})$ are merely vector subtractions; we know how to do those. Limits of vectors are also familiar. What we cannot mimic from the scalar case is the division of $\mathbf{g}(\mathbf{x}) - \mathbf{g}(\mathbf{b})$ by $\mathbf{x} - \mathbf{b}$. Consequently, we need to characterize derivatives in a way that does not require division by the change in the independent variable.

The equation $\lim_{t\to a} \frac{f(t)-f(a)}{(t-a)} = f'(a)$ is equivalent to

$$\lim_{t\to a} \frac{f(t) - f(a) - f'(a)[t - a]}{t - a} = 0.$$

The latter says, in part, that $f(t) \approx f(a) + f'(a)[t-a]$. We call this expression the **first-degree approximation to $f(t)$ near $t = a$**. What is important for us is that $f(a) + f'(a)[t-a]$ is not just close to $f(t)$. It is so close that the difference is small *even in comparison with the small quantity $t - a$*.

It is the last statement that completely describes $f'(a)$. If f is continuous, then

$$f(t) \approx f(a) + k[t-a]$$

holds near $t = a$ for every constant k. However,

$$\frac{f(t) - f(a) - k[t-a]}{t-a} \approx 0$$

holds iff $f'(a)$ exists and $k = f'(a)$.

With this characterization in mind, we make the following definition.

Definition. Assume that there exists a linear transformation $\mathbf{L}\colon \mathbf{R}^n \to \mathbf{R}^m$ such that

$$\lim_{\mathbf{x}\to\mathbf{b}} \frac{\mathbf{g}(\mathbf{x}) - \mathbf{g}(\mathbf{b}) - \mathbf{L}\langle\mathbf{x}-\mathbf{b}\rangle}{\|\mathbf{x}-\mathbf{b}\|} = \mathbf{O}.$$

Then we say **g is differentiable at b**, we call **L** the **derivative** of **g** at **b**, and we write $\mathbf{L} = \mathbf{g}'(\mathbf{b})$.

We will be consistent in using $\mathbf{L}\langle\mathbf{v}\rangle$ to denote the image of $\mathbf{v}$ under the mapping $\mathbf{L}$. This notation distinguishes the dependence of $\mathbf{L} = \mathbf{g}'(\mathbf{b})$ on $\mathbf{b}$, which is not linear in general, from the dependence of $\mathbf{L}\langle\mathbf{v}\rangle = \mathbf{g}'(\mathbf{b})\langle\mathbf{v}\rangle$ on $\mathbf{v}$, which is required to be linear. The notation also suggests multiplication, which is how we will normally calculate $\mathbf{L}\langle\mathbf{v}\rangle$. Right now, we have $\mathbf{L}$ mapping $\mathbf{R}^n$ into $\mathbf{R}^m$. Such a map can always be represented by a matrix, $\mathbf{L} = \begin{bmatrix} a_{11} & \cdots & a_{1n} \\ & \ddots & \\ a_{m1} & \cdots & a_{mn} \end{bmatrix}$. Write $\langle\mathbf{v}\rangle \equiv \begin{bmatrix} v_1 \\ \vdots \\ v_n \end{bmatrix}$ for the column matrix identified with the vector $\mathbf{v} = (v_1, \ldots, v_n)$; we will make no distinction between those two objects. Then the image of $\mathbf{v}$ is a matrix product,

$$\mathbf{L}\langle\mathbf{v}\rangle = \begin{bmatrix} a_{11} & \cdots & a_{1n} \\ & \ddots & \\ a_{m1} & \cdots & a_{mn} \end{bmatrix} \begin{bmatrix} v_1 \\ \vdots \\ v_n \end{bmatrix} = \begin{bmatrix} a_{11}v_1 + \cdots + a_{1n}v_n \\ \ddots \\ a_{m1}v_1 + \cdots + a_{mn}v_n \end{bmatrix},$$

where we view the last column as identical to $\left(a_{11}v_1 + \cdots + a_{1n}v_n, \ldots, a_{m1}v_1 + \cdots + a_{mn}v_n\right)$.

Throughout this book, we will use $\|\mathbf{x}\|$ to denote the Pythagorean norm

$$\|\mathbf{x}\| = \left\|(x_1, \ldots, x_n)\right\| \equiv \left(x_1^2 + \cdots + x_n^2\right)^{1/2}.$$

It will frequently be clear that what we write would make sense under other norms, and even in infinite dimensional spaces. Such is the case with the definition, at least if $\mathbf{L}$ is assumed to be continuous. Nevertheless, we cast everything in the language of our normal habitat, hereby declared to be Euclidean space.

Under our definition, the derivative of a vector function is also a vector function, but of a different character. Thus, $\mathbf{g}(\mathbf{b})$ is a vector in $\mathbf{R}^m$. The derivative $\mathbf{L} = \mathbf{g}'(\mathbf{b})$ is a map from $\mathbf{R}^n$ to $\mathbf{R}^m$. We could view such a map as a member of $\mathbf{R}^{nm}$. Instead, as we said above, we will follow linear algebra's practice and associate $\mathbf{g}'(\mathbf{b})$ with the matrix that represents it relative to the standard bases in $\mathbf{R}^n$ and $\mathbf{R}^m$. With this viewpoint, we see $\mathbf{g}'(\mathbf{b})$ as a member of the vector space M_{mn} of $m \times n$ matrices.

Our interpretation may seem distant from the one-variable case, but the two are easy to reconcile. Because the only linear maps on $\mathbf{R}^1$ are the scalar multiplications, to say that $\frac{f(t)-f(a)-f'(a)[t-a]}{t-a} \approx 0$ is to say that $f(t) - f(a)$ is closely approximated by the image of $t - a$ under the linear transformation $\mathbf{L}$ defined by $\mathbf{L}\langle s\rangle \equiv f'(a)s$.

Example 1. (a) Let $F(x, y) \equiv x^2 + 3y^2$. To determine F' at $\mathbf{c} \equiv (a, b)$, we first note that

$$\begin{aligned} F(\mathbf{x}) - F(\mathbf{c}) &= x^2 + 3y^2 - a^2 - 3b^2 = (x + a)(x - a) + (3y + 3b)(y - b) \\ &= [x + a \quad 3y + 3b] \begin{bmatrix} x - a \\ y - b \end{bmatrix}. \end{aligned}$$

The last expression is a product of matrices, but we make no distinction between the matrix $[z]$ and the real number z. If $(x, y) \approx (a, b)$, then $x + a \approx 2a$ and $3y + 3b \approx 6b$. Therefore, we approximate

$$F(\mathbf{x}) - F(\mathbf{c}) \approx \begin{bmatrix} 2a & 6b \end{bmatrix} \begin{bmatrix} x - a \\ y - b \end{bmatrix} = \mathbf{L}\langle (x, y) - (a, b)\rangle,$$

where $\mathbf{L} \equiv \begin{bmatrix} 2a & 6b \end{bmatrix}$. The last approximation suggests that $\mathbf{L}$ is the derivative of F at $\mathbf{c}$.

To give proof, we must show that $\frac{F(\mathbf{x})-F(\mathbf{c})-\mathbf{L}\langle\mathbf{x}-\mathbf{c}\rangle}{\|\mathbf{x}-\mathbf{c}\|}$ has zero limit. We obtain

$$\begin{aligned} F(\mathbf{x}) - F(\mathbf{c}) - \mathbf{L}\langle \mathbf{x} - \mathbf{c}\rangle &= x^2 + 3y^2 - a^2 - 3b^2 - 2a(x - a) - 6b(y - b) \\ &= (x - a)^2 + 3(y - b)^2. \end{aligned}$$

In this last expression, each term is small compared to $\|\mathbf{x} - \mathbf{c}\|$. For example,

$$\frac{(x - a)^2}{\|\mathbf{x} - \mathbf{c}\|} = |x - a| \frac{|x - a|}{\left[(x - a)^2 + (y - b)^2\right]^{1/2}} \leq |x - a|.$$

(We should remember the relationship

$$\frac{|x - a|}{\left[(x - a)^2 + (y - b)^2\right]^{1/2}} \leq 1,$$

and the corresponding one for $y - b$. It is obvious from algebra, but it is also a geometric statement: The fraction is the cosine of the angle between $\mathbf{x} - \mathbf{c}$ and the x-axis.) Thus,

$$\frac{|F(\mathbf{x}) - F(\mathbf{c}) - \mathbf{L}\langle \mathbf{x} - \mathbf{c}\rangle|}{\|\mathbf{x} - \mathbf{c}\|} \le |x - a| + |y - b|,$$

which goes to zero as $(x, y) \to (a, b)$. Therefore, as required,

$$\frac{F(\mathbf{x}) - F(\mathbf{c}) - \mathbf{L}\langle \mathbf{x} - \mathbf{c}\rangle}{\|\mathbf{x} - \mathbf{c}\|} \to \mathbf{O} \qquad \text{when } (x, y) \to (a, b).$$

(b) Suppose $\mathbf{G}(x, y) \equiv \left(x^2 + 3y^2, e^{x+y}\right)$. Again we examine

$$\mathbf{G}(\mathbf{x}) - \mathbf{G}(\mathbf{c}) = \left([(x + a)(x - a) + (3y + 3b)(y - b)], [e^{x+y} - e^{a+b}]\right).$$

The first component looks the same as in (a). What do we do for the second?

By the mean value theorem, $e^{x+y} - e^{a+b} = e^t(x + y - a - b)$ for some real t between $x + y$ and $a + b$. Rewriting the last as $e^t(x - a) + e^t(y - b)$, we have achieved something essential: We have made $(x-a)$ and $(y-b)$ appear as factors. We now have

$$\begin{aligned}\mathbf{G}(\mathbf{x}) - \mathbf{G}(\mathbf{c}) &= \left([(x + a)(x - a) + (3y + 3b)(y - b)], [e^t(x - a) + e^t(y - b)]\right)\\ &= \begin{bmatrix} x + a & 3y + 3b \\ e^t & e^t \end{bmatrix}\begin{bmatrix} x - a \\ y - b \end{bmatrix} \approx \begin{bmatrix} 2a & 6b \\ e^{a+b} & e^{a+b} \end{bmatrix}\begin{bmatrix} x - a \\ y - b \end{bmatrix},\end{aligned}$$

because $(x, y) \to (a, b)$ forces $x + a \to 2a$, $3y + 3b \to 6b$, and t (which is squeezed between $x + y$ and $a + b$) $\to a + b$.

The approximation gives us a candidate matrix, call it $\mathbf{M}$, for $\mathbf{G}'(a, b)$. This really is the derivative. We have

$$\begin{aligned}&\mathbf{G}(\mathbf{x}) - \mathbf{G}(\mathbf{c}) - \mathbf{M}\langle \mathbf{x} - \mathbf{c}\rangle\\ &\quad = \left([(x - a)^2 + 3(y - b)^2], [e^{x+y} - e^{a+b} - e^{a+b}(x - a) - e^{a+b}(y - b)]\right).\end{aligned}$$

The first component's two terms are small compared to $\|\mathbf{x} - \mathbf{c}\|$, as in (a). For the second component,

$$\begin{aligned}\left|e^{x+y} - e^{a+b} - e^{a+b}(x - a) - e^{a+b}(y - b)\right| &= \left|\left(e^t - e^{a+b}\right)(x - a + y - b)\right|\\ &\le \left|e^t - e^{a+b}\right|\left(\|\mathbf{x} - \mathbf{c}\| + \|\mathbf{x} - \mathbf{c}\|\right),\end{aligned}$$

so that its ratio to $\|\mathbf{x} - \mathbf{c}\|$ is at most $2\left|e^t - e^{a+b}\right| \to 0$ as $(x, y) \to (a, b)$. Accordingly, $\frac{\mathbf{G}(\mathbf{x}) - \mathbf{G}(\mathbf{c}) - \mathbf{M}\langle \mathbf{x} - \mathbf{c}\rangle}{\|\mathbf{x} - \mathbf{c}\|}$ has two components that approach 0, so its limit is $\mathbf{O}$. We conclude that $\mathbf{M} = \mathbf{G}'(\mathbf{c})$.

Example 2. $h(\mathbf{x}) \equiv \|\mathbf{x}\|$ **is never differentiable at** $\mathbf{O}$**.**

Assume that the domain of h is $\mathbf{R}^n$; the range is within $\mathbf{R}$. If h had a derivative at $\mathbf{O}$, then the derivative would be a map $\mathbf{L}\colon \mathbf{R}^n \to \mathbf{R}$. Let $\mathbf{L}$ be any such map, and fix $\mathbf{d} \neq \mathbf{O}$ in $\mathbf{R}^n$. For every $t > 0$,

$$\frac{h(t\mathbf{d}) - h(\mathbf{O}) - \mathbf{L}\langle t\mathbf{d} - \mathbf{O}\rangle}{\|t\mathbf{d}\|} = \frac{\|t\mathbf{d}\| - \mathbf{L}\langle t\mathbf{d}\rangle}{t\|\mathbf{d}\|} = 1 - \mathbf{L}\left\langle \frac{\mathbf{d}}{\|\mathbf{d}\|}\right\rangle.$$

For $t < 0$,

$$\frac{h(t\mathbf{d}) - h(\mathbf{O}) - \mathbf{L}\langle t\mathbf{d} - \mathbf{O}\rangle}{\|t\mathbf{d}\|} = \frac{\|t\mathbf{d}\| - \mathbf{L}\langle t\mathbf{d}\rangle}{(-t)\|\mathbf{d}\|} = 1 + \mathbf{L}\left\langle \frac{\mathbf{d}}{\|\mathbf{d}\|}\right\rangle.$$

If h were differentiable at $\mathbf{O}$, then these two constant expressions would both have zero limits. That clearly cannot happen.

The method in Example 1(b) suggests that we can deal with the derivative of a vector function by working with its components individually. Observe that in the approximation $\mathbf{g}(\mathbf{x}) - \mathbf{g}(\mathbf{b}) \approx \mathbf{L}\langle \mathbf{x} - \mathbf{b}\rangle$, the change $\mathbf{g}(\mathbf{x}) - \mathbf{g}(\mathbf{b})$ in the function reflects no interaction among the components of $\mathbf{g}$, which we write as rows (entries) in a column. Similarly, in $\mathbf{L}\langle \mathbf{x} - \mathbf{b}\rangle$, rows of $\mathbf{L}$ operate on $\mathbf{x} - \mathbf{b}$, but not on each other. As a result, we can do derivatives of vector functions one component at a time.

Theorem 1. *Assume that* $\mathbf{g}\colon \mathbf{R}^n \to \mathbf{R}^m$ *has* $\mathbf{g}(\mathbf{x}) = \big(g_1(\mathbf{x}), \ldots, g_m(\mathbf{x})\big)$. *Then:*

(a) *$\mathbf{g}$ is differentiable at $\mathbf{b}$ iff each component g_j is differentiable at $\mathbf{b}$.*

(b) *If $\mathbf{g}$ is differentiable at $\mathbf{b}$, then* $\mathbf{g}'(\mathbf{b}) = \begin{bmatrix} g_1'(\mathbf{b}) \\ \vdots \\ g_m'(\mathbf{b}) \end{bmatrix}$*; that is, the components (rows) of the derivative (matrix) are the derivatives of the components.*

Proof. $\Rightarrow$ Assume that $\mathbf{g}$ is differentiable at $\mathbf{b}$. Its derivative is some matrix; let us use $\mathbf{g}'(\mathbf{b}) = \begin{bmatrix} \mathbf{R}_1 \\ \vdots \\ \mathbf{R}_m \end{bmatrix}$ to denote the row-structure of $\mathbf{g}'(\mathbf{b})$. Then

$$\begin{aligned}\mathbf{g}(\mathbf{x}) - \mathbf{g}(\mathbf{b}) - \mathbf{g}'(\mathbf{b})\langle \mathbf{x} - \mathbf{b}\rangle &= \begin{bmatrix} g_1(\mathbf{x}) \\ \vdots \\ g_m(\mathbf{x}) \end{bmatrix} - \begin{bmatrix} g_1(\mathbf{b}) \\ \vdots \\ g_m(\mathbf{b}) \end{bmatrix} - \begin{bmatrix} \mathbf{R}_1 \\ \vdots \\ \mathbf{R}_m \end{bmatrix}\langle \mathbf{x} - \mathbf{b}\rangle \\ &= \begin{bmatrix} g_1(\mathbf{x}) - g_1(\mathbf{b}) - \mathbf{R}_1\langle \mathbf{x} - \mathbf{b}\rangle \\ \vdots \\ g_m(\mathbf{x}) - g_m(\mathbf{b}) - \mathbf{R}_m\langle \mathbf{x} - \mathbf{b}\rangle \end{bmatrix}.\end{aligned}$$

Let $\pi_j\left(\begin{bmatrix} v_1 \\ \vdots \\ v_m \end{bmatrix}\right) \equiv v_j$ define the projection π_j. For each j, we have

$$\frac{g_j(\mathbf{x}) - g_j(\mathbf{b}) - \mathbf{R}_j\langle \mathbf{x} - \mathbf{b}\rangle}{\|\mathbf{x} - \mathbf{b}\|} = \pi_j\left(\frac{\mathbf{g}(\mathbf{x}) - \mathbf{g}(\mathbf{b}) - \mathbf{g}'(\mathbf{b})\langle \mathbf{x} - \mathbf{b}\rangle}{\|\mathbf{x} - \mathbf{b}\|}\right).$$

As $\mathbf{x} \to \mathbf{b}$, the argument of π_j approaches $\mathbf{O}$, because $\mathbf{g}$ is differentiable. Since π_j is a continuous function, the whole right side approaches $\pi_j(\mathbf{O}) = 0$. This proves that (a) g_j is differentiable and that (b) $g_j'(\mathbf{b}) = \mathbf{R}_j$ = row number j of $\mathbf{g}'(\mathbf{b})$.

$\Leftarrow$ Assume conversely that every g_k is differentiable. The derivative $g_k'(\mathbf{b})$, being a mapping of $\mathbf{R}^n$ to $\mathbf{R}$, is a row. Stack these rows into a matrix $\mathbf{M}$, and examine

$$\frac{\mathbf{g}(\mathbf{x}) - \mathbf{g}(\mathbf{b}) - \mathbf{M}\langle \mathbf{x} - \mathbf{b}\rangle}{\|\mathbf{x} - \mathbf{b}\|} = \frac{\begin{bmatrix} g_1(\mathbf{x}) \\ \vdots \\ g_m(\mathbf{x}) \end{bmatrix} - \begin{bmatrix} g_1(\mathbf{b}) \\ \vdots \\ g_m(\mathbf{b}) \end{bmatrix} - \begin{bmatrix} g_1'(\mathbf{b}) \\ \vdots \\ g_m'(\mathbf{b}) \end{bmatrix}\langle \mathbf{x} - \mathbf{b}\rangle}{\|\mathbf{x} - \mathbf{b}\|}$$
$$= \begin{bmatrix} \frac{g_1(\mathbf{x}) - g_1(\mathbf{b}) - g_1'(\mathbf{b})\langle \mathbf{x} - \mathbf{b}\rangle}{\|\mathbf{x} - \mathbf{b}\|} \\ \vdots \\ \frac{g_m(\mathbf{x}) - g_m(\mathbf{b}) - g_m'(\mathbf{b})\langle \mathbf{x} - \mathbf{b}\rangle}{\|\mathbf{x} - \mathbf{b}\|} \end{bmatrix}.$$

Because every g_k is differentiable, all the (scalar) entries in the last column approach 0. Therefore, the column approaches $\mathbf{O}$. We conclude that (a) $\mathbf{g}$ is differentiable at $\mathbf{b}$ and (b) $\mathbf{g}'(\mathbf{b}) = \mathbf{M}$. □

Example 3. Find the derivative of $\mathbf{G}(x, y) \equiv (x^3 - y^3, xy)$.
Here $G_1(x, y) \equiv x^3 - y^3$. If $(x, y) \approx (a, b)$, then

$$\begin{aligned} G_1(x, y) - G_1(a, b) &= x^3 - y^3 - a^3 + b^3 \\ &= (x^2 + xa + a^2)(x - a) - (y^2 + yb + b^2)(y - b) \\ &\approx 3a^2(x - a) - 3b^2(y - b). \end{aligned}$$

This suggests that $G_1'(a, b) = \begin{bmatrix} 3a^2 & -3b^2 \end{bmatrix}$.
Also, $G_2(x, y) \equiv xy$ gives

$$\begin{aligned} G_2(x, y) - G_2(a, b) &= xy - ab = xy - ay + ay - ab \\ &= y(x - a) + a(y - b) \\ &\approx b(x - a) + a(y - b), \end{aligned}$$

suggesting that $G_2'(a, b) = \begin{bmatrix} b & a \end{bmatrix}$. Proofs for G_1' and G_2' are left to Exercise 1.

By Theorem 2,

$$\mathbf{G}'(a,b) = \begin{bmatrix} 3a^2 & -3b^2 \\ b & a \end{bmatrix}.$$

We keep talking about "the" derivative. Our next result shows that, at least at the kind of place we need to deal with, "the" and the notation $\mathbf{g}'(\mathbf{b})$ are justified.

Theorem 2. *Let* $\mathbf{g}$ *have domain* D. *Let* $\mathbf{b}$ *be a point and* $N(\mathbf{c}, \delta)$ *a neighborhood with the property that for any* $\mathbf{x} \in N(\mathbf{c}, \delta)$, *the line segment* $\mathbf{bx}$ *is contained in* D. *If* $\mathbf{g}$ *is differentiable at* $\mathbf{b}$, *then the derivative is unique.*

We do not normally put remarks between the statement and proof of a theorem, but some description is appropriate here. We write $N(\mathbf{c}, \delta)$ for the **neighborhood** or **open ball** $\{\mathbf{x}\colon \|\mathbf{x}-\mathbf{c}\| < \delta\}$. The hypothesis states that the (open) shaded region in Figure 1.1 is a subset of D, and $\mathbf{b} \in D$. In $\mathbf{R}^2$, it is natural to call the region a "sector" or "fan"; in $\mathbf{R}^3$ and beyond, we can use "cone."

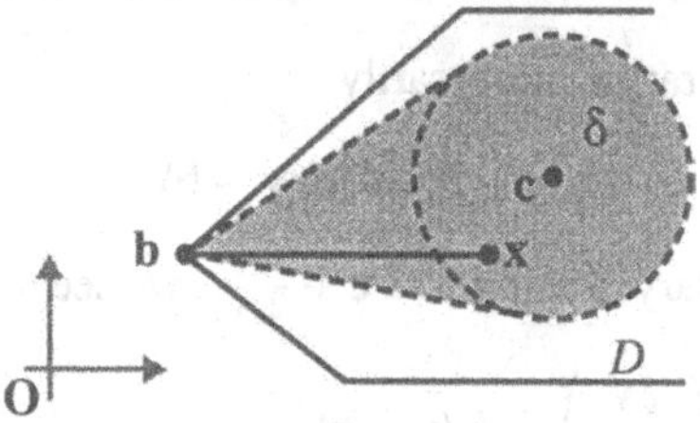

Figure 1.1.

Notice that the definition of derivative does not allow $\mathbf{b}$ to be an isolated point of D. It demands the existence of a limit for a quotient that is undefined at $\mathbf{b}$; if the quotient is also undefined nearby, then no limit can exist. Accordingly, $\mathbf{b}$ has to be at least an accumulation point of D. Our hypothesis may seem odd, but it covers two situations especially important to us, and we will show that it is essential. One situation is that $\mathbf{b}$ is an interior point of D. In that case, some $N(\mathbf{b}, \varepsilon)$ is contained in D, and we may take $\mathbf{c} \equiv \mathbf{b}$ and $\delta \equiv \varepsilon$. The second is that $\mathbf{b}$ is on the boundary and D is a disk or a box. In those cases, any neighborhood $N(\mathbf{c}, \delta) \subseteq D$ fits the requirement, and we get unique derivatives at $\mathbf{b}$.

Proof of Theorem 2. The differentiability hypothesis states that there exists a linear map $\mathbf{L}$ such that

$$\lim_{\mathbf{x}\to\mathbf{b}} \frac{\mathbf{g}(\mathbf{x}) - \mathbf{g}(\mathbf{b}) - \mathbf{L}\langle \mathbf{x} - \mathbf{b}\rangle}{\|\mathbf{x} - \mathbf{b}\|} = \mathbf{O}.$$

We must prove that if $\mathbf{M}$ is another linear map with the same property, then $\mathbf{M}$ must equal $\mathbf{L}$.

First we show that $\mathbf{L}\langle \mathbf{y} - \mathbf{b}\rangle = \mathbf{M}\langle \mathbf{y} - \mathbf{b}\rangle$ in $N(\mathbf{c}, \delta)$. Let $\varepsilon > 0$. By definition of limit, there exist δ_1 and δ_2 such that

$$\|\mathbf{x} - \mathbf{b}\| < \delta_1 \text{ and } \mathbf{x} \in D \Rightarrow \frac{\|\mathbf{g}(\mathbf{x}) - \mathbf{g}(\mathbf{b}) - \mathbf{L}\langle \mathbf{x} - \mathbf{b}\rangle\|}{\|\mathbf{x} - \mathbf{b}\|} < \varepsilon,$$

$$\|\mathbf{x}-\mathbf{b}\| < \delta_2 \text{ and } \mathbf{x} \in D \Rightarrow \frac{\|\mathbf{g}(\mathbf{x}) - \mathbf{g}(\mathbf{b}) - \mathbf{M}\langle \mathbf{x}-\mathbf{b}\rangle\|}{\|\mathbf{x}-\mathbf{b}\|} < \varepsilon.$$

Fix $\mathbf{y} \in N(\mathbf{c}, \delta)$, and let t be the smallest of $\frac{\delta_1}{2\|\mathbf{y}-\mathbf{b}\|}, \frac{\delta_2}{2\|\mathbf{y}-\mathbf{b}\|}, 1$. Then $\mathbf{x} \equiv \mathbf{b} + t(\mathbf{y}-\mathbf{b})$ is on the segment $\mathbf{by}$ and within δ_1 and δ_2 of $\mathbf{b}$. By the bounds and the triangle inequality,

$$\begin{aligned}\|\mathbf{L}\langle \mathbf{x}-\mathbf{b}\rangle - \mathbf{M}\langle \mathbf{x}-\mathbf{b}\rangle\| &= \|\mathbf{g}(\mathbf{x}) - \mathbf{g}(\mathbf{b}) - \mathbf{M}\langle \mathbf{x}-\mathbf{b}\rangle \\ &\quad - \mathbf{g}(\mathbf{x}) + \mathbf{g}(\mathbf{b}) + \mathbf{L}\langle \mathbf{x}-\mathbf{b}\rangle\| \\ &\le \|\mathbf{g}(\mathbf{x}) - \mathbf{g}(\mathbf{b}) - \mathbf{M}\langle \mathbf{x}-\mathbf{b}\rangle\| \\ &\quad + \|\mathbf{g}(\mathbf{x}) - \mathbf{g}(\mathbf{b}) - \mathbf{L}\langle \mathbf{x}-\mathbf{b}\rangle\| \\ &\le 2\varepsilon\|\mathbf{x}-\mathbf{b}\|.\end{aligned}$$

Thus, $\|\mathbf{L}\langle t(\mathbf{y}-\mathbf{b})\rangle - \mathbf{M}\langle t(\mathbf{y}-\mathbf{b})\rangle\| = 2\varepsilon\|t(\mathbf{y}-\mathbf{b})\|$, and by linearity

$$\|\mathbf{L}\langle \mathbf{y}-\mathbf{b}\rangle - \mathbf{M}\langle \mathbf{y}-\mathbf{b}\rangle\| \le 2\varepsilon\|\mathbf{y}-\mathbf{b}\|.$$

This being true for arbitrary ε, necessarily

$$\mathbf{L}\langle \mathbf{y}-\mathbf{b}\rangle = \mathbf{M}\langle \mathbf{y}-\mathbf{b}\rangle.$$

Now let $\mathbf{v} \neq \mathbf{O}$ be arbitrary. Setting $\mathbf{y} \equiv \mathbf{c} + \frac{\delta\mathbf{v}}{2\|\mathbf{v}\|}$, we see that $\mathbf{y} \in N(\mathbf{c}, \delta)$. Hence

$$\begin{aligned}\mathbf{L}\langle \mathbf{c}-\mathbf{b}\rangle + \mathbf{L}\left\langle \frac{\delta\mathbf{v}}{2\|\mathbf{v}\|}\right\rangle &= \mathbf{L}\langle \mathbf{y}-\mathbf{b}\rangle \\ &= \mathbf{M}\langle \mathbf{y}-\mathbf{b}\rangle = \mathbf{M}\langle \mathbf{c}-\mathbf{b}\rangle + \mathbf{M}\left\langle \frac{\delta\mathbf{v}}{2\|\mathbf{v}\|}\right\rangle.\end{aligned}$$

Since $\mathbf{L}\langle \mathbf{c}-\mathbf{b}\rangle = \mathbf{M}\langle \mathbf{c}-\mathbf{b}\rangle$, we conclude that $\mathbf{L}\left\langle \frac{\delta\mathbf{v}}{2\|\mathbf{v}\|}\right\rangle = \mathbf{M}\left\langle \frac{\delta\mathbf{v}}{2\|\mathbf{v}\|}\right\rangle$, and linearity forces $\mathbf{L}\langle \mathbf{v}\rangle = \mathbf{M}\langle \mathbf{v}\rangle$. We have shown that $\mathbf{L}$ and $\mathbf{M}$, which must agree at $\mathbf{O}$, also match for nonzero $\mathbf{v}$. □

If $\mathbf{b}$ does not satisfy the sector assumption—if, roughly speaking, D lies along just one direction from $\mathbf{b}$—then there can be directions in which $\mathbf{L}$ and $\mathbf{M}$ disagree.

Example 4. Consider $f(x, y) \equiv \left(x^2 - y\right)^{3/2} + \left(x^2 + y\right)^{3/2}$. Its domain is given by $-x^2 \le y \le x^2$.

The origin is in there, but because the two parabolas are tangent to the x-axis, every sector with vertex at the origin has points outside the domain. The function is differentiable, with zero derivative, along the edges of its domain, for roughly the same reason that $h(s) \equiv s^{3/2}$ has $h'(0) = 0$. We will not prove this, but we show that f' is not unique at (0, 0).

Fix c, and let $\mathbf{L} \equiv \begin{bmatrix} 0 & c \end{bmatrix}$. Then

$$f(x, y) - f(0, 0) - \mathbf{L}\langle (x, y) - (0, 0)\rangle = \left(x^2 - y\right)^{3/2} + \left(x^2 + y\right)^{3/2} - cy.$$

By the mean value theorem,

$$\left(x^2 - y\right)^{3/2} = 1.5\sqrt{s}\left(x^2 - y\right)$$

and

$$\left(x^2 + y\right)^{3/2} = 1.5\sqrt{t}\left(x^2 + y\right)$$

for appropriate s and t. Since $\left|\frac{x}{\|(x,y)\|}\right| \le 1$ and $\left|\frac{y}{\|(x,y)\|}\right| \le 1$, we have

$$\frac{\left|f(x,y) - f(0,0) - \mathbf{L}\begin{bmatrix} x-0 \\ y-0 \end{bmatrix}\right|}{\|(x,y)-(0,0)\|} \le \frac{3}{2}\sqrt{s}(|x|+1) + \frac{3}{2}\sqrt{t}(|x|+1) + |c|\frac{|y|}{\|(x,y)\|}.$$

As $(x, y) \to (0, 0)$, $\sqrt{s} \to 0$ and $\sqrt{t} \to 0$. The surprise here is that $\frac{|y|}{(x^2+y^2)^{1/2}}$ is not merely less than 1; because $|y| \le x^2$, we see that $\frac{|y|}{(x^2+y^2)^{1/2}} \le |x| \to 0$.

We have shown that $\mathbf{L}$ is a derivative of f at $(0, 0)$, no matter what c is.

We have used the phrase "is small compared to" (as in Example 1). There is a handy abbreviation to indicate relative limiting size. If a vector quantity $\mathbf{h}(\mathbf{y})$ satisfies $\lim_{\mathbf{y}\to\mathbf{O}} \frac{\mathbf{h}(\mathbf{y})}{\|\mathbf{y}\|} = \mathbf{O}$ (equivalently, $\lim_{\mathbf{y}\to\mathbf{O}} \frac{\|\mathbf{h}(\mathbf{y})\|}{\|\mathbf{y}\|} = 0$), then we say $\mathbf{h}(\mathbf{y})$ **is small in comparison with y**, and write $\mathbf{h}(\mathbf{y}) = \mathbf{o}(\mathbf{y})$ (respectively $\|\mathbf{h}(\mathbf{y})\| = o(\mathbf{y})$). This notation simplifies many proofs, but (as do many weapons) demands careful use. We employ it to end the section with the generalization of a familiar result.

Theorem 3. *Suppose* $\mathbf{g}$ *maps from* $\mathbf{R}^n$ *to* $\mathbf{R}^m$. *If* $\mathbf{g}$ *is differentiable at* $\mathbf{b}$, *then* $\mathbf{g}$ *is continuous at* $\mathbf{b}$.

Proof. By definition, there is a linear map $\mathbf{L}\colon \mathbf{R}^n \to \mathbf{R}^m$ with

$$\mathbf{g}(\mathbf{x}) - \mathbf{g}(\mathbf{b}) = \mathbf{L}\langle\mathbf{x}-\mathbf{b}\rangle + \mathbf{o}(\mathbf{x}-\mathbf{b}) = \mathbf{L}\langle\mathbf{x}\rangle - \mathbf{L}\langle\mathbf{b}\rangle + \mathbf{o}(\mathbf{x}-\mathbf{b}).$$

As $\mathbf{x} \to \mathbf{b}$, $\mathbf{o}(\mathbf{x}-\mathbf{b})$ approaches $\mathbf{O}$ by definition, and $\mathbf{L}\langle\mathbf{x}\rangle \to \mathbf{L}\langle\mathbf{b}\rangle$ because every linear map from $\mathbf{R}^n$ to $\mathbf{R}^m$ is a continuous function [Guzman, Section 3.4]. Therefore, $\mathbf{x} \to \mathbf{b}$ implies $\mathbf{g}(\mathbf{x}) \to \mathbf{g}(\mathbf{b})$. □

Theorem 3 is an example of a result that we could establish in general normed spaces. It is clear that the proof is correct as long as the derivative is a continuous linear map. Such continuity can hold between infinite-dimensional spaces, and is automatic if the domain space of $\mathbf{g}$ is finite dimensional, regardless of the image space [Guzman, Section 4.6]. So, the theorem need not be limited to Euclidean space.

Exercises

1. (a) Let $G_1(x, y) \equiv x^3 - y^3$. Prove that $G_1'(a, b) = \begin{bmatrix} 3a^2 & -3b^2 \end{bmatrix}$; that is, show that

$$\frac{G_1(x, y) - G_1(a, b) - [3a^2 \quad -3b^2] \begin{bmatrix} x - a \\ y - b \end{bmatrix}}{\|(x - a, y - b)\|} \to 0$$

as $(x, y) \to (a, b)$.

(b) Do similarly for $G_2(x, y) \equiv xy$: Prove that $G_2'(a, b) = \begin{bmatrix} b & a \end{bmatrix}$.

2. Find the derivative of $\mathbf{F}(x, y) \equiv \left((x + y)^2, (x - y)^2\right)$.

3. Find the derivative of $\mathbf{H}(x, y) \equiv \left(\sin x - \cos y, e^{2x} + e^{3y}\right)$.

4. Show that $h(\mathbf{x}) \equiv \|\mathbf{x}\|$ is differentiable in $\mathbf{R}^2$, except at the origin.

5. Show that differentiation is linear: If $\mathbf{F}$ and $\mathbf{G}$ are differentiable at $\mathbf{b}$, then so is every linear combination $\alpha\mathbf{F} + \beta\mathbf{G}$, and $(\alpha\mathbf{F} + \beta\mathbf{G})' = \alpha\mathbf{F}' + \beta\mathbf{G}'$.

6. Let $\mathbf{A}$ be a (matrix defining a) linear map from $\mathbf{R}^n$ to $\mathbf{R}^m$ and $\mathbf{b} \in \mathbf{R}^m$. Find the derivative of the first-degree function $\mathbf{H}(\mathbf{x}) \equiv \mathbf{A}\langle\mathbf{x}\rangle + \mathbf{b}$.

1.2 Derivatives and Partial Derivatives

As in elementary calculus, the definition of differentiability is a clumsy tool for finding actual derivatives. In this section, we show that for our working purposes, derivatives are easy to calculate, and that the important functions have them.

We are interested in a function $\mathbf{f}$ mapping $\mathbf{x} = (x_1, \ldots, x_n)$ in the domain $D \subseteq \mathbf{R}^n$ to $\mathbf{f}(\mathbf{x}) = \left(f_1(\mathbf{x}), \ldots, f_m(\mathbf{x})\right)$ in $\mathbf{R}^m$, around the place $\mathbf{x} = \mathbf{b} \equiv (b_1, \ldots, b_n)$. To avoid worrying about whether function values are defined near $\mathbf{b}$—as well as to guarantee that derivatives are unique—we restrict our attention to $\mathbf{b}$ in the interior of D. Given that restriction, we may as well make the following rule: When discussing derivatives, *we will assume that the domains of the functions involved are open sets.*

We have seen that the differentiability of $\mathbf{f}$, and its derivative, may be studied in terms of its components. Therefore, we may restrict our attention to any component of $\mathbf{f}$, that is, to a scalar function, which we label $f(\mathbf{x})$. Our tools will be "partial derivatives" of f. These are rates of change with respect to single variables, for which we think of all the other variables as constant. Take a fixed j and define

$$g_j(t) \equiv f\left(b_1, \ldots, b_{j-1}, b_j + t, b_{j+1}, \ldots, b_n\right).$$

If we let $\mathbf{e}_j \equiv (0, \ldots, 0, 1, 0, \ldots, 0)$ (nonzero in position number j) be the unit vector in the x_j-direction, then

$$g_j(t) = f(\mathbf{b} + t\mathbf{e}_j).$$

Definition. The derivative

$$g_j'(0) = \lim_{t\to 0} \frac{f(\mathbf{b}+t\mathbf{e}_j) - f(\mathbf{b})}{t}$$

is called the **partial derivative** of f **with respect to** x_j at $\mathbf{b}$, and denoted by $\frac{\partial f}{\partial x_j}(\mathbf{b})$.

Theorem 1. *Suppose f is differentiable at* $\mathbf{b}$. *Then*:

(a) *Every partial derivative* $\frac{\partial f}{\partial x_j}(\mathbf{b})$ *exists.*

(b) $f'(\mathbf{b})$ *is the row* $\left[\frac{\partial f}{\partial x_1}(\mathbf{b}) \ \cdots \ \frac{\partial f}{\partial x_n}(\mathbf{b})\right]$.

Proof. Assume that f is differentiable at $\mathbf{b}$. Its derivative is some row; say $f'(\mathbf{b}) = [a_1 \cdots a_n]$.

Write $\mathbf{x} \equiv \mathbf{b} + t\mathbf{e}_j$. By definition of differentiability, the function difference $f(\mathbf{x}) - f(\mathbf{b})$ is approximated in terms of $f'(\mathbf{b})$. In detail,

$$f(\mathbf{x}) - f(\mathbf{b}) = f'(\mathbf{b})\langle\mathbf{x} - \mathbf{b}\rangle + o(\mathbf{x} - \mathbf{b}),$$

so that

$$\frac{f(\mathbf{x}) - f(\mathbf{b})}{t} = \frac{f'(\mathbf{b})\langle\mathbf{x}-\mathbf{b}\rangle}{t} + \frac{o(\mathbf{x}-\mathbf{b})}{t} = f'(\mathbf{b})\left\langle\frac{\mathbf{x}-\mathbf{b}}{t}\right\rangle \pm \frac{o(\mathbf{x}-\mathbf{b})}{\|\mathbf{x}-\mathbf{b}\|}.$$

On the right side, the first term is fixed; it is $f'(\mathbf{b})\langle\mathbf{e}_j\rangle$. The second term approaches $\mathbf{O}$ as $\mathbf{x} - \mathbf{b} \to \mathbf{O}$. We conclude that $\frac{f(\mathbf{b}+t\mathbf{e}_j)-f(\mathbf{b})}{t}$ has a limit as $t \to 0$, and the limit is $f'(\mathbf{b})\langle\mathbf{e}_j\rangle$. That is to say, $\frac{\partial f}{\partial x_j}(\mathbf{b})$ exists, and

$$\frac{\partial f}{\partial x_j}(\mathbf{b}) = [a_1 \cdots a_n][0 \cdots 0\,1\,0 \cdots 0]^t = a_j. \qquad \square$$

The last line introduces a space-saving notation: We will often represent the column with entries $c_1, \ldots, c_k$ as the transposed row $[c_1 \cdots c_k]^t$.

We will use $\frac{\partial f}{\partial \mathbf{x}}(\mathbf{b})$ to denote the derivative row-matrix $\left[\frac{\partial f}{\partial x_1}(\mathbf{b}) \ \cdots \ \frac{\partial f}{\partial x_n}(\mathbf{b})\right]$, and call it the **Jacobian matrix** of f. If $\mathbf{b}$ is understood, then we omit it. For a vector function $\mathbf{g}(\mathbf{x}) = (g_1(\mathbf{x}), \ldots, g_m(\mathbf{x}))$, we have (Theorem 1.1:1)

$$\mathbf{g}'(\mathbf{b}) = \begin{bmatrix} g_1'(\mathbf{b}) \\ \ddots \\ g_m'(\mathbf{b}) \end{bmatrix} = \begin{bmatrix} \frac{\partial g_1}{\partial x_1}(\mathbf{b}) \cdots & \frac{\partial g_1}{\partial x_n}(\mathbf{b}) \\ & \ddots \\ \frac{\partial g_m}{\partial x_1}(\mathbf{b}) \cdots & \frac{\partial g_m}{\partial x_n}(\mathbf{b}) \end{bmatrix}.$$

We call the last matrix the Jacobian matrix of $\mathbf{g}$. Its most compact symbol is $\frac{\partial \mathbf{g}}{\partial \mathbf{x}}$, but we sometimes expand $\mathbf{g}$ or $\mathbf{x}$ to write $\frac{\partial \mathbf{g}}{\partial(x_1,\dots,x_n)}$, $\frac{\partial(g_1,\dots,g_m)}{\partial \mathbf{x}}$, or the most standard form $\frac{\partial(g_1,\dots,g_m)}{\partial(x_1,\dots,x_n)}$.

Theorem 1 is important for what it says and for what it does not. It says that if a scalar function has a derivative, then that derivative is easy to express in terms of the most basic ones, namely, one-variable derivatives. In other words, if a multivariable function is differentiable, then we know how to take its derivative. It does not tell us when a function has a derivative. Specifically, the converse statement is false: Existence of the partials does not imply differentiability.

Example 1. Let $f(x, y) \equiv (xy)^{1/3}$. Then f is continuous and has zero-value partial derivatives at $\mathbf{O}$ (Exercises 1(a), (b)).

Suppose f were differentiable. Theorem 1 would apply, and the derivative would have to be $\begin{bmatrix}0 & 0\end{bmatrix}$. Then $f(t,t) - f(0,0) - \begin{bmatrix}0 & 0\end{bmatrix}\langle(t,t) - (0,0)\rangle$ would have to be $o(t)$. This approximation does not hold (Exercise 1(c)). We conclude that f is not differentiable.

Notice how we used Theorem 1 in Example 1. We reiterate that the theorem does not tell us whether a function is differentiable. On the other hand, it limits the field of derivative candidates. It tells us that the only thing that *can* work is the Jacobian matrix. If we use the matrix and it yields the required approximation, then the function has a derivative, and the Jacobian matrix is it. If the matrix does not work, then nothing else needs a tryout. We put these ideas to work in the proof of a theorem that does identify differentiability.

Theorem 2. *Suppose every partial derivative $\frac{\partial f}{\partial x_j}$ in the matrix $\frac{\partial f}{\partial \mathbf{x}}$ is defined near* $\mathbf{b}$ *and continuous at* $\mathbf{b}$. *Then f is differentiable at* $\mathbf{b}$.

Proof. We stated that to prove a function differentiable, we need only show that the Jacobian matrix approximates change in the function. Accordingly, we need to show that $f(\mathbf{x}) - f(\mathbf{b}) - \frac{\partial f}{\partial \mathbf{x}}(\mathbf{b})\langle\mathbf{x} - \mathbf{b}\rangle = o(\mathbf{x} - \mathbf{b})$.

Let $\varepsilon > 0$ be specified. By assumption, the partials of f are defined near $\mathbf{b}$ and continuous at $\mathbf{b}$. Hence there is a neighborhood $N(\mathbf{b}, \delta)$ in which

$$\mathbf{x} \in N(\mathbf{b}, \delta) \Rightarrow \left|\frac{\partial f}{\partial x_j}(\mathbf{x}) - \frac{\partial f}{\partial x_j}(\mathbf{b})\right| < \frac{\varepsilon}{n}, \quad j = 1, 2, \dots, n.$$

For any such $\mathbf{x}$, write

$$\begin{aligned} f(\mathbf{x}) - f(\mathbf{b}) = {} & f(x_1, x_2, \dots, x_n) - f(b_1, x_2, \dots, x_n) \\ & + f(b_1, x_2, \dots, x_n) - f(b_1, b_2, x_3, \dots, x_n) \\ & + \cdots + f(b_1, b_2, \dots, b_{n-1}, x_n) - f(b_1, b_2, \dots, b_n). \end{aligned}$$

If $b_1 \le t \le x_1$, then $\mathbf{v}_1(t) \equiv (t, x_2, \dots, x_n)$ is in $N(\mathbf{b}, \delta)$. This means that the function

$$h_1(t) \equiv f(\mathbf{v}_1(t)) = f(t, x_2, \dots, x_n)$$

is differentiable, $h_1'(t)$ being $\frac{\partial f}{\partial x_1}(\mathbf{v}(t))$. The mean value theorem applies, so

$$f(x_1, x_2, \ldots, x_n) - f(b_1, x_2, \ldots, x_n) = h_1'(t_1)[x_1 - b_1] = \frac{\partial f}{\partial x_1}(\mathbf{w}_1)[x_1 - b_1].$$

We have a similar result for each increment in f, down to

$$\begin{aligned} f(b_1, b_2, \ldots, b_{n-1}, x_n) - f(b_1, b_2, \ldots, b_n) &= h_n'(t_n)[x_n - b_n] \\ &= \frac{\partial f}{\partial x_n}(\mathbf{w}_n)[x_n - b_n]. \end{aligned}$$

Now

$$\begin{aligned} \frac{\partial f}{\partial \mathbf{x}}(\mathbf{b})\langle \mathbf{x} - \mathbf{b}\rangle &= \begin{bmatrix} \frac{\partial f}{\partial x_1}(\mathbf{b}) & \cdots & \frac{\partial f}{\partial x_n}(\mathbf{b}) \end{bmatrix} \begin{bmatrix} x_1 - b_1 \\ \vdots \\ x_n - b_n \end{bmatrix} \\ &= \frac{\partial f}{\partial x_1}(\mathbf{b})[x_1 - b_1] + \cdots + \frac{\partial f}{\partial x_n}(\mathbf{b})[x_n - b_n]. \end{aligned}$$

Substituting and rearranging, we obtain

$$\begin{aligned} f(\mathbf{x}) - f(\mathbf{b}) - \frac{\partial f}{\partial \mathbf{x}}(\mathbf{b})\langle \mathbf{x} - \mathbf{b}\rangle = &\frac{\partial f}{\partial x_1}(\mathbf{w}_1)[x_1 - b_1] - \frac{\partial f}{\partial x_1}(\mathbf{b})[x_1 - b_1] \\ &+ \cdots + \frac{\partial f}{\partial x_n}(\mathbf{w}_n)[x_n - b_n] \\ &- \frac{\partial f}{\partial x_n}(\mathbf{b})[x_n - b_n]. \end{aligned}$$

Since $\left|\frac{\partial f}{\partial x_j}(\mathbf{w}_j) - \frac{\partial f}{\partial x_j}(\mathbf{b})\right| < \frac{\varepsilon}{n}$ and $|x_j - b_j| \le \|\mathbf{x} - \mathbf{b}\|$ for each j, we have

$$\left| f(\mathbf{x}) - f(\mathbf{b}) - \frac{\partial f}{\partial \mathbf{x}}(\mathbf{b})\langle \mathbf{x} - \mathbf{b}\rangle \right| \le n\left(\frac{\varepsilon}{n}\right)\|\mathbf{x} - \mathbf{b}\|.$$

Beginning with $\varepsilon > 0$, we have found a neighborhood of $\mathbf{b}$ in which

$$\left| f(\mathbf{x}) - f(\mathbf{b}) - \frac{\partial f}{\partial \mathbf{x}}(\mathbf{b})\langle \mathbf{x} - \mathbf{b}\rangle \right| \le \varepsilon\|\mathbf{x} - \mathbf{b}\|.$$

This proves that $f(\mathbf{x}) - f(\mathbf{b}) - \frac{\partial f}{\partial \mathbf{x}}(\mathbf{b})\langle \mathbf{x} - \mathbf{b}\rangle = o(\mathbf{x} - \mathbf{b})$; f is differentiable at $\mathbf{b}$. □

Theorem 2 is essential for our work. The elementary one-variable functions are mostly continuously differentiable in the interiors of their domains. Hence a vector-variable function built from elementary functions of the variable's coordinates will typically have continuous partial derivatives. In other words, the functions that are important to us are differentiable multivariable functions on open subsets of their domains.

We should also be aware that the provisions of Theorem 2 are strict.

Example 2. In Theorem 2:

(a) It does not suffice for the partials to be defined in a neighborhood of **b**. For the function

$$h(x, y) \equiv \frac{xy}{x^2 + y^2}, \qquad h(0, 0) \equiv 0,$$

both partials are defined everywhere, but the function is not differentiable at the origin. (Justifications: Exercise 2.)

(b) The converse is false. Thus,

$$G(x, y) \equiv x^2 \sin\left(\frac{1}{x}\right), \quad G(0, y) \equiv 0$$

is differentiable everywhere, but $\frac{\partial G}{\partial x}$ is discontinuous all along the y-axis (Exercise 3).

Exercises

1. Let $f(x, y) \equiv (xy)^{1/3}$. Show that:

   (a) f is continuous at **O**.

   (b) $\frac{\partial f}{\partial x}$ and $\frac{\partial f}{\partial y}$ exist and equal 0 at **O**.

   (c) $f(t, t) - f(0, 0) - \left[\frac{\partial f}{\partial x}(\mathbf{O}) \ \frac{\partial f}{\partial y}(\mathbf{O})\right] [t \ \ t]^t$ is not small compared with $\|(t, t)\|$ around $(0, 0)$.

2. Let $h(x, y) \equiv \frac{xy}{x^2+y^2}$, with $h(0, 0) \equiv 0$. Show that:

   (a) $\frac{\partial h}{\partial x}$ and $\frac{\partial h}{\partial y}$ are defined everywhere in $\mathbf{R}^2$.

   (b) $h(x, y) - h(0, 0) - \frac{\partial h}{\partial \mathbf{x}}\langle(x, y)\rangle$ is not small compared to $\|(x, y)\|$ around $(0, 0)$.

3. Let $G(x, y) \equiv x^2 \sin\left(\frac{1}{x}\right)$, $G(0, y) \equiv 0$. Show that:

   (a) G is differentiable everywhere in $\mathbf{R}^2$.

   (b) $\frac{\partial G}{\partial x}$ is defined everywhere.

   (c) $\frac{\partial G}{\partial x}$ is discontinuous at each $(0, y)$.

4. For each function, decide where it is differentiable and compute its derivative (matrix). Justify your conclusions carefully.

   (a) $\mathbf{f}(x, y) \equiv \left(xy, e^{x+y}, \sin x \cos y\right)$.

   (b) $g(x, y) \equiv x^3 y^3 \sin\left(\frac{1}{xy}\right)$ if $xy \neq 0$, $\equiv 0$ otherwise.

(c) $F(x, y) \equiv x^2y^2 \sin\left(\frac{1}{xy}\right)$ if $xy \neq 0$, $\equiv 0$ otherwise.

(d) $G(x, y) \equiv x^{2/3}y^{2/3}$ (very carefully!).

(e) $H(x, y) \equiv x^{1/3}y^{1/3}$ (same).

5. Let $\mathbf{v} \equiv (v_1, \ldots, v_n)$ and $\mathbf{f}(t) \equiv t\mathbf{v}$. Find $\mathbf{f}'(t)$.

1.3 The Chain Rule

In elementary calculus, the product, quotient, and chain rules are tools for finding derivatives of combinations of functions. Here we do not have such use in mind; we introduce them, primarily the chain rule, for their value in our theoretical work.

Recall that the chain rule deals with composites. Thus, suppose $\mathbf{u} \equiv \mathbf{f}(\mathbf{x})$ is differentiable at $\mathbf{x} = \mathbf{b}$ and $\mathbf{v} \equiv \mathbf{g}(\mathbf{u})$ is differentiable at $\mathbf{u} = \mathbf{f}(\mathbf{b})$. If the rule retains its form, we expect $\mathbf{v} = \mathbf{g}(\mathbf{u}) = \mathbf{g}(\mathbf{f}(\mathbf{x}))$ to satisfy $\frac{\partial \mathbf{v}}{\partial \mathbf{x}} = \frac{\partial \mathbf{v}}{\partial \mathbf{u}}\frac{\partial \mathbf{u}}{\partial \mathbf{x}}$.

Before attempting any proof, let us observe that the equality above makes sense. If $\mathbf{f}$ maps from $\mathbf{R}^n$ to $\mathbf{R}^m$, then so does $\frac{\partial \mathbf{f}}{\partial \mathbf{x}}$. Similarly, $\frac{\partial \mathbf{g}}{\partial \mathbf{u}}$ imitates $\mathbf{g}$ in mapping from $\mathbf{R}^m$ to $\mathbf{R}^k$. With both true, the product $\left(\frac{\partial \mathbf{v}}{\partial \mathbf{u}}\right)\left(\frac{\partial \mathbf{u}}{\partial \mathbf{x}}\right) = \left(\frac{\partial \mathbf{g}}{\partial \mathbf{u}}\right)\left(\frac{\partial \mathbf{f}}{\partial \mathbf{x}}\right)$ maps from $\mathbf{R}^n$ to $\mathbf{R}^k$. This is also how $\frac{\partial \mathbf{v}}{\partial \mathbf{x}}$ has to map, because $\mathbf{g}(\mathbf{f}(\mathbf{x}))$ maps likewise.

We need to introduce the idea of a **norm of a linear map**. Suppose a linear map $\mathbf{L}\colon \mathbf{R}^n \to \mathbf{R}^m$ is given by the standard matrix $[a_{jk}]$. We write

$$\|\mathbf{L}\| \equiv \left(\sum_{1 \le j \le m, 1 \le k \le n} a_{jk}^2 \right)^{1/2}.$$

This quantity is the norm associated with the inner product

$$\mathbf{L} \bullet \mathbf{M} \equiv \sum_{1 \le j \le m, 1 \le k \le n} a_{jk} b_{jk}$$

[refer to Guzman, Section 1.4] on the space M_{mn} of $m \times n$ matrices. We will refer to it as the **Pythagorean** (or **Euclidean**) **norm** of $\mathbf{L}$.

We may use without proof the properties $\|\mathbf{L}\|$ inherits by virtue of being a norm. There is one additional property that we want to highlight.

Theorem 1 (The Operator Inequality). *If* $\mathbf{L}\colon \mathbf{R}^n \to \mathbf{R}^m$ *is linear, then for every* $\mathbf{x} \in \mathbf{R}^n$,

$$\|\mathbf{L}\langle\mathbf{x}\rangle\| \le \|\mathbf{L}\| \, \|\mathbf{x}\|.$$

Proof. Exercise 8.

Theorem 2 (The Chain Rule). *Suppose* $\mathbf{f}\colon \mathbf{R}^n \to \mathbf{R}^m$ *is differentiable at* $\mathbf{x} = \mathbf{b}$ *and* $\mathbf{g}\colon \mathbf{R}^m \to \mathbf{R}^k$ *is differentiable at* $\mathbf{u} = \mathbf{f}(\mathbf{b})$. *Then:*

(a) $\mathbf{h}(\mathbf{x}) \equiv \mathbf{g}(\mathbf{f}(\mathbf{x}))$ *is differentiable at* $\mathbf{x} = \mathbf{b}$, *and*

(b) $\mathbf{h}'(\mathbf{b}) = \mathbf{g}'(\mathbf{f}(\mathbf{b}))\mathbf{f}'(\mathbf{b})$.

Proof. Our rule about open domains requires $\mathbf{f}$ to be defined near $\mathbf{b}$ and $\mathbf{g}$ to be defined near $\mathbf{c} \equiv \mathbf{f}(\mathbf{b})$. However, these provisions do not guarantee that $\mathbf{g}(\mathbf{f})$ is defined near $\mathbf{b}$. (Example?) We must establish this last condition before we discuss differentiability.

For that purpose, observe that by our rules, there is a neighborhood $N(\mathbf{c}, \varepsilon)$ in which $\mathbf{g}$ is defined and one $N(\mathbf{b}, \delta_1)$ is which $\mathbf{f}$ is. The differentiability of $\mathbf{f}$ implies that $\mathbf{f}$ is continuous (Theorem 1.1:3). Hence there is a neighborhood $N(\mathbf{b}, \delta_2)$ such that

$$\mathbf{x} \in N(\mathbf{b}, \delta_2) \Rightarrow \mathbf{f}(\mathbf{x}) \in N(\mathbf{c}, \varepsilon), \quad \text{assuming that } \mathbf{f}(\mathbf{x}) \text{ is defined.}$$

Therefore, $\delta \equiv \min\{\delta_1, \delta_2\}$ gives

$$\mathbf{x} \in N(\mathbf{b}, \delta) \Rightarrow \mathbf{f}(\mathbf{x}) \in N(\mathbf{c}, \varepsilon) \Rightarrow \mathbf{g}(\mathbf{f}(\mathbf{x})) \text{ is defined.}$$

The composite is, indeed, defined near $\mathbf{b}$.

Let $\mathbf{L}_1 \equiv \mathbf{f}'(\mathbf{b})$, $\mathbf{L}_2 \equiv \mathbf{g}'(\mathbf{c})$, and $\mathbf{L} \equiv \mathbf{L}_2\mathbf{L}_1$. Our work in Sections 1.1 and 1.2 makes it clear that to prove that $\mathbf{h}$ is differentiable and that its derivative is $\mathbf{L}$, we need to demonstrate that $\mathbf{h}(\mathbf{x}) - \mathbf{h}(\mathbf{b}) - \mathbf{L}\langle \mathbf{x} - \mathbf{b}\rangle = \mathbf{o}(\mathbf{x} - \mathbf{b})$. Accordingly, we examine

$$\begin{aligned} \mathbf{h}(\mathbf{x}) - \mathbf{h}(\mathbf{b}) - \mathbf{L}\langle \mathbf{x} - \mathbf{b}\rangle &= \mathbf{g}(\mathbf{f}(\mathbf{x})) - \mathbf{g}(\mathbf{f}(\mathbf{b})) - \mathbf{L}_2\mathbf{L}_1\langle \mathbf{x} - \mathbf{b}\rangle \\ &= \big[\mathbf{g}(\mathbf{f}(\mathbf{x})) - \mathbf{g}(\mathbf{c}) - \mathbf{L}_2\langle \mathbf{f}(\mathbf{x}) - \mathbf{c}\rangle\big] \\ &\quad + \big[\mathbf{L}_2\langle \mathbf{f}(\mathbf{x}) - \mathbf{c}\rangle - \mathbf{L}_2\mathbf{L}_1\langle \mathbf{x} - \mathbf{b}\rangle\big]. \end{aligned}$$

To show that this sum is $\mathbf{o}(\mathbf{x} - \mathbf{b})$, we prove the estimate separately for its two terms.

Let a positive s be specified. Because $\mathbf{f}$ is differentiable, there is a neighborhood N_1 of $\mathbf{b}$ in which

$$\|\mathbf{f}(\mathbf{x}) - \mathbf{f}(\mathbf{b}) - \mathbf{L}_1\langle \mathbf{x} - \mathbf{b}\rangle\| \leq s\|\mathbf{x} - \mathbf{b}\|.$$

For such $\mathbf{x}$, Theorem 1 gives

$$\begin{aligned} \|\mathbf{f}(\mathbf{x}) - \mathbf{c}\| &\leq \|\mathbf{f}(\mathbf{x}) - \mathbf{c} - \mathbf{L}_1\langle \mathbf{x} - \mathbf{b}\rangle\| + \|\mathbf{L}_1\langle \mathbf{x} - \mathbf{b}\rangle\| \\ &\leq s\|\mathbf{x} - \mathbf{b}\| + \|\mathbf{L}_1\|\,\|\mathbf{x} - \mathbf{b}\|. \end{aligned}$$

Next, $\mathbf{g}$ being differentiable at $\mathbf{c}$, there is a neighborhood O of $\mathbf{c}$ in which

$$\|\mathbf{g}(\mathbf{u}) - \mathbf{g}(\mathbf{c}) - \mathbf{L}_2\langle \mathbf{u} - \mathbf{c}\rangle\| \leq s\|\mathbf{u} - \mathbf{c}\|.$$

By continuity, there is a neighborhood N_2 of $\mathbf{b}$ such that $\mathbf{f}(N_2) \subseteq O$. Therefore, for any $\mathbf{x} \in N_1 \cap N_2$, we have $\mathbf{f}(\mathbf{x}) \in O$, so that

$$\|\mathbf{g}(\mathbf{f}(\mathbf{x})) - \mathbf{g}(\mathbf{c}) - \mathbf{L}_2\langle \mathbf{f}(\mathbf{x}) - \mathbf{c}\rangle\| \leq s\|\mathbf{f}(\mathbf{x}) - \mathbf{c}\| \leq s(s + \|\mathbf{L}_1\|)\|\mathbf{x} - \mathbf{b}\|.$$

It follows that $\mathbf{g}(\mathbf{f}(\mathbf{x})) - \mathbf{g}(\mathbf{c}) - \mathbf{L}_2\langle \mathbf{f}(\mathbf{x}) - \mathbf{c}\rangle = \mathbf{o}(\mathbf{x} - \mathbf{b})$, taking care of the first term.

The second is $\mathbf{L}_2\langle \mathbf{f}(\mathbf{x}) - \mathbf{c}\rangle - \mathbf{L}_2\mathbf{L}_1\langle \mathbf{x} - \mathbf{b}\rangle = \mathbf{L}_2\langle \mathbf{f}(\mathbf{x}) - \mathbf{c} - \mathbf{L}_1\langle \mathbf{x} - \mathbf{b}\rangle\rangle$, because $\mathbf{L}_2$ is a linear map. For the $s > 0$ specified before and the resulting N_1, Theorem 1 tells us that

$$\begin{aligned}\mathbf{x} \in N_1 \Rightarrow \left\|\mathbf{L}_2\langle \mathbf{f}(\mathbf{x}) - \mathbf{c} - \mathbf{L}_1\langle \mathbf{x} - \mathbf{b}\rangle\rangle\right\| &\leq \left\|\mathbf{L}_2\right\| \left\|\mathbf{f}(\mathbf{x}) - \mathbf{c} - \mathbf{L}_1\langle \mathbf{x} - \mathbf{b}\rangle\right\| \\ &\leq \|\mathbf{L}_2\| s \|\mathbf{x} - \mathbf{b}\|.\end{aligned}$$

We conclude that $\mathbf{L}_2\langle \mathbf{f}(\mathbf{x}) - \mathbf{c}\rangle - \mathbf{L}_2\mathbf{L}_1\langle \mathbf{x} - \mathbf{b}\rangle = \mathbf{o}(\mathbf{x} - \mathbf{b})$. □

Example 1. For a quick check on the theorem, consider

$$g(u, v, w) \equiv u + 2v + 3w,$$

with

$$(u, v, w) \equiv \left(x^2 + y^2, x^2 - y^2, xy\right).$$

The theorem says that

$$\begin{aligned}\frac{\partial g}{\partial(x, y)} = \frac{\partial g}{\partial(u, v, w)} \frac{\partial(u, v, w)}{\partial(x, y)} &= \begin{bmatrix}1 & 2 & 3\end{bmatrix} \begin{bmatrix}2x & 2y \\ 2x & -2y \\ y & x\end{bmatrix} \\ &= \begin{bmatrix}2x + 4x + 3y & 2y - 4y + 3x\end{bmatrix}.\end{aligned}$$

If instead we substitute

$$g = \left(x^2 + y^2\right) + 2\left(x^2 - y^2\right) + 3(xy) = 3x^2 + 3xy - y^2,$$

then we obtain directly the matching result

$$\frac{\partial g}{\partial(x, y)} = \begin{bmatrix}6x + 3y & 3x - 2y\end{bmatrix}.$$

It is clear that the chain rule extends by induction to a composite of any finite number of functions.

Again, we have stated in the context of Euclidean space a theorem that generalizes easily to normed spaces. Other than the definition of differentiability, the essential element of the proof was that the derivatives satisfy conditions of the form

$$\|\mathbf{L}\langle \mathbf{x}\rangle\|_{\text{range}} = K \|\mathbf{x}\|_{\text{domain}}.$$

For a linear map, such a condition ["boundedness"; consult Guzman, Section 3.5] is (a) equivalent to continuity; (b) automatic if the domain has finite dimension; (c) possible even in infinite dimensions. Hence our proof carries over to general normed spaces. Moreover, as long as $\frac{\partial \mathbf{f}}{\partial \mathbf{x}}$ and $\frac{\partial \mathbf{g}}{\partial \mathbf{u}}$ both meet the criterion, their product does likewise, because the product is continuous. Therefore, the inductive argument extends in general spaces to composites $\mathbf{h}(\ldots\mathbf{g}(\mathbf{f}))$ of more than two functions, provided that the required derivatives $\frac{\partial \mathbf{f}}{\partial \mathbf{x}}, \frac{\partial \mathbf{g}}{\partial \mathbf{u}}, \ldots, \frac{\partial \mathbf{h}}{\partial \mathbf{z}}$ are individually continuous.

The relation $\|\mathbf{L}\langle\mathbf{x}\rangle\| \leq K\|\mathbf{x}\|$ is our second involving relative size. It too has a useful shorthand. When such a relation holds for a set of $\mathbf{x}$, we will write $\mathbf{L}\langle\mathbf{x}\rangle = \mathbf{O}(\mathbf{x})$ (equivalently $\|\mathbf{L}\langle\mathbf{x}\rangle\| = O(\mathbf{x})$) and say that $\mathbf{L}\langle\mathbf{x}\rangle$ **is of the order of x** (or $\mathbf{L}\langle\mathbf{x}\rangle$ **is dominated by x**) on that set.

In Example 1, we considered a real-valued composite. This is not an excessive restriction. Suppose $\mathbf{g}(\mathbf{u}) = (g_1(\mathbf{u}), \ldots, g_k(\mathbf{u}))$ and $\mathbf{u}(\mathbf{x}) = (u_1(\mathbf{x}), \ldots, u_m(\mathbf{x}))$. We know that $\frac{\partial \mathbf{g}}{\partial \mathbf{x}} = \frac{\partial \mathbf{g}}{\partial \mathbf{u}}\frac{\partial \mathbf{u}}{\partial \mathbf{x}}$. In the product of the matrices $\frac{\partial \mathbf{g}}{\partial \mathbf{u}}$ and $\frac{\partial \mathbf{u}}{\partial \mathbf{x}}$, any row is the product of the corresponding row in $\frac{\partial \mathbf{g}}{\partial \mathbf{u}}$ and (all of) $\frac{\partial \mathbf{u}}{\partial \mathbf{x}}$. Let us abbreviate the row-structure of

$$\frac{\partial \mathbf{g}}{\partial \mathbf{u}} = \begin{bmatrix} \frac{\partial g_1}{\partial u_1} & \cdots & \frac{\partial g_1}{\partial u_m} \\ & \ddots & \\ \frac{\partial g_k}{\partial u_1} & \cdots & \frac{\partial g_k}{\partial u_m} \end{bmatrix}$$

by $\begin{bmatrix} \frac{\partial g_1}{\partial \mathbf{u}} \\ \ddots \\ \frac{\partial g_k}{\partial \mathbf{u}} \end{bmatrix}$. Then

$$\frac{\partial \mathbf{g}}{\partial \mathbf{x}} = \begin{bmatrix} \frac{\partial g_1}{\partial \mathbf{u}} \\ \ddots \\ \frac{\partial g_k}{\partial \mathbf{u}} \end{bmatrix} \frac{\partial \mathbf{u}}{\partial \mathbf{x}} = \begin{bmatrix} \frac{\partial g_1}{\partial \mathbf{u}}\frac{\partial \mathbf{u}}{\partial \mathbf{x}} \\ \ddots \\ \frac{\partial g_k}{\partial \mathbf{u}}\frac{\partial \mathbf{u}}{\partial \mathbf{x}} \end{bmatrix} = \begin{bmatrix} \frac{\partial g_1}{\partial \mathbf{x}} \\ \ddots \\ \frac{\partial g_k}{\partial \mathbf{x}} \end{bmatrix}$$

by application of the chain rule to each g_j. This tells us that we may always apply the rule componentwise to the final function in the chain. Therefore, for purposes of the chain rule, we may restrict our attention to (ultimately) scalar composites $F(\mathbf{h}(\ldots\mathbf{f}(\mathbf{x})))$.

Example 2. Given a scalar function $G(x, y)$ on $\mathbf{R}^2$, how does G vary with polar coordinates?

We have $(x, y) = (r\cos\theta, r\sin\theta)$. Hence

$$\begin{aligned} \frac{\partial G}{\partial(r,\theta)} &= \frac{\partial G}{\partial(x,y)}\frac{\partial(x,y)}{\partial(r,\theta)} \\ &= \begin{bmatrix} \frac{\partial G}{\partial x} & \frac{\partial G}{\partial y} \end{bmatrix} \begin{bmatrix} \cos\theta & -r\sin\theta \\ \sin\theta & r\cos\theta \end{bmatrix} \\ &= \begin{bmatrix} \frac{\partial G}{\partial x}\cos\theta + \frac{\partial G}{\partial y}\sin\theta & -\frac{\partial G}{\partial x}r\sin\theta + \frac{\partial G}{\partial y}r\cos\theta \end{bmatrix}. \end{aligned}$$

Thus,

$$\frac{\partial G}{\partial r} = \frac{\partial G}{\partial x}\cos\theta + \frac{\partial G}{\partial y}\sin\theta,$$
$$\frac{\partial G}{\partial \theta} = -r\frac{\partial G}{\partial x}\sin\theta + r\frac{\partial G}{\partial y}\cos\theta.$$

Concerning a scalar function $F(\mathbf{x})$, our defining approximation says that

$$F(\mathbf{x}) - F(\mathbf{b}) \approx \frac{\partial F}{\partial \mathbf{x}}(\mathbf{b})\langle \mathbf{x} - \mathbf{b}\rangle.$$

By analogy to the one-variable case, we call the differences $F(\mathbf{x}) - F(\mathbf{b})$ and $\mathbf{x}-\mathbf{b}$ the **increments** (or **changes**) in F and $\mathbf{x}$, respectively. We call

$$dF(\mathbf{b}) \equiv \frac{\partial F}{\partial \mathbf{x}}(\mathbf{b})\langle \mathbf{x} - \mathbf{b}\rangle$$

the **differential** (or **total differential**) of F. Thus

$$dF = \left[\frac{\partial F}{\partial x_1}(\mathbf{b}) \;\cdots\; \frac{\partial F}{\partial x_n}(\mathbf{b})\right][x_1 - b_1 \;\cdots\; x_n - b_n]^t$$
$$= \frac{\partial F}{\partial x_1}(\mathbf{b})[x_1 - b_1] + \cdots + \frac{\partial F}{\partial x_n}(\mathbf{b})[x_n - b_n].$$

Since $dx_j = \frac{\partial x_j}{\partial x_1}[x_1 - b_1] + \cdots + \frac{\partial x_j}{\partial x_n}[x_n - b_n] = x_j - b_j$, the increment and differential of x_j are the same. Therefore, we may write

$$dF = \frac{\partial F}{\partial x_1}dx_1 + \cdots + \frac{\partial F}{\partial x_n}dx_n.$$

Suppose now that $F(\mathbf{x})$ is actually a composite $G(\mathbf{u}(\mathbf{x}))$, with $\mathbf{u}(\mathbf{x}) = (u_1(\mathbf{x}), \ldots, u_m(\mathbf{x}))$. By our characterization, we have

$$dG = \frac{\partial G}{\partial u_1}du_1 + \cdots + \frac{\partial G}{\partial u_m}du_m$$

and

$$du_j = \frac{\partial u_j}{\partial x_1}dx_1 + \cdots + \frac{\partial u_j}{\partial x_n}dx_n, \quad j = 1, \ldots, m.$$

Hence

$$\begin{aligned} dG &= \frac{\partial G}{\partial u_1}\left(\frac{\partial u_1}{\partial x_1}dx_1 + \cdots + \frac{\partial u_1}{\partial x_n}dx_n\right) \\ &\quad + \cdots + \frac{\partial G}{\partial u_m}\left(\frac{\partial u_m}{\partial x_1}dx_1 + \cdots + \frac{\partial u_m}{\partial x_n}dx_n\right) \\ &= \left(\frac{\partial G}{\partial u_1}\frac{\partial u_1}{\partial x_1} + \cdots + \frac{\partial G}{\partial u_m}\frac{\partial u_m}{\partial x_1}\right)dx_1 \\ &\quad + \cdots + \left(\frac{\partial G}{\partial u_1}\frac{\partial u_1}{\partial x_n} + \cdots + \frac{\partial G}{\partial u_m}\frac{\partial u_m}{\partial x_n}\right)dx_n \\ &= \left(\frac{\partial G}{\partial x_1}\right)dx_1 + \cdots + \left(\frac{\partial G}{\partial x_n}\right)dx_n. \end{aligned}$$

The message is that whether H is a function of $\mathbf{x}$ immediately, like $F(\mathbf{x})$, or at the end of a chain, like $G(\mathbf{u}(\mathbf{x}))$, we may write its differential as

$$dH = \frac{\partial H}{\partial x_1}dx_1 + \cdots + \frac{\partial H}{\partial x_n}dx_n.$$

Example 3. We can illustrate differentials, as in elementary calculus, with approximate calculation of functional values close to simple ones. Thus, suppose a right circular cylinder with radius 5 cm and height 8 cm expands 1 mm in each dimension. What is the change in its volume?

From $V = \pi r^2 h$, we write

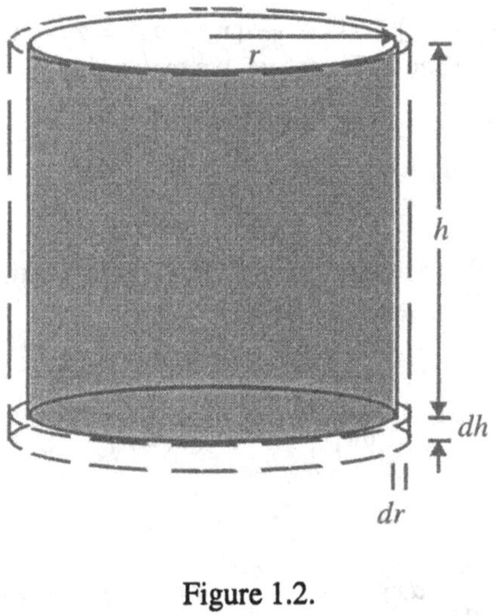

Figure 1.2.

$$\begin{aligned}\Delta V \approx dV &= \frac{\partial V}{\partial r}\,dr + \frac{\partial V}{\partial h}\,dh = 2\pi rh\,dr + \pi r^2\,dh\\ &= 2\pi(5\text{ cm})(8\text{ cm})(0.1\text{ cm}) + \pi(5\text{ cm})^2(0.1\text{ cm})\\ &= 10.5\pi\text{ cm}^3.\end{aligned}$$

It is instructive to look at the geometry of the situation; refer to Figure 1.2. The increase in radius adds to the side of the cylinder a skin dr thick, h high, and between $2\pi r$ and $2\pi(r + dr)$ around, depending on where you measure. The first term in the differential approximates this added volume with $(2\pi r)h\,dr$. Similarly, the increase in height adds to the bottom of the cylinder a disk dh thick that has face area πr^2 or $\pi(r + dr)^2$, depending on whether you change h or r first. The second term in the differential accounts just for the change in h, using $\pi r^2\,dh$ for the added volume.

In our work, $dF = \frac{\partial F}{\partial x_1}\,dx_1 + \cdots + \frac{\partial F}{\partial x_n}\,dx_n$ will serve as a template for differentiation with respect to any set of variables. What we called its message is this: If F is immediately a function of $x_1, \ldots, x_n$, then its (perhaps partial) derivatives relative to other variables are linear combinations of the derivatives of the x_j relative to those others, using the coefficients $\frac{\partial F}{\partial x_j}$.

As a final application of the chain rule, we give the product rule.

Theorem 3. *If $f(\mathbf{x})$ and $g(\mathbf{x})$ are differentiable at* $\mathbf{b}$, *then:*

(a) $h(\mathbf{x}) \equiv f(\mathbf{x})g(\mathbf{x})$ *is differentiable at* $\mathbf{b}$.

(b) $\frac{\partial h}{\partial \mathbf{x}}(\mathbf{b}) = f(\mathbf{b})\frac{\partial g}{\partial \mathbf{x}}(\mathbf{b}) + g(\mathbf{b})\frac{\partial f}{\partial \mathbf{x}}(\mathbf{b})$.

Proof. Write $\mathbf{u}(\mathbf{x}) \equiv (f(\mathbf{x}), g(\mathbf{x}))$. This function has differentiable components, so it is differentiable at $\mathbf{b}$. The function $H(\mathbf{u}) \equiv u_1 u_2$ is a differentiable function of $\mathbf{u}$ everywhere. (Why?) By the chain rule:

(a) $h(\mathbf{x}) = H(\mathbf{u}(\mathbf{x}))$ is differentiable at $\mathbf{x} = \mathbf{b}$.

(b)

$$\begin{aligned}\frac{\partial h}{\partial \mathbf{x}}(\mathbf{b}) &= \frac{\partial H}{\partial \mathbf{u}}(\mathbf{u}(\mathbf{b}))\frac{\partial \mathbf{u}}{\partial \mathbf{x}}(\mathbf{b})\\ &= \begin{bmatrix}\frac{\partial H}{\partial u_1}(\mathbf{b}) & \frac{\partial H}{\partial u_2}(\mathbf{b})\end{bmatrix}\begin{bmatrix}\frac{\partial f}{\partial \mathbf{x}}(\mathbf{b})\\ \frac{\partial g}{\partial \mathbf{x}}(\mathbf{b})\end{bmatrix} = g(\mathbf{b})\frac{\partial f}{\partial \mathbf{x}}(\mathbf{b}) + f(\mathbf{b})\frac{\partial g}{\partial \mathbf{x}}(\mathbf{b}).\end{aligned}$$

□

Exercises

1. Let $g(u, v) \equiv u^2 + v^2$, where $(u, v) = (\sin xy, \cos xy)$.

   (a) Use the chain rule to find $\frac{\partial g}{\partial(x,y)}$.

   (b) Write g in terms of x and y, find $\frac{\partial g}{\partial(x,y)}$, and verify that the result matches (a).

2. A particle's position at time t is given by $x = 10\cos t$, $y = 8\sin t$. Use the chain rule to decide whether its distance from the origin is increasing when it is at the point in quadrant I where $x = 5$.

3. In $\mathbf{R}^3$, **spherical coordinates** (ρ, θ, ϕ) are related to rectangular coordinates by $x = \rho\cos\theta\sin\phi$, $y = \rho\sin\theta\sin\phi$, $z = \rho\cos\phi$. Given $f(x, y, z)$, express $\frac{\partial f}{\partial(\rho,\theta,\phi)}$ in terms of $\frac{\partial f}{\partial x}$, $\frac{\partial f}{\partial y}$, $\frac{\partial f}{\partial z}$ and ρ, θ, ϕ.

4. Suppose $\mathbf{L}\colon \mathbf{R}^n \to \mathbf{R}^m$ and $\mathbf{M}\colon \mathbf{R}^m \to \mathbf{R}^k$ are linear maps. What is the derivative of $\mathbf{ML}$?

5. State and prove the quotient rule for the ratio of two scalar functions of a vector variable.

6. Suppose $\mathbf{u}(\mathbf{x})$ and $\mathbf{v}(\mathbf{x})$ are vector functions with values in $\mathbf{R}^m$. Prove the "dot-product rule":

$$\frac{\partial(\mathbf{u} \bullet \mathbf{v})}{\partial \mathbf{x}} = \text{row}(\mathbf{u})\,\frac{\partial \mathbf{v}}{\partial \mathbf{x}} + \text{row}(\mathbf{v})\frac{\partial \mathbf{u}}{\partial \mathbf{x}}.$$

   Here we have written row($\mathbf{u}$) for the row-matrix $[u_1 \;\cdots\; u_m]$ (which matches $\langle \mathbf{u} \rangle^t$).

7. Suppose $G(x, y)$ actually depends only on distance $r = \left(x^2 + y^2\right)^{1/2}$ from the origin: $G(x, y) = H(r)$. Show that:

   (a) Away from the origin, G is differentiable at (x, y) iff H is differentiable at r.

   (b) At the origin, G is differentiable iff H is differentiable and $H'(0) = 0$.

   (c) Anywhere that G is differentiable, $y\frac{\partial G}{\partial x} = x\frac{\partial G}{\partial y}$.

8. Prove Theorem 1.

   (Hint: Write out $\mathbf{L}\langle \mathbf{x} \rangle$ as the matrix product $\mathbf{L}[x_1 \;\cdots\; x_n]^t$, then apply Cauchy's inequality to $\|\mathbf{L}\langle \mathbf{x} \rangle\|^2$.)

1.4 Higher Derivatives

With vector variables, higher derivatives are a necessary evil. We have to study them, because they relate to two matters of interest to us: maxima/minima and Taylor's theorem. Here we give a brief introduction, concentrating on the specific material of most use to us, namely, second derivatives of scalar functions.

The idea is elementary, in that it mirrors the one-variable case. If $\mathbf{f}$ is a vector-valued function of $\mathbf{x}$, then at any point $\mathbf{y}$, $\mathbf{f}'(\mathbf{y})$ is also a vector, albeit not of the same kind as $\mathbf{f}(\mathbf{x})$. Since we may vary $\mathbf{y}$, $\mathbf{f}'(\mathbf{y})$ is a function of a vector variable, and we may inquire about its derivative. It is this derivative of the derivative that we mean by "second derivative."

Definition. Assume that $\mathbf{f}'(\mathbf{x})$ exists for each $\mathbf{x}$ in a neighborhood of $\mathbf{b}$. Suppose $\mathbf{f}'(\mathbf{x})$ is differentiable at $\mathbf{x} = \mathbf{b}$; that is, suppose there is a linear map Λ such that

$$\mathbf{f}'(\mathbf{x}) - \mathbf{f}'(\mathbf{b}) = \Lambda\langle \mathbf{x} - \mathbf{b}\rangle + \mathbf{o}(\mathbf{x} - \mathbf{b}) \quad \text{as } \mathbf{x} \to \mathbf{b}.$$

Then we say that $\mathbf{f}$ is **twice differentiable** at $\mathbf{b}$ (or **has a second derivative** at $\mathbf{b}$) and call Λ (which we also write as $\mathbf{f}''(\mathbf{b})$, $\frac{\partial^2 \mathbf{f}}{\partial \mathbf{x}^2}(\mathbf{b})$, $\left(\frac{\partial \mathbf{f}}{\partial \mathbf{x}}\right)'(\mathbf{b})$) the **second derivative** of $\mathbf{f}$ at $\mathbf{b}$.

Before proceeding, we need to remember in what sense $\mathbf{f}'(\mathbf{x})$ is a vector in Euclidean space. By definition, it is a linear map from $\mathbf{R}^n$ to $\mathbf{R}^m$. The set of such maps, which we will denote by $L(\mathbf{R}^n, \mathbf{R}^m)$, has a natural vector-space structure. By habit, we think of $\mathbf{f}'(\mathbf{x})$ as an $m \times n$ matrix. In the set M_{mn} of such matrices, the linear structure is a standard linear algebra example. Consequently, working with derivatives, we will utilize the algebra of vectors without further worry.

Because $\mathbf{f}'(\mathbf{x}) - \mathbf{f}'(\mathbf{b})$ is a linear map from $\mathbf{R}^n$ to $\mathbf{R}^m$, Λ has to turn $\mathbf{x} - \mathbf{b}$ into such a map. Thus, $\Lambda\colon \mathbf{R}^n \to L(\mathbf{R}^n, \mathbf{R}^m)$. It becomes a problem to represent Λ by a mere matrix, but we will design a matrix-related structure for the map.

To take derivatives (of derivatives), we must also handle limits. We have already specified that we work under the Pythagorean norm: $\|\mathbf{x}\| \equiv \left(x_1{}^2 + \cdots + x_n{}^2\right)^{1/2}$ for vectors, $\|[a_{jk}]\| \equiv \left(\sum a_{jk}^2\right)^{1/2}$ for matrices, and extensions where necessary. In finite dimensions, though, questions about limits of vector functions reduce to questions about limits of their components [Guzman, Section 4.6], which are (at least ultimately) real functions. Therefore, we will deal componentwise with limits, continuity, and the like.

We consider first the case of a scalar function. Assume that f is scalar and differentiable near $\mathbf{b}$. We want to find an f'' that makes $f'(\mathbf{x}) - f'(\mathbf{b}) \approx f''(\mathbf{b})\langle \mathbf{x} - \mathbf{b}\rangle$. From

$$f'(\mathbf{x}) = \frac{\partial f}{\partial \mathbf{x}}(\mathbf{x}) = \left[\frac{\partial f}{\partial x_1}(\mathbf{x}) \quad \cdots \quad \frac{\partial f}{\partial x_n}(\mathbf{x})\right],$$

we have

$$f'(\mathbf{x}) - f'(\mathbf{b}) = \left[\frac{\partial f}{\partial x_1}(\mathbf{x}) - \frac{\partial f}{\partial x_1}(\mathbf{b}) \quad \cdots \quad \frac{\partial f}{\partial x_n}(\mathbf{x}) - \frac{\partial f}{\partial x_n}(\mathbf{b})\right].$$

If the scalar function $\frac{\partial f}{\partial x_1}$ is differentiable at **b**, then

$$\frac{\partial f}{\partial x_1}(\mathbf{x}) - \frac{\partial f}{\partial x_1}(\mathbf{b}) \approx \left(\frac{\partial f}{\partial x_1}\right)'(\mathbf{b})\langle \mathbf{x}-\mathbf{b}\rangle$$
$$= \left[\frac{\partial(\partial f/\partial x_1)}{\partial x_1}(\mathbf{b}) \;\cdots\; \frac{\partial(\partial f/\partial x_1)}{\partial x_n}(\mathbf{b})\right]\langle \mathbf{x}-\mathbf{b}\rangle;$$

similarly for $\frac{\partial f}{\partial x_2}, \ldots, \frac{\partial f}{\partial x_n}$. Therefore, we approximate

$$f'(\mathbf{x}) - f'(\mathbf{b}) \approx \Bigg[\left[\frac{\partial(\partial f/\partial x_1)}{\partial x_1}(\mathbf{b}) \;\cdots\; \frac{\partial(\partial f/\partial x_1)}{\partial x_n}(\mathbf{b})\right]\langle \mathbf{x}-\mathbf{b}\rangle$$
$$\cdots\left[\frac{\partial(\partial f/\partial x_n)}{\partial x_1}(\mathbf{b}) \;\cdots\; \frac{\partial(\partial f/\partial x_n)}{\partial x_n}(\mathbf{b})\right]\langle \mathbf{x}-\mathbf{b}\rangle\Bigg].$$

This expression is getting wide, so we will abbreviate. Since we must deal with partials of partials, we introduce the notation $\frac{\partial^2 f}{\partial x_j \partial x_k}$ for $\frac{\partial(\partial f/\partial x_j)}{\partial x_k}$. This symbol is called a **mixed partial derivative**. Its two partial differentiations are done in reading (left-to-right) order: first by x_j, then by x_k.

To profit from the wide expression, we need do just two things: Define it as a linear image of $\mathbf{x}-\mathbf{b}$; and make precise the meaning of $\approx$.

Definition. Let $\mathbf{R}_1 \equiv [a_{11} \cdots a_{1n}], \ldots, \mathbf{R}_n \equiv [a_{n1} \cdots a_{nn}]$ be row-matrices. The expression

$$\Lambda \equiv \big[\mathbf{R}_1 \cdots \mathbf{R}_n\big] = \big[[a_{11} \cdots a_{1n}] \cdots [a_{n1} \cdots a_{nn}]\big]$$

is called a **row of rows**, or **row**$^{(2)}$. It represents an operator from $\mathbf{R}^n$ to $L(\mathbf{R}^n, \mathbf{R})$ defined by

$$\Lambda\langle \mathbf{v}\rangle \equiv \big[\mathbf{R}_1\langle \mathbf{v}\rangle \cdots \mathbf{R}_n\langle \mathbf{v}\rangle\big].$$

Notice that the expression $\Lambda\langle \mathbf{v}\rangle$ is no longer a matrix multiplication, but each $\mathbf{R}_j\langle \mathbf{v}\rangle$ is, representing a row times a column. That matrix product is

$$\mathbf{R}_j\langle \mathbf{v}\rangle = \big[a_{j1} \cdots a_{jn}\big]\big[v_1 \cdots v_n\big]^t = a_{j1}v_1 + \cdots + a_{jn}v_n,$$

a real number. Accordingly, $[\mathbf{R}_1\langle \mathbf{v}\rangle \cdots \mathbf{R}_n\langle \mathbf{v}\rangle]$ is a row; it is a member of $L(\mathbf{R}^n, \mathbf{R})$.

With row$^{(2)}$ and mixed-partial notation, our approximation for the change in f' becomes

$$f'(\mathbf{x}) - f'(\mathbf{b}) \approx \Bigg[\left[\frac{\partial^2 f}{\partial x_1 \partial x_1}(\mathbf{b}) \cdots \frac{\partial^2 f}{\partial x_1 \partial x_n}(\mathbf{b})\right]$$
$$\cdots\left[\frac{\partial^2 f}{\partial x_n \partial x_1}(\mathbf{b}) \cdots \frac{\partial^2 f}{\partial x_n \partial x_n}(\mathbf{b})\right]\Bigg]\langle \mathbf{x}-\mathbf{b}\rangle.$$

This suggests that the row$^{(2)}$ on the right is f''. We now make it official.

Theorem 1. *Assume that f is differentiable in a neighborhood of* **b**. *Then*:

(a) *f is twice differentiable at* $\mathbf{b}$ *iff each partial derivative* $\frac{\partial f}{\partial x_j}$ *is differentiable at* $\mathbf{b}$.

(b) *If f is twice differentiable, then $f''(\mathbf{b})$ is the same operator as the row*$^{(2)}$ *whose entries (constituent rows) are the derivatives of the* $\frac{\partial f}{\partial x_j}$*; that is,*

$$f''(\mathbf{b}) = \left[\left[\frac{\partial^2 f}{\partial x_1 \partial x_1}(\mathbf{b}) \cdots \frac{\partial^2 f}{\partial x_1 \partial x_n}(\mathbf{b})\right] \cdots \left[\frac{\partial^2 f}{\partial x_n \partial x_1}(\mathbf{b}) \cdots \frac{\partial^2 f}{\partial x_n \partial x_n}(\mathbf{b})\right]\right].$$

Proof. $\Leftarrow$ Suppose each $\frac{\partial f}{\partial x_j}$ is differentiable at $\mathbf{b}$. For each j,

$$\begin{aligned}\frac{\partial f}{\partial x_j}(\mathbf{x}) - \frac{\partial f}{\partial x_j}(\mathbf{b}) &= \left(\frac{\partial f}{\partial x_j}\right)' \langle \mathbf{x} - \mathbf{b} \rangle + o(\mathbf{x} - \mathbf{b}) \\ &= \left[\frac{\partial^2 f}{\partial x_j \partial x_1}(\mathbf{b}) \ \cdots \ \frac{\partial^2 f}{\partial x_j \partial x_n}(\mathbf{b})\right] \langle \mathbf{x} - \mathbf{b} \rangle + o(\mathbf{x} - \mathbf{b}).\end{aligned}$$

Therefore,

$$\begin{aligned} f'(\mathbf{x}) - f'(\mathbf{b}) &= \left[\frac{\partial f}{\partial x_1}(\mathbf{x}) \ \cdots \ \frac{\partial f}{\partial x_n}(\mathbf{x})\right] - \left[\frac{\partial f}{\partial x_1}(\mathbf{b}) \ \cdots \ \frac{\partial f}{\partial x_n}(\mathbf{b})\right] \\ &= \left[\left[\frac{\partial^2 f}{\partial x_1 \partial x_1}(\mathbf{b}) \ \cdots \ \frac{\partial^2 f}{\partial x_1 \partial x_n}(\mathbf{b})\right] \langle \mathbf{x} - \mathbf{b} \rangle \right. \\ &\quad \left. \cdots \left[\frac{\partial^2 f}{\partial x_n \partial x_1}(\mathbf{b}) \ \cdots \ \frac{\partial^2 f}{\partial x_n \partial x_n}(\mathbf{b})\right] \langle \mathbf{x} - \mathbf{b} \rangle \right] \\ &\quad + \left[o(\mathbf{x} - \mathbf{b}) \ \cdots \ o(\mathbf{x} - \mathbf{b})\right] = \Lambda \langle \mathbf{x} - \mathbf{b} \rangle + \mathbf{o}(\mathbf{x} - \mathbf{b}), \end{aligned}$$

where Λ is the row$^{(2)}$ of mixed partials. This proves simultaneously that (a) f' is differentiable, so f is twice differentiable, and (b) that $f'' = \Lambda$.

$\Rightarrow$ Suppose now that f is twice differentiable at $\mathbf{b}$. By definition of second derivative,

$$f'(\mathbf{x}) - f'(\mathbf{b}) = f''(\mathbf{b}) \langle \mathbf{x} - \mathbf{b} \rangle + \mathbf{o}(\mathbf{x} - \mathbf{b}).$$

The row on the left side is

$$\left[\frac{\partial f}{\partial x_1}(\mathbf{x}) - \frac{\partial f}{\partial x_1}(\mathbf{b}) \ \cdots \ \frac{\partial f}{\partial x_n}(\mathbf{x}) - \frac{\partial f}{\partial x_n}(\mathbf{b})\right].$$

Let Π_j represent the projection defined on row matrices by $\Pi_j([a_1 \cdots a_n]) \equiv a_j$. Then

$$\Pi_j(f'(\mathbf{x}) - f'(\mathbf{b})) = \frac{\partial f}{\partial x_j}(\mathbf{x}) - \frac{\partial f}{\partial x_j}(\mathbf{b}).$$

Since Π_j is linear, applying it to the right side gives

$$\Pi_j(f''(\mathbf{b})\langle \mathbf{x} - \mathbf{b} \rangle + \mathbf{o}(\mathbf{x} - \mathbf{b})) = \Pi_j(f''(\mathbf{b})\langle \mathbf{x} - \mathbf{b} \rangle) + \Pi_j(\mathbf{o}(\mathbf{x} - \mathbf{b})).$$

The expression $\Pi_j\big(\mathbf{o}(\mathbf{x}-\mathbf{b})\big)$ is just another $o(\mathbf{x}-\mathbf{b})$, because clearly for any row $\mathbf{A}$, $\|\Pi_j(\mathbf{A})\| \le \|\mathbf{A}\|$. Let us define

$$\mathbf{L}\langle\mathbf{v}\rangle \equiv \Pi_j\big(f''(\mathbf{b})\langle\mathbf{v}\rangle\big).$$

This $\mathbf{L}$ is linear, because it is a composite of two linear maps. We have arrived at

$$\frac{\partial f}{\partial x_j}(\mathbf{x}) - \frac{\partial f}{\partial x_j}(\mathbf{b}) = \mathbf{L}\langle\mathbf{x}-\mathbf{b}\rangle + o(\mathbf{x}-\mathbf{b}).$$

This says that each $\frac{\partial f}{\partial x_j}$ is differentiable at $\mathbf{b}$, proving the implication we want, and putting us back in the upper half of this argument, so that (b) follows. □

Example 1. Let $f(x, y) \equiv x^2y - xy^2$. Then f is always differentiable (Reason?), with

$$f'(x, y) = \begin{bmatrix}2xy - y^2 & x^2 - 2xy\end{bmatrix}.$$

The quantity $\frac{\partial f}{\partial x} = 2xy - y^2$ is differentiable, with derivative

$$\left(\frac{\partial f}{\partial x}\right)' = \begin{bmatrix}2y & 2x - 2y\end{bmatrix}.$$

Similarly, $\left(\frac{\partial f}{\partial y}\right)' = \begin{bmatrix}2x - 2y & -2x\end{bmatrix}$. Theorem 1 says that

$$f''(a, b) = \big[[2b \quad 2a - 2b] \quad [2a - 2b \quad -2a]\big].$$

To verify, we would need to show that at $\mathbf{c} = (a, b)$, we have

$$f'(\mathbf{x}) - f'(\mathbf{c}) - \big[[2b \quad 2a - 2b] \quad [2a - 2b \quad -2a]\big]\langle\mathbf{x}-\mathbf{c}\rangle = \mathbf{o}(\mathbf{x}-\mathbf{c}).$$

The increment $f'(\mathbf{x}) - f'(\mathbf{c})$ is $\begin{bmatrix}2xy - y^2 & x^2 - 2xy\end{bmatrix} - \begin{bmatrix}2ab - b^2 & a^2 - 2ab\end{bmatrix}$. The other term on the left is (by definition of the operator)

$$\left[[2b \quad 2a - 2b]\begin{bmatrix}x-a\\y-b\end{bmatrix} \quad [2a - 2b \quad -2a]\begin{bmatrix}x-a\\y-b\end{bmatrix}\right]$$
$$= [2b(x-a) + (2a-2b)(y-b) \quad (2a-2b)(x-a) - 2a(y-b)].$$

We leave it to the reader to show that the difference is

$$\left[2(x-a)(y-b) - (y-b)^2 \quad (x-a)^2 - 2(x-a)(y-b)\right],$$

each of whose entries is $o\big((x-a, y-b)\big) = o(\mathbf{x}-\mathbf{c})$.

We stated that in $\frac{\partial^2 f}{\partial x_j \partial x_k}$, the two differentiations are done in reading order. In our work, however, it turns out that the order does not matter.

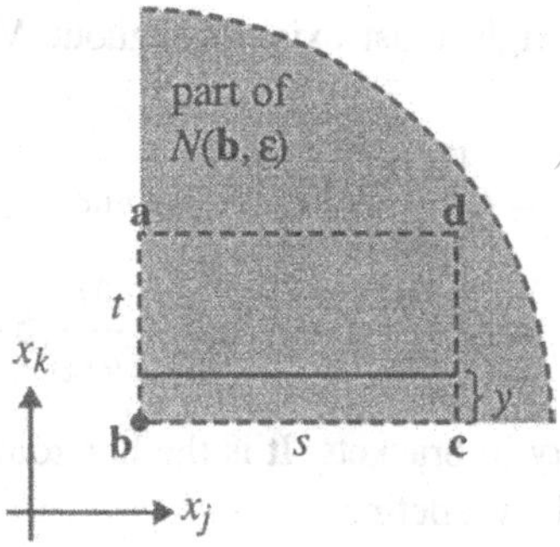

Figure 1.3.

Theorem 2. *Assume that each mixed partial derivative of f is defined near* **b** *and continuous at* **b**. *Then the mixed partials are symmetric; that is,*

$$\frac{\partial^2 f}{\partial x_j \partial x_k}(\mathbf{b}) = \frac{\partial^2 f}{\partial x_k \partial x_j}(\mathbf{b}).$$

Proof. Assume that the mixed partials are defined in $N(\mathbf{b}, \varepsilon)$. Look at the rectangle (dashed in Figure 1.3) with corners at $\mathbf{b}$, $\mathbf{c} \equiv \mathbf{b}+s\mathbf{e}_j$, $\mathbf{d} \equiv \mathbf{c}+t\mathbf{e}_k$, $\mathbf{a} \equiv \mathbf{b}+t\mathbf{e}_k$, where $s^2 + t^2 < \varepsilon^2$ (so all the points are in $N(\mathbf{b}, \varepsilon)$).

The idea of the proof is the following. Derivatives being limits of difference quotients, $\frac{\partial(\partial f/\partial x_j)}{\partial x_k}(\mathbf{b})$ should be roughly $\frac{\frac{\partial f}{\partial x_j}(\mathbf{a})-\frac{\partial f}{\partial x_j}(\mathbf{b})}{t}$. For the same reason, $\frac{\partial f}{\partial x_j}(\mathbf{a}) \approx \frac{f(\mathbf{d})-f(\mathbf{a})}{s}$ and $\frac{\partial f}{\partial x_j}(\mathbf{b}) \approx \frac{f(\mathbf{c})-f(\mathbf{b})}{s}$. Hence

$$\frac{\partial(\partial f/\partial x_j)}{\partial x_k}(\mathbf{b}) \approx \frac{\frac{f(\mathbf{d})-f(\mathbf{a})}{s} - \frac{f(\mathbf{c})-f(\mathbf{b})}{s}}{t};$$

similarly,

$$\frac{\partial(\partial f/\partial x_k)}{\partial x_j}(\mathbf{b}) \approx \frac{\frac{f(\mathbf{d})-f(\mathbf{c})}{t} - \frac{f(\mathbf{c})-f(\mathbf{b})}{t}}{s};$$

and clearly the two complex fractions are equal.

First, draw the horizontal at height y above the bottom of the rectangle, and look at the horizontal change in f (across the rectangle) as a function of y:

$$\Delta(y) \equiv f(\mathbf{c} + y\mathbf{e}_k) - f(\mathbf{b} + y\mathbf{e}_k), \qquad 0 \le y \le t.$$

Notice that $[f(\mathbf{d}) - f(\mathbf{a})] - [f(\mathbf{c}) - f(\mathbf{b})] = \Delta(t) - \Delta(0)$. Also, in each term of Δ, only x_k is varying. By definition of the partial derivative,

$$\frac{d\Delta}{dy} = \frac{\partial f}{\partial x_k}(\mathbf{c} + y\mathbf{e}_k) - \frac{\partial f}{\partial x_k}(\mathbf{b} + y\mathbf{e}_k).$$

Since the partials on the right must exist throughout $N(\mathbf{b}, \varepsilon)$, the mean value theorem applies, and

$$\begin{aligned}\Delta(t) - \Delta(0) &= t\left[\frac{d\Delta}{dy}(y^*)\right] \qquad (\text{some } 0 \le y^* \le t)\\ &= t\left[\frac{\partial f}{\partial x_k}(\mathbf{c} + y^*\mathbf{e}_k) - \frac{\partial f}{\partial x_k}(\mathbf{b} + y^*\mathbf{e}_k)\right].\end{aligned}$$

Next, examine the quantity in brackets. It is the horizontal change in $\frac{\partial f}{\partial x_k}$ at the $x_k = y^*$ level. Accordingly, we define

$$g^*(x) \equiv \frac{\partial f}{\partial x_k}\big(\mathbf{b} + x\mathbf{e}_j + y^*\mathbf{e}_k\big), \qquad 0 \le x \le s.$$

By hypothesis, the partial derivative $\frac{\partial(\partial f/\partial x_k)}{\partial x_j}$ exists in $N(\mathbf{b}, \varepsilon)$. Reasoning as above, we infer that

$$\begin{aligned}\frac{\partial f}{\partial x_k}(\mathbf{c} + y^*\mathbf{e}_k) - \frac{\partial f}{\partial x_k}(\mathbf{b} + y^*\mathbf{e}_k) &= g^*(s) - g^*(0)\\ &= s\left[\frac{dg^*}{dx}(x^*)\right] \qquad (\text{some } 0 \le x^* \le s)\\ &= s\frac{\partial(\partial f/\partial x_k)}{\partial x_j}\big(\mathbf{b} + x^*\mathbf{e}_j + y^*\mathbf{e}_k\big).\end{aligned}$$

Here x^* depends on y^*, but we know this: $\mathbf{b}^* \equiv \mathbf{b}+x^*\mathbf{e}_j+y^*\mathbf{e}_k$ is in the rectangle.

We have arrived at

$$[f(\mathbf{d}) - f(\mathbf{a})] - [f(\mathbf{c}) - f(\mathbf{b})] = ts\,\frac{\partial^2 f}{\partial x_k \partial x_j}(\mathbf{b}^*).$$

Applying similar reasoning to vertical changes, we get

$$[f(\mathbf{d}) - f(\mathbf{c})] - [f(\mathbf{a}) - f(\mathbf{b})] = st\,\frac{\partial^2 f}{\partial x_j \partial x_k}(\mathbf{b}^{**})$$

for some $\mathbf{b}^{**}$ in the rectangle. We conclude that $\frac{\partial^2 f}{\partial x_k \partial x_j}(\mathbf{b}^*) = \frac{\partial^2 f}{\partial x_j \partial x_k}(\mathbf{b}^{**})$.

Finally, observe that the same construction can be carried out in any neighborhood $N(\mathbf{b}, \delta) \subseteq N(\mathbf{b}, \varepsilon)$. Thus, in any such $N(\mathbf{b}, \delta)$, some value $\frac{\partial^2 f}{\partial x_k \partial x_j}(\mathbf{b}^*)$ matches some value $\frac{\partial^2 f}{\partial x_j \partial x_k}(\mathbf{b}^{**})$. Hence we can define two sequences $(\mathbf{b}_i^*)$ and $(\mathbf{b}_i^{**})$ converging to $\mathbf{b}$ with $\frac{\partial^2 f}{\partial x_k \partial x_j}(\mathbf{b}_i^*) = \frac{\partial^2 f}{\partial x_j \partial x_k}(\mathbf{b}_i^{**})$. Because the two mixed partials are continuous at $\mathbf{b}$, it follows that $\frac{\partial^2 f}{\partial x_k \partial x_j}(\mathbf{b}) = \frac{\partial^2 f}{\partial x_j \partial x_k}(\mathbf{b})$. □

Symmetry of the mixed partials allows us to think of second derivatives in either of two ways. We described the rows in the row of rows

$$\left[\left[\frac{\partial^2 f}{\partial x_1 \partial x_1}(\mathbf{b}) \cdots \frac{\partial^2 f}{\partial x_1 \partial x_n}(\mathbf{b})\right] \cdots \left[\frac{\partial^2 f}{\partial x_n \partial x_1}(\mathbf{b}) \cdots \frac{\partial^2 f}{\partial x_n \partial x_n}(\mathbf{b})\right]\right]$$

as the derivative of $\frac{\partial f}{\partial x_1}, \ldots,$ the derivative of $\frac{\partial f}{\partial x_n}$. If we reverse the denominators to read

$$\left[\left[\frac{\partial^2 f}{\partial x_1 \partial x_1}(\mathbf{b}) \cdots \frac{\partial^2 f}{\partial x_n \partial x_1}(\mathbf{b})\right] \cdots \left[\frac{\partial^2 f}{\partial x_1 \partial x_n}(\mathbf{b}) \cdots \frac{\partial^2 f}{\partial x_n \partial x_n}(\mathbf{b})\right]\right],$$

then the rows appear to be $\frac{\partial f'}{\partial x_1}, \frac{\partial f'}{\partial x_2}, \ldots$. In words: To construct the rows that make up f'', we may either write down the (row) derivative of each entry in f', or take the partial derivative of the one row representing f' with respect to each variable x_j.

As we did with differentiable functions, we have discussed twice-differentiable functions without pointing to any class of examples. It should come as no surprise that the elementary functions have second derivatives, as is evident from the next theorem.

Theorem 3. *Suppose that every mixed partial* $\frac{\partial^2 f}{\partial x_j \partial x_k}$ *is defined near* **b** *and continuous at* **b** *(same hypothesis as Theorem 2). Then* f *is twice differentiable at* **b**.

Proof. By hypothesis, the partial derivatives of $\frac{\partial f}{\partial x_j}$, $j = 1, \ldots, n$, are defined near **b** and continuous at **b**. Therefore (Theorem 1.2:2), $\frac{\partial f}{\partial x_j}$ is differentiable at **b**. By Theorem 1.1:3, $\frac{\partial f}{\partial x_j}$ is continuous at **b**, and f, in turn, is differentiable there. Also, f' has differentiable components, so f' is differentiable at **b**. □

Derivatives beyond the second are easy to define and messy to write. For that reason, we limit our treatment to a discussion of how one would use recursion and induction to define higher derivatives and establish their properties.

First, the $(i+1)$th derivative is the derivative of the ith. Thus, if f'' is defined in a neighborhood and differentiable at a point there, then f **is three-times differentiable (has a third derivative)** at the point; and similarly for **order** higher than 3.

Second, f has an $(i+1)$th derivative iff all the partials that go into the ith derivative are differentiable. Where this occurs, the $(i+1)$th derivative is a row$^{(i+1)}$ = row of rows$^{(i)}$ whose rows$^{(i)}$ are derivatives of the rows$^{(i-1)}$ that make up the ith derivative. Thus in $\mathbf{R}^2$,

$$f'' = \left[\left[\frac{\partial^2 f}{\partial x^2}\ \frac{\partial^2 f}{\partial x \partial y}\right]\left[\frac{\partial^2 f}{\partial y \partial x}\ \frac{\partial^2 f}{\partial y^2}\right]\right]$$

leads to

$$f''' = \left[\left[\frac{\partial}{\partial x}\left[\frac{\partial^2 f}{\partial x^2}\ \frac{\partial^2 f}{\partial x \partial y}\right]\ \frac{\partial}{\partial y}\left[\frac{\partial^2 f}{\partial x^2}\ \frac{\partial^2 f}{\partial x \partial y}\right]\right] \right.$$
$$\left.\left[\frac{\partial}{\partial x}\left[\frac{\partial^2 f}{\partial y \partial x}\ \frac{\partial^2 f}{\partial y^2}\right]\ \frac{\partial}{\partial y}\left[\frac{\partial^2 f}{\partial y \partial x}\ \frac{\partial^2 f}{\partial y^2}\right]\right]\right]$$
$$= \left[\left[\left[\frac{\partial^3 f}{\partial x^3}\ \frac{\partial^3 f}{\partial x \partial y \partial x}\right]\left[\frac{\partial^3 f}{\partial x^2 \partial y}\ \frac{\partial^3 f}{\partial x \partial y^2}\right]\right] \text{etc.}\right].$$

Last, one guarantee for f to have an $(i+1)$th derivative at a point is that the **mixed partials of order** $i+1$, that is, $\frac{\partial^{i+1} f}{\partial x_j \partial x_k \cdots}$ (with $i+1$ factors, not necessarily distinct, in the denominator), should be continuous at the point and defined nearby. In the case of such continuity, the mixed partials of order $i+1$ are symmetric.

Example 2. (a) We have seen that $f(x, y) \equiv x^2y - xy^2$ has

$$f''(x, y) = \big[[2y\ \ 2x - 2y]\ \ [2x - 2y\ \ -2x]\big].$$

Hence

$$f'''(x, y) = \left[\big[[0\ \ 2]\ [2\ \ -2]\big]\ \big[[2\ \ -2]\ [-2\ \ 0]\big]\right].$$

To verify this statement, we would have to show that

$$f''(x, y) - f''(a, b) \approx \left[\big[[0\ \ 2]\ [2\ \ -2]\big]\ \big[[2\ \ -2]\ [-2\ \ 0]\big]\right]\langle(x - a, y - b)\rangle.$$

We leave this check to Exercise 4.

(b) It is clear that $f'''' = \mathbf{O}$, but it takes a lot of writing. (We offer an extra exercise: Show that the kth derivative of $f(x_1, \ldots, x_n)$ has n^k entries enclosed by $1 + n + \cdots + n^{k-1}$ pairs of brackets.) Here f is a polynomial of degree 3. As with functions of a single variable, a polynomial of degree k always has zero $(k+1)$th derivative.

We end the section by giving evidence that higher derivatives of a vector function $\mathbf{f}(\mathbf{x}) \equiv \big(f_1(\mathbf{x}), \ldots, f_m(\mathbf{x})\big)$ can be taken, as with the first derivative, componentwise.

Suppose each f_j is twice differentiable. First, each f_j is differentiable. By Theorem 1.1:1, $\mathbf{f}$ is differentiable, and

$$\mathbf{f}'(\mathbf{x}) - \mathbf{f}'(\mathbf{b}) = \begin{bmatrix} f_1'(\mathbf{x}) - f_1'(\mathbf{b}) \\ \vdots \\ f_m'(\mathbf{x}) - f_m'(\mathbf{b}) \end{bmatrix}.$$

Because f_j is twice differentiable,

$$f_j'(\mathbf{x}) - f_j'(\mathbf{b}) = f_j''(\mathbf{b})\langle \mathbf{x} - \mathbf{b}\rangle + \mathbf{o}(\mathbf{x} - \mathbf{b}).$$

It follows that

$$\mathbf{f}'(\mathbf{x}) - \mathbf{f}'(\mathbf{b}) = \begin{bmatrix} f_1''(\mathbf{b})\langle\mathbf{x}-\mathbf{b}\rangle \\ \vdots \\ f_m''(\mathbf{b})\langle\mathbf{x}-\mathbf{b}\rangle \end{bmatrix} + \mathbf{o}(\mathbf{x}-\mathbf{b}).$$

Let us call

$$\Lambda \equiv \begin{bmatrix} f_1''(\mathbf{b}) \\ \vdots \\ f_m''(\mathbf{b}) \end{bmatrix}$$

a **column of rows$^{(2)}$** and define its operation by

$$\Lambda\langle\mathbf{v}\rangle \equiv \begin{bmatrix} f_1''(\mathbf{b})\langle\mathbf{v}\rangle \\ \vdots \\ f_m''(\mathbf{b})\langle\mathbf{v}\rangle \end{bmatrix}.$$

We then have

$$\mathbf{f}'(\mathbf{x}) - \mathbf{f}'(\mathbf{b}) = \Lambda\langle\mathbf{x}-\mathbf{b}\rangle + \mathbf{o}(\mathbf{x}-\mathbf{b}),$$

which tells us that (a) $\mathbf{f}$ is twice differentiable and (b) $\mathbf{f}''$ is the same as the column of rows$^{(2)}$ Λ whose rows$^{(2)}$ are the second derivatives of its components.

Conversely, suppose $\mathbf{f}$ is twice differentiable. Then

$$\mathbf{f}'(\mathbf{x}) - \mathbf{f}'(\mathbf{b}) = \mathbf{f}''(\mathbf{b})\langle\mathbf{x}-\mathbf{b}\rangle + \mathbf{o}(\mathbf{x}-\mathbf{b}).$$

Each side of this equation is a column of rows. (Such a thing is normally called a "matrix"; our name has the advantage of extending to "column of rows$^{(2)}$, ...) Consider the "projection" Π_j defined on columns of rows by

$$\Pi_j\left\langle \begin{bmatrix} \mathbf{R}_1 \\ \cdots \\ \mathbf{R}_m \end{bmatrix} \right\rangle \equiv \mathbf{R}_j.$$

Applying Π_j to the two sides of the previous equation, we get

$$\begin{aligned} f_j'(\mathbf{x}) - f_j'(\mathbf{b}) &= \Pi_j\left\langle \mathbf{f}''(\mathbf{b})\langle\mathbf{x}-\mathbf{b}\rangle\right\rangle + \Pi_j\langle\mathbf{o}(\mathbf{x}-\mathbf{b})\rangle \\ &= \Pi_j\left\langle \mathbf{f}''(\mathbf{b})\langle\mathbf{x}-\mathbf{b}\rangle\right\rangle + \mathbf{o}(\mathbf{x}-\mathbf{b}) \\ &= \Lambda\langle\mathbf{x}-\mathbf{b}\rangle + \mathbf{o}(\mathbf{x}-\mathbf{b}), \end{aligned}$$

where Λ is the linear composite given by $\Lambda\langle\mathbf{v}\rangle \equiv \Pi_j\langle\mathbf{f}''(\mathbf{b})\langle\mathbf{v}\rangle\rangle$. Consequently, f_j' is differentiable, f_j is twice differentiable, and by the previous paragraph, $\mathbf{f}''$ may be represented by the column whose rows$^{(2)}$ are $f_1'', \ldots, f_m''$.

Compare our uses of Π_j here and the similar scalar projections π_j in the proof of Theorem 1.1:1 and Π_j in Theorem 1.4:1. We did not look for an explicit expression for $\left(\frac{\partial f}{\partial x_j}\right)'$ in Theorem 1.4:1, whereas we did get one for g_j' in Theorem

1.1:1, choosing instead an argument whose recursive extension to higher order is revealed in the previous paragraph. In other words, the above argument for second derivatives indicates how we would develop an induction proof for the principle that a vector function is k-times differentiable iff each of its components is k-times differentiable, and the kth derivative of the function is the column of rows$^{(k)}$ that are the kth derivatives of the components.

Exercises

1. For $f(x, y) \equiv \exp(x^2 y)$:

   (a) Confirm that $\frac{\partial^2 f}{\partial x \partial y} = \frac{\partial^2 f}{\partial y \partial x}$.

   (b) Write f''.

2. Let $\mathbf{g}(x, y) \equiv \left(x^2 - y^2, xy\right)$. Write $\mathbf{g}'$ and $\mathbf{g}''$.

3. In Exercises 1 and 2, find the third derivatives f''' and $\mathbf{g}'''$.

4. In Example 2, show that the approximation at the end of (a) is actually an equality.

5. Is it possible to find $h(x, y)$ such that $\frac{\partial h}{\partial x} = y \exp(x^2)$, $\frac{\partial h}{\partial y} = x \sin^2 y$ for all x, y?

6. Assume that $\mathbf{f}$ and $\mathbf{g}$ are vector functions defined near $\mathbf{b}$ and continuous at $\mathbf{b}$. Suppose that in every neighborhood $N(\mathbf{b}, \varepsilon)$, there exist $\mathbf{x}$ and $\mathbf{y}$ such that $\mathbf{g}(\mathbf{x}) = \mathbf{f}(\mathbf{y})$. Prove that $\mathbf{g}(\mathbf{b}) = \mathbf{f}(\mathbf{b})$.

2

Derivatives of Scalar Functions

We have seen that derivatives of vector functions can be taken component by component. Consequently, much of our investigation of such derivatives reduces to the study of derivatives of real-valued functions. In this chapter we study some derivative properties for which reduction to the scalar case has an interesting variety of advantages: from being helpful (directional derivatives) to sensible (the mean value theorem) to necessary (maxima/minima).

2.1 Directional Derivatives and the Gradient

The partial derivatives of a function are rates of change in the directions of the coordinate axes ("coordinate directions"). There is nothing special about these; we select them when we establish a coordinate system. In this section we look at rates of change in arbitrary directions, and use them to introduce an important concept.

Definition. Given $f\colon \mathbf{R}^n \to \mathbf{R}$ defined near $\mathbf{b}$, let $\mathbf{u} \in \mathbf{R}^n$ be a unit vector. Write $g(t) \equiv f(\mathbf{b} + t\mathbf{u})$. If the derivative

$$\partial_{\mathbf{u}} f(\mathbf{b}) \equiv g'(0) = \lim_{t\to 0} \frac{f(\mathbf{b}+t\mathbf{u}) - f(\mathbf{b})}{t}$$

exists, then we call it the **directional derivative along u** (or **u-directional derivative** or **derivative in the direction of u**) of f at $\mathbf{b}$.

Theorem 1. *If f is differentiable at* **b**, *then every directional derivative exists there, and*

$$\partial_{\mathbf{u}} f(\mathbf{b}) = f'(\mathbf{b})\langle\mathbf{u}\rangle.$$

Proof. Writing $\mathbf{x}(t) \equiv \mathbf{b} + t\mathbf{u}$, we have $g(t) = f(\mathbf{x}(t))$.

The function $\mathbf{x}(t)$ is (of first degree and therefore) differentiable. Its derivative (Exercise 1.2:5) is $\langle\mathbf{u}\rangle$ (the column whose entries are $u_1, \ldots, u_n$). By hypothesis, f is differentiable at $\mathbf{b} = \mathbf{x}(0)$. Therefore, the chain rule applies, telling us that $g(t)$ is differentiable at $t = 0$, and

$$g'(0) = f'(\mathbf{x}(0))\mathbf{x}'(0) = f'(\mathbf{b})\langle\mathbf{u}\rangle. \qquad \square$$

Example 1. Consider $f(x, y) \equiv \left(x^2 + y^2\right)^{1/2}$ at (4, 3).

(a) In the direction of (1, 0), the directional derivative is

$$\partial_{\mathbf{i}} f(4,3) = \begin{bmatrix} \frac{\partial f}{\partial x}(4,3) & \frac{\partial f}{\partial y}(4,3) \end{bmatrix} \begin{bmatrix} 1 \\ 0 \end{bmatrix} = \frac{\partial f}{\partial x}(4,3).$$

This is always the case: In the (coordinate) direction of increasing x_j, the directional derivative is $\frac{\partial f}{\partial x_j}$.

(b) In the direction of the unit vector $\mathbf{u} \equiv (u_1, u_2)$,

$$\begin{aligned} \partial_{\mathbf{u}} f(4,3) &= \left[x\left(x^2+y^2\right)^{-1/2} \quad y\left(x^2+y^2\right)^{-1/2} \right] \left[u_1 \quad u_2 \right]^t \\ &= \frac{4}{5}u_1 + \frac{3}{5}u_2. \end{aligned}$$

This relation can be viewed geometrically. Let ϕ be the angle between the line of (4, 3) (the line containing (4.3) and the origin) and the x-axis; see Figure 2.1. Similarly, let **u** have inclination θ, so that $\mathbf{u} = (\cos\theta, \sin\theta)$. Then

$$\partial_{\mathbf{u}} f(4,3) = \frac{4}{5}u_1 + \frac{3}{5}u_2 = \cos\phi\cos\theta + \sin\phi\sin\theta = \cos(\theta - \phi).$$

In the figure, let us walk distance Δs from (4, 3) in the direction specified by **u**. Our distance from the origin increases by the length Δf past (4, 3) intercepted by the origin-centered arc. If Δs is small, then $\frac{\Delta f}{\Delta s} \approx \cos(\theta - \phi) = \partial_{\mathbf{u}} f$. Thus, $\partial_{\mathbf{u}} f$ is always the rate of change of f with respect to distance in the **u**-direction.

Example 2. The converse of Theorem 1 is false.

Let $g(x, y) \equiv \frac{y^3}{x^2+y^2}$, with $g(0, 0) \equiv 0$. If $\mathbf{u} = (\cos\theta, \sin\theta)$, then

$$\lim_{s\to 0} \frac{g(s\mathbf{u}) - g(\mathbf{O})}{s} = \lim_{s\to 0} \sin^3\theta = \sin^3\theta,$$

so all the directional derivatives of g exist at **O**. But g is not differentiable there (Exercise 5).

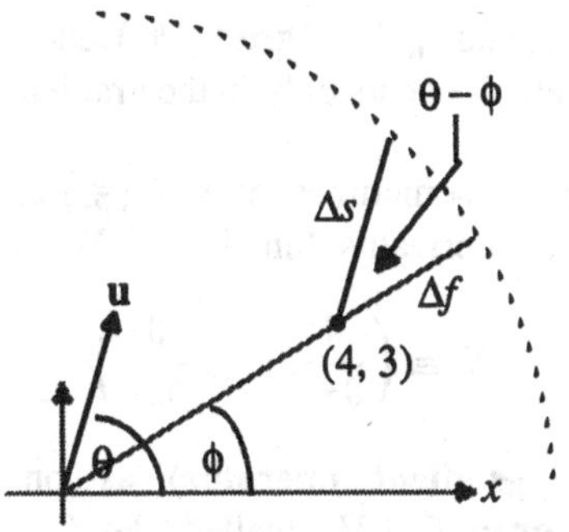

Figure 2.1.

We see in Example 2 that Theorem 1 belongs to a family of similar results. These say (compare Theorem 1.2:1) that where a vector-variable derivative exists, there the related scalar-variable derivatives exist, but not vice versa.

If f is real, then the product

$$f'(\mathbf{b})\langle\mathbf{u}\rangle = \left[\frac{\partial f}{\partial x_1}(\mathbf{b}) \cdots \frac{\partial f}{\partial x_n}(\mathbf{b})\right]\left[u_1 \cdots u_n\right]^t$$

is a 1×1 matrix, but we have identified it with its only entry, the real number $(\frac{\partial f}{\partial x_1})u_1 + \cdots + (\frac{\partial f}{\partial x_n})u_n$. On the other hand, this last quantity is the dot product of the two vectors $(\frac{\partial f}{\partial x_1}, \ldots, \frac{\partial f}{\partial x_n})$ and $(u_1, \ldots, u_n) = \mathbf{u}$.

Definition. The vector **grad** $f \equiv (\frac{\partial f}{\partial x_1}, \ldots, \frac{\partial f}{\partial x_n})$ is called the **gradient** of f.

Theorem 2. *Suppose f is differentiable at* **b**.

(a) *Along any unit vector* **u**, *the directional derivative of f is the component* [**grad** f] • **u** *of* **grad** f *in the direction of* **u**. *In particular,*

(b) *The gradient points in the direction of fastest increase of f, and* ‖**grad** f‖ *is the rate of increase in that direction.*

Proof. (a) Theorem 1 says that the directional derivative is $\partial_{\mathbf{u}} f(\mathbf{b}) = f'(\mathbf{b})\langle\mathbf{u}\rangle$, and we have seen that

$$f'(\mathbf{b})\langle\mathbf{u}\rangle = [(\mathbf{grad}\ f)(\mathbf{b})] \bullet \mathbf{u}.$$

(b) We also have

$$[\mathbf{grad}\ f] \bullet \mathbf{u} = \|\mathbf{grad}\ f\|\ \|\mathbf{u}\| \cos\theta,$$

where θ is the angle between **u** and **grad** f. Assume **grad** $f \neq \mathbf{O}$. If we pick $\mathbf{u}^*$ to point in the direction of the gradient, then $\partial_{\mathbf{u}^*} f = \|\mathbf{grad}\ f\|$. In every other direction, $\cos\theta < 1$, and $\partial_{\mathbf{u}} f < \|\mathbf{grad}\ f\|$. Hence $\|\mathbf{grad}\ f\|$ is the biggest possible rate of change, and it occurs only in the gradient's direction. □

In elementary calculus we sometimes think of $\left(\frac{d}{dt}\right)$ as an operator, which acts on $f = f(t)$ to produce the derivative function f'. We will similarly think of

$$\nabla \equiv \left(\frac{\partial}{\partial x_1}, \ldots, \frac{\partial}{\partial x_n}\right)$$

(the **del operator** or **gradient operator**) as one, whose action on $f(x_1, \ldots, x_n)$ produces $\mathbf{grad}\ f = \nabla f$ (called "del f").

In view of Theorem 2, we will write $\nabla f \bullet \mathbf{u}$ for the derivative of f along $\mathbf{u}$. In fact, since

$$f'(\mathbf{b})\langle\mathbf{v}\rangle = \left[\frac{\partial f}{\partial x_1} \cdots \frac{\partial f}{\partial x_n}\right] \left[v_1 \cdots v_n\right]^t = \nabla f \bullet \mathbf{v}$$

for every vector $\mathbf{v}$, we will frequently adopt the dot product notation for the action of the map $f'(\mathbf{b})$.

Exercises

1. Let $f(x, y) = x^2 - y^2$.
   (a) Find the directional derivative of f along $\mathbf{u} \equiv (\cos\theta, \sin\theta)$ at $(5, 3)$.
   (b) Show that $\partial_{\mathbf{u}} f(5, 3) = 0$ in the direction tangent to the graph of $x^2 - y^2 = 16$. (Hint: That direction is given by $\tan\theta = \frac{dy}{dx}$.)

2. Let $g(x, y) = x^2 + 4y^2$.
   (a) Write $\nabla g(a, b)$.
   (b) Show that at every $(a, b) \neq \mathbf{O}$, $\nabla g(a, b)$ is perpendicular to the tangent to the graph of $g(x, y) = a^2 + 4b^2$.

3. Let $h(x, y) = (x^2 + y^2)^{1/2}$.
   (a) Show that $\|\nabla h\| = 1$.
   (b) Explain (a) geometrically; that is, explain why at every point, the maximal rate of increase of h should work out to 1.

4. Given $F(x_1, \ldots, x_n)$, find the directional derivative of F at $(b_1, \ldots, b_n)$ in the direction of negative x_j.

5. Show that $G(x, y) \equiv \frac{y^3}{x^2+y^2}$, $\equiv 0$ at $(0, 0)$, is not differentiable at the origin.

6. Show that wherever f and g are both differentiable, $\nabla(fg) = g\nabla f + f\nabla g$.

7. Assume that f is differentiable.

   (a) Use Exercise 6 to show that $\nabla(f^2) = 2f\nabla f$.

   (b) Use the chain rule to prove the same relation.

   (c) At a given point, does f^2 increase fastest in the same direction that f does? Verify your answer using $f(x, y) \equiv xy$ at $(5, 3)$ and $(-2, 2)$.

8. Suppose f is twice differentiable near $\mathbf{b}$, and let $\mathbf{u}$, $\mathbf{v}$ be fixed unit vectors. Show that the directional derivative along $\mathbf{u}$ of the directional derivative along $\mathbf{v}$ of f at $\mathbf{b}$ is $[f''(\mathbf{b})\langle\mathbf{v}\rangle]\langle\mathbf{u}\rangle$.

2.2 The Mean Value Theorem

In the form

$$\frac{f(b) - f(a)}{b - a} = f'(t^*),$$

the mean value theorem for a single variable says that the average rate of change of a function is one of the values of the instantaneous rate of change. The form is no good for us, because we cannot divide by change in our vector variable. But

$$f(b) - f(a) = f'(t^*)[b - a]$$

works, and we can directly translate the one-variable theorem.

We work with a scalar function f defined in an open subset O of $\mathbf{R}^n$.

Theorem 1 (Mean Value Theorem). *Assume that the segment* **ab** *from* **a** *to* **b** *(which includes* **a** *and* **b***) is contained in* O. *Suppose* f *is continuous along the segment and differentiable between* **a** *and* **b** *(at the points of the segment other than* **a** *and* **b***). Then there exists* $\mathbf{x}^*$ *on* **ab** *such that*

$$f(\mathbf{b}) - f(\mathbf{a}) = \nabla f(\mathbf{x}^*) \bullet (\mathbf{b} - \mathbf{a}).$$

Proof. Our usual parametrization of the segment is

$$\mathbf{x}(t) = (1 - t)\mathbf{a} + t\mathbf{b}, \qquad 0 \le t \le 1.$$

Define $g(t) \equiv f(\mathbf{x}(t))$. This g is a composite of continuous functions for $t \in [0, 1]$ and of differentiable functions for $t \in (0, 1)$. Hence g is a continuous function on the closed interval, differentiable on the open interval.

By the single-variable mean value theorem,

$$f(\mathbf{b}) - f(\mathbf{a}) = g(1) - g(0) = g'(t^*),$$

for some t^* between 0 and 1. By the chain rule,

$$g'(t^*) = f'(\mathbf{x}(t^*))\mathbf{x}'(t^*) = f'(\mathbf{x}^*)\langle \mathbf{b} - \mathbf{a}\rangle = \nabla f(\mathbf{x}^*) \bullet (\mathbf{b} - \mathbf{a}). \qquad \square$$

The mean value theorem is not a special case of a vector-function result. That is, it is not true that if $\mathbf{g}(\mathbf{x})$ is continuous and differentiable in the right places, then $\mathbf{g}(\mathbf{b}) - \mathbf{g}(\mathbf{a}) = \mathbf{g}'(\mathbf{x}^*)\langle \mathbf{b} - \mathbf{a}\rangle$.

Example 1. Let the position $\mathbf{r}$ of a particle be given by $\mathbf{r}(t) \equiv (\cos t, \sin t)$. Then $\mathbf{r}(2\pi) - \mathbf{r}(0) = \mathbf{O}$, but $\mathbf{r}'(t^*)\langle 2\pi - 0\rangle = 2\pi[-\sin t^* \;\; \cos t^*]^t$ is never $\mathbf{O}$.

The components $x(t)$ and $y(t)$ of $\mathbf{r}$ separately satisfy the theorem. Thus,

$$0 = \frac{x(2\pi) - x(0)}{2\pi - 0} = x'(t^*) \qquad \text{for } t^* = \pi,$$
$$0 = \frac{y(2\pi) - y(0)}{2\pi - 0} = y'(t^{**}) \qquad \text{for } t^{**} = \frac{\pi}{2} \text{ or } \frac{3\pi}{2},$$

although at no time are both derivatives zero.

We may recast the mean value theorem as

$$\frac{f(\mathbf{b}) - f(\mathbf{a})}{\|\mathbf{b} - \mathbf{a}\|} = \nabla f(\mathbf{x}^*) \bullet \frac{\mathbf{b} - \mathbf{a}}{\|\mathbf{b} - \mathbf{a}\|}.$$

It is natural to refer to the left side as the **average rate of change of f (from a to b) with (respect to) distance**. On the right, we have $\nabla f(\mathbf{x}^*) \bullet \mathbf{u}$, where $\mathbf{u} = \frac{(\mathbf{b}-\mathbf{a})}{\|\mathbf{b}-\mathbf{a}\|}$ is the unit vector in the direction from $\mathbf{a}$ to $\mathbf{b}$. By Theorem 2.1:2, $\nabla f(\mathbf{x}^*) \bullet \mathbf{u}$ is the rate of change of f along the (directed) segment. Thus, in agreement with the case of one variable, the multivariable mean value theorem says that the average rate of change of a function from here to there matches some instantaneous directional rate along the trip.

A second rigid element of the theorem is the line segment. That is, we may not replace the segment with a curved path.

Example 2. Let $f(x, y) \equiv x^2y + y^3$. From $\mathbf{a} = (0, -1)$ to $\mathbf{b} = (0, 1)$, the average rate of change in f is $\frac{f(\mathbf{b}) - f(\mathbf{a})}{2} = 1$. The vertical directional is

$$\partial_{\mathbf{j}} f(x, y) = \left[2xy \;\; x^2 + 3y^2\right][0 \;\; 1]^t = x^2 + 3y^2.$$

At a point $(0, y)$ along the segment, $\partial_{\mathbf{j}} f(0, y) = 3y^2$, which matches 1 at $\left(0, \pm\frac{1}{\sqrt{3}}\right)$. But along the ellipse given by $x^2 + 3y^2 = 3$, whose right half also joins the points, $\partial_{\mathbf{j}} f(x, y) = 3$ never matches 1.

Example 2 shows us that the vector-variable mean value theorem cannot look around corners. This is not a handicap in our most frequent use of the mean value theorem, which occurs within a neighborhood ("locally"). There, the segment joining two points is necessarily a subset of the neighborhood. (What is that property called?) To deal with the entirety of an open set ("globally"), we have to hope that it suffices to break up the trip from one place to another into a finite number of straight pieces.

An example of such partitionable trips is the theorem that represents the first important application of the real mean value theorem: Zero derivative implies constant function. To prove it, we need a topological result—which will give us some mileage later, as well—having to do with straight-piece approximation.

Definition. In a subset $S \subseteq \mathbf{R}^n$, a **broken line** (or **polygonal path**) **from a to b** is a union $\mathbf{p}_0\mathbf{p}_1 \ldots \mathbf{p}_k \equiv \mathbf{p}_0\mathbf{p}_1 \cup \ldots \cup \mathbf{p}_{k-1}\mathbf{p}_k$ of segments contained in S with $\mathbf{a} = \mathbf{p}_0$, $\mathbf{b} = \mathbf{p}_k$.

Theorem 2. *Assume that* **a** *and* **b** *are points in a connected open set O.*

(a) *There is within O a broken line from* **a** *to* **b**.

(b) *Any arc within O from* **a** *to* **b** *can be approximated by a broken line. Specifically, if an $\varepsilon > 0$ and an arc are specified, then there is within O a broken line from* **a** *to* **b** *such that each of its points is at distance less than ε from the arc and each point of the arc is less than ε from the broken line.*

Proof. (a) Every connected open set is arc-connected [Guzman, Section 5.1]. This means that there exists at least one arc within O from **a** to **b**. Consequently, if we demonstrate (b), then (a) follows.

(b) Assume that we are given $\varepsilon > 0$ and a continuous (arc) $\mathbf{f}\colon [r, s] \to O$ with $\mathbf{f}(r) = \mathbf{a}$ and $\mathbf{f}(s) = \mathbf{b}$. For each $t \in [r, s]$, $\mathbf{f}(t) \in O$. Since O is open, there is a neighborhood $N(t) \equiv N(\mathbf{f}(t), \Delta(t)) \subseteq O$. We may assume every $\Delta(t) < \varepsilon$. Because **f** is continuous, there is a neighborhood $I(t) \equiv \big(t - \delta(t), t + \delta(t)\big)$ whose image is contained in $N(t)$ and therefore in O.

The intervals $I(t)$ cover $[r, s]$. By the Heine–Borel theorem, there is a finite subcollection $I(t_1), \ldots, I(t_k)$ that still covers $[r, s]$. We will use some of the images $\mathbf{f}(t_j)$ as vertices for the broken line.

Let $\mathbf{p}_0 \equiv \mathbf{a} = \mathbf{f}(r)$. Among the intervals $I(t_j)$, one must hold r; call it $I(t_1{}^*)$. Since $\mathbf{f}(I(t_1{}^*)) \subseteq N(t_1{}^*)$, both $\mathbf{p}_0$ and $\mathbf{p}_1 \equiv \mathbf{f}(t_1{}^*)$ are in $N(t_1{}^*)$, and so the segment $\mathbf{p}_0\mathbf{p}_1$ is a subset of $N(t_1{}^*) \subseteq O$. (See Figure 2.2, which is contained in O.)

If $I(t_1{}^*)$ reaches rightward past s, then $\mathbf{f}(s)$ is also in $N(t_1{}^*)$, and we can finish with $\mathbf{p}_2 \equiv \mathbf{f}(s) = \mathbf{b}$. If not, then the end $t_1{}^* + \delta(t_1{}^*)$ of $I(t_1{}^*)$ is in some interval $I(t_2{}^*)$. The two intervals overlap, so the two neighborhoods $N(t_1{}^*)$ and $N(t_2{}^*)$ overlap. That being the case, the segment joining their centers is contained in their union (Exercise 9). Taking $\mathbf{p}_2 \equiv \mathbf{f}(t_2{}^*)$, we have $\mathbf{p}_1\mathbf{p}_2 \subseteq N(t_1{}^*) \cup N(t_2{}^*) \subseteq O$.

We cannot continue indefinitely this march through $[r, s]$. We eventually reach $t_m^* + \delta(t_m^*) > s$, so that $\mathbf{p}_{m+1} \equiv \mathbf{f}(s) = \mathbf{b}$ and $\mathbf{p}_m \equiv \mathbf{f}(t_m^*)$ are in the last

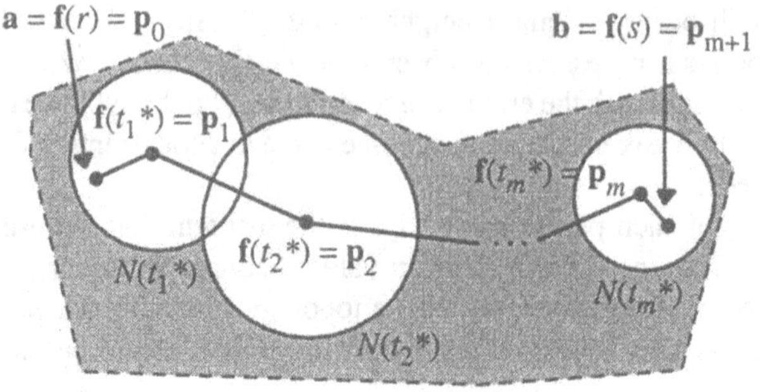

Figure 2.2.

neighborhood $N(t_m^*)$. Then $\mathbf{p}_0\mathbf{p}_1\ldots\mathbf{p}_{m+1}$ is a polygonal path within O from **a** to **b**. It remains only to show that it is close to the arc.

By construction, the image $\mathbf{f}([r, t_1^*])$ is a subset of $N(t_1^*)$. Therefore, every point of that part of the arc is less than ε from $\mathbf{f}(t_1^*)$, which is on the broken line, and similarly with $\mathbf{f}([t_m^*, s])$ within $N(t_m^*)$. For the in-between pieces, the image of each $[t_j^*, t_{j+1}^*]$ (or $[t_{j+1}^*, t_j^*]$, whichever is the right order) is a subset of the union of the overlapping neighborhoods $N(t_j^*)$ and $N(t_{j+1}^*)$ (recall Figure 2.2). Every member of that union is less than ε from one of $\mathbf{f}(t_j^*)$ and $\mathbf{f}(t_{j+1}^*)$, both of which are vertices of the broken line. We conclude that each point of the curve lies less than ε from the broken line.

On the other hand, the broken line is also a subset of the union of the neighborhoods $N(t_j^*)$, and the points $\mathbf{f}(t_j^*)$ are on the arc. Therefore, no point of the polygonal path is as much as ε from the arc. □

Theorem 3. *Assume that* $g'(\mathbf{x})$ *is* **O** *throughout a connected open set. Then* g *is constant there.*

Proof. Suppose O is the set and **a** and **b** are in O. By Theorem 2, there is a polygonal path $\mathbf{p}_0\mathbf{p}_1\ldots\mathbf{p}_k$ from **a** to **b** within O. By hypothesis, g is differentiable along each segment $\mathbf{p}_j\mathbf{p}_{j+1}$.

Applying Theorem 1 to each segment, we have

$$g(\mathbf{p}_{j+1}) - g(\mathbf{p}_j) = \nabla g(\mathbf{x}_j^*) \bullet (\mathbf{p}_{j+1} - \mathbf{p}_j) = 0,$$

because ∇g is **O** everywhere. Hence $g(\mathbf{p}_j) = g(\mathbf{p}_{j+1})$, $j = 0, 1, \ldots, k-1$, and we conclude that $g(\mathbf{a}) = g(\mathbf{b})$. □

Exercises

1. Find a place **c** on the segment **ab** at which $f(\mathbf{b}) - f(\mathbf{a}) = \nabla f(\mathbf{c}) \bullet (\mathbf{b} - \mathbf{a})$:

(a) $f(x, y) \equiv \left(x^2 + y^2\right)^{1/2}$, $\mathbf{a} = (3, -4)$, $\mathbf{b} = (3, 4)$.

(b) $f(x, y) \equiv e^{x+y}$, $\mathbf{a} = (1, 0)$, $\mathbf{b} = (2, 1)$.

2. Give an example of a function that is differentiable and has zero derivative throughout an open set, but is not constant there.

3. (a) Give an example of a function f continuous everywhere and differentiable for all but two points $\mathbf{a}$ and $\mathbf{b}$ in a convex open set.

   (b) For your function, find $\mathbf{c}$ on the segment $\mathbf{ab}$ with $f(\mathbf{b}) - f(\mathbf{a}) = \nabla f(\mathbf{c}) \bullet (\mathbf{b} - \mathbf{a})$.

4. Assume that $\mathbf{F}\colon \mathbf{R}^n \to \mathbf{R}^m$ is defined on a connected open set. Assume that $\mathbf{F}' = \mathbf{O}$ throughout the set. Prove that $\mathbf{F}$ is constant there.

5. Assume that $\mathbf{h}\colon \mathbf{R}^n \to \mathbf{R}^m$ is twice differentiable and $\mathbf{h}'' = \mathbf{O}$ throughout a connected open set O. Show that $\mathbf{h}$ is of first degree: There is a linear map $\mathbf{M}$ and vector $\mathbf{b}$ such that $\mathbf{h}(\mathbf{x}) = \mathbf{M}(\mathbf{x}) + \mathbf{b}$ for all $\mathbf{x} \in O$.

6. Assume that $\mathbf{g}\colon \mathbf{R}^n \to \mathbf{R}^m$ is defined on a connected open set O. We say that $\mathbf{g}$ is **Lipschitz** (or **Lipschitz of degree** r) if there exists M such that $\|\mathbf{g}(\mathbf{x}) - \mathbf{g}(\mathbf{y})\| \le M\|\mathbf{x} - \mathbf{y}\|$ (respectively $\le M\|\mathbf{x} - \mathbf{y}\|^r$) for $\mathbf{x}, \mathbf{y} \in O$.

   (a) Show that if $\mathbf{g}$ is Lipschitz of degree r for some $r > 0$, then $\mathbf{g}$ is uniformly continuous.

   (b) Show that if $\mathbf{g}$ is Lipschitz of degree $r > 1$, then $\mathbf{g}$ is constant.

   (c) Suppose O is convex, $\mathbf{g}$ is differentiable on O, and $\mathbf{g}'$ is a bounded function. (In symbols, the Pythagorean norm

$$\|\mathbf{g}'\| \equiv \left(\sum_{1 \le j \le m, 1 \le k \le n} \left[\frac{\partial g_j}{\partial x_k}\right]^2 \right)^{1/2}$$

   is a bounded real function.) Show that $\mathbf{g}$ is Lipschitz on O.

   (d) Suppose $\mathbf{g}$ is differentiable on O and $\mathbf{g}'$ is continuous at $\mathbf{b}$. Show that $\mathbf{g}$ is Lipschitz in some neighborhood of $\mathbf{b}$ ("$\mathbf{g}$ is locally Lipschitz at $\mathbf{b}$").

7. On the open interval $(-1, 1)$, give examples of functions that are:

   (a) Lipschitz;

   (b) not Lipschitz;

   (c) Lipschitz of degree $\frac{1}{2}$, but not of any bigger degree (Justify!);

   (d) Lipschitz, but not everywhere differentiable;

   (e) locally Lipschitz (for each b, there is $M(b)$ such that $|f(x) - f(y)| < M(b)|x - y|$ near b) but not Lipschitz;

(f) *uniformly continuous, but not Lipschitz of any positive degree (showing that the converse of 6(a) is false).

8. (a) *Show that convexity is essential in 6(c): Construct a function that has bounded derivative on some (necessarily nonconvex) connected open set but is not Lipschitz there.

   (b) Show that continuity is essential in 6(d): Give an example of a function that is differentiable throughout an open set but is at some point not locally Lipschitz (has a point for which it fails to be Lipschitz in every neighborhood).

9. Show that if two neighborhoods have nonempty intersection, then the segment joining their centers is a subset of their union. (Hint: If $N(\mathbf{b}, r)$ and $N(\mathbf{c}, s)$ overlap, then $r + s$ exceeds the distance from $\mathbf{b}$ to $\mathbf{c}$. [See pictures: Guzman, Theorem 2.2.])

10. ("differentiation under the integral sign") Suppose $f(x, y, z)$ is differentiable on the unit cube, and f' is continuous there.

   (a) Show that for fixed x and y, $f(x, y, z)$ and $\frac{\partial f}{\partial x}(x, y, z)$ are Riemann integrable functions of z.

   (b) Show that $\frac{\partial}{\partial x}\left(\int_0^1 f(x, y, z)\, dz\right) = \int_0^1 \frac{\partial f}{\partial x}(x, y, z)\, dz$.

   (c) Show that $\int_0^1 f(x, y, z)\, dz$ is a differentiable function of (x, y).

2.3 Extreme Values and the Derivative

The derivatives of a function help identify maxima and minima. Although the extra degrees of freedom possessed by a vector variable add complications, the one-variable results have close cousins in higher dimensions. In this section we generalize the extreme-value results related to the (first) derivative.

Definition. We say that f **has a minimum** at $\mathbf{b}$, or $f(\mathbf{b})$ **is a minimum,** if there is a neighborhood $N(\mathbf{b}, \varepsilon)$ in which $\mathbf{x} \in N(\mathbf{b}, \varepsilon) \Rightarrow f(\mathbf{x}) \geq f(\mathbf{b})$. The minimum is **strict** if $f(\mathbf{x}) > f(\mathbf{b})$ for $\mathbf{x} \neq \mathbf{b}$ in $N(\mathbf{b}, \varepsilon)$. **Maximum** and **strict maximum** are defined similarly, substituting $f(\mathbf{x}) \leq f(\mathbf{b})$ and $f(\mathbf{x}) < f(\mathbf{b})$.

We lump maxima and minima together under the name **extremes.** In this section, all the extremes we discuss are local (synonym: "relative"). We make no attempt to describe global ("absolute") extremes, since working mainly within open sets, we do not even know whether our functions are bounded.

Theorem 1. *Assume that f has an extreme at a place $\mathbf{b}$ where its derivative is defined. Then $f'(\mathbf{b}) = \mathbf{O}$.*

Proof. It suffices to deal with a minimum: Assume that $f(\mathbf{x}) \geq f(\mathbf{b})$ in a neighborhood $N(\mathbf{b}, \varepsilon)$.

Let $\mathbf{u}$ be any unit vector. By hypothesis, $\nabla f(\mathbf{b}) = f'(\mathbf{b})$ exists. By Theorem 2.1:2(a), the directional derivative $\partial_{\mathbf{u}} f(\mathbf{b})$ exists and is $\nabla f(\mathbf{b}) \bullet \mathbf{u}$. By definition,

$$\partial_{\mathbf{u}} f(\mathbf{b}) = \lim_{t \to 0} \frac{f(\mathbf{b} + t\mathbf{u}) - f(\mathbf{b})}{t}.$$

For $0 < t < \varepsilon$, we have $\frac{f(\mathbf{b}+t\mathbf{u}) - f(\mathbf{b})}{t} \geq 0$, because $f(\mathbf{b} + t\mathbf{u}) \geq f(\mathbf{b})$. Of necessity,

$$\partial_{\mathbf{u}} f(\mathbf{b}) = \lim_{t \to 0^+} \frac{f(\mathbf{b} + t\mathbf{u}) - f(\mathbf{b})}{t} \geq 0.$$

If instead $-\varepsilon < t < 0$, then $f(\mathbf{b} + t\mathbf{u}) \geq f(\mathbf{b})$ still holds, so $\frac{f(\mathbf{b}+t\mathbf{u}) - f(\mathbf{b})}{t} \leq 0$. Hence

$$\partial_{\mathbf{u}} f(\mathbf{b}) = \lim_{t \to 0^-} \frac{f(\mathbf{b} + t\mathbf{u}) - f(\mathbf{b})}{t} \leq 0.$$

We conclude that $\partial_{\mathbf{u}} f(\mathbf{b}) = 0$.

We have found that $\nabla f(\mathbf{b}) \bullet \mathbf{u} = 0$ for every unit vector. It follows that $\nabla f(\mathbf{b}) = \mathbf{O}$. □

If $\mathbf{b}$ is in the (open) domain of f and $f'(\mathbf{b})$ is either undefined or zero, we call $\mathbf{b}$ a **critical point** of f. Theorem 1 says that an extreme can occur only at a critical point.

We know from elementary calculus that the converse is false. Hence we need criteria to decide whether a given critical point is really an extreme. We proceed by analogy to the elementary case, beginning with an equivalent of the first derivative test.

Theorem 2. *Assume that f is continuous at $\mathbf{b}$. Suppose that for $\mathbf{x} \neq \mathbf{b}$ in some neighborhood $N(\mathbf{b}, \varepsilon)$, the derivative $\partial_{\mathbf{u}} f(\mathbf{x})$ exists in the direction of $\mathbf{u} \equiv \frac{\mathbf{x}-\mathbf{b}}{\|\mathbf{x}-\mathbf{b}\|}$. If every such* **outward directional derivative** *is nonnegative (respectively nonpositive, positive, negative), then $f(\mathbf{b})$ is a minimum (respectively maximum, strict minimum, strict maximum).*

Proof. Under the hypothesis that the outward directionals exist in $N(\mathbf{b}, \varepsilon)$, fix $\mathbf{c} \neq \mathbf{b}$ in $N(\mathbf{b}, \varepsilon)$. Let $r \equiv \|\mathbf{c} - \mathbf{b}\|$ and $\mathbf{u} \equiv \frac{\mathbf{c}-\mathbf{b}}{r}$.

We will look at f along the segment $\mathbf{bc}$. Write $\mathbf{x}(t) \equiv (1-t)\mathbf{b} + t\mathbf{c}$, and define

$$g(t) \equiv f(\mathbf{x}(t)), \qquad 0 \leq t \leq 1.$$

By the chain rule, g is a differentiable composite, with

$$g'(t) = f'(\mathbf{x}(t))\mathbf{x}'(t) = f'(\mathbf{x}(t))\langle \mathbf{c} - \mathbf{b} \rangle = r\, f'(\mathbf{x}(t))\langle \mathbf{u} \rangle = r\, \partial_{\mathbf{u}} f(\mathbf{x}(t)).$$

Now $f(\mathbf{c}) - f(\mathbf{b}) = g(1) - g(0)$. By the real mean value theorem,

$$g(1) - g(0) = g'(t^*), \quad \text{for some } 0 < t^* < 1.$$

Therefore, $f(\mathbf{c}) - f(\mathbf{b}) = r\,\partial_{\mathbf{u}} f(\mathbf{x}(t^*))$. It follows that if $\partial_{\mathbf{u}} f \ge 0$ throughout $N(\mathbf{b}, \varepsilon)$, then likewise $f(\mathbf{c}) \ge f(\mathbf{b})$, making $f(\mathbf{b})$ a minimum. The other cases correspond to $\partial_{\mathbf{u}} f \le 0$, $\partial_{\mathbf{u}} f > 0$, $\partial_{\mathbf{u}} f < 0$. □

Example 1. Let $f(x, y) \equiv \exp([x - y]^2)$.

(a) We obtain

$$\nabla f = \left(2[x - y]e^{[x-y]^2},\ -2[x - y]e^{[x-y]^2}\right).$$

This gradient is always defined, and $\nabla f = \mathbf{O}$ iff $x = y$. Therefore, the set of critical points is the line $x = y$.

(b) Pick $\mathbf{b} = (t, t)$ along the line. If $\mathbf{x} \equiv (x, y)$, then

$$\mathbf{u} \equiv \frac{\mathbf{x} - \mathbf{b}}{\|\mathbf{x} - \mathbf{b}\|} = \frac{(x - t, y - t)}{\|\mathbf{x} - \mathbf{b}\|}.$$

Therefore,

$$\begin{aligned}\partial_{\mathbf{u}} f(x, y) &= \nabla f \bullet \mathbf{u} = 2[x - y]e^{[x-y]^2} \frac{[x - t] - [y - t]}{\|\mathbf{x} - \mathbf{b}\|} \\ &= 2e^{(x-y)^2} \frac{(x - y)^2}{\sqrt{(x - t)^2 + (y - t)^2}}.\end{aligned}$$

Clearly, $\partial_{\mathbf{u}} f(x, y) \ge 0$, and we conclude that f has a loose (not strict) minimum at (t, t).

In Theorem 2, the hypothesis that the outward directional derivatives exist is not hard to satisfy. Frequently, we work with functions that are differentiable all around $\mathbf{b}$, but maybe not at $\mathbf{b}$. In that case, existence of the nearby directionals is guaranteed. Even if every neighborhood of $\mathbf{b}$ has places where the derivative is undefined, it may still turn out that the outward directionals exist.

Example 2. Let $g(x, y) \equiv x^{1/3}y^{1/3}$. Away from the coordinate axes,

$$\nabla g = \left(\frac{x^{-2/3}y^{1/3}}{3}, \frac{x^{1/3}y^{-2/3}}{3}\right) = \frac{1}{3}x^{-2/3}y^{-2/3}(y, x)$$

is always defined and is never zero. On the axes, g is not differentiable (Exercise 1.2:4e), so the critical points constitute the two axes.

(a) Suppose $\mathbf{x} \equiv (x, y)$ is near $\mathbf{a} \equiv (a, 0)$, $a \ne 0$. Let $\mathbf{u} \equiv \frac{\mathbf{x}-\mathbf{a}}{\|\mathbf{x}-\mathbf{a}\|}$. If $\mathbf{x}$ is on the x-axis, then $\mathbf{u} = \pm(1, 0)$. Hence $\partial_{\mathbf{u}} g(\mathbf{x}) = \pm\frac{\partial g}{\partial x}(\mathbf{x}) = 0$, because g is identically zero along the axis. Thus, the outward $\mathbf{u}$-directional at $\mathbf{x}$ exists. If instead $\mathbf{x}$ is not on the x-axis, then $\nabla g(x, y)$ exists, because $x \approx a$ means that (x, y) is off the y-axis as well. Therefore,

$$\partial_{\mathbf{u}} g(\mathbf{x}) = \nabla g(x, y) \bullet \mathbf{u} = \frac{1}{3}x^{-2/3}y^{-2/3}\frac{(y[x - a] + xy)}{\|\mathbf{x} - \mathbf{a}\|};$$

again the outward **u**-directional exists, and

$$\partial_{\mathbf{u}} g(\mathbf{x}) \approx \frac{a^{1/3} y^{1/3}}{3\|\mathbf{x} - \mathbf{a}\|}.$$

For negative a and $x \approx a$, the outward directionals are positive below the x-axis ($y < 0$) and negative above ($y > 0$). By the reasoning in Theorem 2—in other words, by the mean value theorem—$g(x, y) > g(a, 0)$ below $(a, 0)$ and $g(x, y) < g(a, 0)$ above. We conclude that $g(a, 0)$ is neither a maximum nor a minimum.

An analogous treatment shows that there are no extremes along the positive x-axis, negative y-axis, or positive y-axis.

(b) Suppose now that $\mathbf{x} \equiv (x, y)$ is near $(0, 0)$. There, $\mathbf{u} = \frac{\mathbf{x}}{\|\mathbf{x}\|}$. As in (a), if $\mathbf{x}$ is on the x-axis (or y-axis), then $\partial_{\mathbf{u}} g(\mathbf{x}) = \pm\frac{\partial g}{\partial x}(x, 0) = 0$ (respectively, $\partial_{\mathbf{u}} g(\mathbf{x}) = \pm\frac{\partial g}{\partial y}(0, y) = 0$). If instead $\mathbf{x}$ is in a quadrant, then

$$\partial_{\mathbf{u}} g(\mathbf{x}) = \frac{1}{3} \frac{x^{-2/3} y^{-2/3} (y, x) \bullet (x, y)}{\|(x, y)\|} = \frac{2}{3} \frac{x^{1/3} y^{1/3}}{\|(x, y)\|}.$$

These directionals are positive precisely in Quadrants I and III, so $g(x, y)$ exceeds $g(0, 0)$ to the upper right and lower left, and the opposite in the perpendicular directions. We conclude that $(0, 0)$ is not an extreme of g.

The situation in Example 2(b) is of separate interest to us. There, along any line going up to the right, like $y = x$, the function value at the origin is the smallest possible. Along a line with negative slope, like $y = -x$, the value is biggest.

Definition. The function f has a **saddle point** at **b** if $f(\mathbf{b})$ is a strict maximum of $f(\mathbf{x})$ along one line and a strict minimum along another line through **b**.

In symbols, **b** is a saddle point if there are $\varepsilon > 0$ and unit vectors $\mathbf{u}_1$ and $\mathbf{u}_2$ such that $f(\mathbf{b} + t\mathbf{u}_1) < f(\mathbf{b})$ and $f(\mathbf{b} + t\mathbf{u}_2) > f(\mathbf{b})$ for $0 < |t| < \varepsilon$.

Example 3. $h(x, y) \equiv x^2 - y^2$ has a saddle point at $(0, 0)$ (Exercise 1(e)).

Consider the surface given by $z = h(x, y)$, which we illustrate in Figure 2.3. For fixed y, its cross-sections perpendicular to the y-axis are parabolas opening upward, two of which are shown. For fixed x, the parabolas have equations of the form $z = d^2 - y^2$, so they open downward. For a fixed positive b, the horizontal plane $z = b$ cuts the graph in the hyperbola with equation $\frac{x^2}{b} - \frac{y^2}{b} = 1$; we see its left branch. If instead b is negative, then the standard form shifts to $\frac{y^2}{(-b)} - \frac{x^2}{(-b)} = 1$, putting the transverse axis along the y-axis. These cross sections give a good indication of the graph's shape, and earn it the name "hyperbolic paraboloid." The reason for the term "saddle point" is obvious.

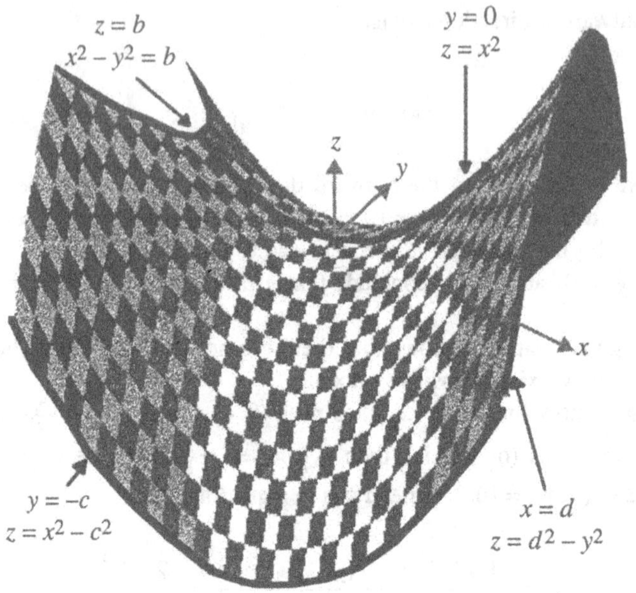

Figure 2.3.

Exercises

1. For each function, find all the critical points, then use the outward directional derivatives (as in Theorem 2 or Examples 1 and 2) to decide whether the points represent strict extremes, loose extremes, saddle points, or none of these.

   (a) $f(x, y) \equiv x^2 + y^2$

   (b) $g(x, y) \equiv x^2 y^2$

   (c) $h(x, y) \equiv \left(x^2 + y^2\right)^{1/2}$

   (d) $F(x, y) \equiv x^4 y^3$

   (e) $G(x, y) \equiv x^2 - y^2$

   (f) $H(x, y) \equiv x^3 y^3$

   (g) $K(x, y) \equiv x^{2/3} y^{2/3}$

2. *Show that the converse of Theorem 2 is false. That is, find an example of a function that is differentiable and has a strict minimum at some point but does not, in any neighborhood of the point, have all its outward directionals positive.

2.4 Extreme Values and the Second Derivative

In analyzing critical points, the first derivative has the disadvantage of forcing us to study the components of a variable gradient throughout a multidimensional neighborhood. We next want to develop a principle that, like the second derivative test, uses signs at just the critical point.

Suppose $\mathbf{b}$ is a critical point of f, and f is differentiable in $N(\mathbf{b}, \varepsilon)$; necessarily, $f'(\mathbf{b}) = \mathbf{O}$. Fix a unit vector $\mathbf{u}$ and look at our usual directional function

$$g(t) \equiv f(\mathbf{b} + t\mathbf{u}), \qquad -\varepsilon < t < \varepsilon.$$

By the chain rule, g is differentiable, with $g'(t) = f'(\mathbf{b} + t\mathbf{u})\langle\mathbf{u}\rangle$. In particular, $g'(0) = 0$. If g' is differentiable at $t = 0$ and $g''(0) > 0$, then the second derivative test tells us that $g(0) = f(\mathbf{b})$ is a minimum compared to the values of f along the line $\{\mathbf{b} + t\mathbf{u}\}$. We will turn this line-related information into n-dimensional information.

Theorem 1. *Suppose f is differentiable near* $\mathbf{b}$ *and twice differentiable at* $\mathbf{b}$. *Then for any direction* $\mathbf{u}$, $g(t) \equiv f(\mathbf{b}+t\mathbf{u})$ *is twice differentiable at $t = 0$, and* $g''(0) = (f''(\mathbf{b})\langle\mathbf{u}\rangle)\langle\mathbf{u}\rangle$.

Proof. We are hypothesizing that f is differentiable in a neighborhood $N(\mathbf{b}, \varepsilon)$ and that f' is differentiable at $\mathbf{b}$.

From $g'(t) = f'(\mathbf{b} + t\mathbf{u})\langle\mathbf{u}\rangle$, we have

$$\begin{aligned}\frac{g'(t) - g'(0)}{t} &= \frac{f'(\mathbf{b} + t\mathbf{u})\langle\mathbf{u}\rangle - f'(\mathbf{b})\langle\mathbf{u}\rangle}{t}\\ &= \frac{f'(\mathbf{b} + t\mathbf{u}) - f'(\mathbf{b})}{t}\langle\mathbf{u}\rangle \qquad \text{(algebra of operators)}.\end{aligned}$$

Because $f''(\mathbf{b})$ exists, the fraction in the last line satisfies

$$\begin{aligned}\frac{f'(\mathbf{b} + t\mathbf{u}) - f'(\mathbf{b})}{t} &= \frac{f''(\mathbf{b})\langle t\mathbf{u}\rangle + \mathbf{o}(t\mathbf{u})}{t}\\ &= f''(\mathbf{b})\langle\mathbf{u}\rangle + \frac{\mathbf{o}(t\mathbf{u})}{t} \to f''(\mathbf{b})\langle\mathbf{u}\rangle \qquad \text{as } t \to 0.\end{aligned}$$

It follows that $\frac{g'(t)-g'(0)}{t}$ has a limit, and the limit is $(f''(\mathbf{b})\langle\mathbf{u}\rangle)\langle\mathbf{u}\rangle$. □

(Compare the conclusion of Theorem 1 with Exercise 2.1:8.)

We will abbreviate $(f''(\mathbf{b})\langle\mathbf{u}\rangle)\langle\mathbf{u}\rangle$ by, naturally enough, $f''(\mathbf{b})\langle\mathbf{u}\rangle^2$.

Theorem 2. *Suppose f is differentiable near* $\mathbf{b}$, *is twice differentiable at* $\mathbf{b}$, *and* $f'(\mathbf{b}) = \mathbf{O}$.

(a) *If $f''(\mathbf{b})\langle\mathbf{v}\rangle^2 > 0$ (or < 0) for every $\mathbf{v} \neq \mathbf{O}$, then $f(\mathbf{b})$ is a strict minimum (respectively, strict maximum).*

(b) *If* $f''(\mathbf{b})\langle\mathbf{v}\rangle^2 > 0$ *for some vectors* $\mathbf{v}$ *and* < 0 *for others, then* $\mathbf{b}$ *is a saddle point of* f.

Proof. (a) Let us assume that $f''(\mathbf{b})\langle\mathbf{v}\rangle^2 > 0$ for all $\mathbf{v} \neq \mathbf{O}$. We will derive our conclusion from the outward directional derivatives of f around $\mathbf{b}$.

Recall (compare the discussion leading to Theorem 1.4:1) that the action of $f''(\mathbf{b})$ looks like

$$(f''(\mathbf{b})\langle\mathbf{v}\rangle)\langle\mathbf{v}\rangle = \left[[a_{11}\cdots a_{1n}]\langle\mathbf{v}\rangle \;\cdots\; [a_{n1}\cdots a_{nn}]\langle\mathbf{v}\rangle\right]\langle\mathbf{v}\rangle$$
$$= \sum_{1\le j\le n, 1\le k\le n} a_{kj} v_j v_k.$$

Thus, $f''(\mathbf{b})\langle\mathbf{v}\rangle^2$ is a second-degree polynomial in $(v_1, \ldots, v_n)$. Such a function, being continuous in $\mathbf{R}^n$, achieves a minimum value $c \equiv f''(\mathbf{b})\langle\mathbf{u}_0\rangle^2$ on the (compact) unit sphere. Since by hypothesis all its values there are positive, we conclude that $f''(\mathbf{b})\langle\mathbf{u}\rangle^2 \ge c > 0$ for every unit vector $\mathbf{u}$.

By definition of the second derivative, corresponding to $\varepsilon \equiv \frac{c}{2}$ there must be a neighborhood $N(\mathbf{b}, \delta)$ in which

$$\|f'(\mathbf{x}) - f'(\mathbf{b}) - f''(\mathbf{b})\langle\mathbf{x}-\mathbf{b}\rangle\| \le \frac{c}{2}\|\mathbf{x}-\mathbf{b}\|.$$

Since $f'(\mathbf{b}) = \mathbf{O}$,

$$\mathbf{x} \in N(\mathbf{b}, \delta) \Rightarrow \|f'(\mathbf{x}) - f''(\mathbf{b})\langle\mathbf{x}-\mathbf{b}\rangle\| \le \frac{c\|\mathbf{x}-\mathbf{b}\|}{2}.$$

Now, for any fixed $\mathbf{x} \neq \mathbf{b}$ in $N(\mathbf{b}, \delta)$, write $\mathbf{u} \equiv \frac{\mathbf{x}-\mathbf{b}}{\|\mathbf{x}-\mathbf{b}\|}$. Then

$$\partial_{\mathbf{u}} f(\mathbf{x}) = f'(\mathbf{x})\langle\mathbf{u}\rangle = [f''(\mathbf{b})\langle\mathbf{x}-\mathbf{b}\rangle]\langle\mathbf{u}\rangle + [f'(\mathbf{x}) - f''(\mathbf{b})\langle\mathbf{x}-\mathbf{b}\rangle]\langle\mathbf{u}\rangle.$$

For the first term, we have

$$[f''(\mathbf{b})\langle\mathbf{x}-\mathbf{b}\rangle]\langle\mathbf{u}\rangle = [f''(\mathbf{b})\langle\|\mathbf{x}-\mathbf{b}\|\mathbf{u}\rangle]\langle\mathbf{u}\rangle = \|\mathbf{x}-\mathbf{b}\| f''(\mathbf{b})\langle\mathbf{u}\rangle^2 \ge \|\mathbf{x}-\mathbf{b}\|c.$$

For the second, the absolute value satisfies

$$\left|[f'(\mathbf{x}) - f''(\mathbf{b})\langle\mathbf{x}-\mathbf{b}\rangle]\langle\mathbf{u}\rangle\right| \le \|f'(\mathbf{x}) - f''(\mathbf{b})\langle\mathbf{x}-\mathbf{b}\rangle\| \, \|\mathbf{u}\| \quad \text{(operator inequality)}$$
$$\le \frac{c}{2}\|\mathbf{x}-\mathbf{b}\| \; 1.$$

Therefore,

$$\partial_{\mathbf{u}} f(\mathbf{x}) \ge c\|\mathbf{x}-\mathbf{b}\| - \frac{c\|\mathbf{x}-\mathbf{b}\|}{2} > 0.$$

We have shown that the outward directionals are positive in a neighborhood of $\mathbf{b}$. By Theorem 2.3:2, $f(\mathbf{b})$ is a strict minimum.

(b) Assume that $f''(\mathbf{b})\langle\mathbf{v}\rangle^2 > 0$ and $f''(\mathbf{b})\langle\mathbf{w}\rangle^2 < 0$. Then the same relations hold for $\frac{\mathbf{v}}{\|\mathbf{v}\|}$ and $\frac{\mathbf{w}}{\|\mathbf{w}\|}$, so we may assume that $\mathbf{v}$ and $\mathbf{w}$ are unit vectors.

Write $g(t) \equiv f(\mathbf{b} + t\mathbf{v})$. By Theorem 1, $g''(0) = f''(\mathbf{b})\langle\mathbf{v}\rangle^2 > 0$. By the second derivative test, $g(0)$ is a strict minimum of g. By similar reasoning, $h(t) \equiv f(\mathbf{b} + t\mathbf{w})$ has a strict maximum at $t = 0$. By definition, $\mathbf{b}$ is a saddle point of f. □

Example 1. (a) Let $f(x, y) \equiv x^2 + y^2$. Then $f'(x, y) = \begin{bmatrix} 2x & 2y \end{bmatrix}$, so that $(0, 0)$ is the unique critical point. Also, $f''(x, y) = [[2\ \ 0]\ \ [0\ \ 2]]$, so that

$$f''(0,0)\langle\mathbf{v}\rangle^2 = \left[\begin{bmatrix} 2 & 0 \end{bmatrix} \begin{bmatrix} v_1 \\ v_2 \end{bmatrix} \quad \begin{bmatrix} 0 & 2 \end{bmatrix} \begin{bmatrix} v_1 \\ v_2 \end{bmatrix} \right] \begin{bmatrix} v_1 \\ v_2 \end{bmatrix} = 2v_1^2 + 2v_2^2 > 0$$

for $\mathbf{v} \neq \mathbf{O}$. Hence the origin is a strict minimum of f.

(b) Let $g(x, y) \equiv x^2 - y^2$. Again $(0, 0)$ is the lone critical point, and

$$g'' = [[2\ \ \ 0]\ \ [0\ \ \ -2]],$$

giving

$$g''(0,0)\langle\mathbf{v}\rangle^2 = 2v_1^2 - 2v_2^2.$$

Clearly, $g''(0, 0)\langle\mathbf{v}\rangle^2$ is positive if $\mathbf{v}$ is horizontal, and negative if $\mathbf{v}$ is vertical. Hence $(0, 0)$ is a saddle point of g. (Compare Example 2.3:(3) and Exercise 2.3:1e.)

Second derivatives of scalar functions map from $\mathbf{R}^n$ to $L(\mathbf{R}^n, \mathbf{R})$. If such an operator $\mathbf{T}$ has $\mathbf{T}\langle\mathbf{v}\rangle^2 > 0$ for all $\mathbf{v} \neq \mathbf{O}$, then $\mathbf{T}$ is called **positive definite.** If $\mathbf{T}\langle\mathbf{v}\rangle^2 < 0$ for all $\mathbf{v} \neq \mathbf{O}$, then $\mathbf{T}$ is **negative definite.** If instead $\mathbf{T}\langle\mathbf{v}\rangle^2 > 0$ for some $\mathbf{v}$ and < 0 for others, then $\mathbf{T}$ is **indefinite**. This language is unfortunate, because "indefinite" is not the negation of "definite." If $\mathbf{T}$ is not indefinite, then either $\mathbf{T}\langle\mathbf{v}\rangle^2 \geq 0$ for all $\mathbf{v}$ ("$\mathbf{T}$ is **positive semidefinite**") or $\mathbf{T}\langle\mathbf{v}\rangle^2 \leq 0$ for all $\mathbf{v}$ ("$\mathbf{T}$ is **negative semidefinite**"); and conversely. Thus, "indefinite" is the negation of "semidefinite."

Theorem 2 has the same weaknesses as the second derivative test. First, it does not apply at a critical point where f' is undefined or undifferentiable. Second, even where $f' = \mathbf{O}$ and f'' is defined, it remains inapplicable if f'' is just semidefinite (Exercises 4 and 5). Finally, as the same exercises show, its conditions are sufficient but not necessary for the existence of extremes or saddle points.

Nevertheless, it gives criteria that are often easy to check analytically, and can always be checked algebraically. To end this section, we explain the latter idea.

The second-degree function $Q(\mathbf{x}) \equiv \sum_{1\le j\le n, 1\le k\le n} a_{jk}x_jx_k$, from the proof of Theorem 2, is always expressible in terms of the matrix $\mathbf{A} \equiv [a_{jk}]$. Thus,

$$\sum_{1\le j\le n, 1\le k\le n} a_{jk}x_jx_k = \begin{bmatrix} a_{11}x_1 + \cdots + a_{n1}x_n & \cdots & a_{1n}x_1 + \cdots + a_{nn}x_n \end{bmatrix} \begin{bmatrix} x_1 \\ \vdots \\ x_n \end{bmatrix}$$

$$= \begin{bmatrix} x_1 & \cdots & x_n \end{bmatrix} \begin{bmatrix} a_{11} & \cdots & a_{1n} \\ & \ddots & \\ a_{n1} & \cdots & a_{nn} \end{bmatrix} \begin{bmatrix} x_1 \\ \vdots \\ x_n \end{bmatrix}$$

$$= \langle\mathbf{x}\rangle^t \mathbf{A} \langle\mathbf{x}\rangle \qquad \text{(meaning matrix product).}$$

If $\mathbf{A}$ is symmetric ($\mathbf{A} = \mathbf{A}^t$), then $Q(\mathbf{x})$ is called a **quadratic form** in $\mathbf{x}$. The form is called positive or negative definite, positive or negative semidefinite, or indefinite, according to whether $\mathbf{A}$ is any of those.

The form we want to analyze is

$$[f''(\mathbf{b})\langle\mathbf{v}\rangle]\langle\mathbf{v}\rangle = \left[\left[\frac{\partial^2 f}{\partial x_1 \partial x_1} \cdots \frac{\partial^2 f}{\partial x_1 \partial x_n} \right] \begin{bmatrix} v_1 \\ \vdots \\ v_n \end{bmatrix} \right.$$

$$\left. \cdots \left[\frac{\partial^2 f}{\partial x_n \partial x_1} \cdots \frac{\partial^2 f}{\partial x_n \partial x_n} \right] \begin{bmatrix} v_1 \\ \vdots \\ v_n \end{bmatrix} \right] \langle\mathbf{v}\rangle$$

$$= \left[v_1 \frac{\partial^2 f}{\partial x_1 \partial x_1} + \cdots + v_n \frac{\partial^2 f}{\partial x_1 \partial x_n} \right.$$

$$\left. \cdots v_1 \frac{\partial^2 f}{\partial x_n \partial x_1} + \cdots + v_n \frac{\partial^2 f}{\partial x_n \partial x_n} \right] \langle\mathbf{v}\rangle$$

$$= \langle\mathbf{v}\rangle^t \begin{bmatrix} \frac{\partial^2 f}{\partial x_1 \partial x_1} & \cdots & \frac{\partial^2 f}{\partial x_n \partial x_1} \\ & \ddots & \\ \frac{\partial^2 f}{\partial x_1 \partial x_n} & \cdots & \frac{\partial^2 f}{\partial x_n \partial x_n} \end{bmatrix} \langle\mathbf{v}\rangle,$$

all these partials being evaluated at $\mathbf{b}$. The matrix in the last line is called the **matrix of second partials** or **Hessian matrix** of f. Because it is symmetric—assuming that the mixed partials behave—there is a simple numerical way to decide its definiteness.

Let the matrix $\mathbf{A} \equiv [a_{jk}]$ be symmetric. Define the **upper-left subdeterminants** of $\mathbf{A}$ by

$$\Delta_j \equiv \begin{vmatrix} a_{11} & \cdots & a_{1j} \\ & \ddots & \\ a_{j1} & \cdots & a_{jj} \end{vmatrix}, \qquad j = 1, \ldots, n.$$

By a theorem from linear algebra, $\mathbf{A}$ is positive definite (or positive semidefinite) iff every Δ_j is positive (respectively, nonnegative). It is easy to see that $\mathbf{A}$ is negative definite (or semidefinite) iff $-\mathbf{A}$ is positive definite (respectively semidefinite). Therefore $\mathbf{A}$ is negative-definite iff each Δ_j has the sign of $(-1)^j$; similarly $\mathbf{A}$ is negative-semidefinite iff $(-1)^j \Delta_j \geq 0$. Finally, for $\mathbf{A}$ to be indefinite, it must violate both conditions "all $\Delta_j \geq 0$" and "all $(-1)^j \Delta_j \geq 0$." It follows that $\mathbf{A}$ is indefinite iff either an even-numbered Δ_j is negative or there are two odd-numbered Δ_j of opposite signs. (Check the last sentence's logic.)

[One reference for the linear-algebra theorem is Mirsky, Section 13.3 in Chapter 13. The theorem is difficult to find in current texts. This omission is surprising, because it can be proved with small weapons: row operations, their elementary matrices, inverses, transposes, and their relations to determinants.]

Exercises

1. For the functions in Example 1:
   (a) Write down the Hessian matrices of $f(x, y)$ and $g(x, y)$ at the origin.
   (b) Verify that the one for f has positive subdeterminants and the one for g has a negative even-numbered subdeterminant.
2. Let $h(x, y) \equiv xy$. Use Theorem 2 to test whether $(0, 0)$ is an extreme or a saddle point for h.
3. Let $F(x, y) \equiv \sin x \sin y$.
   (a) Characterize the critical points of F. (Hint: Sketch.)
   (b) Use Theorem 2 to decide which ones are maxima, minima, and saddle points.
4. Give examples of functions with zero first and second derivatives at the origin and:
   (a) a strict minimum there;
   (b) a loose (not strict) minimum there;
   (c) a saddle point there;
   (d) no extreme and no saddle there.

5. Give examples of functions G for which $G'(0,0) = \mathbf{O}$ and $G''(0,0)$ is semidefinite without being either zero or definite—say $G''(0,0)\langle\mathbf{v}\rangle^2$ is ≥ 0 for all $\mathbf{v}$, > 0 for some $\mathbf{w}$, and $= 0$ for some nonzero $\mathbf{u}$—exhibiting the four behaviors (a)–(d) in Exercise 4.

6. (a) (Taylor's theorem) Assume that f has $k+1$ derivatives in the neighborhood $N(\mathbf{a},\delta)$. Prove that for each $\mathbf{b} \in N(\mathbf{a},\delta)$, there exists $\mathbf{c}$ along the segment $\mathbf{ab}$ such that

$$f(\mathbf{b}) = f(\mathbf{a}) + f'(\mathbf{a})\langle\mathbf{b}-\mathbf{a}\rangle + \cdots + \frac{f^{(k)}(\mathbf{a})\langle\mathbf{b}-\mathbf{a}\rangle^k}{k!} + \frac{f^{(k+1)}(\mathbf{c})\langle\mathbf{b}-\mathbf{a}\rangle^{k+1}}{(k+1)!}.$$

Here $f^{(j)}(\mathbf{a})$ is a $\text{row}^{(j)}$, so $f^{(j)}(\mathbf{a})\langle\mathbf{v}\rangle^j$ abbreviates

$$\Big[[f^{(j)}(\mathbf{a})\langle\mathbf{v}\rangle]\langle\mathbf{v}\rangle \ldots\Big]\langle\mathbf{v}\rangle;$$

such operators and their actions are defined in Section 1.4. (Hint: You may assume the corresponding one-variable theorem [Ross, Theorem 31.3]: If $g(t)$ has $k+1$ derivatives near $t = 0$, then nearby $g(s) = g(0) + g'(0)s + \cdots + \frac{g^{(k)}(0)s^k}{k!} + \frac{g^{(k+1)}(c)s^{k+1}}{(k+1)!}$ for some c between s and 0.)

(b) The function $f(\mathbf{a}) + f'(\mathbf{a})\langle\mathbf{x}-\mathbf{a}\rangle + \cdots + \frac{f^{(k)}(\mathbf{a})\langle\mathbf{x}-\mathbf{a}\rangle^k}{k!}$ is the **Taylor polynomial of degree k for f near a**. Find the Taylor Polynomial of degree 3 for $F(x,y) \equiv x\,e^y$ near $(0,0)$.

(c) The end term $\frac{f^{(k+1)}(\mathbf{c})\langle\mathbf{x}-\mathbf{a}\rangle^{k+1}}{(k+1)!}$ is the (Taylor) **remainder for $f(\mathbf{x})$ near a**. Show that the remainder for $F(x,y) \equiv x\,e^y$ near $(0,0)$ tends to zero as $k \to \infty$ for every fixed $\mathbf{x} = (x,y)$.

(d) From (c), we conclude that $F(x,y) = \lim_{k\to\infty}$ (polynomial of degree k at (x,y)). Does that agree with the one-variable Taylor series for e^y?

(e) Use Taylor's theorem to prove this weaker version of Theorem 2: Assume that f is twice differentiable near $\mathbf{b}$, $f'(\mathbf{b}) = \mathbf{O}$, and f'' is continuous and positive definite at $\mathbf{b}$; then $f(\mathbf{b})$ is a strict minimum.

2.5 Implicit Scalar Functions

In algebra, we often write an equation relating some quantities, then ask whether it is possible to solve the equation for one quantity in terms of the others. Even if we are unable to express the one quantity as an elementary function of the others ("explicitly"), we sometimes assume that the equation implies some such relationship, and use the equation to deduce properties of the correspondence. Indeed, "implicit differentiation" allows us to get information about derivatives, even when we cannot explicitly specify the function. In this section, we produce a sufficient condition for the existence of implicit functions, together with a characterization of their derivatives.

An equation in $x_1, \ldots, x_n$ can always be written in the form $f(x_1, \ldots, x_n) = 0$. The solution set is some subset of the domain of f in $\mathbf{R}^n$. A subset of $\mathbf{R}^n$ is called a **relation** in $x_1, \ldots, x_n$. The relation determined by $f(\mathbf{x}) = 0$ is **functional** (or is **a function**) **with respect to** $x_1, \ldots, x_{n-1}$ if for each fixed $(n-1)$-tuple $(a_1, \ldots, a_{n-1})$, there is no more than one value of x_n such that $f(a_1, \ldots, a_{n-1}, x_n) = 0$.

It is useful to introduce some abbreviations. For any $\mathbf{x} = (x_1, \ldots, x_n) \in \mathbf{R}^n$, we use $\mathbf{x}^\#$ for the "projection" $(x_1, \ldots, x_{n-1}) \in \mathbf{R}^{n-1}$. Then we write $\mathbf{x} = (\mathbf{x}^\#, x_n)$. We can handle the latter ordered-pair symbol much as we do vectors in $\mathbf{R}^2$. For example,

$$\left(\mathbf{x}^\#, x_n\right) + \left(\mathbf{y}^\#, y_n\right) = \mathbf{x} + \mathbf{y} = \left(\mathbf{x}^\# + \mathbf{y}^\#, x_n + y_n\right)$$

and

$$\left\|\left(\mathbf{x}^\#, x_n\right)\right\|^2 = x_1^2 + \cdots + x_{n-1}^2 + x_n^2 = \left\|\mathbf{x}^\#\right\|^2 + x_n^2.$$

With this notation, assume that $f(\mathbf{x}) = 0$ is functional with respect to $\mathbf{x}^\#$. Then we say that $f(\mathbf{x}) = 0$ **determines**, or **can be solved for**, x_n as the function g of $\mathbf{x}^\#$ defined by

$$\text{domain of } g = D^\# \equiv \left\{\mathbf{x}^\#\colon f(\mathbf{x}^\#, y) = 0 \text{ for some (necessarily unique) } y\right\};$$
$$g(\mathbf{x}^\#) \equiv y \text{ for } \mathbf{x}^\# \in D^\#.$$

Theorem 1 (Scalar Form of the Implicit Function Theorem). *Assume that F is differentiable near* **b**, *with F' continuous at* **b**. *Suppose $\frac{\partial F}{\partial x_n}(\mathbf{b}) \neq 0$. Then near* **b**, *the equation $F(\mathbf{x}) = F(\mathbf{b})$ can be solved for x_n as a differentiable function of the other variables. In detail, there is a* **box** $[\mathbf{a}, \mathbf{c}] \equiv \{\mathbf{x}\colon a_1 \le x_1 \le c_1, \ldots, a_n \le x_n \le c_n\}$ *centered at* **b** *in $\mathbf{R}^n$ and a differentiable function $g(\mathbf{x}^\#)$ defined in the box $[\mathbf{a}^\#, \mathbf{c}^\#]$ in $\mathbf{R}^{n-1}$ such that for all $\mathbf{x} \in [\mathbf{a}, \mathbf{c}]$*:

(a) $\left(\mathbf{x}^\#, g\left(\mathbf{x}^\#\right)\right) \in [\mathbf{a}, \mathbf{c}]$.

(b) $F(\mathbf{x}) = F(\mathbf{b})$ *iff* $x_n = g\left(\mathbf{x}^\#\right)$.

(c) *If* $\mathbf{x} = \left(\mathbf{x}^\#, g\left(\mathbf{x}^\#\right)\right)$*—in other words, if* $x_n = g(\mathbf{x}^\#)$*—then* $g'(\mathbf{x}^\#)$ *is given by*

$$\frac{\partial x_n}{\partial(x_1, \ldots, x_{n-1})}\left(\mathbf{x}^\#\right) = \frac{-\partial F/\partial(x_1, \ldots, x_{n-1})(\mathbf{x})}{\partial F/\partial x_n(\mathbf{x})}.$$

Proof. Write $f(\mathbf{x}) \equiv F(\mathbf{x}) - F(\mathbf{b})$. Throughout the proof, we work with f, turning the original equation into $f(\mathbf{x}) = 0$.

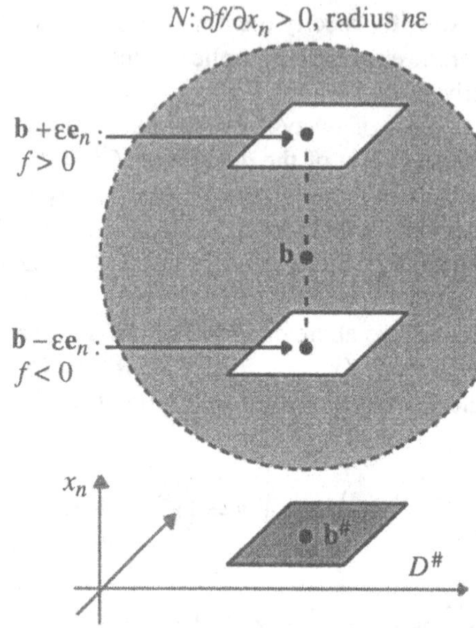

Figure 2.4.

For definiteness, we assume that $\frac{\partial f}{\partial x_n}(\mathbf{b}) = \frac{\partial F}{\partial x_n}(\mathbf{b}) > 0$. By hypothesis, all the partials of f are continuous at $\mathbf{b}$. Therefore, there is a neighborhood $N \equiv N(\mathbf{b}, n\varepsilon)$ in which $\frac{\partial f}{\partial x_n} > 0$. (In this neighborhood, the equation is already functional; by the mean value theorem, we cannot have $f(a_1, \ldots, a_{n-1}, x_n) = 0$ for two different values of x_n if $\frac{\partial f}{\partial x_n}$ is never zero.)

We are going to examine f around the two points $\mathbf{b} \pm \varepsilon \mathbf{e}_n = (b_1, \ldots, b_{n-1}, b_n \pm \varepsilon) = (\mathbf{b}^{\#}, b_n \pm \varepsilon)$. In Figure 2.4, we suggest $\mathbf{R}^n$ by showing the x_n-axis vertically and the $x_1 x_2 \ldots x_{n-1}$-hyperplane horizontally. Our two points are the ends of a vertical segment (dashed), which is easily within N.

By the (multivariable) mean value theorem, we have

$$f(\mathbf{b} + \varepsilon \mathbf{e}_n) = f(\mathbf{b} + \varepsilon \mathbf{e}_n) - f(\mathbf{b}) = f'(\mathbf{b} + t\mathbf{e}_n)\langle \varepsilon \mathbf{e}_n \rangle$$

$$\text{(for some } t \text{ between 0 and 1)}$$

$$= \varepsilon \frac{\partial f}{\partial x_n}(\mathbf{b} + t\mathbf{e}_n) > 0.$$

Because f is continuous, there is a neighborhood $N_{\text{up}} \equiv N(\mathbf{b} + \varepsilon \mathbf{e}_n, n\Delta_{\text{up}})$ (not drawn) in which $f(\mathbf{x})$ stays positive. In particular, f is positive throughout the "hyperrectangle" given by

$$b_1 - \Delta_{\text{up}} \le x_1 \le b_1 + \Delta_{\text{up}}, \ldots, b_{n-1} - \Delta_{\text{up}} \le x_{n-1} \le b_{n-1} + \Delta_{\text{up}},\ x_n = b_n + \varepsilon,$$

which looks to our eyes like the parallelogram drawn uppermost in the figure.

By similar reasoning, we obtain the lower hyperrectangle

$$b_1 - \Delta_{\text{lo}} \le x_1 \le b_1 + \Delta_{\text{lo}}, \ldots, b_{n-1} - \Delta_{\text{lo}} \le x_{n-1} \le b_{n-1} + \Delta_{\text{lo}},\ x_n = b_n - \varepsilon,$$

in which $f(\mathbf{x})$ is negative. We set δ to be the smallest of Δ_{up}, Δ_{lo}, and ε.

Let $D^{\#} \equiv \{\mathbf{x}^{\#}: |x_1 - b_1| \leq \delta, \ldots, |x_{n-1} - b_{n-1}| \leq \delta\}$. This set is a box in $\mathbf{R}^{n-1}$, and therefore not part of the picture, but we may pretend that it is the projection (lowest parallelogram in the figure) of the two hyperrectangles onto the $x_1 x_2 \ldots x_{n-1}$-hyperplane. For any fixed $\mathbf{x}^{\#} \in D^{\#}$, write

$$h(t) \equiv f\left(\mathbf{x}^{\#}, b_n + t\right), \qquad -\varepsilon \leq t \leq \varepsilon.$$

In words, h tracks the values of f along any vertical segment from the lower hyperrectangle (middle parallelogram) to the upper. Clearly, h is a continuous function of t with $h(-\varepsilon) < 0$ and $h(\varepsilon) > 0$. By the intermediate value theorem, there is t^* for which $h(t^*) = 0$. Because $h'(t) = \frac{\partial f}{\partial x_n}\left(\mathbf{x}^{\#}, b_n + t\right) > 0$, this t^* is unique. Hence each $\mathbf{x}^{\#}$ gives rise to one and only one $x_n = g\left(\mathbf{x}^{\#}\right) \equiv b_n + t^*$ between $b_n - \varepsilon$ and $b_n + \varepsilon$ for which $f\left(\mathbf{x}^{\#}, x_n\right) = 0$.

Now define $[\mathbf{a}, \mathbf{c}] \equiv \left[(b_1 - \delta, \ldots, b_{n-1} - \delta, b_n - \varepsilon), (b_1 + \delta, \ldots, b_{n-1} + \delta, b_n + \varepsilon)\right]$ (whose top is the upper, bottom is the middle, parallelogram in the figure) in $\mathbf{R}^n$, and let $\mathbf{x} \in [\mathbf{a}, \mathbf{c}]$.

(a) By assumption, $a_j \leq x_j \leq c_j$ for $1 \leq j \leq n-1$. Hence $\mathbf{x}^{\#} \in \left[\mathbf{a}^{\#}, \mathbf{c}^{\#}\right] = D^{\#}$, and $g\left(\mathbf{x}^{\#}\right)$ is defined. By the construction of g, $a_n \leq g\left(\mathbf{x}^{\#}\right) \leq c_n$. Therefore, $\left(\mathbf{x}^{\#}, g\left(\mathbf{x}^{\#}\right)\right) \in [\mathbf{a}, \mathbf{c}]$.

(b) From $\mathbf{x} \in [\mathbf{a}, \mathbf{c}]$, we know that $\mathbf{x}^{\#} \in D^{\#}$ and that x_n is between $b_n - \varepsilon$ and $b_n + \varepsilon$. Hence

$$f(\mathbf{x}) = 0 \Leftrightarrow f\left(\mathbf{x}^{\#}, x_n\right) = 0 \Leftrightarrow x_n = g\left(\mathbf{x}^{\#}\right).$$

(c) By hypothesis, f' is continuous at $\mathbf{b}$. Consequently, there is a neighborhood of $\mathbf{b}$ in which $f'(\mathbf{x})$ is bounded, say $\|f'(\mathbf{x})\| < M$. Also, $\frac{\partial f}{\partial x_n}(\mathbf{x})$ is continuous and positive at $\mathbf{x} = \mathbf{b}$, so in some neighborhood $\frac{\partial f}{\partial x_n}(\mathbf{x})$ stays bigger than $K \equiv \frac{\partial f}{\partial x_n}(\mathbf{b})/2$. We may assume that the construction in (a) takes place within these neighborhoods, so that $\|f'(\mathbf{x})\| < M$ and $\frac{\partial f}{\partial x_n}(\mathbf{x}) > K$ throughout the box $[\mathbf{a}, \mathbf{c}]$.

Suppose $\mathbf{x}^{\#}$ and $\mathbf{v}^{\#}$ are in $D^{\#}$. (Refer to Figure 2.5.) Write $y = g\left(\mathbf{x}^{\#}\right)$, $w = g(\mathbf{v}^{\#})$. By the mean value thoerem,

$$\begin{aligned}
0 &= f\left(\mathbf{x}^{\#}, y\right) - f\left(\mathbf{v}^{\#}, w\right) \\
&= \frac{\partial f}{\partial \mathbf{x}}(\mathbf{z})\left\langle \left(\mathbf{x}^{\#}, y\right) - \left(\mathbf{v}^{\#}, w\right)\right\rangle \quad \text{(for some } \mathbf{z} \text{ in the box)} \\
&= \left[\frac{\partial f}{\partial x_1}(\mathbf{z}) \cdots \frac{\partial f}{\partial x_{n-1}}(\mathbf{z})\ \frac{\partial f}{\partial x_n}(\mathbf{z})\right]\left[x_1 - v_1 \cdots x_{n-1} - v_{n-1}\ \ y - w\right]^t \\
&= \frac{\partial f}{\partial x_1}(\mathbf{z})[x_1 - v_1] + \cdots + \frac{\partial f}{\partial x_{n-1}}(\mathbf{z})[x_{n-1} - v_{n-1}] + \frac{\partial f}{\partial x_n}(\mathbf{z})[y - w].
\end{aligned}$$

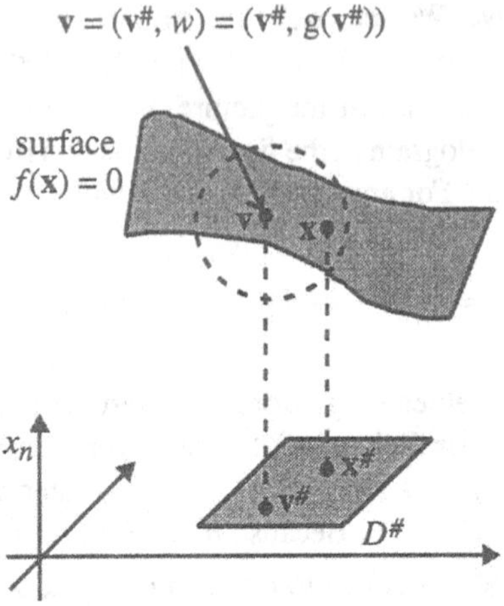

Figure 2.5.

Solving for $[y - w]$, we obtain

$$y - w = \frac{-\frac{\partial f}{\partial x_1}(\mathbf{z})[x_1 - v_1] - \cdots - \frac{\partial f}{\partial x_{n-1}}(\mathbf{z})[x_{n-1} - v_{n-1}]}{\frac{\partial f}{\partial x_n}(\mathbf{z})},$$

from which we conclude that

$$|y - w| \leq \left\| \left(\frac{\partial f}{\partial x_1}(\mathbf{z}), \ldots, \frac{\partial f}{\partial x_{n-1}}(\mathbf{z}) \right) \right\| \frac{\|(x_1 - v_1, \ldots, x_{n-1} - v_{n-1})\|}{K}$$
$$\leq \frac{M \|\mathbf{x}^\# - \mathbf{v}^\#\|}{K}. \qquad \text{(Reasons?)}$$

Notice that this last line tells us that g is continuous.

Let $\mathbf{v}^\#$ be fixed in the interior of $D^\#$, that is, $a_1 < v_1 < c_1, \ldots, a_{n-1} < v_{n-1} < c_{n-1}$. Let r be an arbitrary positive number. The corresponding point on the surface is $\mathbf{v} \equiv (\mathbf{v}^\#, w)$. Because f is differentiable at $\mathbf{v}$, there exists a neighborhood $N(\mathbf{v}, s)$ (dashed circle in Figure 2.5) in which $\mathbf{z} \in N(\mathbf{v}, s)$ implies

$$|f(\mathbf{z}) - f(\mathbf{v}) - f'(\mathbf{v})\langle \mathbf{z} - \mathbf{v} \rangle| \leq \frac{r\|\mathbf{z} - \mathbf{v}\| K^2}{K + M}.$$

Let $t = \min\left\{\frac{s}{2}, \frac{sK}{2M}\right\}$, and suppose $\mathbf{x}^\#$ is any point in $D^\#$ within the open ball in $\mathbf{R}^{n-1}$ given by

$$\|\mathbf{x}^\# - \mathbf{v}^\#\| < t.$$

Above it on the surface is $\mathbf{x} \equiv (\mathbf{x}^\#, y)$. Then

$$\|\mathbf{x}^\# - \mathbf{v}^\#\| < \frac{s}{2} \quad \text{and} \quad |y - w| \leq \frac{M\|\mathbf{x}^\# - \mathbf{v}^\#\|}{K} < \frac{s}{2},$$

so

$$\|\mathbf{x}-\mathbf{v}\| = \left\|\left(\mathbf{x}^{\#}, y\right) - \left(\mathbf{v}^{\#}, w\right)\right\| < \frac{s}{\sqrt{2}} < s.$$

In other words, **x** is close enough to **v** to make

$$|f(\mathbf{x}) - f(\mathbf{v}) - f'(\mathbf{v})\langle\mathbf{x}-\mathbf{v}\rangle| \le \frac{r\|\mathbf{x}-\mathbf{v}\|K^2}{K+M}.$$

Since $f(\mathbf{x}) = f(\mathbf{v}) = 0$, we have

$$|f'(\mathbf{v})\langle\mathbf{x}-\mathbf{v}\rangle| \le \frac{r\|\mathbf{x}-\mathbf{v}\|K^2}{K+M}.$$

It is this last inequality, derived from the statement that **x** and **v** are on the surface, that decides the differentiability of g. On the left side, we see the expression

$$\begin{aligned}
&f'(\mathbf{v})\langle\mathbf{x}-\mathbf{v}\rangle \\
&= \frac{\partial f}{\partial x_1}(\mathbf{v})[x_1 - v_1] + \cdots + \frac{\partial f}{\partial x_{n-1}}(\mathbf{v})[x_{n-1} - v_{n-1}] + \frac{\partial f}{\partial x_n}(\mathbf{v})[y - w] \\
&= \left[\frac{\partial f}{\partial x_1}(\mathbf{v}) \cdots \frac{\partial f}{\partial x_{n-1}}(\mathbf{v})\right]\left\langle\mathbf{x}^{\#} - \mathbf{v}^{\#}\right\rangle + \frac{\partial f}{\partial x_n}(\mathbf{v})\left[g(\mathbf{x}^{\#}) - g(\mathbf{v}^{\#})\right] \\
&= \frac{\partial f}{\partial x_n}(\mathbf{v})\left[g(\mathbf{x}^{\#}) - g(\mathbf{v}^{\#}) - \mathbf{L}\left\langle\mathbf{x}^{\#} - \mathbf{v}^{\#}\right\rangle\right],
\end{aligned}$$

where we have written **L** for the operator $\frac{-\partial f/\partial(x_1,\ldots,x_{n-1})(\mathbf{v})}{\partial f/\partial x_n(\mathbf{v})}$. On the right side of the inequality,

$$\|\mathbf{x}-\mathbf{v}\| = \left[\left\|\mathbf{x}^{\#} - \mathbf{v}^{\#}\right\|^2 + (y-w)^2\right]^{1/2} \le \left\|\mathbf{x}^{\#} - \mathbf{v}^{\#}\right\|\left[1 + \frac{M}{K}\right].$$

Therefore,

$$\begin{aligned}
\left|g\left(\mathbf{x}^{\#}\right) - g\left(\mathbf{v}^{\#}\right) - \mathbf{L}\left\langle\mathbf{x}^{\#} - \mathbf{v}^{\#}\right\rangle\right| &\le r\|\mathbf{x}-\mathbf{v}\|\frac{K^2}{K+M}\frac{1}{\partial f/\partial x_n(\mathbf{v})} \\
&\le r\left\|\mathbf{x}^{\#} - \mathbf{v}^{\#}\right\|\left[1 + \frac{M}{K}\right]\frac{K^2}{K+M}\frac{1}{K} \\
&= r\left\|\mathbf{x}^{\#} - \mathbf{v}^{\#}\right\|.
\end{aligned}$$

We have shown that near $\mathbf{v}^{\#}$, $g\left(\mathbf{x}^{\#}\right) - g\left(\mathbf{v}^{\#}\right)$ is approximated to within $\mathbf{o}(\mathbf{x}^{\#} - \mathbf{v}^{\#})$ by $\mathbf{L}\left\langle\mathbf{x}^{\#} - \mathbf{v}^{\#}\right\rangle$. The conclusion follows. □

Example 1. At what points does Theorem 1 guarantee that $36x^2+9y^2+4z^2 = 36$ determines z as a differentiable function of x and y?

By Theorem 1, $F(x, y, z) \equiv 36x^2 + 9y^2 + 4z^2 = 36$ can be solved for z as long as $\frac{\partial F}{\partial z} = 8z \neq 0$. Therefore, anywhere except along the equatorial belt, the surface is given by $z = G(x, y)$. In fact, wherever $z > 0$, we have $z = 0.5\sqrt{36 - 36x^2 - 9y^2}$ for (x, y) in the elliptical region of the xy-plane given by $x^2/1 + y^2/4 < 1$; and the opposite sign for $z < 0$. In either half,

$$\frac{\partial z}{\partial(x,y)} = 0.5\left[\frac{(1/2)(-72x)}{\pm\sqrt{36-36x^2-9y^2}} \quad \frac{(1/2)(-18y)}{\pm\sqrt{36-36x^2-9y^2}}\right]$$
$$= \frac{\begin{bmatrix}-72x & -18y\end{bmatrix}}{8z} = \frac{-\partial F/\partial(x,y)}{\partial F/\partial z}.$$

Considering our affection for Leibniz's notation, it is sad to see it injured by the negative sign in $\frac{\partial x_n}{\partial \mathbf{x}^\#} = \frac{-\partial F/\partial \mathbf{x}^\#}{\partial F/\partial x_n}$. Unfortunately, we had to expect it there. If $\frac{\partial F}{\partial x_n}$ is positive, then a change in $\mathbf{x}^\#$ that increases F must decrease x_n, to drive F back to its constant value along the surface; therefore, $\frac{\partial F}{\partial \mathbf{x}^\#}$ and $\frac{\partial x_n}{\partial \mathbf{x}^\#}$ have opposite signs. By analogous reasoning, if $\frac{\partial F}{\partial x_n}$ is negative, then $\frac{\partial F}{\partial \mathbf{x}^\#}$ and $\frac{\partial x_n}{\partial \mathbf{x}^\#}$ have like signs.

We should pay special attention to the case $n = 2$. Suppose $F(x, y)$ is continuously differentiable at (a, b) and $\frac{\partial F}{\partial y}(a, b) \neq 0$. Then in some neighborhood of (a, b), the graph of $F(x, y) = c \equiv F(a, b)$ is the graph of $y = g(x)$ for some real-variable function g, with $g'(x) = \frac{-\partial F/\partial x(x,y)}{\partial F/\partial y(x,y)}$.

Example 2. Can $x^2 + y^2 - 4y = 5$ be solved for y as a function of x?

(a) By the quadratic formula, y satisfies

$$y = \frac{4 \pm \sqrt{16 - 4(x^2 - 5)}}{2} = 2 \pm \sqrt{9 - x^2}.$$

The relation is not functional.

If we investigate analytically rather than algebraically, we look to solve

$$0 = F(x, y) \equiv x^2 + y^2 - 4y - 5.$$

We observe that $\frac{\partial F}{\partial y} = 2y - 4 = 0$ iff $y = 2$. By Theorem 1, the relation is functional in the vicinity of any point (a, b) with $b \neq 2$. The meaning of this prediction is clear in Figure 2.6. If $b > 2$, then we can draw within the open strip defined by $|x| < 3$, $y > 2$ (shaded) numerous rectangles in which each x gives precisely one y with $F(x, y) = 0$, namely $y = 2 + \sqrt{9 - x^2}$; and correspondingly for $b < 2$, strip $y < 2$, $y = 2 - \sqrt{9 - x^2}$.

(b) The condition $\frac{\partial F}{\partial y}(a, b) \neq 0$—in general, $\frac{\partial F}{\partial x_n}(\mathbf{b}) \neq 0$—is essential. Around the bad point $(3, 2)$, it is impossible to define an interval $3 < x < 3 + \delta$ in which $F(x, y) = 0$ has any solutions, or an interval $3 - \delta < x < 3$ in which the solution is unique. The same thing, except mirror-imaged, occurs around $(-3, 2)$. Thus, the conclusion of Theorem 1 may be untenable near a bad point.

On the other hand, the condition is not necessary for the existence of a solution; see Exercise 3.

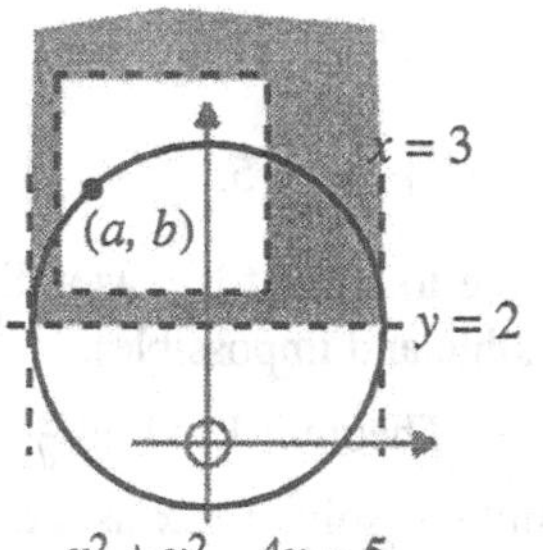

Figure 2.6.

(c) Theorem 1 also predicts that

$$\frac{dy}{dx} = \frac{-\partial F/\partial x}{\partial F/\partial y} = \frac{-(2x)}{2y-4}.$$

Notice that this result agrees with the answer from implicit differentiation:

$$2x + 2yy' - 4y' = 0, \quad \text{yielding } y' = \frac{-x}{y-2}.$$

Notice also that it applies to either half of the circle. On the upper semicircle, $y = 2 + \sqrt{9-x^2}$ gives $y' = \frac{-x}{\sqrt{9-x^2}} = \frac{-x}{y-2}$. For the lower, $y = 2 - \sqrt{9-x^2}$ gives $y' = \frac{x}{\sqrt{9-x^2}} = \frac{x}{2-y}$.

Suppose the equation $f(x, y) = 0$ is equivalent to one with the form $y = g(x)$. If we can also solve it for x as a function $h(y)$, then

$$y = g(x) \Leftrightarrow f(x, y) = 0 \Leftrightarrow x = h(y)$$

says that h is the inverse of g. From Theorem 1, we easily obtain the following elementary result about derivatives of inverse functions:

Theorem 2. *Suppose $g(x)$ is differentiable near $x = b$ and g' is continuous and nonzero at* **b**. *Then in some closed interval $[b - \delta, b + \delta]$:*

(a) *g is invertible.*

(b) *Its inverse $h \equiv g^{-1}$ satisfies $h'(g(x)) = \frac{1}{g'(x)}$.*

Proof. Exercise 5.

Exercises

1. Along the graph of $x - y^2 + 4y = 5$:

   (a) Where is it possible to solve for y as a differentiable function of x? Justify both (possible and impossible).

   (b) Where possible, use Theorem 1 to find $\frac{dy}{dx}$.

   (c) Where is it possible to solve for x as a differentiable function of y? Find $\frac{dx}{dy}$ there.

2. Along the graph of $e^{xy} = 1$:

   (a) Where is it possible to solve for one variable as a function of the other?

   (b) Of those places, where is the resulting function differentiable?

3. (a) Where along the graph of $e^x\left(x^3 + y^3\right) - 1 = 0$ can one solve for y as a function of x?

   (b) Where is the resulting function differentiable?

   (c) Find the derivative by Theorem 1 and explicitly. Do they agree?

4. Consider the graph of $z^2 = x^2 + y^2$:

   (a) At which points does Theorem 1 guarantee a solution for x in terms of y and z?

   (b) At the points in (a), find $\frac{\partial x}{\partial(y,z)}$.

   (c) At the points not in (a), decide whether there is a solution, even though Theorem 1 does not promise one.

5. Use Theorem 1 to prove Theorem 2. (Caution: Theorem 1 has to do with differentiable functions of *vector* variables.)

2.6 Curves, Surfaces, Tangents, and Normals

In this section we give analytical meanings to some familiar geometric concepts. These include curves and surfaces, together with their tangents and normals. We enlarge later upon this material. The limited version that we have of the implicit function theorem enables us to introduce the material now, and doing so will add to our knowledge of gradients.

Definition. A differentiable function $\mathbf{g}\colon [a, c] \to \mathbf{R}^n$ is called a **curve in $\mathbf{R}^n$**. A curve is **smooth at $\mathbf{b}$** $\equiv \mathbf{g}(t_0)$ if $\mathbf{g}'$ is continuous and nonzero at $t = t_0$. In that case, we say the curve **has a tangent** at $\mathbf{b}$, and we call the vector $\mathbf{g}'(t_0)$ the **tangent** and the line $\mathbf{b} + \langle\langle \mathbf{g}'(t_0) \rangle\rangle \equiv \{\mathbf{b} + \alpha\mathbf{g}'(t_0) \colon \alpha \in \mathbf{R}\}$ the **tangent line** to the curve. We say that the **curve is smooth** if it is smooth at all its points.

The same way we abuse "arc," we will allow the word "curve" to signify both the function and the set of points that is the function's range. It should be clear, however, that our definition attaches to the curve some properties that do not belong to the set of points. (It is usual to say of such properties that they are "not geometric," or "not intrinsic to the point set.") One of these is **orientation**. Because the interval $[a, c]$ has an order, the curve "goes from $\mathbf{g}(a)$ to $\mathbf{g}(c)$" (the endpoints of the curve) and the tangents point one way and not the other. Thus, the parametrization $\mathbf{G}(s) \equiv \mathbf{g}(-s)$, $-c \le s \le -a$, describes the same set of points as $\mathbf{g}$, but turns the trip and the tangents around. More important, the existence of "tangents" is also a nongeometric property.

Example 1. (a) Existence and values of tangents are characteristics of parametrizations.

The curve $(x, y, z) = \mathbf{g}(t) \equiv (\cos t, \sin t, t)$, $t \in [-1, 1]$, describes part of a helix in $\mathbf{R}^3$. We find that

$$\mathbf{g}'(t) = \begin{bmatrix} -\sin t \\ \cos t \\ 1 \end{bmatrix},$$

so that $\|\mathbf{g}'(t)\| = \sqrt{2}$. Hence the curve is smooth, and there is always a tangent as we have defined it.

But the same helix is described by $\mathbf{G}(t) \equiv \left(\cos t^3, \sin t^3, t^3\right)$. For this parametrization,

$$\mathbf{G}'(t) = \left[-3t^2 \sin t^3 \;\; 3t^2 \cos t^3 \;\; 3t^2\right]^t.$$

Consequently, $\mathbf{G}' = \mathbf{O}$ at the origin, and the curve—that is, the parametrization—lacks a tangent there.

(b) Where there is a tangent in our sense, there is also a tangent in the geometric sense, meaning a limit of the secants. (See Exercise 6(a) for precise formulations.) However, it is clear from (a) that the converse is false.

(c) Some irregularities *are* intrinsic to the point set. These always show up as "singularities" ($\mathbf{g}' = \mathbf{O}$ or $\mathbf{g}'$ undefined) of parametrizations.

We know from calculus that the graph of $y = x^{2/3}$ has a kink at the origin, although it still has a (vertical) tangent in the geometric sense. The graph can be given parametrically by $(x, y) = \mathbf{H}(t) \equiv \left(t^3, t^2\right)$. Here $\mathbf{H}'(0) = \mathbf{O}$. Is it possible to find a parametrization of this curve that is smooth at the origin? We will see (Exercise 6(b)) that none is possible. The singularity is an intrinsic feature of the graph.

The graph of $y = |x|$ is different: It does not have a geometric tangent at the origin. It, too, can be given by a continuously differentiable parametrization, for example

$$(x, y) = \left(\sin^3 t, \left[\frac{1 - \cos 2t}{2}\right]^{3/2}\right).$$

Clearly, the origin is singular for this function. Could we, with harder thought, find a different function that gives a smooth curve? Part (b) already says "no"; we will see (Exercise 6(c)) how to identify this kind of point.

Example 2. Describe the tangent line at $\mathbf{b} = \left(\frac{\sqrt{3}}{2}, \frac{1}{2}, \frac{\pi}{6}\right)$ to the curve in Example 1(a).

(a) For $\mathbf{g}(t) = (\cos t, \sin t, t)$, we have $\mathbf{g}'\left(\frac{\pi}{6}\right) = \left[-\frac{1}{2} \;\; \frac{\sqrt{3}}{2} \;\; 1\right]^t$. Hence the tangent line is

$$\mathbf{b} + \left\langle\!\left\langle \mathbf{g}'\left(\frac{\pi}{6}\right) \right\rangle\!\right\rangle \equiv \left\{ \begin{bmatrix} \frac{\sqrt{3}}{2} \\ \frac{1}{2} \\ \frac{\pi}{6} \end{bmatrix} + \alpha \begin{bmatrix} -\frac{1}{2} \\ \frac{\sqrt{3}}{2} \\ 1 \end{bmatrix} : \alpha \in \mathbf{R} \right\}$$

$$= \left\{ \begin{bmatrix} \frac{\sqrt{3}-\alpha}{2} \\ \frac{1+\sqrt{3}\alpha}{2} \\ \frac{\pi}{6} + \alpha \end{bmatrix} : \alpha \in \mathbf{R} \right\}.$$

(b) An equation is a more familiar description. The vector $\mathbf{x} = (x, y, z)$ is on the tangent line

$$\Leftrightarrow \mathbf{x} = \mathbf{b} + \alpha \mathbf{g}'\left(\frac{\pi}{6}\right) \Leftrightarrow \left(x - \frac{\sqrt{3}}{2}, y - \frac{1}{2}, z - \frac{\pi}{6}\right) = \left(-\frac{\alpha}{2}, \frac{\alpha\sqrt{3}}{2}, \alpha\right).$$

Therefore, the line is characterized by the system

$$\frac{x - \sqrt{3}/2}{-1/2} = \frac{y - 1/2}{\sqrt{3}/2} = \frac{z - \pi/6}{1}.$$

A curve is the generalization of a line segment in the world of objects we will describe as "one-dimensional." The next step up is a generalization of rectangles.

Definition. Let $R = \left[(a, c), (b, d)\right] \equiv \{(r, s) \colon a \le r \le b, c \le s \le d\}$ be a rectangle in $\mathbf{R}^2$. Suppose $\mathbf{F} \colon R \to \mathbf{R}^n$ is differentiable. We call $\mathbf{F}$ a **surface** in $\mathbf{R}^n$. We say that $\mathbf{F}$ is **smooth at b** $\equiv \mathbf{F}(r, s)$ if $\frac{\partial \mathbf{F}}{\partial r}$ and $\frac{\partial \mathbf{F}}{\partial s}$ are continuous and linearly independent at (r, s); otherwise, $\mathbf{b}$ is a **singular point**, or **singularity**, of $\mathbf{F}$. At a place where $\mathbf{F}$ is smooth, the plane

$$\mathbf{b} + \left\langle\!\left\langle \frac{\partial \mathbf{F}}{\partial r}, \frac{\partial \mathbf{F}}{\partial s} \right\rangle\!\right\rangle \equiv \left\{ \mathbf{b} + \alpha \frac{\partial \mathbf{F}}{\partial r} + \beta \frac{\partial \mathbf{F}}{\partial s} \right\}$$

is the **tangent (plane)** to the surface.

Notice that $\frac{\partial \mathbf{F}}{\partial r} = \begin{bmatrix} \frac{\partial F_1}{\partial r} \\ \vdots \\ \frac{\partial F_n}{\partial r} \end{bmatrix}$ and $\frac{\partial \mathbf{F}}{\partial s} = \begin{bmatrix} \frac{\partial F_1}{\partial s} \\ \vdots \\ \frac{\partial F_n}{\partial s} \end{bmatrix}$ are the columns of $\frac{\partial \mathbf{F}}{\partial (r,s)} = \mathbf{F}'(r, s)$. The tangent plane is a translate of the plane that those columns span.

To justify the name "tangent" plane, we will characterize the plane in terms of tangents to curves. The next theorem gives half that characterization; the other half comes in the next chapter.

Theorem 1. *Assume that* $\mathbf{p} \equiv (r, s)$ *is interior to* R *and that* $\mathbf{F}$ *is smooth at* $\mathbf{b} \equiv \mathbf{F}(\mathbf{p})$. *Then every nonzero linear combination of* $\mathbf{F}_r \equiv \frac{\partial \mathbf{F}}{\partial r}(\mathbf{p})$ *and* $\mathbf{F}_s \equiv \frac{\partial \mathbf{F}}{\partial s}(\mathbf{p})$ *is the tangent at* $\mathbf{b}$ *to some curve that lies on the surface.*

Proof. Assume that $\mathbf{v} = \alpha\mathbf{F}_r + \beta\mathbf{F}_s$. The rectangle contains some of the line through (r, s) parallel to (α, β); see all these in Figure 2.7. The curve we want is the image of the line.

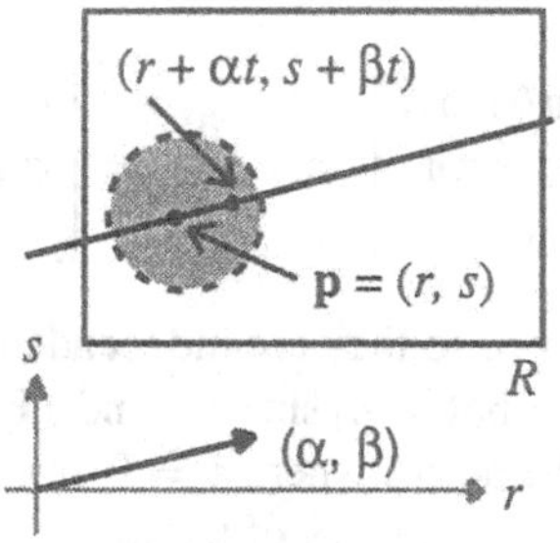

Figure 2.7.

In detail, because $\mathbf{p} \in \text{int}(R)$, there is a neighborhood $N(\mathbf{p}, 2\varepsilon) \subseteq R$. If $|t| = \frac{\varepsilon}{(a^2+b^2)^{1/2}}$, then the point $(r + \alpha t, s + \beta t)$ is (in $N(\mathbf{p}, 2\varepsilon)$ and so) in R. For such t, we set

$$\mathbf{g}(t) \equiv \mathbf{F}(r + \alpha t, s + \beta t).$$

Each $\mathbf{g}(t)$ is on the surface. By the chain rule,

$$\mathbf{g}'(t) = \frac{\partial F}{\partial r}(r + \alpha t, s + \beta t)\frac{d(\alpha t)}{dt} + \frac{\partial F}{\partial s}(t + \alpha t, s + \beta t)\frac{d(\beta t)}{dt},$$

which is defined for all t and continuous at $t = 0$. In fact, $\mathbf{g}'(0) = \alpha\mathbf{F}_r + \beta\mathbf{F}_s = \mathbf{v}$. Since $\mathbf{v}$ is by hypothesis nonzero, $\mathbf{g}$ has a tangent at $\mathbf{g}(0) = \mathbf{b}$, and $\mathbf{v}$ is it. □

Example 3. Let $\mathbf{F} = (x, y, z)$ be the function of θ and ϕ given by

$$x = a\cos\theta\sin\phi, \; y = a\sin\theta\sin\phi, \; z = a\cos\phi, \quad -\pi \le \theta \le \pi, \; 0 \le \phi \le \pi.$$

This surface is the sphere of radius a (constant), described in terms of spherical coordinates. (We urge the reader to examine an Earth globe as we consider $\mathbf{F}$.)

(a) First, in words: It should be clear that at any point on the surface, θ can be viewed as longitude measured eastward (so that Rome has $\theta \approx \frac{12\pi}{180}$, New York $\approx \frac{-74\pi}{180}$) and ϕ as colatitude ($\frac{\pi}{2}$ − latitude, so that the North Pole at latitude 90° has $\phi = 0$, Rio de Janeiro at latitude −23° has $\phi \approx \frac{113\pi}{180}$). If we hold $\phi = \phi_0$

constant, then the curve $\mathbf{g}(\theta) \equiv \mathbf{F}(\theta, \phi_0)$ is a parallel of latitude, and the tangent $\mathbf{g}'(\theta) = \frac{\partial \mathbf{F}}{\partial \theta}(\theta, \phi_0)$ points east along a small circle, except that the curves and tangent referred to in this sentence degenerate at both poles. If instead we fix $\theta = \theta_0$, then the resulting curve $\mathbf{h}(\phi) \equiv \mathbf{F}(\theta_0, \phi)$ is a meridian of longitude, its tangent $\mathbf{h}'(\phi) = \frac{\partial \mathbf{F}}{\partial \phi}(\theta_0, \phi)$ is the southward-pointing tangent to a (half) great circle, and again we must think separately about the poles, where "southward" is not unique. Away from the poles, the parallels and meridians meet (the way the Tropic of Cancer, $\phi \approx \frac{67\pi}{180}$, meets the International Date Line, $\theta = \pi$) at right angles, so their tangents are not only independent, they are orthogonal. Finally, at any point, the tangent plane is the union of the lines that are perpendicular to the sphere's radius there.

(b) In symbols: We have

$$\frac{\partial \mathbf{F}}{\partial \theta} = \begin{bmatrix} -a \sin\theta \sin\phi \\ a \cos\theta \sin\phi \\ 0 \end{bmatrix} \quad \text{and} \quad \frac{\partial \mathbf{F}}{\partial \phi} = \begin{bmatrix} a \cos\theta \cos\phi \\ a \sin\theta \cos\phi \\ -a \sin\phi \end{bmatrix}.$$

These are orthogonal (Verify!), so they are independent unless one of them is $\mathbf{O}$. From $\|\mathbf{F}_\theta\| = a \sin\phi$ (Why not $= a|\sin\phi|$?) and $\|\mathbf{F}_\phi\| = a$, we see that $\mathbf{F}$ is smooth everywhere but at the places where $\phi = 0$ or π, namely the poles. (Again we find a singularity that is a peculiarity of the function, not of the sphere.)

Because $\mathbf{F}_\theta$ and $\mathbf{F}_\phi$ are orthogonal, the equation $\mathbf{v} = \alpha\mathbf{F}_\theta + \beta\mathbf{F}_\phi$ means that $\alpha\mathbf{F}_\theta$ is the projection of $\mathbf{v}$ onto $\mathbf{F}_\theta$, and similarly with $\beta\mathbf{F}_\phi$. In particular, if $\mathbf{v}$ is a unit vector, then $\mathbf{v} \bullet \left(\frac{\mathbf{F}_\phi}{a}\right) = \beta a$ is the cosine of the "deviation from south" of $\mathbf{v}$ (the angle $\mathbf{v}$ makes with true south).

We see also that $\mathbf{F}_\theta$ and $\mathbf{F}_\phi$ are both orthogonal to (x, y, z), which is the radius vector at any point. Hence each of their linear combinations is orthogonal to (x, y, z); the tangent plane is perpendicular to the radius.

(c) The observation that $\mathbf{F}_\theta$ and $\mathbf{F}_\phi$ are the columns of the Jacobian (derivative) matrix gives us a practical way to decide smoothness and to characterize the tangent plane.

Consider the point $\mathbf{b} \equiv \left(\frac{a}{2}, \frac{a}{2}, \frac{a}{\sqrt{2}}\right)$, located at latitude = longitude = $\frac{\pi}{4}$. There,

$$\frac{\partial \mathbf{F}}{\partial(\theta, \phi)} = \begin{bmatrix} -\frac{a}{2} & \frac{a}{2} \\ \frac{a}{2} & \frac{a}{2} \\ 0 & -\frac{a}{\sqrt{2}} \end{bmatrix}.$$

It is easy to tell when two vectors are independent. But even if there were three or more columns, we could determine their independence by row-reducing the matrix. In this matrix, the row-echelon form is $\begin{bmatrix} 1 & -1 \\ 0 & 1 \\ 0 & 0 \end{bmatrix}$. Hence the rank is 2, the same as the number of columns; the original columns are independent.

As for the tangent plane, a vector $\mathbf{x}$ is in the plane $\Leftrightarrow$ it has the form $\mathbf{b}+\alpha\mathbf{F}_\theta+\beta\mathbf{F}_\phi \Leftrightarrow \mathbf{x}-\mathbf{b}$ is a combination of $\mathbf{F}_\theta$ and $\mathbf{F}_\phi \Leftrightarrow$ the determinant

$$\left|\mathbf{F}_\theta\ \mathbf{F}_\phi\ \mathbf{x}-\mathbf{b}\right| = \begin{vmatrix} -\frac{a}{2} & \frac{a}{2} & x-\frac{a}{2} \\ \frac{a}{2} & \frac{a}{2} & y-\frac{a}{2} \\ 0 & -\frac{a}{\sqrt{2}} & z-\frac{a}{\sqrt{2}} \end{vmatrix} = 0.$$

The determinant equation yields $x+y+z\sqrt{2}-2a=0$.

Notice that the argument in Theorem 1 still works, toward one side, at any edge of R, and in a 90-degree sector from any corner. That possibility leads us to talk about derivatives at the edges of regions. Some discussion is essential: Almost all of our theorems about differentiability and derivatives, but not the definition, operate at interior points of the domain. What happens at the boundary?

Suppose O is a bounded open set in $\mathbf{R}^n$, equivalently, O is the interior of a bounded set. If g is differentiable and g' is *uniformly* continuous on O, then [and only then; consult Guzman, Section 5.5] g' can be extended uniquely to a continuous (row-valued) function $\mathbf{G}(\mathbf{x})$ on the closure $\mathrm{cl}(O)$. Suppose now $\mathbf{b}$ is a boundary point of O at the vertex of a sector opening into O, the situation described in Theorem 1.1:2. Then for $\mathbf{x} \in O$ in the sector, the mean value theorem applies, and

$$\begin{aligned}\|g(\mathbf{x})-g(\mathbf{b})-\mathbf{G}(\mathbf{b})\langle\mathbf{x}-\mathbf{b}\rangle\| &= \|\mathbf{G}(\mathbf{x}^*)\langle\mathbf{x}-\mathbf{b}\rangle-\mathbf{G}(\mathbf{b})\langle\mathbf{x}-\mathbf{b}\rangle\| \quad (\text{some } \mathbf{x}^*)\\ &\le \|\mathbf{G}(\mathbf{x}^*)-\mathbf{G}(\mathbf{b})\|\ \|\mathbf{x}-\mathbf{b}\|.\end{aligned}$$

Because $\mathbf{G}$ is continuous at $\mathbf{b}$, we see that $g(\mathbf{x})-g(\mathbf{b})-\mathbf{G}(\mathbf{b})\langle\mathbf{x}-\mathbf{b}\rangle = o(\mathbf{x}-\mathbf{b})$, so that $\mathbf{G}$ is the derivative of g on the boundary as well.

Thus, the condition that (surface) $\mathbf{F}$ have continuous derivatives on the box amounts to requiring that $\mathbf{F}'$ be uniformly continuous on the interior of the box.

Now it is easy to generalize to higher dimensions. We list the needed definitions, together with results corresponding to Theorem 2 and the method of Example 3. Their proofs and interpretations are identical in form to the theorem and example.

A **surface of dimension** j, $3 \le j \le n-1$, is a differentiable mapping $\mathbf{H}$ from a box in $\mathbf{R}^j$ to $\mathbf{R}^n$. The mapping $\mathbf{H}$ is **smooth** at any place $\mathbf{b} \equiv \mathbf{H}(v_1, \ldots, v_j)$ where the vectors $\frac{\partial\mathbf{H}}{\partial v_1}, \ldots, \frac{\partial\mathbf{H}}{\partial v_j}$—that is, the columns of $\mathbf{H}'$—are continuous and independent, at which place the translate $\mathbf{b}+\left(\frac{\partial\mathbf{H}}{\partial v_1}, \ldots, \frac{\partial\mathbf{H}}{\partial v_j}\right)$ is the **tangent plane** to the surface. Where $\mathbf{H}$ is continuously differentiable, it will be smooth if $\frac{\partial\mathbf{H}}{\partial\mathbf{v}}$ has rank j. If a point $\mathbf{x} \ne \mathbf{b}$ is in the tangent plane, then $\mathbf{x}-\mathbf{b}$ is tangent at $\mathbf{b}$ to some curve that lies on the surface.

We will always use the name "curve" for the case $j = 1$ and "surface" for $j = 2$. There is no need for a name to cover the cases $j = 3$ through $j = n-2$, but we will reserve the names **hypersurface** and **tangent hyperplane** for the specific case $j = n - 1 > 2$.

This last case is important for two reasons, one of them being what we have said about implicit scalar functions.

Theorem 2. *Assume that f is differentiable near* **b**, *with ∇f continuous and nonzero at* **b**. *Then near* **b**, *the solutions of $f(\mathbf{x}) = f(\mathbf{b})$ constitute a hypersurface that is smooth at* **b**.

Proof. Because $\nabla f(\mathbf{b}) \neq \mathbf{O}$, one of the partials of f is nonzero; say $\frac{\partial f}{\partial x_n}(\mathbf{b}) \neq 0$. By the implicit function theorem (Theorem 2.5:1), there is a box B surrounding **b** in which $f(\mathbf{x}) = f(\mathbf{b})$ iff $x_n = g(x_1, \ldots, x_{n-1})$, where g is a differentiable function of $x_1, \ldots, x_{n-1}$. Thus, the nearby solution set is parametrized by

$$x_1 = t_1, \ldots, x_{n-1} = t_{n-1}, \; x_n = g(t_1, \ldots, t_{n-1}),$$

for $\mathbf{t} \equiv (t_1, \ldots, t_{n-1})$ in the box $B^{\#}$.

By the theorem, the hypersurface $\mathbf{G} \equiv (x_1, \ldots, x_n)$ given by the parametrization has

$$\frac{\partial \mathbf{G}}{\partial t_1} = \begin{bmatrix} 1 \\ 0 \\ \vdots \\ 0 \\ \frac{\partial x_n}{\partial t_1} \end{bmatrix} = \begin{bmatrix} 1 \\ 0 \\ \vdots \\ 0 \\ \frac{-\partial f/\partial x_1}{\partial f/\partial x_n} \end{bmatrix},$$

and similarly for

$$\frac{\partial \mathbf{G}}{\partial t_2} = \begin{bmatrix} 0 & 1 & \cdots & 0 & \dfrac{-\partial f/\partial x_2}{\partial f/\partial x_n} \end{bmatrix}^t$$

through

$$\frac{\partial \mathbf{G}}{\partial t_{n-1}} = \begin{bmatrix} 0 & 0 & \cdots & 1 & \dfrac{-\partial f/\partial x_{n-1}}{\partial f/\partial x_n} \end{bmatrix}^t.$$

These are evidently independent. The hypothesis guarantees that they exist in $B^{\#}$ and are continuous at $(b_1, \ldots, b_{n-1})$. Hence **G** is a smooth hypersurface at **b**. □

We will call the set of points where $f(\mathbf{x}) = f(\mathbf{b})$ a **level hypersurface** ("level curve" in $\mathbf{R}^2$, "level surface" in $\mathbf{R}^3$) of f.

Example 4. The proof indicates that wherever the hypersurface is given by $x_n = g(x_1, \ldots, x_{n-1})$, the vectors $\left(1, \ldots, 0, \frac{\partial x_n}{\partial x_1}\right), \ldots, \left(0, \ldots, 1, \frac{\partial x_n}{\partial x_{n-1}}\right)$ define the tangent hyperplane; that is, they give a basis for the associated subspace.

Consider a picture of the graph of $z = g(x, y)$, as in Figure 2.8. The plane of the paper has $y = y_0$ and cuts the surface along a curve given by $z = f(x) \equiv g(x, y_0)$. The tangent line to the curve at $\big(x, y_0, g(x, y_0)\big)$ has slope $\frac{df}{dx} = \frac{\partial g}{\partial x}$. If we move horizontally one unit, the tangent rises $\frac{\partial g}{\partial x}$. Therefore, the tangent line

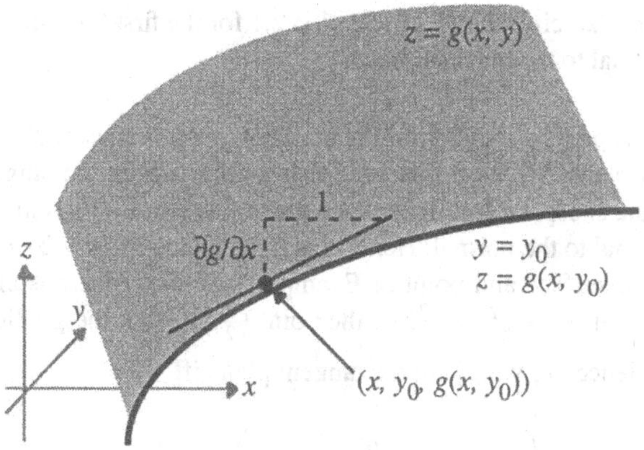

Figure 2.8.

follows the vector $\left(1, 0, \frac{\partial g}{\partial x}\right)$. This is precisely the first vector from the proof of Theorem 2.

Similarly, $\left(0, 1, \frac{\partial g}{\partial y}\right)$ is the tangent to the curve $z = g(x_0, y)$ intercepted along the surface by the plane $x = x_0$.

The other factor behind the importance of hypersurfaces and their tangents is that a hyperplane has a unique perpendicular direction. More precisely, a hyperplane is the translate of a subspace of dimension $n - 1$. The orthogonal complement of that subspace has dimension 1. Therefore, there is a vector $\mathbf{v}$ such that every vector orthogonal to the subspace is a multiple of $\mathbf{v}$. Returning to geometric language, the lines perpendicular to the hyperplane are translates of the line of $\mathbf{v}$.

Definition. Suppose a surface is smooth at $\mathbf{b}$ and the tangent plane is $\mathbf{b} + T$. The **normal plane** to the surface at $\mathbf{b}$ is the translate $\mathbf{b} + N$ of the orthogonal complement N of T. For a hypersurface, if $N = \langle\langle \mathbf{v} \rangle\rangle$, then we call $\mathbf{v}$ the **normal** to the hypersurface.

Theorem 3. *Assume that f is differentiable near* $\mathbf{b}$, *with ∇f continuous and nonzero at* $\mathbf{b}$. *Then $\nabla f(\mathbf{b})$ is the normal at* $\mathbf{b}$ *to the hypersurface given by $f(\mathbf{x}) = f(\mathbf{b})$.*

Proof. Again, some partial of f is nonzero; say $\frac{\partial f}{\partial x_n}(\mathbf{b}) \neq 0$.

In Theorem 2, we showed that the solutions of $f(\mathbf{x}) = f(\mathbf{b})$ make a hypersurface, whose tangent hyperplane is

$$\mathbf{b} + \left\langle\!\!\left\langle \left(1, \ldots 0, \frac{\partial x_n}{\partial x_1}\right), \ldots, \left(0, \ldots, 1, \frac{\partial x_n}{\partial x_{n-1}}\right) \right\rangle\!\!\right\rangle.$$

Each of these vectors is orthogonal to $\nabla f(\mathbf{b})$ (Exercise 7). Therefore, $\nabla f(\mathbf{b})$ is orthogonal to their span. It follows that the normal line is spanned by $\nabla f(\mathbf{b})$. $\square$

Look back at Exercises 1 and 2 in Section 2.1 for the first hints that a function's gradient is normal to the level surfaces.

Example 5. The perpendicularity of the gradient gives us an easy characterization of the tangent plane. We have observed that a point $\mathbf{x}$ is in the tangent plane iff $\mathbf{x} - \mathbf{b}$ lies in the subspace that defines the plane. In turn, $\mathbf{x} - \mathbf{b}$ is in the subspace iff it is orthogonal to the normal. Hence $\mathbf{x}$ is in the plane iff $(\mathbf{x} - \mathbf{b}) \bullet \nabla f(\mathbf{b}) = 0$.

Return to the surface and point of Example 3(c). The sphere is also given by $f(x, y, z) \equiv x^2 + y^2 + z^2 = a^2$. At the point $\left(\frac{a}{2}, \frac{a}{2}, \frac{a}{\sqrt{2}}\right)$, the gradient is $\nabla f = \left(a, a, a\sqrt{2}\right)$. Hence (x, y, z) is in the tangent plane iff

$$\begin{aligned} 0 &= \left(x - \frac{a}{2}, y - \frac{a}{2}, z - \frac{a}{\sqrt{2}}\right) \bullet \left(a, a, a\sqrt{2}\right) \\ &= a\left[\left(x - \frac{a}{2}\right) + \left(y - \frac{a}{2}\right) + \left(z\sqrt{2} - a\right)\right]; \end{aligned}$$

equivalently, $0 = x + y + z\sqrt{2} - 2a$, which is what we found before.

It is worthwhile, as in Section 2.5, to make special mention of the case $n = 2$. If $\frac{\partial f}{\partial y} \neq 0$ at a point (a, b) where f is continuously differentiable, then the tangent to the graph of $f(x, y) = f(a, b)$ is not vertical. The reason is that $\frac{\partial f}{\partial y}$ is the vertical component of ∇f; if ∇f has nonzero vertical component, then the tangent line, being perpendicular to ∇f, has nonzero horizontal component. The converse is also true: If the tangent is not vertical, then the normal is not horizontal, and its component $\frac{\partial f}{\partial y}$ is nonzero. At these points, since the tangent is not vertical and the graph is approximately the tangent, the graph passes the vertical line test. Geometrically speaking, that is why the graph goes with a function $y = g(x)$. Indeed, in general, $\frac{\partial F}{\partial x_k}(\mathbf{b}) = 0$ exactly where the tangent hyperplane to the hypersurface $F(\mathbf{x}) = F(\mathbf{b})$ is parallel to (or contains) the x_k-axis.

The case $n = 2$ also relates to two familiar kinds of map. Figure 2.9 is a standard part of weather reports. It has an outline of the contiguous USA, and shows the level curves of air pressure $p = f(x, y)$. Thus, at all the points of the curve above Chicago, the pressure is 30.4 inches of mercury, the highest value shown. At any point on such an "isobar," the pressure's gradient ∇p is perpendicular to the (tangent to the) curve, pointing toward the side of higher pressure. For example, at the point on the 30.0 isobar directly above the "L" in Las Vegas, ∇p points northeast (upper right). From there, the pressure drops fastest in the direction of $-\nabla p$. Ignoring other factors, we might figure that the wind, seeking to fill in the places of lower pressure, would blow in the $-\nabla p$ direction. The magnitude of the gradient corresponds to the spacing of the curves. Where the isobars cluster together, the pressure changes rapidly in relation to distance along the land. So, from Orlando to Atlanta, the pressure goes from under 29.8 inches to about 30.1 over a distance of about 420 miles. There, the magnitude of the gradient, the rate of change in the direction of fastest increase, is relatively big: roughly 0.07 inch

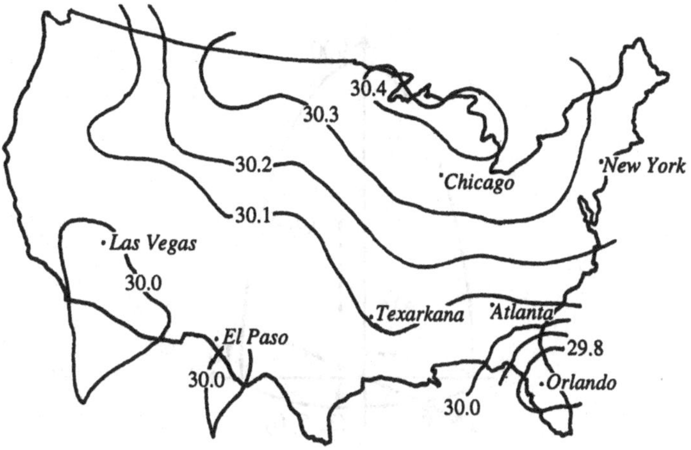

Figure 2.9.

per 100 miles. Where the curves are spaced far apart, the pressure has small (average) rate of change, and the gradients are short: from El Paso at the western end of Texas to Texarkana at the east, the average gradient norm is 0.1 in/720mi ≈ 0.01 in/100 mi.

A similar description applies to "contour maps." On those, the level curves of altitude $h = h(x, y)$ above sea level are plotted—though usually over an extent of thousands of meters rather than thousands of kilometers—projected onto level ground. If you are on a hill and walk along a curve of constant altitude ("isocline"), then you walk around the hill without going up or down. If you stop and turn 90°, to face the horizontal direction ("azimuth") normal to the curve, then the route of steepest ascent lies before or behind you, with steepest descent the opposite way. Walking in either of those two directions, you would gain altitude at a rate $\pm\|\nabla h\|$ (vertical meters per horizontal meter walked, for example). If you walk the hill along a path whose projection onto sea level makes an acute angle θ with $\pm\nabla h$, then you are climbing (up or down) the hill the way roads do, by spiraling around it with a climb rate $\pm\|\nabla h\|\cos\theta$ less severe than the two extremes. On the map itself, a place where the isoclines crowd together has long ∇p, steep terrain, and the opposite where the contours are widely spaced.

Example 6. Consider the mountain rising from an infinite plain (suggested by Figure 2.10) that is the graph of

$$z = \max\left\{0, 900 - 25x^2 - 9y^2\right\}.$$

If we stand at $P(3, 5, 450)$, in which direction is the climb up the mountain steepest?

There are two separate things to consider. One is that our movement is confined to the surface; we may not tunnel under nor fly above it, so that we must travel on

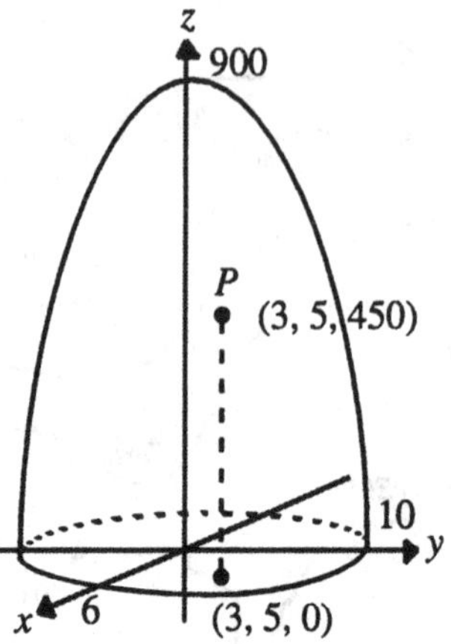

Figure 2.10.

a path in $\mathbf{R}^3$ tangent to the hill. The other is that steepness on the hill is the rate of change of z, a function of two variables whose derivatives are calculated in $\mathbf{R}^2$.

To characterize the directions tangent to the hill, we may use the normal. Consider the mountain first as the level surface $F(x, y, z) \equiv 25x^2 + 9y^2 + z = 900$. By Theorem 3, the normal is the gradient $\nabla F(3, 5, 450) = (50x, 18y, 1) = (150, 90, 1)$. Every vector $\mathbf{v}$ tangent to the mountain at P must have $0 = \mathbf{v} \bullet \nabla F = 150v_1 + 90v_2 + v_3$.

Now, to decide which such vector gives the steepest ascent, think of z as the function $f(x, y) \equiv 900 - 25x^2 - 9y^2$. The vector $\mathbf{v}$ projects onto (v_1, v_2) in $\mathbf{R}^2$, which we identify with the xy-plane. If $\mathbf{v}$ is to maximize the rate of increase of f, then (v_1, v_2) must point along $\nabla f(3, 5) = (-150, -90)$. It suffices to take $(v_1, v_2) = (-5, -3)$. From the earlier equation, we get $v_3 = -150v_1 - 90v_2 = 1020$. Thus, $\mathbf{v} \equiv (-5, -3, 1020)$ gives the direction we must travel from P.

Note that this vector does not project onto $(-3, -5)$, so it does not point toward the z-axis. From P, you do not get the fastest rate of climb by sighting the flagpole at the top of the mountain and walking towards it.

Exercises

1. Give a vector description and an equation for the tangent line to each curve at the indicated point:

   (a) $x = 2\cos\theta$, $y = 2\sin\theta$, $z = e^\theta$, at $\left(1, \sqrt{3}, e^{\pi/3}\right)$.

   (b) $y = x^2$ in $\mathbf{R}^2$, at $(2, 4)$.

2. Show that the graph of each equation is smooth at all its points, and find an equation for the tangent plane at the point (a, b, c):

(a) $z = x^2 + y^2$ (paraboloid);

(b) $x^2 + y^2 = 4$, z arbitrary (cylinder);

(c) $x^2 + y^2 + z^2 = r^2$ (sphere);

(d) $z = (x - y)^{1/3}$.

3. On the paraboloid $z = x^2 + y^2$, $\mathbf{v} \equiv (1, 2, 0)$ is a vector in the tangent plane at the origin (see Exercise 2(a)). Find a curve lying on the surface for which $\mathbf{v}$ is the tangent.

4. Find a "basis for the tangent plane" to the surface given by $x - y + z^2 = 1$ at a point (a, b, c). (Find $\mathbf{u}$ and $\mathbf{v}$ such that the tangent plane is $(a, b, c) + \langle\langle \mathbf{u}, \mathbf{v} \rangle\rangle$.)

5. Assume that $F(x)$ is continuously differentiable near $x = a$. Use Theorem 3 to find an equation of the tangent line to the graph of $y = F(x)$ at $(a, F(a))$.

6. (a) Prove that wherever a curve is smooth, the **forward-pointing secants** have a limiting direction: in symbols, if $\mathbf{g}$ is smooth at $\mathbf{g}(t)$, then there is a unit vector $\mathbf{u}$ such that $\lim_{s \to t^+} \frac{\mathbf{g}(s)-\mathbf{g}(t)}{\|\mathbf{g}(s)-\mathbf{g}(t)\|}$ and $\lim_{s \to t^-} \frac{\mathbf{g}(t)-\mathbf{g}(s)}{\|\mathbf{g}(t)-\mathbf{g}(s)\|}$ are both $\mathbf{u}$.

   (b) Show that if $\mathbf{g}$ is one-to-one and its range is $\{(x, y) \colon -1 \le x \le 1, y = x^{2/3}\}$ (Example 1(c)), then $\mathbf{g}$ is not smooth at $(0, 0)$. (Hint: Prove that the two limits in part (a) work out to be opposites at the origin. This opposition of the limits is the usual behavior at a cusp.)

   (c) Suppose $\mathbf{g}$ is one-to-one and has range $\{(x, y) \colon -1 \le x \le 1, y = |x|\}$. Show that at the origin, the two limits in part (a) are not in the same line.

7. Assume ∇f exists near $\mathbf{b}$ and is continuous at $\mathbf{b}$, and $\frac{\partial f}{\partial x_n}(\mathbf{b}) \neq 0$. Show that $\nabla f(\mathbf{b})$ is orthogonal to each vector

$$\left(1, 0, \ldots, 0, \frac{\partial x_n}{\partial x_1}(\mathbf{b}^\#)\right), \ldots, \left(0, \ldots, 0, 1, \frac{\partial x_n}{\partial x_{n-1}}(\mathbf{b}^\#)\right).$$

3

Derivatives of Vector Functions

In this chapter we study derivative-related results that are not reducible to the scalar case.

The central concepts are inverses and implicit functions, with inverses leading the way. Inverses of vector functions cannot be explained component by component. For example, existence of the function inverse (unlike existence of the derivative) is not equivalent to existence of component inverses. Indeed, it will be clear that if a nonscalar differentiable function has a differentiable inverse, then its components cannot be one-to-one. Thus, for analysis of inverses of vector functions, the interplay among components makes a difference; we must deal with the function as a vector.

3.1 Contractions

The result we want about inverses and their derivatives is almost invariably proved by reference to a theorem that has nothing to do with inverses or derivatives. Nevertheless, the theorem is so useful in analysis, and belongs to a family of such importance, that it merits a section of its own.

Definition. A function $\mathbf{f}$ mapping a set $D \subseteq \mathbf{R}^n$ into itself is a **contraction** if there exists $K < 1$ such that $\|\mathbf{f}(\mathbf{x}) - \mathbf{f}(\mathbf{y})\| \leq K\|\mathbf{x} - \mathbf{y}\|$ for all $\mathbf{x}, \mathbf{y} \in D$.

Contraction is a metric notion, so the definition and the principle below carry over to any set in which a metric is defined. Using $\mathbf{R}^n$ as the setting merely reflects our current interest. The function does not have to be defined on the whole space, but it is essential for the range to be a subset of the domain.

Example 1. (a) A function with a small enough derivative will be a contraction.

Let $f(x) \equiv \cos x$, $0 \le x \le 1$. Observe that $f\colon [0,1] \to [0,1]$. Also, $|f'(x)| = \sin x$ is between 0 and $\sin 1 \approx 0.8$. By the mean value theorem,

$$|f(x) - f(y)| = |f'(t)|\,|x-y| \le 0.9|x-y|,$$

and f is a contraction.

(b) It does not suffice to demand that the derivative be less than 1. Write $g(x) \equiv \sqrt{1+x^2}$, $0 \le x$. Then $g\colon [0,\infty) \to [0,\infty)$, and

$$|g(x) - g(y)| = |g'(t)|\,|x-y| = t(1+t^2)^{-1/2}|x-y| < |x-y|,$$
$$\text{except if } x = y.$$

[In the language of Guzman, Section 2.3, g is "contractive."] But g is not a contraction, because there is no fixed $K < 1$ such that $|g(x) - g(y)| \le K|x-y|$.

To justify the last statement, we put it differently. For this function, $\sup\left|\frac{g(x)-g(y)}{x-y}\right| = 1$, because

$$\lim_{x\to\infty} \frac{g(2x)-g(x)}{2x-x} = \lim_{x\to\infty} \frac{\sqrt{1+4x^2}-\sqrt{1+x^2}}{x} = 1.$$

For a contraction, the ratio must be bounded away from 1; in other words, the sup must be strictly less than 1.

Theorem 1 (The Contraction Principle). *Assume that* $\mathbf{f}$ *is a contraction on a nonempty closed set* $D \subseteq \mathbf{R}^n$. *Then* $\mathbf{f}$ *has a unique fixed point in* D*; that is, there is one and only one* $\mathbf{x} \in D$ *such that* $\mathbf{f}(\mathbf{x}) = \mathbf{x}$.

Proof. Let $\mathbf{b}$ be any member of D, and examine the sequence $\mathbf{y}_0 \equiv \mathbf{b}$, $\mathbf{y}_1 \equiv \mathbf{f}(\mathbf{y}_0) = \mathbf{f}(\mathbf{b})$, $\mathbf{y}_2 \equiv \mathbf{f}(\mathbf{y}_1) = \mathbf{f}\big(\mathbf{f}(\mathbf{b})\big), \ldots$. Its consecutive terms are squeezed together, in that

$$\|\mathbf{y}_{i+1} - \mathbf{y}_i\| \le K\|\mathbf{y}_i - \mathbf{y}_{i-1}\| \le \cdots \le K^i\|\mathbf{y}_1 - \mathbf{y}_0\|.$$

Consequently, terms are not far from the initial term:

$$\begin{aligned}\|\mathbf{y}_{i+1} - \mathbf{y}_0\| &\le \|\mathbf{y}_{i+1} - \mathbf{y}_i\| + \cdots + \|\mathbf{y}_1 - \mathbf{y}_0\| \\ &\le \big(K^i + \cdots + K + 1\big)\|\mathbf{y}_1 - \mathbf{y}_0\| \le \frac{\|\mathbf{f}(\mathbf{b}) - \mathbf{b}\|}{1-K}.\end{aligned}$$

As a result, all its terms are squeezed. Let i and j be separate indices. Then

$$\begin{aligned}\|\mathbf{y}_{i+j} - \mathbf{y}_i\| &\le K\|\mathbf{y}_{i+j-1} - \mathbf{y}_{i-1}\| \\ &\le \cdots \le K^i\|\mathbf{y}_j - \mathbf{y}_0\| \le \frac{K^i\|\mathbf{f}(\mathbf{b}) - \mathbf{b}\|}{1-K}.\end{aligned}$$

With $K < 1$, we see that $\|\mathbf{y}_{i+j} - \mathbf{y}_i\| \to 0$ as $i \to \infty$. Hence the sequence is Cauchy.

Because every Cauchy sequence in $\mathbf{R}^n$ converges, $(\mathbf{y}_i)$ has a limit $\mathbf{y}$; and this limit is in D, since D is closed. Therefore, $\mathbf{f}(\mathbf{y})$ is defined, and

$$\begin{aligned}\mathbf{f}(\mathbf{y}) &\equiv \mathbf{f}\left(\lim_{i\to\infty} \mathbf{y}_i\right) = \lim_{i\to\infty} \mathbf{f}(\mathbf{y}_i) \quad \text{(Reason? See Exercise 4(a).)}\\ &= \lim_{i\to\infty} \mathbf{y}_{i+1} = \mathbf{y}.\end{aligned}$$

We have found a fixed point. The proof of uniqueness is Exercise 5. □

Example 2. Pick any real number r and use a calculator (in RADIAN mode) to evaluate $\cos r$, then $\cos(\cos r)$, $\cos(\cos(\cos r))$, ... The second term $\cos(\cos r)$ is necessarily in $[0, 1]$; your sequence of values should reach roughly 0.74 within about twelve terms.

Example 1(a) demonstrated that $\cos x$ is a contraction in $[0, 1]$. By Theorem 2, it has a fixed point θ. The calculations show that $\theta \approx 0.739$.

One of the important features of the proof of the contraction principle is that it does not matter where you start; any $\mathbf{b}$ in the domain leads to the fixed point. Try the same calculations beginning with different values of r. In this case, you may even start with r outside $[0, 1]$. (How come it works anyway? See Exercises 2 and 3 for similar questions.)

Example 3. The contraction principle underlies a simple method that approximates solutions of one-variable equations. Consider $x^2 - 10x + 21 = 0$. If the equation can be recast into the form $g(x) = x$ so that g is a contraction, then the solutions can be found by **iteration**.

(a) We can immediately get the form by adding x to both sides: $g(x) \equiv x^2 - 9x + 21 = x$. The function has $|g'(x)| \le 1$ only in the interval $[4, 5]$, which g does not map to itself. Hence this reformulation is unsuccessful.

(b) Suppose we rewrite the equation as $x^2 + 21 = 10x$, or

$$h(x) \equiv \frac{x^2}{10} + 2.1 = x.$$

Then $|h'(x)| < 1$ for $|x| < 5$, so that h is a contraction on, say, $[-4.9, 4.9]$. Beginning with $x_0 \equiv 0$, we get $x_1 = h(x_0) = 2.1$, $x_2 = h(x_1) = 2.541$, $x_3 \approx 2.75, \ldots$, approaching the root $x = 3$.

(c) Let us reformulate to $10x - 21 = x^2$, or

$$G(x) \equiv 10 - \frac{21}{x} = x.$$

Then $0 < G'(x) < 1$ for $|x| > \sqrt{21}$; G is a contraction on $[5, \infty)$. Here $x = 0$ is an illegal substitution, but $x_0 = 1000$ gives $x_1 = 9.979$, $x_2 \approx 7.9$, $x_3 \approx 7.3, \ldots$, leading quickly toward the other root, $x = 7$.

It is important to note that in the proof of Theorem 1 there were just two key requirements for $\mathbf{R}^n$ and D. First, every Cauchy sequence in $\mathbf{R}^n$ has a limit in $\mathbf{R}^n$. Hence the proof works in any metric space having that property, that is, in any complete metric space. Second, a convergent sequence from D has its limit in D. Accordingly, the proof succeeds as long as D is a closed subset of the space.

The contraction principle belongs to the family of "fixed-point theorems," which guarantee that certain mappings have fixed points within their domains. This family is fabulously useful in the theoretical development of analysis and topology, as well as in the applications of mathematics to areas, like physics and economics, in which "stability" and "equilibrium" are important concepts. To illustrate its reach, we will use the Principle to prove a standard theorem about differential equations.

The simplest differential equation relates the derivative $\frac{dy}{dx}$ to the independent variable x and the function y itself, via an equation $f(x, y, y') = 0$, which we assume may be solved for $y' = F(x, y)$. (If either x or y is missing from F, then the question reduces to an antidifferentiation.) An **initial value problem** asks for the solution of this equation satisfying the "initial condition" $y = y_0$ when $x = x_0$. (Why is such a solution necessarily unique?) Our next theorem shows that if F is decent—for example, a continuously differentiable F will do—then its initial value problems admit solutions.

Theorem 2. *Suppose F is continuous on the box in $\mathbf{R}^2$ given by $a - \Delta \le x \le a + \Delta$, $b - \varepsilon \le y \le b + \varepsilon$. Assume that F is also Lipschitz relative to y; that is (Exercises 2.2:6–8), assume that there exists K such that*

$$|F(x, y) - F(x, z)| \le K|y - z|$$

in the box. Then there exists $\delta > 0$ and a (unique) function f such that

$$f(a) = b \quad \textit{and} \quad f'(x) = F\big(x, f(x)\big) \;\; \textit{for } a - \delta \le x \le a + \delta.$$

Proof. Being continuous, F must be bounded on the box; say $|F(x, y)| \le M$. The problem is trivial if $M = 0$, so we assume that M is positive. For δ, we may use any positive number not exceeding Δ or $\frac{\varepsilon}{M}$ and strictly smaller than $\frac{1}{K}$.

We are going to employ facts about the normed linear space $C_0[a - \delta, a + \delta]$ of continuous functions on the closed interval $I \equiv [a - \delta, a + \delta]$, under the norm $\|g\|_0 \equiv \max\{|g(x)| : x \in I\}$. In this space, the vector sequence (g_i) converges to g iff the function sequence (g_i) converges uniformly to g, and every Cauchy sequence converges. The subset $D \equiv \{g : b - \varepsilon \le g(x) \le b + \varepsilon \text{ for } x \in I\}$—in words, the functions whose graphs lie in the box—is the ball of radius ε centered at the constant function $h_0 \equiv b$, so D is closed. [Consult the corresponding statements for $C_0[0, 1]$ in Guzman, Section 2.4.]

For $g \in D$, we define a function $\phi(g)$ by

$$[\phi(g)](x) \equiv b + \int_a^x F\big(t, g(t)\big)\, dt, \qquad x \in I.$$

We must first note that the integral is defined. Since g is continuous and has values between $b-\varepsilon$ and $b+\varepsilon$, the integrand $F\big(t, g(t)\big)$ is a continuous function of t, and therefore integrable. Second, by the fundamental theorem of calculus, the integral is a continuous function of x. Hence $\phi(g)$ is a member of $C_0[a-\delta, a+\delta]$. Third, for every x,

$$\big|[\phi(g)](x) - b\big| = \left|\int_a^x F\big(t, g(t)\big)\,dt\right| \le |x-a| \max\{|F(t,y)|\} \le \delta M \le \varepsilon.$$

Hence $\phi(g)$ is also in D, and ϕ maps D to itself. Fourth, if $g, h \in D$, then

$$\begin{aligned}\|\phi(g) - \phi(h)\|_0 &\equiv \max_{x\in I}\left|\int_a^x \big[F(t, g(t)) - F(t, h(t))\big]\,dt\right| \\ &\le \delta \max_{t\in I} |F(t, g(t)) - F(t, h(t))| \\ &\le \delta K \max |g(t) - h(t)| = \delta K \|g-h\|_0.\end{aligned}$$

This last property means that ϕ is a contraction, because $\delta K < 1$.

By the contraction principle, ϕ must have a fixed point in D. Thus, there is a continuous f such that $f = \phi(f)$, or

$$f(x) = b + \int_a^x F\big(t, f(t)\big)\,dt, \qquad a-\delta \le x \le a+\delta.$$

Here $f(a) = b$, and by the fundamental theorem,

$$f'(x) = F\big(x, f(x)\big). \qquad \square$$

The mapping defined in the proof is not a product of magic. If $f'(x) = F(x, f(x))$, then necessarily

$$f(x) = b + \int_a^x f'(t)\,dt = b + \int_a^x F\big(t, f(t)\big)\,dt.$$

The thing to remember is that the fixed point of a contraction can be approximated by iterating the contraction. Hence the (at least local) solution of an initial value problem can be approximated uniformly by iterated integration; see Exercise 6.

Exercises

1. In Example 1(b), we showed that $g(x) \equiv \sqrt{1+x^2}$, as a mapping in $[0, \infty)$, is not a contraction. Show that g does not have a fixed point there.

2. (a) Put a calculator in DEGREE mode, pick a number r, and calculate $\cos r$, $\cos(\cos r)$, ... Is there a fixed point?

(b) According to Example 2, $\theta \approx 0.739$ has $\theta = \cos\theta$, and 0.739 measures an angle of about 42°. Why is the answer to (a) not 42°?

(c) If $\theta = \cos\theta$, then $\cos^{-1}\theta = \cos^{-1}(\cos\theta) = \theta$. (Our θ is an acute angle.) Thus, θ is a fixed point of $\cos^{-1}$. Calculate $\cos^{-1}\theta$, $\cos^{-1}(\cos^{-1}\theta), \ldots$, and explain the result.

(d) Calculate $\sin r$, $\sin(\sin r), \ldots$. Explain the result.

3. In Example 3(c):

(a) Show that the iteration $x_1 = G(x_0)$, $x_2 = G(x_1), \ldots$, converges toward $x = 7$ for all $x_0 > 3$. (Hint: Show that x_{i+1} is "significantly" closer to 7 than x_i is.) This behavior is odd in two ways: First, G is not a contraction for $x > 3$; second, even if you start with x_0 very close to 3, which is one root of the equation, the iterations move you toward the other one.

(b) Show that the same is true for all $x_0 < 2.1$, except 0.

(c) Show that the iterations lead to $x_k = 0$ eventually (that is, for some k) if x_0 is any term from a certain sequence of numbers in $[2.1, 3)$.

(d) Show that the iterations converge to $x = 7$ for the numbers in $[2.1, 3)$ other than the ones in (c). (Hint: Show that the iterations eventually go below 2.1.)

4. Assume that **f** is a contraction. On its domain:

(a) Need **f** be continuous?

(b) Need **f** be differentiable?

(c) Suppose **f** is known to be differentiable. Need $\|\mathbf{f}'\|$ be small?

5. Show that a contraction cannot have more than one fixed point.

6. Consider the initial value problem

$$\frac{dy}{dx} = y, \qquad y(0) = 1.$$

(a) Let $f_0 \equiv 0$ and

$$f_{i+1}(x) \equiv 1 + \int_0^x f_i(t)\,dt, \qquad i = 0, 1, 2, \ldots.$$

Express $f_i(x)$ as an elementary function of x.

(b) What is the function $f(x) \equiv \lim_{i\to\infty} f_i(x)$?

(c) Perform the same iteration starting from $g_0(x) \equiv \cos x$. Verify that the resulting limit is the same as in (b). (Hint: The pattern becomes clear by g_5.)

3.2 The Inverse Function Theorem

Inverses are important anywhere that functions are. Here we deal with inverses in the context of differentiable functions.

We begin by showing that you cannot have a function and its inverse mapping differentiably from one dimension to a different one.

Theorem 1. *If the differentiable function* $\mathbf{f}\colon \mathbf{R}^n \to \mathbf{R}^m$ *has a differentiable inverse, then* $n = m$.

Proof. Write $\mathbf{y} = \mathbf{f}(\mathbf{x})$, so that $\mathbf{x} = \mathbf{f}^{-1}(\mathbf{y})$.

We have $\mathbf{x} = \mathbf{f}^{-1}(\mathbf{f}(\mathbf{x}))$ throughout the domain of $\mathbf{f}$. Since both $\mathbf{f}$ and $\mathbf{f}^{-1}$ are differentiable, the chain rule applies. Thus, $\left[\frac{\partial \mathbf{f}^{-1}}{\partial \mathbf{y}}\right]\left[\frac{\partial \mathbf{f}}{\partial \mathbf{x}}\right] = \frac{\partial \mathbf{x}}{\partial \mathbf{x}}$ is the identity mapping on $\mathbf{R}^n$. But $\frac{\partial \mathbf{f}}{\partial \mathbf{x}}$ maps $\mathbf{R}^n$ to $\mathbf{R}^m$; if a product of linear maps has the form $\mathbf{L}\left[\frac{\partial \mathbf{f}}{\partial \mathbf{x}}\right]$, then the range has dimension m or less. Therefore, $n = \dim(\mathbf{R}^n) \le m$. Symmetrically, $m \le n$. □

Theorem 1 allows us to restrict our attention to functions mapping $\mathbf{R}^n$ to itself.

The theorem we are approaching says that under appropriate conditions, a differentiable function has locally a differentiable inverse, and the derivative of the inverse is the inverse of the derivative. We will prove it in installments.

Any function that is one-to-one has an inverse from its range back to its domain. In the one-variable theory, a function continuous on an interval is one-to-one iff it is strictly monotone. Assuming that the function is differentiable, it is strictly monotone if, for example, the derivative is of one sign. If the derivative is continuous, then it has just one sign iff it is never zero.

Reasoning by analogy with the one-variable case, we might think that we need a vector function whose derivative is continuous and never zero. It should immediately be clear that merely nonzero is insufficient: $\mathbf{F}(x, y) \equiv (x, x)$ has nonzero derivative, but is evidently many-to-one in every neighborhood. However, if we recall that derivatives are linear maps, then our thinking improves.

What is special about a nonzero scalar derivative is that it gives an invertible map. Accordingly, suppose $\mathbf{f}'(\mathbf{b})$ is invertible. Our fundamental approximation

$$\mathbf{f}(\mathbf{x}) - \mathbf{f}(\mathbf{b}) \approx \mathbf{f}'(\mathbf{b})\langle \mathbf{x} - \mathbf{b}\rangle$$

suggests

$$\mathbf{f}'(\mathbf{b})^{-1}\langle \mathbf{f}(\mathbf{x}) - \mathbf{f}(\mathbf{b})\rangle \approx \mathbf{x} - \mathbf{b}$$

or

$$\mathbf{x} \approx \mathbf{b} - \mathbf{f}'(\mathbf{b})^{-1}\langle \mathbf{f}(\mathbf{b}) - \mathbf{f}(\mathbf{x})\rangle.$$

That is, if we know $\mathbf{y} \equiv \mathbf{f}(\mathbf{x})$, then we can more or less retrieve $\mathbf{x}$ from

$$\mathbf{x} \approx \mathbf{b} - \mathbf{f}'(\mathbf{b})^{-1}\langle \mathbf{f}(\mathbf{b}) - \mathbf{y}\rangle.$$

Retrieving $\mathbf{x}$ is precisely what we need to do.

Theorem 2. *A mapping of* $\mathbf{R}^n$ *to itself is one-to-one near any place that is not a singularity. In detail: Let* $\mathbf{f}\colon \mathbf{R}^n \to \mathbf{R}^n$ *be differentiable near* $\mathbf{b}$, *with* $\mathbf{f}'(\mathbf{x})$ *continuous at* $\mathbf{b}$. *Assume that* $\mathbf{f}'(\mathbf{b})$ *is an invertible linear map. (*$\mathbf{f}'(\mathbf{x})$ *also maps* $\mathbf{R}^n$ *to itself. We may call* $\mathbf{f}$ *"smooth at* $\mathbf{b}$*" because* $\mathbf{f}'(\mathbf{b})$*, being invertible, must have independent columns.) Then there are neighborhoods* $N(\mathbf{b}, \delta)$ *and* $N(\mathbf{f}(\mathbf{b}), \varepsilon)$ *such that* $\mathbf{f}$ *maps an open subset of* $N(\mathbf{b}, \delta)$ *one-to-one onto* $N(\mathbf{f}(\mathbf{b}), \varepsilon)$.

Proof. Let $\mathbf{L}$ represent $\mathbf{f}'(\mathbf{b})$ and $M \equiv \|\mathbf{L}^{-1}\|$. By the continuity of $\mathbf{f}'(\mathbf{x})$, there is a neighborhood of $\mathbf{b}$ in which $\|\mathbf{f}'(\mathbf{x}) - \mathbf{L}\| < \frac{1}{2Mn}$. By the nature of $\mathbf{L}$, there is a neighborhood of $\mathbf{b}$ in which

$$\|\mathbf{f}(\mathbf{x}) - \mathbf{f}(\mathbf{b}) - \mathbf{L}\langle \mathbf{x} - \mathbf{b}\rangle\| \le \frac{1}{2M}\|\mathbf{x} - \mathbf{b}\|.$$

Assume that the ball $B(\mathbf{b}, \delta)$ is contained in both neighborhoods. We will establish properties of f by defining a contraction on this ball.

Take $\varepsilon \equiv \frac{\delta}{2M}$ and fix $\mathbf{y} \in N(\mathbf{f}(\mathbf{b}), \varepsilon)$. Examine the function

$$\mathbf{g}(\mathbf{x}) \equiv \mathbf{x} - \mathbf{L}^{-1}\langle \mathbf{f}(\mathbf{x}) - \mathbf{y}\rangle, \qquad \mathbf{x} \in B(\mathbf{b}, \delta).$$

For any such $\mathbf{x}$,

$$\begin{aligned}\mathbf{g}(\mathbf{x}) - \mathbf{b} &= \mathbf{x} - \mathbf{b} - \mathbf{L}^{-1}\langle \mathbf{f}(\mathbf{x}) - \mathbf{f}(\mathbf{b})\rangle + \mathbf{L}^{-1}\langle \mathbf{y} - \mathbf{f}(\mathbf{b})\rangle \\ &= \mathbf{L}^{-1}\big\langle \mathbf{L}\langle \mathbf{x} - \mathbf{b}\rangle - [\mathbf{f}(\mathbf{x}) - \mathbf{f}(\mathbf{b})]\big\rangle + \mathbf{L}^{-1}\langle \mathbf{y} - \mathbf{f}(\mathbf{b})\rangle.\end{aligned}$$

By the operator and triangle inequalities,

$$\|\mathbf{g}(\mathbf{x}) - \mathbf{b}\| \le M\left(\frac{1}{2M}\right)\|\mathbf{x} - \mathbf{b}\| + M\|\mathbf{y} - \mathbf{f}(\mathbf{b})\| < \frac{\delta}{2} + M\varepsilon = \delta.$$

That is, $\mathbf{g}(\mathbf{x}) \in N(\mathbf{b}, \delta) \subseteq B(\mathbf{b}, \delta)$; $\mathbf{g}$ maps the ball to itself. Further, if $\mathbf{x}$ and $\mathbf{v}$ are in $B(\mathbf{b}, \delta)$, then

$$\mathbf{g}(\mathbf{x}) - \mathbf{g}(\mathbf{v}) = \mathbf{x} - \mathbf{v} - \mathbf{L}^{-1}\langle \mathbf{f}(\mathbf{x}) - \mathbf{f}(\mathbf{v})\rangle = \mathbf{L}^{-1}\big\langle [\mathbf{L}\langle \mathbf{x} - \mathbf{v}\rangle] - [\mathbf{f}(\mathbf{x}) - \mathbf{f}(\mathbf{v})]\big\rangle.$$

The first term in brackets is

$$\mathbf{L}\langle \mathbf{x} - \mathbf{v}\rangle = \begin{bmatrix} f_1'(\mathbf{b})\langle \mathbf{x} - \mathbf{v}\rangle \\ \vdots \\ f_n'(\mathbf{b})\langle \mathbf{x} - \mathbf{v}\rangle \end{bmatrix}.$$

For the second term, we apply the mean value theorem to the individual components to get

$$\mathbf{f}(\mathbf{x}) - \mathbf{f}(\mathbf{v}) \equiv \begin{bmatrix} f_1(\mathbf{x}) - f_1(\mathbf{v}) \\ \vdots \\ f_n(\mathbf{x}) - f_n(\mathbf{v}) \end{bmatrix} = \begin{bmatrix} f_1'(\mathbf{c}_1)\langle \mathbf{x} - \mathbf{v}\rangle \\ \vdots \\ f_n'(\mathbf{c}_n)\langle \mathbf{x} - \mathbf{v}\rangle \end{bmatrix},$$

each $\mathbf{c}_j$ lying on the segment from $\mathbf{x}$ to $\mathbf{v}$. In the difference of the two terms, each row has form $f_j'(\mathbf{b})\langle\mathbf{x}-\mathbf{v}\rangle - f_j'(\mathbf{c}_j)\langle\mathbf{x}-\mathbf{v}\rangle$. This expression satisfies

$$|f_j'(\mathbf{b})\langle\mathbf{x}-\mathbf{v}\rangle - f_j'(\mathbf{c}_j)\langle\mathbf{x}-\mathbf{v}\rangle| \le \|f_j'(\mathbf{b}) - f_j'(\mathbf{c}_j)\|\,\|\mathbf{x}-\mathbf{v}\|$$
$$\le \|\mathbf{f}'(\mathbf{b}) - \mathbf{f}'(\mathbf{c}_j)\|\,\|\mathbf{x}-\mathbf{v}\| \le \left(\frac{1}{2Mn}\right)\|\mathbf{x}-\mathbf{v}\|,$$

because each $\mathbf{c}_j$ is in the ball. Hence the difference of the terms satisfies

$$\|\mathbf{L}\langle\mathbf{x}-\mathbf{v}\rangle - [\mathbf{f}(\mathbf{x}) - f(\mathbf{v})]\| \le \sqrt{n}\left(\frac{1}{2Mn}\right)\|\mathbf{x}-\mathbf{v}\|,$$

and

$$\|\mathbf{g}(\mathbf{x}) - \mathbf{g}(\mathbf{v})\| \le M\sqrt{n}\left(\frac{1}{2Mn}\right)\|\mathbf{x}-\mathbf{v}\| \le \frac{1}{2}\|\mathbf{x}-\mathbf{v}\|.$$

The last inequality says that $\mathbf{g}$ is a contraction on $B(\mathbf{b},\delta)$.

By the contraction principle, there is a unique solution to $\mathbf{g}(\mathbf{x}) = \mathbf{x}$. Because $\mathbf{L}$ is invertible,

$$\mathbf{x} = \mathbf{g}(\mathbf{x}) = \mathbf{x} - \mathbf{L}^{-1}\langle\mathbf{f}(\mathbf{x}) - \mathbf{y}\rangle$$

is equivalent to $\mathbf{y} = \mathbf{f}(\mathbf{x})$. We conclude that there is exactly one $\mathbf{x} \in B(\mathbf{b},\delta)$ with $\mathbf{f}(\mathbf{x}) = \mathbf{y}$. This $\mathbf{x}$ comes from the interior; we have seen that $\mathbf{x} = \mathbf{g}(\mathbf{x}) \in N(\mathbf{b},\delta)$. By continuity of $\mathbf{f}$, the inverse image $O \equiv \{\mathbf{x} \in N(\mathbf{b},\delta) : \mathbf{f}(\mathbf{x}) \in N(\mathbf{f}(\mathbf{b}),\varepsilon)\}$ has to be open. It follows that $\mathbf{f}$ maps O one-to-one onto $N(\mathbf{f}(\mathbf{b}),\varepsilon)$. □

Demanding that $\mathbf{f}'(\mathbf{b})$ be invertible amounts to requiring the determinant

$$\det(\mathbf{f}'(\mathbf{b})) = \begin{vmatrix} \frac{\partial f_1}{\partial x_1}(\mathbf{b}) & \cdots & \frac{\partial f_1}{\partial x_n}(\mathbf{b}) \\ & \ddots & \\ \frac{\partial f_n}{\partial x_1}(\mathbf{b}) & \cdots & \frac{\partial f_n}{\partial x_n}(\mathbf{b}) \end{vmatrix}$$

of the Jacobian matrix to be nonzero. [Consult, for example, Lay, Chapter 3, on determinants and their relation to matrix inverses.] We call $\det(\mathbf{f}'(\mathbf{b}))$ the **Jacobian** of $\mathbf{f}$. We therefore do have an analogy to the real case: Where the Jacobian of a continuously differentiable function is nonzero, the function is locally one-to-one.

Example 1. Local one-to-one-ness is the best we can do.

Consider the open three-quarter ring given in polar coordinates by $1 < r < 2$, $0 < \theta < \frac{3\pi}{2}$. The function $\mathbf{f}(r,\theta) \equiv (r, 2\theta)$ becomes

$$\mathbf{h}(x,y) \equiv (r\cos 2\theta, r\sin 2\theta) = \left(r\left[\cos^2\theta - \sin^2\theta\right], 2r\sin\theta\cos\theta\right)$$
$$= \left(\frac{x^2-y^2}{\sqrt{x^2+y^2}}, \frac{2xy}{\sqrt{x^2+y^2}}\right).$$

The Jacobian $\det(\mathbf{h}')$ is nonzero throughout the open set (Exercise 1), so $\mathbf{h}$ is locally one-to-one. But $\mathbf{h}$ is not (globally) one-to-one on its domain, because $\mathbf{h}(1, 1) = \mathbf{h}(-1, -1) = (0, \sqrt{2})$.

We now know that $\mathbf{f}$ maps an open set surrounding $\mathbf{b}$ one-one onto a neighborhood of $\mathbf{f}(\mathbf{b})$. The inverse $\mathbf{f}^{-1}$ is therefore defined near $\mathbf{f}(\mathbf{b})$, and it is fair to discuss its differentiability.

Theorem 3. *Under the hypothesis of Theorem 2, the local inverse is differentiable, and the derivative of the inverse is the inverse of the derivative: There is a neighborhood of* $\mathbf{f}(\mathbf{b})$ *in which* $\mathbf{f}^{-1}$ *is differentiable and* $\left(\mathbf{f}^{-1}\right)'(\mathbf{f}(\mathbf{x})) = \mathbf{f}'(\mathbf{x})^{-1}$.

Proof. We first look at the Jacobian

$$\Delta(\mathbf{x}) \equiv \begin{vmatrix} \frac{\partial f_1}{\partial x_1}(\mathbf{x}) & \cdots & \frac{\partial f_1}{\partial x_n}(\mathbf{x}) \\ & \ddots & \\ \frac{\partial f_n}{\partial x_1}(\mathbf{x}) & \cdots & \frac{\partial f_n}{\partial x_n}(\mathbf{x}) \end{vmatrix} = \det(\mathbf{f}'(\mathbf{x}))$$

as $\mathbf{x}$ varies near $\mathbf{b}$. By definition of the determinant, $\Delta(\mathbf{x})$ is a sum of terms of the form

$$\pm \frac{\partial f_1}{\partial x_i} \frac{\partial f_2}{\partial x_j} \cdots \frac{\partial f_n}{\partial x_k}.$$

That is, it is a sum of products of the functions $\frac{\partial f_l}{\partial x_m}$. By hypothesis, each of these partials is continuous at $\mathbf{b}$. Therefore, $\Delta(\mathbf{x})$ is a continuous real function at $\mathbf{b}$. Also by hypothesis, $\mathbf{L} \equiv \mathbf{f}'(\mathbf{b})$ is invertible. Hence $\Delta(\mathbf{b}) \neq 0$; say $\Delta(\mathbf{b}) > 0$. Since $\Delta(\mathbf{x})$ is continuous at $\mathbf{b}$, there is a neighborhood of $\mathbf{b}$ in which $\Delta(\mathbf{x})$ stays bigger than $c \equiv \frac{\Delta(\mathbf{b})}{2}$ (implying that $\mathbf{f}'(\mathbf{x})$ is invertible).

Next, look at the inverse $\mathbf{f}'(\mathbf{x})^{-1}$ as a function of $\mathbf{x}$. Linear algebra tells us that the inverse of a matrix is its adjoint divided by its determinant:

$$\mathbf{f}'(\mathbf{x})^{-1} = \frac{\operatorname{adj}(\mathbf{f}'(\mathbf{x}))}{\Delta(\mathbf{x})}.$$

Here $\Delta(\mathbf{x})$ is a scalar divisor, and we have already argued that it is continuous at $\mathbf{b}$ and nonzero near there. The adjoint is a matrix of subdeterminants from $\mathbf{f}'(\mathbf{x})$, each such subdeterminant being a sum of products of partials $\frac{\partial f_l}{\partial x_m}$. The adjoint is therefore a matrix function with continuous entries, and is itself continuous at $\mathbf{b}$. Hence $\operatorname{adj}(\mathbf{f}'(\mathbf{x}))$ is bounded near $\mathbf{b}$. We have, say, $\|\operatorname{adj}(\mathbf{f}'(\mathbf{x}))\| \leq Mc$, from which $\|\mathbf{f}'(\mathbf{x})^{-1}\| \leq M$ in a neighborhood of $\mathbf{b}$.

Now, going back to the beginning of the proof of Theorem 2, let us additionally assume that the ball $B(\mathbf{b}, \delta)$ is contained in the neighborhood where $\|\mathbf{f}'(\mathbf{x})^{-1}\| \leq M$. The proof establishes an open subset of $B(\mathbf{b}, \delta)$ that is mapped one-to-one onto some neighborhood $N\left(\mathbf{f}(\mathbf{b}), \varepsilon\right)$. Let $\mathbf{y}$ be a fixed vector in $N\left(\mathbf{f}(\mathbf{b}), \varepsilon\right)$ and

$\mathbf{x} = \mathbf{f}^{-1}(\mathbf{y})$. We will show that $\mathbf{f}^{-1}$ is differentiable at $\mathbf{y}$, and simultaneously that $(\mathbf{f}^{-1})'(\mathbf{y}) = \mathbf{f}'(\mathbf{x})^{-1}$, by proving that

$$\mathbf{f}^{-1}(\mathbf{w}) - \mathbf{f}^{-1}(\mathbf{y}) - \mathbf{f}'(\mathbf{x})^{-1}\langle \mathbf{w} - \mathbf{y}\rangle = \mathbf{o}(\mathbf{w} - \mathbf{y}) \qquad \text{as } \mathbf{w} \to \mathbf{y}.$$

In the earlier proof, we matched $\mathbf{x}$ to $\mathbf{y}$ via the fixed point of a contraction:

$$\mathbf{x} = \mathbf{g}(\mathbf{x}) \equiv \mathbf{x} - \mathbf{L}^{-1}\langle \mathbf{f}(\mathbf{x}) - \mathbf{y}\rangle.$$

Let $\mathbf{w}$ vary within $N(\mathbf{f}(\mathbf{b}), \varepsilon)$. Then $\mathbf{v} \equiv \mathbf{f}^{-1}(\mathbf{w})$ is in the (ball) domain of $\mathbf{g}$, and

$$\mathbf{g}(\mathbf{v}) = \mathbf{v} - \mathbf{L}^{-1}\langle \mathbf{f}(\mathbf{v}) - \mathbf{y}\rangle.$$

Consequently,

$$\mathbf{v} - \mathbf{x} = \mathbf{g}(\mathbf{v}) - \mathbf{g}(\mathbf{x}) + \mathbf{L}^{-1}\langle \mathbf{f}(\mathbf{v}) - \mathbf{f}(\mathbf{x})\rangle.$$

By the triangle inequality,

$$\begin{aligned}\|\mathbf{v} - \mathbf{x}\| &\le \|\mathbf{g}(\mathbf{v}) - \mathbf{g}(\mathbf{x})\| + \|\mathbf{L}^{-1}\|\,\|\mathbf{f}(\mathbf{v}) - \mathbf{f}(\mathbf{x})\| \\ &\le \frac{1}{2}\|\mathbf{v} - \mathbf{x}\| + M\|\mathbf{f}(\mathbf{v}) - \mathbf{f}(\mathbf{x})\|.\end{aligned}$$

Hence

$$\|\mathbf{v} - \mathbf{x}\| \le 2M\|\mathbf{f}(\mathbf{v}) - \mathbf{f}(\mathbf{x})\|.$$

This is an odd-looking equation, except that if we rewrite it as

$$\|\mathbf{f}^{-1}(\mathbf{w}) - \mathbf{f}^{-1}(\mathbf{y})\| \le 2M\|\mathbf{w} - \mathbf{y}\|,$$

it tells us that $\mathbf{f}^{-1}$ is continuous.

Finally, we look at

$$\begin{aligned}&\|\mathbf{f}^{-1}(\mathbf{w}) - \mathbf{f}^{-1}(\mathbf{y}) - \mathbf{f}'(\mathbf{x})^{-1}\langle \mathbf{w} - \mathbf{y}\rangle\| \\ &\quad= \|\mathbf{v} - \mathbf{x} - \mathbf{f}'(\mathbf{x})^{-1}\langle \mathbf{f}(\mathbf{v}) - \mathbf{f}(\mathbf{x})\rangle\| \\ &\quad\le \|\mathbf{f}'(\mathbf{x})^{-1}\|\,\|\mathbf{f}'(\mathbf{x})\langle \mathbf{v} - \mathbf{x}\rangle - [\mathbf{f}(\mathbf{v}) - \mathbf{f}(\mathbf{x})]\|.\end{aligned}$$

We know that $\|\mathbf{f}'(\mathbf{x})^{-1}\| \le M$. Because $\mathbf{f}$ is differentiable at $\mathbf{x}$, each positive r gives rise to an s such that

$$\|\mathbf{v} - \mathbf{x}\| < s \Rightarrow \|[\mathbf{f}(\mathbf{v}) - \mathbf{f}(\mathbf{x})] - \mathbf{f}'(\mathbf{x})\langle \mathbf{v} - \mathbf{x}\rangle\| \le \frac{r}{2M^2}\|\mathbf{v} - \mathbf{x}\|.$$

If we restrict $\mathbf{w}$ to the neighborhood $N\left(\mathbf{y}, \frac{s}{2M}\right)$, then

$$\|\mathbf{v} - \mathbf{x}\| \le 2M\|\mathbf{w} - \mathbf{y}\| < s,$$

so that

$$\begin{aligned}\|\mathbf{f}^{-1}(\mathbf{w}) - \mathbf{f}^{-1}(\mathbf{y}) - \mathbf{f}'(\mathbf{x})^{-1}\langle \mathbf{w} - \mathbf{y}\rangle\| &\le M\|\mathbf{f}'(\mathbf{x})\langle \mathbf{v} - \mathbf{x}\rangle - [\mathbf{f}(\mathbf{v}) - \mathbf{f}(\mathbf{x})]\| \\ &\le M\left(\frac{r}{2M^2}\right)\|\mathbf{v} - \mathbf{x}\| \\ &\le M\left(\frac{r}{2M^2}\right)2M\|\mathbf{w} - \mathbf{y}\| = r\|\mathbf{w} - \mathbf{y}\|.\end{aligned}$$

This shows that $\mathbf{f}^{-1}(\mathbf{w}) - \mathbf{f}^{-1}(\mathbf{y}) - \mathbf{f}'(\mathbf{x})^{-1}\langle \mathbf{w} - \mathbf{y}\rangle = \mathbf{o}(\mathbf{w} - \mathbf{y})$. □

We will refer to the union of Theorems 2 and 3 as the **inverse function theorem.**

Example 2. In the transformation to polar coordinates, we have $x = r\cos\theta$, $y = r\sin\theta$. The Jacobian

$$\left|\frac{\partial(x,y)}{\partial(r,\theta)}\right| = \begin{vmatrix}\cos\theta & -r\sin\theta \\ \sin\theta & r\cos\theta\end{vmatrix} = r\cos^2\theta + r\sin^2\theta = r$$

is zero only at the origin. Hence the transformation $(x, y) \leftrightarrow (r, \theta)$ is locally invertible away from $(0, 0)$.

By a familiar formula,

$$\left[\frac{\partial(x,y)}{\partial(r,\theta)}\right]^{-1} = \frac{1}{r}\begin{bmatrix} r\cos\theta & r\sin\theta \\ -\sin\theta & \cos\theta\end{bmatrix}.$$

By Theorem 3, $\left[\frac{\partial(x,y)}{\partial(r,\theta)}\right]^{-1} = \frac{\partial(r,\theta)}{\partial(x,y)}$. (Notice again the effectiveness of Leibniz's notation.) This suggests $\frac{\partial r}{\partial x} = \cos\theta$, $\frac{\partial r}{\partial y} = \sin\theta$, $\frac{\partial\theta}{\partial x} = \frac{-\sin\theta}{r}$, $\frac{\partial\theta}{\partial y} = \frac{\cos\theta}{r}$. Do these formulas agree with the direct differentiation?

From $r = \sqrt{x^2 + y^2}$, we have $\frac{\partial r}{\partial x} = \frac{x}{\sqrt{x^2+y^2}} = \frac{x}{r} = \cos\theta$ and $\frac{\partial r}{\partial y} = \frac{y}{r} = \sin\theta$. The angle needs more attention. In the right half-plane, $\theta = \tan^{-1}\left(\frac{y}{x}\right)$, so

$$\frac{\partial\theta}{\partial x} = \frac{1}{1 + y^2/x^2}\,\frac{-y}{x^2} = \frac{-y}{(x^2+y^2)} = \frac{-\sin\theta}{r},$$

and the latter formulas work even if $x = 0$, as long as y is not simultaneously zero. In the left-hand half, $\theta = \pi + \tan^{-1}\left(\frac{y}{x}\right)$, so we may use the same derivative. Similarly,

$$\frac{\partial\theta}{\partial y} = \frac{1}{1 + y^2/x^2}\,\frac{1}{x} = \frac{x}{x^2+y^2} = \frac{\cos\theta}{r}.$$

Exercises

1. Show that $\mathbf{h}(x, y) \equiv \left(\frac{x^2-y^2}{\sqrt{x^2+y^2}}, \frac{2xy}{\sqrt{x^2+y^2}}\right)$ (Example 1) has nonzero Jacobian everywhere it is defined.

2. For spherical coordinates (Exercise 3 of Section 1.3), $x = \rho\cos\theta\sin\phi$, $y = \rho\sin\theta\sin\phi$, $z = \rho\cos\phi$.

   (a) Write the Jacobian matrix $\frac{\partial(x,y,z)}{\partial(\rho,\theta,\phi)}$.

   (b) Find the inverse to the answer in (a), and state the inverse's domain.

(c) Given $F = f(\rho, \theta, \phi)$, express $\frac{\partial F}{\partial(x,y,z)}$ in terms of functions of ρ, θ, and ϕ and partials of f.

3. (a) Show that Theorem 2 gives a sufficient but not necessary condition for the existence of an inverse: A continuously differentiable function may be one-to-one near a place where its Jacobian is zero.

   (b) Show that if **f** is one-to-one near a place **b** where its Jacobian is zero, then its inverse cannot be differentiable at **f**(**b**).

3.3 The Implicit Function Theorem

In Section 2.5 we investigated the possibility of solving an equation for one of its variables. We now escalate to the question of solving a system of equations for an equal number of variables.

A system of $k < n$ equations involving $x_1, \ldots, x_n$ can be put into the form

$$f_1(x_1, \ldots, x_n) = \alpha_1,$$
$$\ldots$$
$$f_k(x_1, \ldots, x_n) = \alpha_k.$$

We say that the system **determines**, or **can be solved for**, $x_1, \ldots, x_k$ in terms of $x_{k+1}, \ldots, x_n$ if there are functions $g_1, \ldots, g_k$ such that the system is equivalent to

$$x_1 = g_1(x_{k+1}, \ldots, x_n),$$
$$\ldots$$
$$x_k = g_k(x_{k+1}, \ldots, x_n),$$

at least for suitably restricted $(x_1, \ldots, x_n)$.

As before, we may abbreviate using vectors. The system becomes the single equation

$$\mathbf{f}(\mathbf{x}) \equiv (f_1(\mathbf{x}), \ldots, f_k(\mathbf{x})) = (\alpha_1, \ldots, \alpha_k).$$

If for each appropriate $\mathbf{x}^\# \equiv (x_{k+1}, \ldots, x_n)$, there is precisely one $\mathbf{v} \equiv (x_1, \ldots, x_k)$ with

$$\mathbf{f}(\mathbf{v}, \mathbf{x}^\#) = (\alpha_1, \ldots, \alpha_k),$$

then we say that the vector equation **can be solved for v as a function of $\mathbf{x}^\#$**.

Theorem 1 (The Implicit Function Theorem). *Assume that* $\mathbf{f}(\mathbf{x}) \equiv \big(f_1(\mathbf{x}), \ldots, f_k(\mathbf{x})\big)$ *is differentiable near* $\mathbf{x} = \mathbf{b} \equiv (b_1, \ldots, b_n)$, *with* $\mathbf{f}'$ *continuous at* $\mathbf{b}$. *Suppose the Jacobian*

$$\det\left(\frac{\partial \mathbf{f}}{\partial(x_1, \ldots, x_k)}(\mathbf{b})\right) \equiv \begin{vmatrix} \frac{\partial f_1}{\partial x_1}(\mathbf{b}) & \cdots & \frac{\partial f_1}{\partial x_k}(\mathbf{b}) \\ & \ddots & \\ \frac{\partial f_k}{\partial x_1}(\mathbf{b}) & \cdots & \frac{\partial f_k}{\partial x_k}(\mathbf{b}) \end{vmatrix}$$

is nonzero. Then near **b**, *the equation* $\mathbf{f}(\mathbf{x}) = \mathbf{f}(\mathbf{b})$ *can be solved for* $x_1, \ldots, x_k$ *as differentiable functions of the other variables. In symbols, there is a box* $[\mathbf{a}, \mathbf{c}]$ *surrounding* **b** *in* $\mathbf{R}^n$ *and a differentiable function* $\mathbf{g}\colon \mathbf{R}^{n-k} \to \mathbf{R}^k$ *defined in the box*

$$\left[\mathbf{a}^{\#}, \mathbf{c}^{\#}\right] \equiv \left[(a_{k+1}, \ldots, a_n), (c_{k+1}, \ldots, c_n)\right]$$

in $\mathbf{R}^{n-k}$ *such that the following are true for every* $\mathbf{x} = (x_1, \ldots, x_n) \in [\mathbf{a}, \mathbf{c}]$:

(a) $\left(\mathbf{g}\left(\mathbf{x}^{\#}\right), \mathbf{x}^{\#}\right) \in [\mathbf{a}, \mathbf{c}]$; *equivalently,* $a_j \le g_j\left(\mathbf{x}^{\#}\right) \le c_j$ *for every* j, $1 \le j \le k$.

(b) $\mathbf{f}(\mathbf{x}) = \mathbf{f}(\mathbf{b})$ *iff* $(x_1, \ldots, x_k) = g(x_{k+1}, \ldots, x_n)$.

(c) *If* $(x_1, \ldots, x_k) = \mathbf{g}(x_{k+1}, \ldots, x_n)$, *then* $\mathbf{g}'$ *is given by*

$$\frac{\partial \mathbf{g}}{\partial(x_{k+1}, \ldots, x_n)}(\mathbf{x}^{\#}) = -\left[\frac{\partial \mathbf{f}}{\partial(x_1, \ldots, x_k)}(\mathbf{x})\right]^{-1} \frac{\partial \mathbf{f}}{\partial(x_{k+1}, \ldots, x_n)}(\mathbf{x}).$$

Proof. Near **b**, examine the function $\mathbf{G}\colon \mathbf{R}^n \to \mathbf{R}^n$ given by

$$\mathbf{G}(\mathbf{x}) \equiv \left(\mathbf{f}(\mathbf{x}), \mathbf{x}^{\#}\right) = \left(f_1(\mathbf{x}), \ldots, f_k(\mathbf{x}), x_{k+1}, \ldots, x_n\right).$$

Observe that **G** is differentiable near **b** and continuously differentiable at **b**, because each of its components is. Indeed, from the components, we see that its Jacobian matrix has the block structure

$$\mathbf{G}'(\mathbf{x}) = \begin{bmatrix} \dfrac{\partial(f_1, \ldots, f_k)}{\partial(x_1, \ldots, x_k)}(\mathbf{x}) & \dfrac{\partial(f_1, \ldots, f_k)}{\partial(x_{k+1}, \ldots, x_n)}(\mathbf{x}) \\ O_{n-k,k} & I_{n-k} \end{bmatrix}.$$

Here I_{n-k} represents the size-$(n-k)$ identity and $O_{n-k,k}$ the zero matrix of $n-k$ rows and k columns. For a matrix of this structure, the determinant is that of the upper-left block,

$$\det\left(\frac{\partial(f_1, \ldots, f_k)}{\partial(x_1, \ldots, x_k)}(\mathbf{x})\right),$$

which by hypothesis is nonzero at **b**. Hence **G** satisfies the hypothesis of the inverse function theorem.

By the theorem, there is an open set O surrounding **b** that is mapped invertibly onto a neighborhood of $\mathbf{G}(\mathbf{b}) = \left(\mathbf{f}(\mathbf{b}), \mathbf{b}^{\#}\right)$. Let $N(\mathbf{b}, \varepsilon)$ be a neighborhood contained in O. The image $\mathbf{G}\left(N(\mathbf{b}, \varepsilon)\right)$ is an open set around $\mathbf{G}(\mathbf{b})$, because an image under **G** is an inverse image under $\mathbf{G}^{-1}$. Let $N\left(\mathbf{G}(\mathbf{b}), n\delta\right) \subseteq \mathbf{G}\left(N(\mathbf{b}, \varepsilon)\right)$. The box we need is delimited by

$$\mathbf{a} \equiv (b_1 - \varepsilon, \ldots, b_k - \varepsilon, b_{k+1} - \delta, \ldots, b_n - \delta)$$

and

$$\mathbf{c} \equiv (b_1 + \varepsilon, \ldots, b_k + \varepsilon, b_{k+1} + \delta, \ldots, b_n + \delta).$$

(Compare this argument, so far, with that in the scalar form, Theorem 2.5:1. The ε and δ produced here have the same significance as the ones in the earlier proof. All we have to do is to pinpoint, for each $\mathbf{x}^{\#}$, the lone place where $\mathbf{f}$ hits $\mathbf{f}(\mathbf{b})$.)

Let $\Pi : \mathbf{R}^n \to \mathbf{R}^k$ be the "projection" map that picks out the leading k coordinates:

$$\Pi(x_1, \ldots, x_n) \equiv (x_1, \ldots, x_k).$$

The function we need, mapping $\mathbf{R}^{n-k}$ to $\mathbf{R}^k$, is given by

$$\mathbf{g}\left(\mathbf{x}^{\#}\right) \equiv \Pi\left(\mathbf{G}^{-1}\left(\mathbf{f}(\mathbf{b}), \mathbf{x}^{\#}\right)\right), \qquad \mathbf{x}^{\#} \in \left[\mathbf{a}^{\#}, \mathbf{c}^{\#}\right].$$

To prove statements (a)–(c), assume now that $\mathbf{x} \in [\mathbf{a}, \mathbf{c}]$.

(a) We have

$$\begin{aligned}\left\|(\mathbf{f}(\mathbf{b}), \mathbf{x}^{\#}) - \mathbf{G}(\mathbf{b})\right\| &= \left\|(\mathbf{f}(\mathbf{b}), \mathbf{x}^{\#}) - (\mathbf{f}(\mathbf{b}), \mathbf{b}^{\#})\right\| \\ &= \left\|\mathbf{x}^{\#} - \mathbf{b}^{\#}\right\| \le \delta\sqrt{n-k}.\end{aligned}$$

Hence $\left(\mathbf{f}(\mathbf{b}), \mathbf{x}^{\#}\right) \in N(\mathbf{G}(\mathbf{b}), n\delta)$. This shows, first, that $\mathbf{G}^{-1}\left(\mathbf{f}(\mathbf{b}), \mathbf{x}^{\#}\right)$ is defined, so $\mathbf{g}\left(\mathbf{x}^{\#}\right)$ is meaningful. Also, since $N(\mathbf{G}(\mathbf{b}), n\delta)$ is part of the image $\mathbf{G}(N(\mathbf{b}, \varepsilon))$, $\mathbf{G}^{-1}\left(\mathbf{f}(\mathbf{b}), \mathbf{x}^{\#}\right) \in N(\mathbf{b}, \varepsilon)$. It follows that no coordinate of $\mathbf{G}^{-1}(\mathbf{f}(\mathbf{b}), \mathbf{x}^{\#})$ can be as much as ε from the corresponding one of $\mathbf{b}$:

$$(\mathbf{b}_1 - \varepsilon, \ldots, b_k - \varepsilon) \le \Pi\left(\mathbf{G}^{-1}(\mathbf{f}(\mathbf{b}), \mathbf{x}^{\#})\right) \le (b_1 + \varepsilon, \ldots, b_k + \varepsilon).$$

This proves (a).

(b) Suppose $\mathbf{f}(\mathbf{x}) = \mathbf{f}(\mathbf{b})$. Then

$$\mathbf{G}(\mathbf{x}) \equiv \left(\mathbf{f}(\mathbf{x}), \mathbf{x}^{\#}\right) = \left(\mathbf{f}(\mathbf{b}), \mathbf{x}^{\#}\right).$$

Therefore, $\mathbf{x} = \mathbf{G}^{-1}\left(\mathbf{f}(\mathbf{b}), \mathbf{x}^{\#}\right)$, and

$$(x_1, \ldots, x_k) = \Pi(\mathbf{x}) = \Pi\left(\mathbf{G}^{-1}(\mathbf{f}(\mathbf{b}), \mathbf{x}^{\#})\right) \equiv \mathbf{g}(\mathbf{x}^{\#}).$$

Suppose, conversely, that $(x_1, \ldots, x_k) = \mathbf{g}\left(\mathbf{x}^{\#}\right)$, the latter being the leading part of $\mathbf{G}^{-1}\left(\mathbf{f}(\mathbf{b}), \mathbf{x}^{\#}\right)$. Then

$$\mathbf{G}^{-1}\left(\mathbf{f}(\mathbf{b}), \mathbf{x}^{\#}\right) = (x_1, \ldots, x_k, w_{k+1}, \ldots, w_n)$$

for some $(w_{k+1}, \ldots, w_n)$. Therefore,

$$\begin{aligned}\left(\mathbf{f}(\mathbf{b}), \mathbf{x}^{\#}\right) &= \mathbf{G}(x_1, \ldots, x_k, w_{k+1}, \ldots, w_n) \\ &\equiv \left(\mathbf{f}(x_1, \ldots, x_k, w_{k+1}, \ldots, w_n), w_{k+1}, \ldots, w_n\right).\end{aligned}$$

We conclude simultaneously that $\mathbf{x}^\# = (w_{k+1}, \dots, w_n)$ and

$$\mathbf{f}(\mathbf{b}) = \mathbf{f}(x_1, \dots, x_k, w_{k+1}, \dots, w_n) = \mathbf{f}(x_1, \dots, x_k, x_{k+1}, \dots, x_n) = \mathbf{f}(\mathbf{x}).$$

We have shown that $\mathbf{f}(\mathbf{x}) = \mathbf{f}(\mathbf{b})$ iff $(x_1, \dots, x_k) = \mathbf{g}\left(\mathbf{x}^\#\right)$.

(c) Working now in $\mathbf{R}^{n-k}$, let $\mathbf{x}^\#$ vary within $\left[\mathbf{a}^\#, \mathbf{c}^\#\right]$. By definition,

$$\mathbf{g}(\mathbf{x}^\#) = \Pi\left(\mathbf{G}^{-1}(\mathbf{f}(\mathbf{b}), \mathbf{x}^\#)\right).$$

Since Π, $\mathbf{G}^{-1}$, and $\left(\mathbf{f}(\mathbf{b}), \mathbf{x}^\#\right)$ are differentiable where we are working, the chain rule tells us that $\mathbf{g}$ is differentiable.

Write $\mathbf{x} \equiv \left(\mathbf{g}\left(\mathbf{x}^\#\right), \mathbf{x}^\#\right)$ and $\mathbf{H}\left(\mathbf{x}^\#\right) \equiv \mathbf{f}(\mathbf{x})$. By part (b), $\mathbf{H}$ has constant value $\mathbf{f}(\mathbf{b})$, so $\mathbf{H}'\left(\mathbf{x}^\#\right) = \mathbf{O}$. On the other hand, $\mathbf{H}$ is a differentiable composite, with

$$\mathbf{H}'\left(\mathbf{x}^\#\right) = \frac{\partial \mathbf{f}}{\partial(x_1, \dots, x_n)}(\mathbf{x}) \frac{\partial \mathbf{x}}{\partial \mathbf{x}^\#}\left(\mathbf{x}^\#\right).$$

Since $\frac{\partial \mathbf{x}}{\partial \mathbf{x}^\#}(\mathbf{x})$ has the block structure

$$\begin{bmatrix} \dfrac{\partial \mathbf{g}}{\partial(x_{k+1}, \dots, x_n)}(\mathbf{x}^\#) \\ \\ I_{n-k} \end{bmatrix},$$

it is convenient to break up $\frac{\partial \mathbf{f}}{\partial(x_1, \dots, x_n)}(\mathbf{x})$ into blocks

$$\left[\frac{\partial \mathbf{f}}{\partial(x_1, \dots, x_k)}(\mathbf{x}) \quad \frac{\partial \mathbf{f}}{\partial(x_{k+1}, \dots, x_n)}(\mathbf{x})\right].$$

The product of the two matrices is then

$$\mathbf{O} = \mathbf{H}'\left(\mathbf{x}^\#\right) = \frac{\partial \mathbf{f}}{\partial(x_1, \dots, x_k)}(\mathbf{x}) \frac{\partial \mathbf{g}}{\partial(x_{k+1}, \dots, x_n)}\left(\mathbf{x}^\#\right) + \frac{\partial \mathbf{f}}{\partial(x_{k+1}, \dots, x_n)}(\mathbf{x}).$$

The first matrix on the right side is invertible, because its determinant matches the Jacobian of the invertible map $\mathbf{G}(\mathbf{x})$. Multiplying both sides by its inverse, we get

$$\mathbf{O} = \frac{\partial \mathbf{g}}{\partial(x_{k+1}, \dots, x_n)}\left(\mathbf{x}^\#\right) + \left[\frac{\partial \mathbf{f}}{\partial(x_1, \dots, x_k)}(\mathbf{x})\right]^{-1} \frac{\partial \mathbf{f}}{\partial(x_{k+1}, \dots, x_n)}(\mathbf{x}). \quad \square$$

Example 1. Can the system

$$x^2 + y^2 + z^2 = 9, \qquad z^2 = 1 + x^2 + y^2$$

be solved for two of its variables in terms of the other?

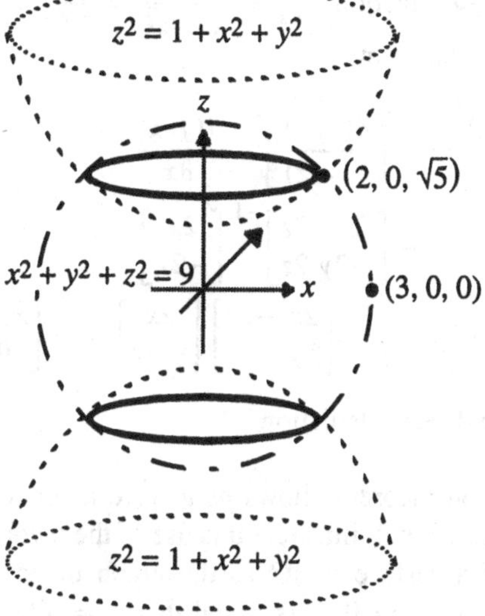

Figure 3.1.

(a) Write $f(x, y, z) \equiv x^2 + y^2 + z^2$, $g(x, y, z) \equiv z^2 - x^2 - y^2$. Since

$$\left|\frac{\partial(f, g)}{\partial(x, y)}\right| = \begin{vmatrix} 2x & 2y \\ -2x & -2y \end{vmatrix}$$

is always zero, it appears that we can never solve for x and y in terms of z.

Figure 3.1 shows why. The hyperboloid cuts the sphere along two horizontal circles, at the levels $z = \pm\sqrt{5}$. On both those, z is constant; we cannot expect x or y to be a function of z.

(b) To solve for y and z, we would need

$$\left|\frac{\partial(f, g)}{\partial(y, z)}\right| = \begin{vmatrix} 2y & 2z \\ -2y & 2z \end{vmatrix} = 8yz \neq 0.$$

We must avoid $z = 0$, which is easy because no point of the intersection—in fact, no point of the hyperboloid—has $z = 0$. We must also avoid $y = 0$, which happens at the places $\left(\pm 2, 0, \pm\sqrt{5}\right)$ where the circles cross the xz-plane. Near those places, x and y are related by $x^2 + y^2 = 4$, and there is no neighborhood in which y is a function of x.

Around any other place, the system determines y and z as functions of x. Thus, let $\left(a, -b, \sqrt{5}\right)$ be in the nearer (to us) half of the upper circle. In the $\mathbf{R}^3$ box given by

$$|x| \leq \sqrt{4 - b^2/4}, \quad -2 \leq y \leq -\frac{b}{2}, \quad 2 \leq z \leq 3,$$

the vector (x, y, z) solves the system—it is on the intersection of the surfaces—exactly if $y = -\sqrt{4 - x^2}$, $z = \sqrt{5}$.

Observe that for these functions, $\frac{dy}{dx} = \frac{x}{\sqrt{4-x^2}} = \frac{x}{-y}$ and $\frac{dz}{dx} = 0$. According to the implicit function theorem,

$$\begin{aligned}\frac{\partial(y,z)}{\partial x} &= -\left[\frac{\partial(f,g)}{\partial(y,z)}\right]^{-1}\frac{\partial(f,g)}{\partial x}\\ &= -\begin{bmatrix}2y & 2z\\ -2y & 2z\end{bmatrix}^{-1}\begin{bmatrix}2x\\ -2x\end{bmatrix}\\ &= -\frac{1}{8yz}\begin{bmatrix}2z & -2z\\ 2y & 2y\end{bmatrix}\begin{bmatrix}2x\\ -2x\end{bmatrix} = -\begin{bmatrix}x/y\\ 0\end{bmatrix},\end{aligned}$$

in agreement with the direct calculation.

The implicit function theorem allows us to give analytic demonstrations for many geometric ideas. We will illustrate this use of the theorem in three results.

We often think of a surface in $\mathbf{R}^3$ as the graph of an equation like either $z = f(x, y)$ or $F(x, y, z) = 0$. It is trivial that $z = f(x, y)$ can be recast as $G(x, y, z) \equiv z - f(x, y) = 0$, with

$$\nabla G = \left(-\frac{\partial f}{\partial x}, -\frac{\partial f}{\partial y}, 1\right) \neq \mathbf{O}.$$

Conversely, by the scalar form of the implicit function theorem (Theorem 2.5:1), the graph of $F(x, y, z) = 0$ is given near any place where $\nabla F \neq \mathbf{O}$ by an equation of the form (one variable) = (function of the other two). Hence the two forms of equation for the surface are equivalent.

The principle extends in an obvious way to arbitrary dimensions: In $\mathbf{R}^n$, every smooth hypersurface that can be defined by $x_j = f(x_1, \dots, x_{j-1}, x_{j+1}, \dots, x_n)$ can also be defined by $F(x_1, \dots, x_n) = 0$, and vice versa.

That the two equation forms define a hypersurface is the content of Theorem 2.6:2. We next show that the definition of hypersurface given in Section 2.6 is also equivalent to these two, by establishing the converse of Theorem 2.6:2.

Theorem 2. *Every smooth hypersurface is (locally) a level hypersurface for some differentiable function. In painful detail: Assume that* $\mathbf{G}$ *is a differentiable function of* $\mathbf{t} = (t_1, \dots, t_{n-1})$, $\mathbf{t}$ *belonging to a box* $B^\# \in \mathbf{R}^{n-1}$, *with* $\frac{\partial \mathbf{G}}{\partial t_1}(\mathbf{t}), \dots, \frac{\partial \mathbf{G}}{\partial t_{n-1}}(\mathbf{t})$ *continuous and independent at* $\mathbf{t}_0$*; then there exists a differentiable function* F, *defined near* $\mathbf{b} \equiv \mathbf{G}(\mathbf{t}_0)$ *in* $\mathbf{R}^n$, *for which* $\nabla F(\mathbf{b}) \neq \mathbf{O}$ *and the hypersurface* $\mathbf{G}$ *is given by* $F(\mathbf{x}) = 0$.

Proof. Under the stated assumption, define $\mathbf{H}\colon \mathbf{R}^{2n-1} \to \mathbf{R}^n$ by

$$\mathbf{H}(y_1, \dots, y_{2n-1}) \equiv \mathbf{G}(y_1, \dots, y_{n-1}) - (y_n, \dots, y_{2n-1}).$$

Using our usual abbreviation $(\mathbf{t}, \mathbf{x}) = (t_1, \dots, t_{n-1}, x_1, \dots, x_n)$, we have $\mathbf{H}(\mathbf{t}, \mathbf{x}) = \mathbf{G}(\mathbf{t}) - \mathbf{x}$. If $(\mathbf{t}, \mathbf{x}) \approx (\mathbf{t}_0, \mathbf{b})$, then $\mathbf{t} \approx \mathbf{t}_0$ and $\mathbf{x} \approx \mathbf{b}$, forcing $\mathbf{H}(\mathbf{t}, \mathbf{x}) \approx$

$\mathbf{G}(\mathbf{t}_0) - \mathbf{b} = \mathbf{O}$. Thus, $\mathbf{H}$ maps the vicinity of $(\mathbf{t}_0, \mathbf{b})$ in $\mathbf{R}^{2n-1}$ to the vicinity of the origin in $\mathbf{R}^n$. Let us consider the vector equation $\mathbf{H}(\mathbf{t}, \mathbf{x}) = \mathbf{O}$.

With what has become our usual way of looking at block matrices, we obtain

$$\frac{\partial \mathbf{H}}{\partial(\mathbf{t}, \mathbf{x})} = \begin{bmatrix} \dfrac{\partial \mathbf{G}}{\partial \mathbf{t}} & -I_n \end{bmatrix}.$$

The right-hand block $-I_n$, having n independent columns, guarantees that the n rows of this Jacobian matrix are independent. Hence the matrix has rank n. [Refer to Lay, Section 4.6.] By hypothesis, the block $\frac{\partial \mathbf{G}}{\partial \mathbf{t}}$ has $n-1$ independent columns $\frac{\partial \mathbf{G}}{\partial t_1}, \dots, \frac{\partial \mathbf{G}}{\partial t_{n-1}}$ at $\mathbf{t} = \mathbf{t}_0$. If each column of $-I_n$ were a combination of the columns of $\frac{\partial \mathbf{G}}{\partial \mathbf{t}}(\mathbf{t}_0)$, then the rank of the Jacobian matrix at $(\mathbf{t}, \mathbf{x}) = (\mathbf{t}_0, \mathbf{b})$ would be only $n-1$. Hence one of the columns of $-I_n$ is independent of the columns of $\frac{\partial \mathbf{G}}{\partial \mathbf{t}}(\mathbf{t}_0)$. For definiteness, we assume that column number 1 is such a column. (Can there be others?) Rewrite

$$\frac{\partial \mathbf{H}}{\partial(\mathbf{t}, \mathbf{x})}(\mathbf{t}_0, \mathbf{b}) = \begin{bmatrix} \dfrac{\partial \mathbf{G}}{\partial \mathbf{t}}(\mathbf{t}_0) & \dfrac{\partial \mathbf{H}}{\partial x_1} & \cdots & \dfrac{\partial \mathbf{H}}{\partial x_n} \end{bmatrix}.$$

(The rightmost n blocks are constants; $\frac{\partial \mathbf{H}}{\partial x_j}$ is always $\langle \mathbf{e}_j \rangle$.) The leftmost two blocks form an $n \times n$ matrix with independent columns. Therefore, its determinant

$$\left| \frac{\partial \mathbf{G}}{\partial t_1}(\mathbf{t}_0) \cdots \frac{\partial \mathbf{G}}{\partial t_{n-1}}(\mathbf{t}_0) \; \frac{\partial \mathbf{H}}{\partial x_1} \right| = \det\left(\frac{\partial \mathbf{H}}{\partial(t_1, \dots, t_{n-1}, x_1)}(\mathbf{t}_0, \mathbf{b}) \right)$$

is nonzero. By the implicit function theorem, there are differentiable functions $F_1, \dots, F_n$ such that near $(\mathbf{t}_0, \mathbf{b})$, $\mathbf{H}(\mathbf{t}, \mathbf{x}) = \mathbf{O}$ iff

$$t_1 = F_1(x_2, \dots, x_n), \dots, t_{n-1} = F_{n-1}(x_2, \dots, x_n), x_1 = F_n(x_2, \dots, x_n).$$

Suppose now that $\mathbf{x} = \mathbf{G}(\mathbf{t})$ is on the hypersurface with $\mathbf{t} \approx \mathbf{t}_0$. Then $(\mathbf{t}, \mathbf{x})$ is close to $(\mathbf{t}_0, \mathbf{b})$, and $\mathbf{H}(\mathbf{t}, \mathbf{x}) = \mathbf{O}$, forcing $x_1 = F_n(x_2, \dots, x_n)$. Conversely, suppose $\mathbf{x}$ is near $\mathbf{b}$ with $x_1 = F_n(x_2, \dots, x_n)$. Then

$$\mathbf{t} \equiv (F_1(x_2, \dots, x_n), \dots, F_{n-1}(x_2, \dots, x_n))$$

has $\mathbf{H}(\mathbf{t}, \mathbf{x}) = \mathbf{O}$, forcing $\mathbf{x} = \mathbf{G}(\mathbf{t})$; $\mathbf{x}$ is on the hypersurface. Hence near $\mathbf{b}$, the hypersurface is determined by $F(\mathbf{x}) \equiv x_1 - F_n(x_2, \dots, x_n) = 0$, where clearly $\nabla F(\mathbf{b}) \neq \mathbf{O}$. □

In view of Theorem 2, every hypersurface in $\mathbf{R}^n$ is locally the graph of a real function of $n-1$ variables.

The second geometric result generalizes the idea that in $\mathbf{R}^3$, two surfaces intersect in a curve. In the same way that Theorem 2 characterizes hypersurfaces as graphs of equations, this next theorem characterizes surfaces as the graphs of systems.

Theorem 3. *In* $\mathbf{R}^n$:

(a) *Suppose the smooth hypersurfaces given by* $F(\mathbf{x}) = \alpha$ *and* $G(\mathbf{x}) = \beta$ *are not tangent (have different tangent hyperplanes) at a common point. Then nearby, their intersection is a surface of dimension* $n-2$.

(b) *More generally, if* $F_1, \ldots, F_k$ *are differentiable near* $\mathbf{b}$, *with* $\nabla F_1(\mathbf{x}), \ldots, \nabla F_k(\mathbf{x})$ *continuous and independent at* $\mathbf{x} = \mathbf{b}$, *then near* $\mathbf{b}$ *the intersection of the hypersurfaces* $F_1(\mathbf{x}) = F_1(\mathbf{b}), \ldots, F_k(\mathbf{x}) = F_k(\mathbf{b})$ *is a smooth surface of dimension* $n-k$.

(c) *Conversely, any surface of dimension* $n-k$ *smooth at* $\mathbf{b}$ *is the intersection of the level hypersurfaces* $f_1(\mathbf{x}) = f_1(\mathbf{b}), \ldots, f_k(\mathbf{x}) = f_k(\mathbf{b})$ *for some set of functions* f_j *that are differentiable near* $\mathbf{b}$ *and have* $\nabla f_1(\mathbf{x}), \ldots, \nabla f_k(\mathbf{x})$ *continuous and independent at* $\mathbf{b}$.

Proof. (a) If the tangent hyperplanes are not the same at $\mathbf{b}$, then the normals are in different lines. In other words, $\nabla F(\mathbf{b})$ and $\nabla G(\mathbf{b})$ are independent vectors. Hence

$$\frac{\partial(F,G)}{\partial \mathbf{x}}(\mathbf{b}) = \begin{bmatrix} \dfrac{\partial F}{\partial(x_1,\ldots,x_n)}(\mathbf{b}) \\[2ex] \dfrac{\partial G}{\partial(x_1,\ldots,x_n)}(\mathbf{b}) \end{bmatrix}$$

has independent rows. It must therefore also have two independent columns.

Say two such columns are

$$\begin{bmatrix} \frac{\partial F}{\partial x_1}(\mathbf{b}) \\[2ex] \frac{\partial G}{\partial x_1}(\mathbf{b}) \end{bmatrix} \quad \text{and} \quad \begin{bmatrix} \frac{\partial F}{\partial x_2}(\mathbf{b}) \\[2ex] \frac{\partial G}{\partial x_2}(\mathbf{b}) \end{bmatrix}.$$

Then the 2×2 Jacobian $\det\left[\frac{\partial(F,G)}{\partial(x_1,x_2)}(\mathbf{b})\right]$, having independent columns, is nonzero. By the implicit function theorem, the intersection of the hypersurfaces is given near $\mathbf{b}$ by

$$x_1 = g_1(x_3, \ldots, x_n),\ x_2 = g_2(x_3, \ldots, x_n),\ x_3 = x_3, \ldots, x_n = x_n.$$

This last is an $n-2$ parametrization, in other words, what we called in Section 2.6 a surface of dimension $n-2$. (Why is it smooth?)

(b) and (c) Exercises 5 and 6. □

The last geometric theorem will settle a debt we owe. In Section 2.6 we suggested that the tangent plane to a surface is the union of tangent lines to curves on the surface. Theorem 2.6:1 there pays half the bill; it shows that if a nonzero vector lies along the tangent plane at some point, then it is tangent at that point to some curve contained in the surface. Here we pick up the other half by proving the converse.

Theorem 4. *At a point where a surface is smooth, if a vector is the tangent to some curve that lies on the surface, then it is tangent to the surface.*

Proof. We will take "surface" literally, meaning a two-dimensional locus in $\mathbf{R}^n$. It is straightforward to extend the proof to dimension $k < n$.

Let S be a surface smooth at $\mathbf{b}$. By Theorem 3(b), there is a neighborhood of $\mathbf{b}$ in which the surface is the graph of some system

$$F_1(x_1, \ldots, x_n) = 0, \ldots, F_{n-2}(x_1, \ldots, x_n) = 0.$$

By Theorem 1, there is a box surrounding $\mathbf{b}$ in which the system can be solved for $n-2$ of the variables in terms of the other two, say

$$x_1 = f_1(x_{n-1}, x_n), \ldots, x_{n-2} = f_{n-2}(x_{n-1}, x_n).$$

For this parametrization, the tangent plane at $\mathbf{b}$ is spanned by

$$\mathbf{v} \equiv \frac{\partial \mathbf{x}}{\partial x_{n-1}}(\mathbf{b}^\#) = \begin{bmatrix} \frac{\partial f_1}{\partial x_{n-1}}(b_{n-1}, b_n) \\ \vdots \\ \frac{\partial f_{n-2}}{\partial x_{n-1}}(b_{n-1}, b_n) \\ 1 \\ 0 \end{bmatrix} \text{ and } \mathbf{w} \equiv \frac{\partial \mathbf{x}}{\partial x_n}(\mathbf{b}^\#) = \begin{bmatrix} \frac{\partial f_1}{\partial x_n}(b_{n-1}, b_n) \\ \vdots \\ \frac{\partial f_{n-2}}{\partial x_n}(b_{n-1}, b_n) \\ 0 \\ 1 \end{bmatrix}.$$

(Compare Example 4 of Section 2.6.)

Suppose now that $\mathbf{x} = \mathbf{g}(t)$ defines a smooth curve with $\mathbf{b} = \mathbf{g}(t_0)$. For t near t_0, $\mathbf{g}(t)$ must (by continuity) be in the box. If the curve lies on the surface, then the coordinates $x_1 = g_1(t), \ldots, x_n = g_n(t)$ of $\mathbf{g}(t)$ must satisfy the parametric equations

$$\begin{aligned} g_1(t) &= f_1(x_{n-1}, x_n) = f_1\big(g_{n-1}(t), g_n(t)\big), \ldots, \\ g_{n-2}(t) &= f_{n-2}(x_{n-1}, x_n) = f_{n-2}\big(g_{n-1}(t), g_n(t)\big). \end{aligned}$$

If $\mathbf{g}$ is smooth at $\mathbf{b}$, then there the chain rule gives

$$\begin{aligned} g_1{}'(t_0) &= \frac{\partial f_1}{\partial x_{n-1}}(b_{n-1}, b_n)\frac{dx_{n-1}}{dt}(t_0) + \frac{\partial f_1}{\partial x_n}(b_{n-1}, b_n)\frac{dx_n}{dt}(t_0) \\ &= v_1 g_{n-1}{}'(t_0) + w_1 g_n{}'(t_0), \end{aligned}$$

and similarly for $g_2{}'(t_0), \ldots, g_{n-2}{}'(t_0)$. Since obviously

$$g_{n-1}{}'(t_0) = 1g_{n-1}{}'(t_0) + 0g_n{}'(t_0) = v_{n-1}g_{n-1}{}'(t_0) + w_{n-1}g_n{}'(t_0)$$

and

$$g_n{}'(t_0) = 0g_{n-1}{}'(t_0) + 1g_n{}'(t_0) = v_n g_{n-1}{}'(t_0) + w_n g_n{}'(t_0),$$

we have

$$[g_1{}'(t_0) \cdots g_n{}'(t_0)]^t = \mathbf{v}g_{n-1}{}'(t_0) + \mathbf{w}g_n{}'(t_0).$$

This last line says that the tangent $\mathbf{g}'(t_0)$ is in the plane spanned by $\mathbf{v}$ and $\mathbf{w}$. □

It pays to invest some time in discussion of the geometry that we claim is involved in these theorems.

Example 2. Consider in $\mathbf{R}^3$ the sphere $x^2 + y^2 + (z-2)^2 = 4$ and paraboloid $z = x^2 + y^2$. Refer to Figure 3.2.

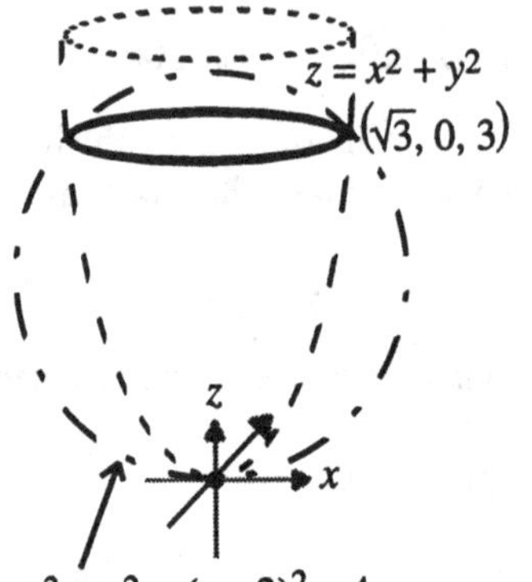

Figure 3.2.

(a) Both surfaces sit atop the xy-plane, making us think of them as meeting tangentially at the origin. This mental picture is consistent with the normals at the origin: $\nabla(x^2+y^2+(z-2)^2) = (2x, 2y, 2z-4) = (0, 0, -4)$ and $\nabla(x^2+y^2-z) = (2x, 2y, -1) = (0, 0, -1)$ are in the same line, so the surfaces have the same tangent plane. At $(0, 0, 0)$, the surfaces do not satisfy the hypothesis of Theorem 3(a). Sure enough, their intersection around there is not a curve.

(b) The other common points are given by $z = 3$, $x^2 + y^2 = 3$. At any such point, the gradients are $\nabla_1 = (2x, 2y, 2)$ and $\nabla_2 = (2x, 2y, -1)$, which are independent. (Justification?) Therefore, the surfaces are not tangent at, say, $(\sqrt{3}, 0, 3)$. By Theorem 3(a), the intersection near there is a curve. (It is usual to say that surfaces are **transversal** or **meet transversally** at a common point where the lower-dimensional tangent is not a subset of the higher.)

(c) Where the meeting is transversal, the tangent to the intersection is the intersection of the tangents. Thus, along the intersection circle, the tangent line is also tangent to the two surfaces; Theorem 4 says so. This observation gives us an easy characterization of the tangent line to the intersection: $\mathbf{x}$ is in the tangent at $\mathbf{b}$ iff $\mathbf{x}-\mathbf{b}$ is orthogonal to both gradients, that is, $(\mathbf{x}-\mathbf{b})\bullet\nabla_1(\mathbf{b}) = (\mathbf{x}-\mathbf{b})\bullet\nabla_2(\mathbf{b}) = 0$.

At $(\sqrt{3}, 0, 3)$, for instance, the system becomes $(x - \sqrt{3}, y - 0, z - 3) \bullet (2\sqrt{3}, 0, 2) = 0 = (x - \sqrt{3}, y - 0, z - 3) \bullet (2\sqrt{3}, 0, -1)$, which reduces to $x = \sqrt{3}$, y arbitrary, $z = 3$.

(d) What happens if we add a third surface? If the third surface's normal is independent of the first two, then Theorem 3(b) speaks; otherwise, the situation is unpredictable.

Suppose we make the third surface the plane $y = x - \sqrt{3}$. This surface has normal $(1, -1, 0)$, which is independent of ∇_1 and ∇_2 at the point $(\sqrt{3}, 0, 3)$. (Justify!) By Theorem 3(b), the three surfaces intersect in a set of dimension $3 - 3 = 0$. (Wait! The intersection of the three surfaces consists of two points. Is that a contradiction?) Also, the tangent to the circle will not be orthogonal to the third normal; that is, it will not lie in the tangent plane to the third surface. Hence the third surface is transversal not only to each of the others, but to their intersection as well.

Suppose instead that the third surface's normal ∇_3 is a linear combination of ∇_1 and ∇_2 at $(\sqrt{3}, 0, 3)$. We can say one thing about the tangents. The tangent to the circle, being orthogonal to both ∇_1 and ∇_2, is also orthogonal to ∇_3. Hence the tangent to the circle is a subset of the tangent plane to the third surface. But that is about the most we can say. The third surface could be the plane $x = \sqrt{3}$, whose normal is $\nabla_1 + 2\nabla_2$ (Check!). This plane is not tangent to the sphere, and it is not tangent to the paraboloid, but it is tangent to the circle of intersection. That is, it is clearly tangent in the geometric sense, and the tangent line to the circle (part (c)) is contained in the (self-tangent) plane. If instead the third surface is the plane $z = 3$, whose normal is $\nabla_1 - \nabla_2$, then we would not describe it geometrically as tangent to the circle, but again the tangent line to the circle is a subset of the third tangent plane.

Although (d) above makes statements about three surfaces in $\mathbf{R}^3$, the principle extends to any number of hypersurfaces in $\mathbf{R}^n$. Thus, if hypersurfaces $S_1, \ldots, S_j$ have independent normals, then the tangent (less-than-hyper-) plane to their intersection is the intersection of their tangent hyperplanes; and if S_{j+1} has normal independent of the others, then S_{j+1} is transversal to the intersection of (any number of) the others. If our application of the principle to the intersection of three surfaces in $\mathbf{R}^3$ seems feeble, remember that improvement is hard, because our brains have difficulty visualizing the intersection of three or more hypersurfaces in dimension exceeding 3.

Exercises

1. Give an example of functions $f(x, y, u, v)$ and $g(x, y, u, v)$ such that $\frac{\partial(f,g)}{\partial(u,v)} = \mathbf{O}$ at some point, but it is still possible near that point to solve the system $f(x, y, u, v) = g(x, y, u, v) = 0$ for u and v as functions of x and y.

2. (a) According to the implicit function theorem, what condition on the constant coefficients a, b, c, e, f, g, i, j, k will guarantee that the system

$$ax + by + cz = d, \quad ex + fy + gz = h, \quad ix + jy + kz = l$$

has a solution for x, y, z?

(b) Show by an example that the condition in (a) is not necessary.

3. (a) Near what points in the plane can $u = x^2 - y^2$, $v = x^2 + y^2$, be solved for x and y as functions of u and v? (Remember that when the hypotheses of Theorem 1 are not satisfied, the theorem does not deny that a solution is possible.)

   (b) Interpret the answer in terms of transversal and tangential intersections.

4. Find equations for the tangent line to the intersection of the cone $z^2 = x^2 + y^2$ and plane $z = x + 4$ at the point in the intersection where $y = 8$.

5. Prove Theorem 3(b).

6. Prove Theorem 3(c). (Hint: Adapt the argument in Theorem 2.)

3.4 Lagrange's Method

There are numerous reasons for investigating the values of a function on a lower-dimensional subset of $\mathbf{R}^n$. For example, we looked at the values of a function along a line for our discussion of directional derivatives. In this section, we analyze maxima/minima of functions on surfaces (in contrast to Sections 2.3 and 2.4, which treated extreme values relative to neighborhoods).

One use of such inquiry is to address the question of absolute extremes. A function continuous on a compact set has to reach a maximum and a minimum. Either extreme must occur at an interior point or on the boundary. The extreme interior value is identifiable by the methods of Chapter 2. We need a way to pick out the extremes on the boundary, which is typically a union of hypersurfaces.

Another use is in optimization problems. In an optimization problem, we seek the extreme value of some quantity, represented by a function, subject to certain requirements. The requirements restrict the candidates to some "feasible set," represented by a subset of Euclidean space. Our job is to find the max/min value of the function on the part of its domain contained in the subset.

We give the name **constrained extremes** to questions with the following analytic standard form: Among the vectors $\mathbf{x} \in \mathbf{R}^n$ that satisfy $g_1(\mathbf{x}) = g_2(\mathbf{x}) = \cdots = g_k(\mathbf{x}) = 0$, find the one that makes $f(\mathbf{x})$ as small as possible. We begin with the case $k = 1$.

Theorem 1. *Assume that f and g are differentiable near* $\mathbf{b}$, $g(\mathbf{b}) = 0$, *and* $\nabla g(\mathbf{x})$ *is continuous and nonzero at* $\mathbf{x} = \mathbf{b}$. *Suppose that in some neighborhood* $N(\mathbf{b}, \varepsilon)$, $f(\mathbf{b}) \le f(\mathbf{x})$ *for every* $\mathbf{x}$ *satisfying* $g(\mathbf{x}) = 0$. *Then* $\nabla f(\mathbf{b})$ *is a scalar multiple of* $\nabla g(\mathbf{b})$.

Proof. From $\nabla g(\mathbf{b}) \neq \mathbf{O}$, we conclude that one of the partials of g is nonzero at $\mathbf{b}$. Assume $\frac{\partial g}{\partial x_n}(\mathbf{b}) \neq 0$. By the implicit function theorem—here it suffices to use the scalar form from Section 2.5—there is a function $G\left(\mathbf{x}^{\#}\right) \equiv G(x_1, \ldots, x_{n-1})$ and a box B contained in $N(\mathbf{b}, \varepsilon)$ surrounding $\mathbf{b}$ in which

$$g(x_1, \ldots, x_n) = 0 \text{ iff } x_n = G(x_1, \ldots, x_{n-1}).$$

This G is defined in a box containing $\mathbf{b}^{\#}$, always has $\left(\mathbf{x}^{\#}, G\left(\mathbf{x}^{\#}\right)\right)$ in B, and satisfies

$$\frac{\partial G}{\partial x_j}(\mathbf{x}^{\#}) = -\frac{\partial g/\partial x_j(\mathbf{x})}{\partial g/\partial x_n(\mathbf{x})} \qquad \text{for } 1 \le j \le n-1$$

at each place $\mathbf{x} = \left(\mathbf{x}^{\#}, G\left(\mathbf{x}^{\#}\right)\right)$ where $g(\mathbf{x}) = 0$.

Consider

$$F(x_1, \ldots, x_{n-1}) \equiv f\left(x_1, \ldots, x_{n-1}, G(x_1, \ldots, x_{n-1})\right).$$

Clearly, F is defined near $\mathbf{b}^{\#}$, and takes the values of f at precisely the points $\mathbf{x}$ whose last coordinate is G of the others, in other words, along the hypersurface $g = 0$. For that reason, $F(\mathbf{b}^{\#})$ is a minimum in some neighborhood of $\mathbf{b}^{\#}$:

$$\begin{aligned} F\left(\mathbf{b}^{\#}\right) &\equiv f\left(b_1, \ldots, b_{n-1}, G(b_1, \ldots, b_{n-1})\right) = f(b_1, \ldots, b_{n-1}, b_n) = f(\mathbf{b}) \\ &\le f\left(\mathbf{x}^{\#}, G(\mathbf{x}^{\#})\right) \equiv F(\mathbf{x}^{\#}) \qquad \text{for } \mathbf{x}^{\#} \text{ near } \mathbf{b}^{\#}. \end{aligned}$$

Because F is a differentiable composite, Theorem 2.3:1 tells us that $\nabla F(\mathbf{b}^{\#}) = \mathbf{O}$. That is,

$$\frac{\partial F}{\partial x_1}\left(\mathbf{b}^{\#}\right) = \cdots = \frac{\partial F}{\partial x_{n-1}}\left(\mathbf{b}^{\#}\right) = 0.$$

Applying the chain rule to the definition of F, we obtain

$$\frac{\partial F}{\partial x_j}\left(\mathbf{b}^{\#}\right) = \frac{\partial f}{\partial x_1}(\mathbf{b})\frac{\partial x_1}{\partial x_j}\left(\mathbf{b}^{\#}\right) + \cdots + \frac{\partial f}{\partial x_n}(\mathbf{b})\frac{\partial x_n}{\partial x_j}\left(\mathbf{b}^{\#}\right)$$

for $1 \le j \le n-1$. On the right, each partial $\frac{\partial x_1}{\partial x_j}\left(\mathbf{b}^{\#}\right)$ through $\frac{\partial x_{n-1}}{\partial x_j}\left(\mathbf{b}^{\#}\right)$ is zero, except for $\frac{\partial x_j}{\partial x_j}\left(\mathbf{b}^{\#}\right) = 1$; and $\frac{\partial x_n}{\partial x_j}\left(\mathbf{b}^{\#}\right)$ means $\frac{\partial G}{\partial x_j}\left(\mathbf{b}^{\#}\right)$. Hence

$$0 = \frac{\partial f}{\partial x_j}(\mathbf{b}) + \frac{\partial f}{\partial x_n}(\mathbf{b})\left[-\frac{\partial g/\partial x_j(\mathbf{b})}{\partial g/\partial x_n(\mathbf{b})}\right]$$

and

$$\frac{\partial f}{\partial x_j}(\mathbf{b}) = \frac{\partial f/\partial x_n(\mathbf{b})}{\partial g/\partial x_n(\mathbf{b})}\frac{\partial g}{\partial x_j}(\mathbf{b}), \qquad j = 1, \ldots, n-1.$$

Since this relation is trivial for $j = n$, we conclude that

$$\left(\frac{\partial f}{\partial x_1}(\mathbf{b}), \ldots, \frac{\partial f}{\partial x_n}(\mathbf{b})\right) = \frac{\partial f/\partial x_n(\mathbf{b})}{\partial g/\partial x_n(\mathbf{b})}\left(\frac{\partial g}{\partial x_1}(\mathbf{b}), \ldots, \frac{\partial g}{\partial x_n}(\mathbf{b})\right). \qquad \square$$

Example 1. Which point on the ellipse given by $9x^2 + 4y^2 = 36$ is closest to the origin?

(a) In our standard form, the problem is to find the solution of

$$g(x, y) \equiv 9x^2 + 4y^2 - 36 = 0$$

that minimizes $\sqrt{x^2 + y^2}$, equivalent to minimizing $f(x, y) \equiv x^2+y^2$. To pursue the standard solution, we calculate $\nabla f = (2x, 2y)$, $\nabla g = (18x, 8y)$. Note that $\nabla g \neq \mathbf{O}$ everywhere on the ellipse. For $(2x, 2y)$ to be a multiple $\lambda(18x, 8y)$, we need

$$2x = 18\lambda x, \qquad 2y = 8\lambda y.$$

Since $9x^2 + 4y^2 = 36$ is also required, we have three equations from which to extract x, y, and λ. If $x \neq 0$, then necessarily $\lambda = \frac{1}{9}$, $y = 0$, and $x = \pm 2$; if $y \neq 0$, then $\lambda = \frac{1}{4}$, $x = 0$, and $y = \pm 3$. Hence the extreme points are among four candidates, $(0, \pm 3)$ and $(\pm 2, 0)$. We simply check that the latter give a minimum, the former a maximum.

(b) This problem has an easy algebraic solution, which we may use as verification. If $9x^2+4y^2-36 = 0$, then $4x^2+4y^2 = 36-5x^2$ and $9x^2+9y^2 = 36+5y^2$. From the second equation, $x^2 + y^2$ is smallest when $y = 0$, which occurs at $(\pm 2, 0)$. From the first, $x^2 + y^2$ is greatest when $x = 0$, at $(0, \pm 3)$.

(c) There is also a geometric interpretation. Examine Figure 3.3. In the geometry of the ellipse, the circle spanned by the minor axis is just small enough to be inscribed within the ellipse, and the one spanned by the major axis is just big enough to be circumscribed. Hence where $f(x, y)$ is a minimum, the graph $f(x, y) = 4$ is tangent to $g(x, y) = 0$, and similarly with the graph of $f(x, y) = 9$ at the maximum.

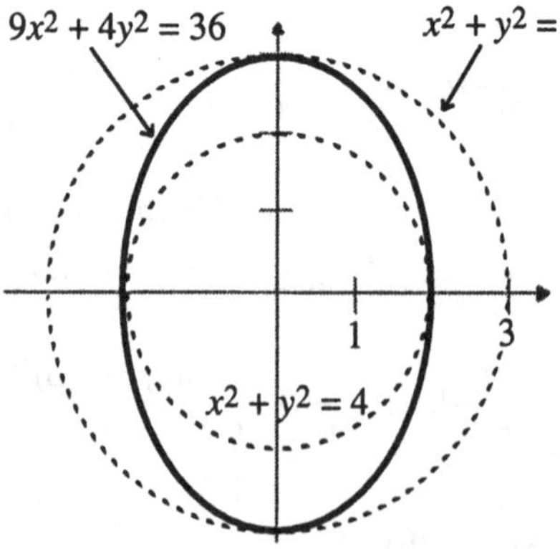

Figure 3.3.

The remark in (c) is an instance of a general principle: If $f(\mathbf{b})$ is the minimum of $f(\mathbf{x})$ on the hypersurface $g(\mathbf{x}) = 0$, then the level hypersurface $f(\mathbf{x}) = f(\mathbf{b})$ (assuming that it is a hypersurface, meaning $\nabla f(\mathbf{b}) \neq \mathbf{O}$) is tangent to $g(\mathbf{x}) = 0$.

This principle makes sense in terms of instantaneous rate of change. If $f(\mathbf{b})$ is minimal, then $\mathbf{b}$ is a stationary point for f on the g-hypersurface. That is, in the directions from $\mathbf{b}$ that hug this hypersurface—in other words, along the vectors lying in the tangent hyperplane—f must have zero (directional) derivative. Therefore, $\nabla f(\mathbf{b})$ has zero component in these directions. That means that $\nabla f(\mathbf{b})$ is orthogonal to the $g = 0$ tangent. Since $\nabla f(\mathbf{b})$ is also orthogonal to the $f(\mathbf{x}) = f(\mathbf{b})$ tangent (Theorem 2.6:3), we conclude that the two hypersurfaces have the same tangent hyperplane.

The principle is, in fact, what Theorem 1 says. To say that $\nabla f(\mathbf{b}) = \lambda \nabla g(\mathbf{b}) \neq \mathbf{O}$ is to say that the normals to the hypersurfaces $f(\mathbf{x}) = f(\mathbf{b})$ and $g(\mathbf{x}) = 0$ are in the same line. Hence the orthogonal complements of the normals are the same. That makes the tangents (the $\mathbf{b}$-translates of those complements) the same.

Part (a) of Example 1 illustrates a method, suggested by Theorem 1, for the solution of problems with a single constraint. The theorem and method extend easily to an extreme constrained by a system of equations.

The system $g_1(\mathbf{x}) = \cdots = g_k(\mathbf{x}) = 0$ abbreviates to the single vector equation

$$\mathbf{g}(\mathbf{x}) \equiv \big(g_1(\mathbf{x}), \dots, g_k(\mathbf{x})\big) = \mathbf{O}.$$

Usually, the system describes the intersection of k hypersurfaces. In view of Theorem 3.3:3(b), this intersection will be an $(n-k)$-dimensional surface. We want $f(\mathbf{b})$ to be a minimum on this locus. The next theorem says that for the minimum to happen, $\nabla f(\mathbf{b})$ must be a linear combination of $\nabla g_1(\mathbf{b}), \dots, \nabla g_k(\mathbf{b})$.

Theorem 2. *Assume that f and $\mathbf{g}$ are differentiable near $\mathbf{b}$, $\mathbf{g}(\mathbf{b}) = \mathbf{O}$, and $\mathbf{g}'(\mathbf{x})$ is continuous at $\mathbf{x} = \mathbf{b}$. Suppose that the gradients $\nabla g_1(\mathbf{b}), \dots, \nabla g_k(\mathbf{b})$ are independent vectors. If in some neighborhood of $\mathbf{b}$, $f(\mathbf{b}) \le f(\mathbf{x})$ for every $\mathbf{x}$ satisfying $\mathbf{g}(\mathbf{x}) = \mathbf{O}$, then there are scalars $\lambda_1, \dots, \lambda_k$ such that*

$$\nabla f(\mathbf{b}) = \lambda_1 \nabla g_1(\mathbf{b}) + \cdots + \lambda_k \nabla g_k(\mathbf{b}).$$

Proof. We give a direct extension of the argument from Theorem 1, employing our now-familiar arguments related to the rank of a matrix.

Assume that the gradients are independent. Then the Jacobian matrix $\mathbf{g}'(\mathbf{b}) = \frac{\partial \mathbf{g}}{\partial \mathbf{x}}(\mathbf{b})$ has k independent rows, so it possesses equally many independent columns. Say the columns $\frac{\partial \mathbf{g}}{\partial x_1}(\mathbf{b}), \dots, \frac{\partial \mathbf{g}}{\partial x_k}(\mathbf{b})$ are independent. Hence

$$\det\left[\frac{\partial \mathbf{g}}{\partial(x_1, \dots, x_k)}(\mathbf{b})\right] = \begin{vmatrix} \frac{\partial g_1}{\partial x_1}(\mathbf{b}) & \cdots & \frac{\partial g_1}{\partial x_k}(\mathbf{b}) \\ & \ddots & \\ \frac{\partial g_k}{\partial x_1}(\mathbf{b}) & \cdots & \frac{\partial g_k}{\partial x_k}(\mathbf{b}) \end{vmatrix}$$

is a nonzero (Jacobian) subdeterminant from $\mathbf{g}'(\mathbf{b})$. By the implicit function theorem, there is a differentiable function $\mathbf{G} = (G_1, \dots, G_k)$ such that near $\mathbf{b}$,

$$\mathbf{g}(\mathbf{x}) = \mathbf{O} \quad \text{iff} \quad (x_1, \dots, x_k) = \mathbf{G}(x_{k+1}, \dots, x_n).$$

Write $\mathbf{x}^* \equiv (x_{k+1}, \dots, x_n)$, and examine the function $F\colon \mathbf{R}^{n-k} \to \mathbf{R}$ given near $\mathbf{b}^*$ by

$$F(\mathbf{x}^*) \equiv f\big(\mathbf{G}(\mathbf{x}^*), \mathbf{x}^*\big).$$

The function F is constructed to track the values of f on the surface $\mathbf{g}(\mathbf{x}) = \mathbf{O}$. We know that $F(\mathbf{b}^*)$ is a minimum. Therefore,

$$\mathbf{O} = \frac{\partial F}{\partial \mathbf{x}^*}(\mathbf{b}^*) = \frac{\partial f}{\partial \mathbf{x}}(\mathbf{b})\frac{\partial(\mathbf{G}(\mathbf{x}^*), \mathbf{x}^*)}{\partial \mathbf{x}^*}(\mathbf{b}^*).$$

Write the row $\frac{\partial f}{\partial \mathbf{x}}(\mathbf{b})$ as

$$\begin{bmatrix} \dfrac{\partial f}{\partial(x_1, \dots, x_k)}(\mathbf{b}) & \dfrac{\partial f}{\partial(x_{k+1}, \dots, x_n)}(\mathbf{b}) \end{bmatrix}$$

and the Jacobian matrix $\frac{\partial(\mathbf{G}(\mathbf{x}^*), \mathbf{x}^*)}{\partial \mathbf{x}^*}(\mathbf{b}^*)$ as

$$\begin{bmatrix} \frac{\partial \mathbf{G}}{\partial \mathbf{x}^*}(\mathbf{b}^*) \\ \\ I_{n-k} \end{bmatrix}.$$

Then

$$\begin{aligned} \mathbf{O} &= \begin{bmatrix} \dfrac{\partial f}{\partial(x_1, \dots, x_k)}(\mathbf{b}) & \dfrac{\partial f}{\partial(x_{k+1}, \dots, x_n)}(\mathbf{b}) \end{bmatrix} \begin{bmatrix} \frac{\partial \mathbf{G}}{\partial \mathbf{x}^*}(\mathbf{b}^*) \\ \\ I_{n-k} \end{bmatrix} \\ &= \frac{\partial f}{\partial(x_1, \dots, x_k)}(\mathbf{b})\frac{\partial \mathbf{G}}{\partial \mathbf{x}^*}(\mathbf{b}^*) + \frac{\partial f}{\partial(x_{k+1}, \dots, x_n)}(\mathbf{b}). \end{aligned}$$

The implicit function theorem gives us

$$\frac{\partial \mathbf{G}}{\partial \mathbf{x}^*}(\mathbf{b}^*) = -\left(\frac{\partial \mathbf{g}}{\partial(x_1, \dots, x_k)}(\mathbf{b})\right)^{-1} \frac{\partial \mathbf{g}}{\partial(x_{k+1}, \dots, x_n)}(\mathbf{b}).$$

We conclude that

$$\begin{aligned} \frac{\partial f}{\partial(x_{k+1}, \dots, x_n)}(\mathbf{b}) = &\left[\frac{\partial f}{\partial(x_1, \dots, x_k)}(\mathbf{b})\left(\frac{\partial \mathbf{g}}{\partial(x_1, \dots, x_k)}(\mathbf{b})\right)^{-1}\right] \\ &\times \frac{\partial \mathbf{g}}{\partial(x_{k+1}, \dots, x_n)}(\mathbf{b}). \end{aligned}$$

The analogous relation is trivial for the block $\frac{\partial f}{\partial(x_1, \dots, x_k)}(\mathbf{b})$:

$$\begin{aligned} \frac{\partial f}{\partial(x_1, \dots, x_k)}(\mathbf{b}) = &\left[\frac{\partial f}{\partial(x_1, \dots, x_k)}(\mathbf{b})\left(\frac{\partial \mathbf{g}}{\partial(x_1, \dots, x_k)}(\mathbf{b})\right)^{-1}\right] \\ &\times \frac{\partial \mathbf{g}}{\partial(x_1, \dots, x_k)}(\mathbf{b}). \end{aligned}$$

The expression in big brackets has size $(1 \times k) \times (k \times k)^{-1} = 1 \times k$. We may write it as a row $\Lambda \equiv [\lambda_1 \cdots \lambda_k]$. Putting together the last two equations, we obtain

$$\begin{aligned}
f'(\mathbf{b}) &= \left[\frac{\partial f}{\partial(x_1, \ldots, x_k)}(\mathbf{b}) \quad \frac{\partial f}{\partial(x_{k+1}, \ldots, x_n)}(\mathbf{b}) \right] \\
&= \left[\Lambda \frac{\partial \mathbf{g}}{\partial(x_1, \ldots, x_k)}(\mathbf{b}) \quad \Lambda \frac{\partial \mathbf{g}}{\partial(x_{k+1}, \ldots, x_n)}(\mathbf{b}) \right] \\
&= \left[\lambda_1 \frac{\partial g_1}{\partial x_1}(\mathbf{b}) + \cdots + \lambda_k \frac{\partial g_k}{\partial x_1}(\mathbf{b}) \quad \cdots \quad \lambda_1 \frac{\partial g_1}{\partial x_n}(\mathbf{b}) + \cdots + \lambda_k \frac{\partial g_k}{\partial x_n}(\mathbf{b}) \right] \\
&= \lambda_1 \left[\frac{\partial g_1}{\partial x_1}(\mathbf{b}) \quad \cdots \quad \frac{\partial g_1}{\partial x_n}(\mathbf{b}) \right] + \cdots + \lambda_k \left[\frac{\partial g_k}{\partial x_1}(\mathbf{b}) \quad \cdots \quad \frac{\partial g_k}{\partial x_n}(\mathbf{b}) \right] \\
&= \lambda_1 g_1'(\mathbf{b}) + \cdots + \lambda_k g_k'(\mathbf{b}),
\end{aligned}$$

which says that the gradient of f is a combination of the other gradients. □

It is both easy and important to see Theorem 2 and its proof in terms of tangency and orthogonality. First, the hypothesis asks for $\nabla g_1(\mathbf{b}), \ldots, \nabla g_k(\mathbf{b})$ to be independent in $\mathbf{R}^n$. We have seen (end of Section 3) that this amounts to demanding that the hypersurfaces $g_j = 0$ intersect transversally. By prohibiting tangency, this condition guarantees that none of the hypersurfaces is too much like the intersection of any others. Accordingly, it guarantees that none of the equations is implied by a subset of the others. In the language of optimization, it says that none of the equations is "redundant."

Next, if $f(\mathbf{b})$ is to be a minimum, then f must be stationary at $\mathbf{b}$ relative to the intersection of the hypersurfaces. Thus, along any vector tangent to the intersection surface, f has zero directional derivative. Hence the tangent to the surface must be orthogonal to $\nabla f(\mathbf{b})$. That means that the tangent to the surface is a subset of the tangent hyperplane to $f(\mathbf{x}) = f(\mathbf{b})$: The f-hypersurface is tangent to the intersection surface. For that reason, $\nabla f(\mathbf{b})$ has to be in the orthogonal complement of the intersection tangent, which complement is spanned by the gradients ∇g_j. Consequently, $\nabla f(\mathbf{b})$ has to be a linear combination of the other gradients. That is the conclusion of Theorem 2.

Example 2. What point on the intersection of $x^2+y^2+z^2 = 9$ and $z^2 = 1+x^2+y^2$ makes $x + 3y$ minimal?

(a) First, $g_1(x, y, z) \equiv x^2 + y^2 + z^2 - 9$ and $g_2(x, y, z) \equiv x^2 + y^2 + 1 - z^2$ have

$$\frac{\partial(g_1, g_2)}{\partial(x, y, z)} = \begin{bmatrix} 2x & 2y & 2z \\ 2x & 2y & -2z \end{bmatrix}.$$

Except where $z = 0$ or $x = y = 0$, the rank of this matrix is 2, so that the gradients are independent. No such point is in the intersection (Example 1 in the previous section). Hence where $f(x, y, z) \equiv x + 3y$ is minimal, we have

$$(1, 3, 0) = \nabla f = \lambda_1 \nabla g_1 + \lambda_2 \nabla g_2 = \lambda_1(2x, 2y, 2z) + \lambda_2(2x, 2y, -2z).$$

We are led to the equations

$$1 = 2\lambda_1 x + 2\lambda_2 x, \quad 3 = 2\lambda_1 y + 2\lambda_2 y, \quad 0 = 2\lambda_1 z - 2\lambda_2 z,$$

together with the intersection characterization $x^2 + y^2 = 4$, $z = \pm\sqrt{5}$. These yield

$$x = \frac{\pm 2}{\sqrt{10}}, \quad y = \frac{\pm 6}{\sqrt{10}} \quad \text{(same sign as } x\text{)},$$
$$z = \pm\sqrt{5} \quad \text{(independent sign)}.$$

That leaves us four suspects: $\left(\frac{2}{\sqrt{10}}, \frac{6}{\sqrt{10}}, \pm\sqrt{5}\right)$ and $\left(\frac{-2}{\sqrt{10}}, \frac{-6}{\sqrt{10}}, \pm\sqrt{5}\right)$. We need only check that the latter two give the smaller value of $f = \frac{-20}{\sqrt{10}}$.

(b) Geometric interpretation? See Exercise 5.

In Example 2, it was easy to eliminate λ_1 and λ_2. Whether they are easy to calculate or not, there is generally no reason to find them. They are irrelevant. For evidence, consider that if we multiply each g_j by a nonzero constant, then the resulting problem is equivalent to the original, but has a rescaled set of λ_j.

The letter λ and the name attached to the method implied by Theorem 2 honor the immortal Joseph Louis Lagrange. It was typical of Lagrange's brilliance to conceive a method—with cousins in modern mathematics plus in physics, economics, and other areas—that simplifies a problem by first making it worse. The **Lagrange multipliers**, after all, are k new unknowns $\lambda_1, \ldots, \lambda_k$ thrown into the problem. With them, however, come n equations: $\nabla f = \lambda_1 \nabla g_1 + \cdots + \lambda_k \nabla g_k$ is equivalent to

$$\frac{\partial f}{\partial x_1} = \lambda_1 \frac{\partial g_1}{\partial x_1} + \cdots + \lambda_k \frac{\partial g_k}{\partial x_1}, \ldots, \frac{\partial f}{\partial x_n} = \lambda_1 \frac{\partial g_1}{\partial x_n} + \cdots + \lambda_k \frac{\partial g_k}{\partial x_n}.$$

The original question of minimizing f subject to k equations in $n > k$ unknowns is traded for a problem of solving $k + n$ equations in $n + k$ unknowns.

Exercises

1. Maximize $f(x, y) \equiv xy$ subject to the condition $x + 2y = 5$. Interpret the result geometrically (in terms of tangent graphs).

2. Find the minimal distance from the point (a, b) to the line $cx + dy = e$. (Hint: Solve for the multiplier, rather than for the minimizing point.)

3. Find the maximal and minimal value of $x + 2y + 3z$ on the sphere $x^2 + y^2 + z^2 = 1$.

4. On the intersection of the cone $z^2 = x^2 + y^2$ and the plane $z = x + 4$, find the point closest to:

(a) $(0, 0, 2)$;

(b) a general z-axis point $(0, 0, b)$.

5. In Example 2, show that the graph of $x + 3y = \frac{-20}{\sqrt{10}}$ is tangent to the intersection of $x^2 + y^2 + z^2 = 9$ and $z^2 = 1 + x^2 + y^2$ at the point of the intersection where $x + 3y$ is minimum. Compare Figure 3.1.

6. Show that the shortest path from a point to a smooth surface is perpendicular to the surface. In symbols: Assume that $\nabla f_1(\mathbf{x}), \ldots, \nabla f_k(\mathbf{x})$ are continuous and independent throughout $\mathbf{R}^n$, the product $f_1(\mathbf{b}) \cdots f_k(\mathbf{b})$ is nonzero, and $\mathbf{c}$ is the point of the surface $f_1(\mathbf{x}) = \cdots = f_k(\mathbf{x}) = 0$ closest to $\mathbf{b}$; show that $\mathbf{b} - \mathbf{c}$ is normal to the tangent plane at $\mathbf{c}$. (Compare Example 1. There, both the shortest and the *longest* segments from the origin to the ellipse are perpendicular to the ellipse.)

4

Integrability of Multivariable Functions

It is time to turn our attention to integrals. As is our habit, we will define them by analogy to the single-variable case. Also, we will organize the material as we did with derivatives. In the earlier chapters, we separated the definitions of differentiability and derivatives from discussion of their properties. We will do similarly with integrals, covering integrability in this chapter and properties of integrals in the next.

4.1 Partitions

For functions of scalar variables, integrals are defined by reference to partitions of a closed interval. We will talk about partitions of boxes in $\mathbf{R}^n$.

Definition. (a) Two sets in $\mathbf{R}^n$ are said to **overlap** if they have a common *interior* point (their interiors have nonempty intersection).

(b) Suppose $S_1 \equiv [\mathbf{a}_1, \mathbf{b}_1], \dots, S_J \equiv [\mathbf{a}_J, \mathbf{b}_J]$ make up a finite collection of (pairwise) nonoverlapping boxes whose union is $[\mathbf{a}, \mathbf{b}]$. We call $\mathcal{P} \equiv \{S_1, \dots, S_J\}$ a **partition** of $[\mathbf{a}, \mathbf{b}]$. The S_j are the **subintervals** of $\mathcal{P}$. The biggest diagonal $\|\mathbf{b}_j - \mathbf{a}_j\|$ is the **norm** (or **fineness**) of $\mathcal{P}$, denoted by $\|\mathcal{P}\|$.

In this section and the next we will make some constructions requiring us to label multiple points along the coordinate axes. Rather than using double subscripts there, we will write x for x_1, (so that $x_{11}, x_{12}, \dots$ become $x_1, x_2, \dots$), y for x_2 (simply to define patterns), and z for x_n.

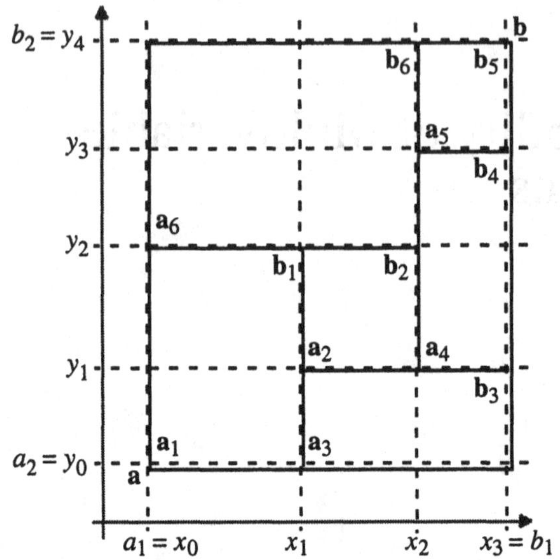

Figure 4.1.

Consider a partition $\mathcal{P} \equiv \{S_1, \ldots, S_J\}$. Suppose that the two ends of a typical subinterval S_j are $\mathbf{a}_j = (a_{j1}, \ldots, a_{jn})$ and $\mathbf{b}_j = (b_{j1}, \ldots, b_{jn})$. For each $k = 1$ to n, write $X_k \equiv \{a_{1k}, b_{1k}, \ldots, a_{Jk}, b_{Jk}\}$. In words, X_1 is the set of x-coordinates of the ends, left or right, of the subintervals of $\mathcal{P}$; X_2 is the set of y-coordinates; and so on. Rearrange the members of X_1 in the order

$$X_1 = \{a_1 = x_0 < x_1 < \cdots < x_{\alpha(1)} = b_1\},$$

and similarly with

$$X_2 = \{a_2 = y_0 < y_1 < \cdots < y_{\alpha(2)} = b_2\}$$

and the others. The number $\alpha(k) + 1$ of elements of X_k is unpredictable, but the first and last are always the kth coordinates of $\mathbf{a}$ and $\mathbf{b}$. Thus, Figure 4.1 shows a partition in $\mathbf{R}^2$ with $J = 6$ subintervals bounded by solid lines, giving rise to four x-ends (members of X_1) and five y-ends (X_2). The boxes of the form $[(x_i, y_m, \ldots, z_l), (x_{i+1}, y_{m+1}, \ldots, z_{l+1})]$—in the figure, they are bordered by dashed lines, $0 \le i \le 2$, $0 \le m \le 3$, and there is no z—constitute the **cross-partition associated with** $\mathcal{P}$.

We will identify a cross-partition by its markers along the coordinate axes, and write it as

$$\{a_1 = x_0 < x_1 < \cdots < x_{\alpha(1)} = b_1\} \otimes \cdots \otimes \{a_n = z_0 < z_1 < \cdots < z_{\alpha(n)} = b_n\}.$$

Our construction of a cross-partition used the subintervals $S_1, \ldots, S_J$ of an existing partition, but the same process can be applied to any finite set of boxes, no matter how they are related. Thus, if $\{[\mathbf{c}_1, \mathbf{d}_1], \ldots, [\mathbf{c}_M, \mathbf{d}_M]\}$ is an arbitrary finite collection of boxes—even overlapping boxes—then $X_k \equiv \{c_{1k}, d_{1k}, \ldots, c_{Mk}, d_{Mk}\}$, $k = 1$ to n, marks along the x_k-axis a cross-partition of the box from

$$\mathbf{c} \equiv \left(\min\{c_{11}, \ldots, c_{M1}\}, \ldots, \min\{c_{1n}, \ldots, c_{Mn}\}\right)$$

to

$$\mathbf{d} \equiv \left(\max\{d_{11}, \ldots, d_{M1}\}, \ldots, \max\{d_{1n}, \ldots, d_{Mn}\}\right);$$

and each of the original boxes is the union of some of the subintervals from this cross-partition. Indeed, if the box $[\mathbf{c}_j, \mathbf{d}_j]$ overlaps none of the others, then substituting $[\mathbf{c}_j, \mathbf{d}_j]$ for the subintervals that add up to it gives us a partition that has $[\mathbf{c}_j, \mathbf{d}_j]$ among its subintervals.

Definition. Let $\mathcal{P} \equiv \{S_1, \ldots, S_J\}$ and $\mathcal{Q} \equiv \{T_1, \ldots, T_K\}$ be partitions. We say that $\mathcal{Q}$ is a **refinement of** (or **refines**) $\mathcal{P}$ if each T_k is a subset of some S_j.

Notice two things. First, the definition of refinement does not mention the boxes that $\mathcal{P}$ and $\mathcal{Q}$ partition. It is unnecessary to specify the box, since it is implied by a partition. So, we will frequently omit explicit reference to the box. Second, by leaving them unstated, this definition permits two different boxes for $\mathcal{P}$ and $\mathcal{Q}$. Generally, we are concerned with refinements within a single box. Still, unequal boxes are allowed. In any case, since each box is the union of its partition's subintervals, the box in which $\mathcal{Q}$ is defined has to be a subset of the one housing $\mathcal{P}$.

It is clear that the cross-partition associated with $\mathcal{P}$ is a refinement of $\mathcal{P}$. One use for the cross-partition is the second part of the next theorem.

Theorem 1. *Let $\mathcal{P} \equiv \{S_1, \ldots, S_J\}$ and $\mathcal{Q} \equiv \{T_1, \ldots, T_K\}$ be partitions of box $[\mathbf{a}, \mathbf{b}]$.*

(a) *If $\mathcal{Q}$ is a refinement of $\mathcal{P}$, then each S_j is actually the union of those T_k that are contained in S_j.*

(b) *The volume of the box is the sum of the subinterval volumes from $\mathcal{P}$ or $\mathcal{Q}$.*

Proof. (a) For simplicity, suppose S_1 contains exactly T_1, T_2, T_3, as in Figure 4.2. It is trivial that $T_1 \cup T_2 \cup T_3 \subseteq S_1$.

Let $\mathbf{x} \in S_1$. In any box, the segment from $\mathbf{x}$ to the center of the box is interior to the box, except perhaps for $\mathbf{x}$. (Why?) Hence $\mathbf{x}$ is the limit of a sequence of interior points of S_1. These interior points cannot come from, say, T_4. After all, T_4 is contained in, say, S_6. Interior points of S_1 cannot be anywhere in S_6; those nonoverlapping boxes can only share points on their boundaries (Exercise 2c). Hence the interior points that approach $\mathbf{x}$ do not come from $T_4, T_5, \ldots, T_K$. However, they have to come from some T's, because the T's fill up $[\mathbf{a}, \mathbf{b}]$. Hence the interior points come from $T_1 \cup T_2 \cup T_3$. Thus, $\mathbf{x}$ is the limit of a sequence from

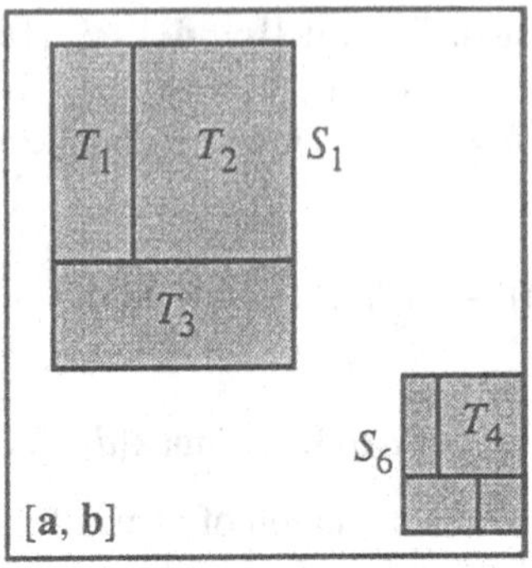

Figure 4.2.

the closed set $T_1 \cup T_2 \cup T_3$. Therefore, $\mathbf{x} \in T_1 \cup T_2 \cup T_3$. We have shown that $S \subseteq T_1 \cup T_2 \cup T_3$, and the equality follows.

(See Exercise 4 for an extension of this result.)

(b) The statement is obvious if the subintervals are uniformly stacked. Thus, assume that

$$\mathcal{Q} = \{a_1 = x_0 < x_1 < \cdots < x_k = b_1\} \otimes \cdots \otimes \{a_n = z_0 < z_1 < \cdots < z_m = b_n\}$$

is the cross-partition associated with $\mathcal{P}$. Then

$$\begin{aligned} V([\mathbf{a}, \mathbf{b}]) &= (x_k - x_0)(y_l - y_0) \cdots (z_m - z_0) \\ &= (x_k - x_{k-1} + \cdots + x_1 - x_0) \cdots (z_m - z_{m-1} + \cdots + z_1 - z_0) \\ &= \sum_{p,q,r} (x_{p+1} - x_p)(y_{q+1} - y_q) \cdots (z_{r+1} - z_r) \end{aligned}$$

says that the volume of the box is the sum of the volumes of $\mathcal{Q}$'s subintervals.

For the same reason, the sum of the volumes of the T_k from $\mathcal{Q}$ contained in any S_j is $V(S_j)$. No T_k is contained in two different S_j (Why?), and each T_k is contained in some S_j. It follows that $V(S_1) + \cdots + V(S_J)$ is a rearrangement of $V(T_1) + \cdots + V(T_K)$, and we conclude that $V(S_1) + \cdots + V(S_J) = V([\mathbf{a}, \mathbf{b}])$. □

(See Exercise 3 for an illustrative calculation.)

The statement in (a) above, that the T_k in a refinement can be gathered into groups whose unions are the S_j, tells us that from a fixed partition, a refinement is produced by subdividing one or more of its subintervals.

One other helpful construction is the simplest partition that is a refinement of two others. Given $\mathcal{P} \equiv \{S_1, \ldots, S_J\}$ and $\mathcal{R} \equiv \{T_1, \ldots, T_K\}$, provided their boxes overlap, let

$$\mathcal{PR} \equiv \{S_j \cap T_k : 1 \le j \le J, 1 \le k \le K, \text{ and } S_j \text{ and } T_k \text{ overlap}\}.$$

Then $\mathcal{PR}$ is a partition (Exercise 5), and we refer to it as the **intersection-partition** of $\mathcal{P}$ and $\mathcal{R}$. Since each subinterval in $\mathcal{PR}$ is a subset of one in $\mathcal{P}$ and one in $\mathcal{R}$, the intersection-partition refines each of $\mathcal{P}$ and $\mathcal{R}$.

Exercises

1. Give an example in $\mathbf{R}^n$ of two boxes that have common points but do not overlap.

2. Let $B \equiv [\mathbf{a}, \mathbf{b}]$ and $C \equiv [\mathbf{c}, \mathbf{d}]$ be boxes in $\mathbf{R}^n$. Show that:

   (a) $B \cap C$ consists of those $\mathbf{x}$ with $\max\{a_j, c_j\} \le x_j \le \min\{b_j, d_j\}$, $j = 1, \ldots, n$.

   (b) $B \cap C$ is a box (with nonempty interior) iff B and C overlap.

   (c) If B and C do not overlap, then any point in their intersection is on the boundary of each box.

   (d) Part (c) may fail if we replace B and C by sets that are not boxes.

   (e) Two sets overlap iff their intersection has nonempty interior.

3. In Figure 4.1, use the cross-partition to show that the volumes (areas) $V([\mathbf{a}_1, \mathbf{b}_1]), \ldots, V([\mathbf{a}_6, \mathbf{b}_6])$ of the subintervals in the original partition add up to the volume of the box $[\mathbf{a}, \mathbf{b}]$.

4. Suppose partition $\mathcal{Q} \equiv \{T_1, \ldots, T_K\}$ of box C refines $\mathcal{P} \equiv \{S_1, \ldots, S_J\}$ of box B, with C a proper subset of B. Show that if S_j overlaps C, then the piece $S_j \cap C$ is the union of the subintervals T_k that are contained in S_j. (Like the remark following Theorem 1, this statement tells us that refinements of $\mathcal{P}$ come from subdividing some of the S_j.)

5. Given partitions $\mathcal{P} \equiv \{S_1, \ldots, S_J\}$ of box B and $\mathcal{R} \equiv \{T_1, \ldots, T_K\}$ of C, with B and C overlapping, prove that $\mathcal{PR}$ is a partition; that is:

   (a) Each intersection in $\mathcal{PR}$ is a box (having nonempty interior).

   (b) No two boxes in $\mathcal{PR}$ overlap.

   (c) The union of the boxes in $\mathcal{PR}$ is $B \cap C$.

   In what sense is $\mathcal{PR}$ the "simplest" refinement of $\mathcal{P}$ and $\mathcal{R}$?

4.2 Integrability in a Box

We proceed to define integrability on one box. In this section f is a fixed function, defined and bounded on the box $B \equiv [\mathbf{a}, \mathbf{b}]$. The box is also fixed, and every partition discussed resides in it.

Given a partition $\mathcal{P} \equiv \{S_1, \ldots, S_J\}$, we associate certain sums with $\mathcal{P}$. Assume that

$$S_j = [(x_j, y_j, \ldots, z_j), (u_j, v_j, \ldots, w_j)],$$

and write

$$\begin{aligned} V(S_j) &\equiv (u_j - x_j) \cdots (w_j - z_j) = \text{volume of } S_j, \\ M(S_j) &\equiv \sup\{f(\mathbf{x}) : \mathbf{x} \in S_j\}, \\ m(S_j) &\equiv \inf\{f(\mathbf{x}) : \mathbf{x} \in S_j\}, \qquad j = 1, \ldots, J. \end{aligned}$$

Then

$$u(f, \mathcal{P}) \equiv M(S_1)V(S_1) + \cdots + M(S_J)V(S_J)$$

is the **upper sum for f on $\mathcal{P}$**. Similarly,

$$l(f, \mathcal{P}) \equiv m(S_1)V(S_1) + \cdots + m(S_J)V(S_J)$$

is the **lower sum for f on $\mathcal{P}$**.

Theorem 1. *For a bounded function f on B:*

(a) *Each partition's lower sum is no more than the upper: $l(f, \mathcal{P}) \le u(f, \mathcal{P})$.*

(b) *Refinement squeezes the sums together: if $\mathcal{Q}$ is a refinement of $\mathcal{P}$, then*

$$l(f, \mathcal{P}) \le l(f, \mathcal{Q}) \le u(f, \mathcal{Q}) \le u(f, \mathcal{P}).$$

(c) *Refinement squeezes the sums gradually. Specifically, suppose refinement $\mathcal{Q}$ is produced by breaking some subintervals of $\mathcal{P}$ along the hyperplane $x_j = c$. Then*

$$(0 \le)\ l(f, \mathcal{Q}) - l(f, \mathcal{P}) \le 4M \|\mathbf{b} - \mathbf{a}\|^{n-1} \|\mathcal{P}\|$$

and

$$(0 \le)\ u(f, \mathcal{P}) - u(f, \mathcal{Q}) \le 4M \|\mathbf{b} - \mathbf{a}\|^{n-1} \|\mathcal{P}\|,$$

where $M \equiv \sup\{|f(\mathbf{x})| : \mathbf{x} \in [\mathbf{a}, \mathbf{b}]\}$.

(d) *No lower sum exceeds any upper sum: For any partitions $\mathcal{P}$ and $\mathcal{R}$, $l(f, \mathcal{P}) \le u(f, \mathcal{R})$.*

(e) $\sup\{l(f, \mathcal{P}) :$ *all partitions* $\mathcal{P}\} \le \inf\{u(f, \mathcal{R}) :$ *all partitions* $\mathcal{R}\}$.

Proof. (a) Each term in $l(f, \mathcal{P})$ is less than or equal to the corresponding term in $u(f, \mathcal{P})$, because $m(S_j) \le M(S_j)$ and $V(S_j) > 0$.

(b) We observed that a refinement comes from repartitioning subintervals of the original partition. Suppose $\mathcal{Q}$ is made from $\mathcal{P}$ by partitioning S_1 into $T_1, \ldots, T_K$. Then

$$u(f, \mathcal{Q}) = M(T_1)V(T_1) + \cdots + M(T_K)V(T_K) + M(S_2)V(S_2) + \cdots + M(S_J)V(S_J).$$

Each T_k is a subset of S_1, so

$$M(T_k) \equiv \sup\{f(\mathbf{x}) : \mathbf{x} \in T_k\} \le \sup\{f(\mathbf{x}) : \mathbf{x} \in S_1\} = M(S_1).$$

Therefore,

$$\begin{aligned} M(T_1)V(T_1) + \cdots + M(T_K)V(T_K) &\le M(S_1)\big(V(T_1) + \cdots + V(T_K)\big) \\ &= M(S_1)V(S_1) \end{aligned}$$

(the volume of S_1 being the sum of the constituent volumes). Substituting into $u(f, \mathcal{Q})$, we find that $u(f, \mathcal{Q}) \le u(f, \mathcal{P})$.

Similarly, $T_k \subseteq S_1$ implies $m(T_k) \ge m(S_1)$, leading to $l(f, \mathcal{Q}) \ge l(f, \mathcal{P})$. In view of (a), we have

$$l(f, \mathcal{P}) \le l(f, \mathcal{Q}) \le u(f, \mathcal{Q}) \le u(f, \mathcal{P}).$$

By induction, it follows that if $\mathcal{Q}$ is produced by any number of repartitionings applied to $\mathcal{P}$ and its refinements, then the sums of $\mathcal{Q}$ are between those of $\mathcal{P}$.

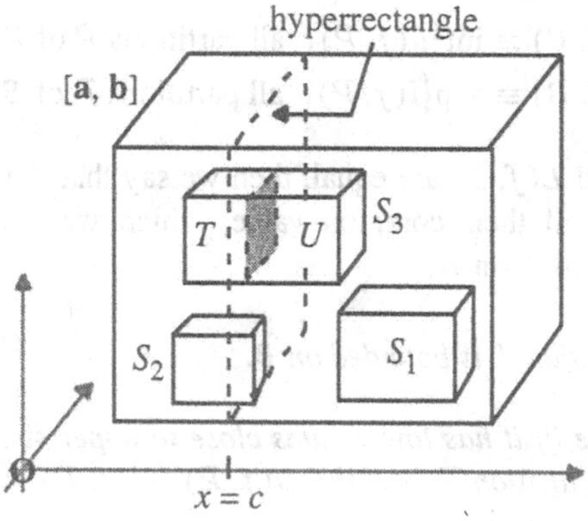

Figure 4.3.

(c) In Figure 4.3 we see the box $[\mathbf{a}, \mathbf{b}]$, intersecting the hyperplane $x = c$ in a hyperrectangle, together with three kinds of subintervals of $\mathcal{P}$. We examine those subintervals' separate contributions to $u(f, \mathcal{P}) - u(f, \mathcal{Q})$.

Subinterval S_1 does not intersect $x = c$, and subinterval S_2 has only boundary points on the hyperplane. Hence each is a subinterval of $\mathcal{Q}$ as well as of $\mathcal{P}$. The terms $M(S_1)V(S_1)$ and $M(S_2)V(S_2)$ appear in both $u(f, \mathcal{P})$ and $u(f, \mathcal{Q})$, so they do not contribute to $u(f, \mathcal{P}) - u(f, \mathcal{Q})$. Subinterval S_3 has points on opposite sides of ("straddles") the hyperrectangle. From $S_3 = T \cup U$, the contribution is

$$\begin{aligned} M(S_3)V(S_3) - [M(T)V(T) + M(U)V(U)] &\le MV(S_3) + M[V(T) + V(U)] \\ &= 2MV(S_3). \end{aligned}$$

Consequently, $u(f, \mathcal{P}) - u(f, \mathcal{Q}) \le 2MV$, where V is the sum of the volumes of those subintervals that straddle the hyperplane and are actually cut in making the refinement. It remains only to show that $V \le 2\|\mathbf{b} - \mathbf{a}\|^{n-1}\|\mathcal{P}\|$.

For every subinterval, the width (left to right) is less than the diagonal, which is no more than $\|\mathcal{P}\|$. Therefore, the straddling subintervals lie in the space between the parallel hyperplanes $x = c - \|\mathcal{P}\|$ and $x = c + \|\mathcal{P}\|$. The part of the box between those hyperplanes has volume at most $2\|\mathcal{P}\|(b_2 - a_2)\cdots(b_n - a_n)$. Since the subintervals do not overlap, we conclude that

$$V \le 2\|\mathcal{P}\|(b_2 - a_2)\cdots(b_n - a_n) < 2\|\mathbf{b} - \mathbf{a}\|^{n-1}\|\mathcal{P}\|.$$

(d) Let $\mathcal{P}$ and $\mathcal{R}$ be partitions. Then $\mathcal{Q} \equiv \mathcal{PR}$ is a refinement of each of $\mathcal{P}$ and $\mathcal{R}$. Applying parts (a) and (b), we get

$$l(f, \mathcal{P}) \le l(f, \mathcal{PR}) \le u(f, \mathcal{PR}) \le u(f, \mathcal{R}).$$

(e) Exercise 9. □

Definition. (a) The **upper integral** and **lower integral** of f **on** B are

$$U(f, B) \equiv \inf\{u(f, \mathcal{P}): \text{ all partitions } \mathcal{P} \text{ of } B\},$$
$$L(f, B) \equiv \sup\{l(f, \mathcal{P}): \text{ all partitions } \mathcal{P} \text{ of } B\}.$$

(b) If $U(f, B)$ and $L(f, B)$ are equal, then we say that f is **(Riemann) integrable on** B and call their common value, which we denote by $\int_B f$, the **(Riemann) integral of** f **on** B.

Theorem 2. *Assume that f is bounded on B.*

(a) *f is integrable iff it has lower sums close to upper sums: for any $\varepsilon > 0$, there exists a partition $\mathcal{P}$ such that $u(f, \mathcal{P}) - l(f, \mathcal{P}) < \varepsilon$.*

(b) *If f is continuous (throughout B), then f is integrable.*

Proof. The arguments are identical to those for a one-variable function on a closed interval. [Compare Ross, Theorems 32.5, 33.2.]

(a) ⇒ Assume that f is integrable. Let $\varepsilon > 0$. By definition, $C \equiv U(f, B) = L(f, B)$. Since C is the infimum of the upper sums, there is some upper sum $u(f, \mathcal{R}) < C + \frac{\varepsilon}{2}$. Also, C is the supremum of the lower sums, so there is some lower sum $l(f, \mathcal{Q}) > C - \frac{\varepsilon}{2}$.

Let $\mathcal{P} = \mathcal{RQ}$. Then $\mathcal{P}$ is a refinement of $\mathcal{R}$ and of $\mathcal{Q}$. By Theorem 1(b),

$$C - \frac{\varepsilon}{2} < l(f, \mathcal{Q}) \le l(f, \mathcal{P}) \le u(f, \mathcal{P}) \le u(f, \mathcal{R}) < C + \frac{\varepsilon}{2}.$$

That tells us that $l(f, \mathcal{P})$ is within ε of $u(f, \mathcal{P})$.

$\Leftarrow$ Assume that each ε has a corresponding $\mathcal{P}$ with $u(f,\mathcal{P}) - l(f,\mathcal{P}) < \varepsilon$. Then

$$\begin{aligned} U(f,B) &\equiv \inf\{u(f,\mathcal{Q})\} \le u(f,\mathcal{P}) < l(f,\mathcal{P}) + \varepsilon \\ &\le \sup\{l(f,\mathcal{Q})\} + \varepsilon \equiv L(f,B) + \varepsilon. \end{aligned}$$

This being true for arbitrary ε, we conclude that $U(f,B) \le L(f,B)$. Equality follows from Theorem 1(e); f is integrable.

(b) Assume that f is continuous on B. Then f is uniformly continuous. If we name $\varepsilon > 0$, then there exists $\delta(\varepsilon)$ such that $\|\mathbf{x} - \mathbf{y}\| < \delta(\varepsilon) \Rightarrow |f(\mathbf{x}) - f(\mathbf{y})| < \frac{\varepsilon}{2V(B)}$ throughout the box.

Given $\varepsilon > 0$, let $\mathcal{P}$ be any partition with $\|\mathcal{P}\| < \delta(\varepsilon)$. In any subinterval S_j of $\mathcal{P}$, we can find sequences $(\mathbf{x}_i)$ with $f(\mathbf{x}_i) \to M(S_j)$ and $(\mathbf{y}_i)$ with $f(\mathbf{y}_i) \to m(S_j)$. Since any $\|\mathbf{x}_i - \mathbf{y}_i\| \le$ diagonal of $S_j \le \|\mathcal{P}\| < \delta(\varepsilon)$,

$$M(S_j) - m(S_j) = \lim_{i\to\infty} [f(\mathbf{x}_i) - f(\mathbf{y}_i)] \le \frac{\varepsilon}{2V(B)}.$$

Hence

$$\begin{aligned} u(f,\mathcal{P}) &= M(S_1)V(S_1) + \cdots + M(S_J)V(S_J) \\ &\le m(S_1)V(S_1) + \cdots + m(S_J)V(S_J) + \big(V(S_1) + \cdots + V(S_J)\big)\frac{\varepsilon}{2V(B)} \\ &= l(f,\mathcal{P}) + \frac{\varepsilon}{2}. \end{aligned}$$

We have found a partition whose sums are within ε. By part (a), f is integrable. □

Example 1. Dirichlet's function, defined by

$$F(\mathbf{x}) \equiv \begin{cases} 1 & \text{if the coordinates } x_1, \ldots, x_n \text{ are all rational,} \\ 0 & \text{otherwise,} \end{cases}$$

is not integrable on any box. (Exercise 6)

Theorem 3 (Darboux's Lemma). *The upper (or lower) integral is the limit of the upper (respectively lower) sums as the partition fineness goes to zero. In symbols, for any $\varepsilon > 0$, there is $\delta > 0$ with the property that $\|\mathcal{P}\| < \delta \Rightarrow U(f,B) \le u(f,\mathcal{P}) < U(f,B) + \varepsilon$.*

Proof. Let $\varepsilon > 0$ be specified. By definition, $U(f,B) = \inf\{u(f,\mathcal{P})\}$, so there exists a partition $\mathcal{Q}$ with $u(f,\mathcal{Q}) < U(f,B) + \frac{\varepsilon}{2}$. Since $\mathcal{Q}$'s cross-partition has even smaller upper sum, we may simply assume that $\mathcal{Q}$ is a cross-partition. Writing

$$\mathcal{Q} = \{a_1 = x_0 < \cdots < x_J = b_1\} \otimes \cdots \otimes \{a_n = z_0 < \cdots < z_K = b_n\},$$

let $L \equiv \max\{J, \ldots, K\} - 1$. Thus, along the x_j-axis, $\mathcal{Q}$ is marked by a_j at one end, b_j at the other, and no more than L other points in between.

Write $M = \sup\{|f(\mathbf{x})| : \mathbf{x} \in B\}$. The fineness we require is

$$\delta \equiv \frac{\varepsilon}{8M\|\mathbf{b}-\mathbf{a}\|^{n-1}nL}.$$

Suppose $\mathcal{P}$ is any partition with $\|\mathcal{P}\| < \delta$. Then the refinement $\mathcal{PQ}$ is produced by breaking the subintervals of $\mathcal{P}$ at the hyperplanes $x = x_1, \ldots, x = x_{J-1}$, $y = y_1, \ldots, y = y_{N-1}, \ldots, z = z_1, \ldots, z = z_{K-1}$, a total of at most nL cuts. By extension of Theorem 1(c),

$$u(f, \mathcal{P}) - u(f, \mathcal{PQ}) \le nL\,4M\|\mathbf{b}-\mathbf{a}\|^{n-1}\|\mathcal{P}\| < \frac{\varepsilon}{2}.$$

Since $\mathcal{PQ}$ also refines $\mathcal{Q}$, we have $u(f, \mathcal{PQ}) \le u(f, \mathcal{Q})$. Hence

$$u(f, \mathcal{P}) < u(f, \mathcal{PQ}) + \frac{\varepsilon}{2} \le u(f, \mathcal{Q}) + \frac{\varepsilon}{2} < U(f, B) + \varepsilon. \qquad \square$$

Darboux's lemma allows us to connect our definitions of integrability and integral to the limit-of-sums definitions. Suppose $\mathcal{P} \equiv \{S_1, \ldots, S_J\}$ is a partition and $\mathbf{x}_1 \in S_1, \ldots, \mathbf{x}_J \in S_J$. The expression $f(\mathbf{x}_1)V(S_1) + \cdots + f(\mathbf{x}_J)V(S_J)$ is called a **Riemann sum for f over $\mathcal{P}$**. From $m(S_j) \le f(\mathbf{x}_j) \le M(S_j)$, it is clear that every Riemann sum is between its partition's upper and lower sums. Therefore, if f is integrable, then its Riemann sums are, like the upper and lower sums, forced by $\|\mathcal{P}\| \to 0$ toward the integral. Our next result provides the details, along with the converse.

(The name "Darboux's lemma" was applied by the late Jesse Douglas, a famous mathematician and extraordinary teacher. It is not in standard use, but accords with the account [Kline, p. 959] of Darboux's contribution to Riemann's theory.)

Theorem 4. *A function is integrable iff its Riemann sums have a limit as the partition norm tends to zero, in which case this limit is the integral.*

Proof. $\Rightarrow$ Assume that f is integrable, with $C \equiv U(f, B) = L(f, B)$, and let $\varepsilon > 0$.

By Theorem 3, there exists δ_1 with $\|\mathcal{P}\| < \delta_1 \Rightarrow u(f, \mathcal{P}) < C + \varepsilon$, along with δ_2 having $\|\mathcal{P}\| < \delta_2 \Rightarrow l(f, \mathcal{P}) > C - \varepsilon$. Set $d \equiv \min\{\delta_1, \delta_2\}$. If $\mathcal{P}$ is any partition with $\|\mathcal{P}\| < \delta$ and $\sigma \equiv f(\mathbf{x}_1)V(S_1) + \cdots + f(\mathbf{x}_J)V(S_J)$ any Riemann sum for f over $\mathcal{P}$, then

$$C - \varepsilon < l(f, \mathcal{P}) \le \sigma \le u(f, \mathcal{P}) < C + \varepsilon.$$

We have proved that $\lim_{\|\mathcal{P}\| \to 0} \sigma = C$.

$\Leftarrow$ Assume that for each ε, there is $\delta(\varepsilon)$ such that whenever $\|\mathcal{P}\| < \delta(\varepsilon)$, all of the Riemann sums over $\mathcal{P}$ are within $\frac{\varepsilon}{4}$ of the limit C. We note that this forces

f to be bounded; if f had arbitrarily large values, then you could define large Riemann sums from arbitrarily fine partitions. Write V for the volume of B.

Fix $\varepsilon > 0$, and let $\mathcal{P} \equiv \{S_1, \ldots, S_J\}$ be one partition with $\|\mathcal{P}\| < \delta(\varepsilon)$. In the subinterval S_j, $M(S_j) = \sup\{f(\mathbf{x}) : \mathbf{x} \in S_j\}$ means that there is $\mathbf{x}_j \in S_j$ with $f(\mathbf{x}_j) > M(S_j) - \frac{\varepsilon}{4V}$. Thus,

$$\begin{aligned}\sigma &\equiv f(\mathbf{x}_1)V(S_1) + \cdots + f(\mathbf{x}_J)V(S_J)\\ &> \left[M(S_1) - \frac{\varepsilon}{4V}\right]V(S_1) + \cdots + \left[M(S_J) - \frac{\varepsilon}{4V}\right]V(S_J)\\ &= u(f, \mathcal{P}) - \frac{\varepsilon}{4}.\end{aligned}$$

Since $C - \frac{\varepsilon}{4} < \sigma < C + \frac{\varepsilon}{4}$, we have $u(f, \mathcal{P}) < \sigma + \frac{\varepsilon}{4} < C + \frac{\varepsilon}{2}$.

Similarly, we can find a Riemann sum ρ with $\rho < l(f, \mathcal{P}) + \frac{\varepsilon}{4}$. Since ρ is also between $C - \frac{\varepsilon}{4}$ and $C + \frac{\varepsilon}{4}$, we have $l(f, \mathcal{P}) > \rho - \frac{\varepsilon}{4} > C - \frac{\varepsilon}{2}$.

Beginning with $\varepsilon > 0$, we have found a partition $\mathcal{P}$ with

$$u(f, \mathcal{P}) - l(f, \mathcal{P}) < C + \frac{\varepsilon}{2} - \left(C - \frac{\varepsilon}{2}\right) = \varepsilon.$$

By Theorem 2(a), f is integrable on B. □

Exercises

1. Show that $f(\mathbf{x}) \equiv 1$ is integrable on any box B and that $\int_B f$ is the volume of B.

2. Let $g(x, y) \equiv x$ and $\mathcal{P} \equiv \{0 = x_0 < \cdots < x_J = 3\} \otimes \{0 = y_0 < \cdots < y_K = 4\}$ be a cross-partition of the box from $(0, 0)$ to $(3, 4)$.

   (a) Find an expression for $u(g, \mathcal{P})$.

   (b) Find the limit as $\|\mathcal{P}\| \to 0$ of the expression in (a).

   (c) Sketch the graph of $z = g(x, y)$ and interpret the result in (b) geometrically.

3. Suppose $h(x, y) \equiv 1$ if there is an integer $k \geq 1$ with $\frac{1}{2^k} \leq x \leq \frac{1}{2^{k-1}}$ and $\frac{1}{2^k} \leq y \leq \frac{1}{2^{k-1}}$, $h(x, y) \equiv 0$ otherwise. Show that h is integrable on the unit square, and find its integral.

4. Prove that integrals in B are:

   (a) Linear: If f and g are integrable, then so is any combination $\alpha f + \beta g$, and

$$\int_B (\alpha f + \beta g) = \alpha \int_B f + \beta \int_B g.$$

(b) Function-monotonic: If f and g are integrable and $f \geq g$ throughout B, then

$$\int_B f \geq \int_B g.$$

(c) Partition-additive: Let $\mathcal{P} \equiv \{S_1, \ldots, S_J\}$ be a partition of B; then f is integrable on B iff f is integrable on each S_j, and $\int_B f = \int_{S_1} f + \cdots + \int_{S_J} f$.

(d) Box-monotonic: If B is a subset of box D and $f \geq 0$ is integrable on D, then f is integrable on B and $\int_B f \leq \int_D f$.

(e) Bounded operators: If f is integrable on B, then

$$\inf\{f(\mathbf{x}) : \mathbf{x} \in B\} V(B) \leq \int_B f \leq \sup\{f(\mathbf{x}) : \mathbf{x} \in B\} V(B).$$

5. Use any of the theorems or exercises from Sections 1 and 2 (only) to find $\int(\alpha x + \beta y + \delta z)$ on the box from $\mathbf{a} \equiv (a_1, a_2, a_3)$ to $\mathbf{b} \equiv (b_1, b_2, b_3)$.

6. Prove that Dirichlet's function is not integrable on any box.

7. (a) Show that if f is integrable on B, then so is $|f|$.

   (b) Show that the converse of (a) is false.

8. Prove the "average value theorem": If f is continuous on B, then there exists $\mathbf{c} \in B$ such that

$$\int_B f = f(\mathbf{c}) V(B).$$

9. Prove Theorem 1(e).

4.3 Domains of Integrability

We introduced integrability and integrals on boxes, with the intention of advancing to more general sets. In this section we begin the advance by extending the notions to sets exhibiting a special kind of good behavior.

To avoid inconveniences related to domains, we will deal with functions that are defined everywhere. Accordingly, if f is defined on a domain D, then we extend f to a new function (with the same name) defined on $\mathbf{R}^n$ by setting $f(\mathbf{x}) \equiv 0$ for $\mathbf{x} \notin D$.

Let A be a fixed bounded set, and assume that f is bounded on A. Given that we can integrate on boxes, there is a natural way to define $\int_A f$. Suppose B is a box containing A. Define f_{AB} by

$$f_{AB}(\mathbf{x}) \equiv \begin{cases} f(\mathbf{x}) & \text{if } \mathbf{x} \in A, \\ 0 & \text{if } \mathbf{x} \in B - A. \end{cases}$$

This new function is bounded on B. If f_{AB} is integrable on B, then $\int_A f$ should match $\int_B f_{AB}$. This description is a standard mathematical device, and it has a standard weakness: It seems to depend on B. To employ it as a definition, we must show that B does not introduce any ambiguity.

Theorem 1. *Suppose B and C are two boxes containing A. Then f_{AB} is integrable on B iff f_{AC} is integrable on C, in which case $\int_B f_{AB} = \int_C f_{AC}$.*

Proof. Suppose $B = [\mathbf{a}, \mathbf{b}]$ and $C = [\mathbf{c}, \mathbf{d}]$. We have seen (Exercises 4.1:2a, b) that $B \cap C$ is flat or empty (collectively, "degenerate"), or else it is a box.

If $B \cap C$ is degenerate, then it and A are subsets of the boundary of B, in fact, of just one wall of B (Exercise 1). Therefore (Exercise 2), f_{AB} is integrable on B, and $\int_B f_{AB} = 0$. Similarly, f_{AC} is integrable on C, with zero integral, and the conclusion is established.

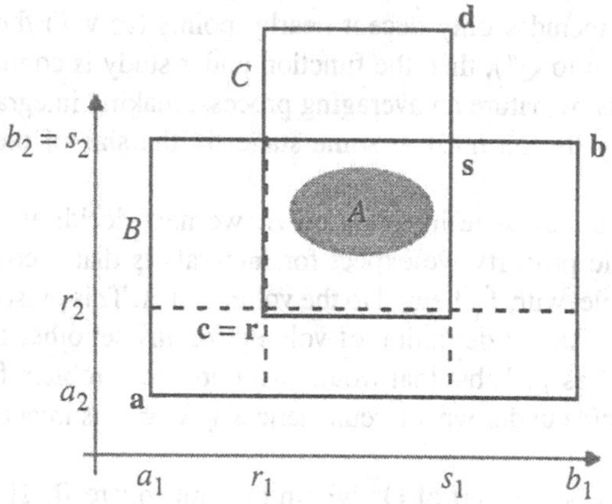

Figure 4.4.

Suppose instead that $B \cap C$ is not degenerate, say $B \cap C = [\mathbf{r}, \mathbf{s}]$, as illustrated in Figure 4.4. Then $B \cap C$ is one subinterval in the cross-partition

$$\mathcal{P} \equiv \{a_1 \le r_1 < s_1 \le b_1\} \otimes \cdots \otimes \{a_n \le r_n < s_n \le b_n\}$$

(dashed lines in the figure) of B. By partition additivity (Exercise 4.2:4c), f_{AB} is integrable on B iff it is integrable on all the subintervals of $\mathcal{P}$. Now, $A \subseteq B \cap C$, so A does not overlap any of the other subintervals. Consequently, $f_{AB} = 0$ in the interiors of those others, and (Exercise 2 again) it is automatic that f_{AB} is integrable and has zero integral on the others. Hence f_{AB} is integrable on B iff its restriction $f_{A(B\cap C)}$ to $B \cap C$ is integrable on $B \cap C$, in which case $\int_B f_{AB} = \int_{B\cap C} f_{A(B\cap C)}$.

By similar reasoning, we establish that f_{AC} is integrable on C iff $f_{A(B\cap C)}$ is integrable on $B \cap C$, with

$$\int_C f_{AC} = \int_{B\cap C} f_{A(B\cap C)}.$$

The desired conclusion follows. □

A useful construct is the function that tells us whether a vector is in a set. For any set S, the **characteristic function** χ_S is defined by $\chi_S(\mathbf{x}) \equiv 1$ if $\mathbf{x} \in S$, $\equiv 0$ if $\mathbf{x} \notin S$. We have already met $\chi_{\mathbf{Q}^n}$, which is Dirichlet's function. If f is defined on $\mathbf{R}^n$, then $f\chi_A$ is likewise, and $f\chi_A$ has precisely the values we want in defining $\int_A f$.

Definition. We say that f **is integrable on** A if $f\chi_A$ is integrable on every box containing A, and set $\int_A f$ to be the single value $\int_B f\chi_A$ has on all such boxes.

It is worth noting the different natures of integrability and continuity. We have already said that continuity is a local property. You study it in terms of arbitrarily small regions. If there are no nearby members (an isolated point) of the domain, or if the domain includes only decent nearby points (as with the restriction of Dirichlet's function to $\mathbf{Q}^n$), then the function under study is continuous. In contrast, integration is by nature an averaging process, making integrability a global property. To study it, you must at some stage fix the size of the region under consideration.

Knowing how to calculate integrals on A, we next decide whether we want to. The most basic property we expect for integrals is that a constant function should be integrable, with $\int_A 1$ equal to the volume of A. This presents a difficulty, because we do not have a definition of volume for any set other than a box. We could define $V(A)$ as $\int_A 1$, but that would just trade one problem for another; we would need to decide under what circumstances $g(\mathbf{x}) \equiv 1$ is integrable on A.

Example 1. Let A be the part of $\mathbf{Q}^2$ within the unit square B. Then $1\chi_A = \chi_A$, matching Dirichlet's function on B, is not integrable on B. Hence 1 is not integrable on A.

We will treat the two questions—volume and integrability—together geometrically. We call, naturally, on the method of the greatest geometer of them all.

In Figure 4.5, we see an irregular bounded set A (shaded), contained in a box B, to which a partition $\mathcal{P}$ has been applied. Some subintervals (slanting lines) contain only interior points of A. These are subsets of A's interior, which we denote by int(A). Some (at lower right) are entirely exterior to A; they are subsets of the exterior, hereafter ext(A). How should we describe the others? The others are precisely those with some points from the boundary of A, written bd(A). That is, if S is a subinterval with boundary points, then S is not a subset of either int(A) or ext(A). Conversely, if S is not contained in int(A) or ext(A), then

$$S = [S \cap \operatorname{int}(A)] \cup [S \cap \operatorname{ext}(A)]$$

is impossible. If that equation were true, then it would describe a disconnection of S, which is an impossibility [Guzman, Sections 5.4 and 4.5]. Accordingly, S must

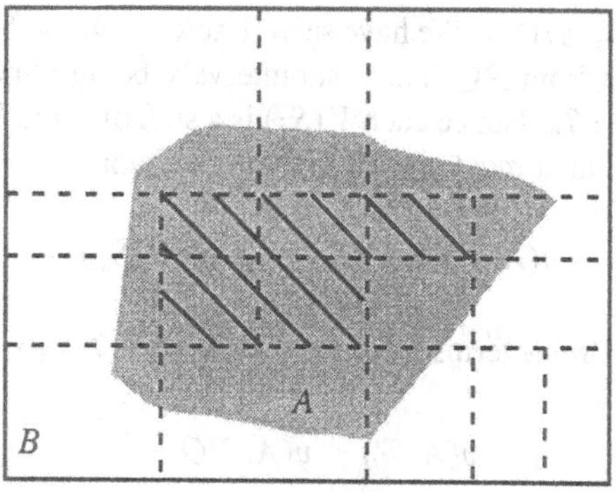

Figure 4.5.

have points that are neither interior nor exterior; it must have boundary points of A.

Look at the sum $v(A, \mathcal{P})$ of the volumes of the subintervals interior to A. This sum could be zero: There might be no such intervals by accident, because the subintervals of $\mathcal{P}$ are misplaced or too big, or by necessity, because A has empty interior. The supremum of all such $v(A, \mathcal{P})$, taken over all partitions $\mathcal{P}$ of all boxes B containing A, is the **inscribed volume** $v^*(A)$ of A. Observe that the boundedness of A guarantees that this supremum is finite. At the other extreme, if A has empty interior, then every $v(A, \mathcal{P}) = 0$, forcing $v^*(A) = 0$; the converse is Exercise 4.

Examine next the boxes that are not exterior to A. These encompass the interior ones, together with those that have points of $\mathrm{bd}(A)$. We may describe these as the ones that intersect the closure of A, for which we write $\mathrm{cl}(A)$. (Compare Exercise 4.4:15(c).) Unless A is empty, there must be some such subintervals, making the sum $V(A, \mathcal{P})$ of their volumes positive; and clearly $V(A, \mathcal{P}) \le V(B)$. The infimum of all such $V(A, \mathcal{P})$ is the **circumscribed volume** $V^*(A)$ of A.

Inscribed and circumscribed volumes are close cousins to lower and upper sums. We could write a series of results corresponding to Theorems 4.2:1 and 4.2:3. Instead, we isolate one that is especially significant for us.

Theorem 2. *If $\mathcal{P}$ and $\mathcal{Q}$ are partitions that cover A, then the intersection-partition $\mathcal{PQ}$ (defined in Section 4.1) satisfies*

$$v(A, \mathcal{P}) \le v(A, \mathcal{PQ}) \le V(A, \mathcal{PQ}) \le V(A, \mathcal{Q}).$$

In particular, a set's inscribed volume does not exceed the circumscribed volume.

Proof. Say $\mathcal{P}$ partitions box B, $\mathcal{Q}$ partitions C, and $D \equiv B \cap C$. We may assume that D is not degenerate. Otherwise, A has empty interior, $v(A, \mathcal{P}) = v(A, \mathcal{PQ}) = V(A, \mathcal{PQ}) = 0$, and the inequalities are trivial.

Suppose $S_1, \ldots, S_J$ are the subintervals from $\mathcal{P}$ that are interior to A, and $T_1, \ldots, T_K$ are the interior ones from $\mathcal{PQ}$. Because $S_j \subseteq A$, each S_j is also a subset of D, and $S_j = S_j \cap D$. We have seen (Exercises 4.1:4, 5) that $S_j \cap D$ is the union of subintervals from $\mathcal{PQ}$. These subintervals, being of necessity interior to A, must be among the T_k. Hence each $V(S_j)$ is a sum of some $V(T_k)$, and a given $V(T_k)$ is a summand in at most one $V(S_j)$. In other words,

$$v(A, \mathcal{P}) \equiv V(S_1) + \cdots + V(S_J)$$

is a rearrangement of *some* terms from $v(A, \mathcal{PQ}) = V(T_1) + \cdots + V(T_K)$. Therefore,

$$v(A, \mathcal{P}) \le v(A, \mathcal{PQ}).$$

$v(A, \mathcal{PQ}) \le V(A, \mathcal{PQ})$ is trivial, since every term on the left appears also on the right.

In a kind of mirror image to the argument two paragraphs back, suppose $T_1, \ldots, T_M$ $(M \ge K)$ are the subintervals from $\mathcal{PQ}$ that touch the closure of A, and $U_1, \ldots, U_L$ the ones from $\mathcal{Q}$. Each T_m is a subset of a unique subinterval from $\mathcal{Q}$, and that subinterval, having points of cl(A), must be one of the U_l. For a given U_l, $U_l \cap D$ is the union of all the subintervals from $\mathcal{PQ}$ contained in U_l. Hence $V(U_1 \cap D) + \cdots + V(U_L \cap D)$ can be broken up into a sum whose terms include all of $V(T_1), \ldots, V(T_M)$, and

$$\begin{aligned} V(A, \mathcal{Q}) &\equiv V(U_1) + \cdots + V(U_L) \\ &\ge V(U_1 \cap D) + \cdots + V(U_L \cap D) \\ &\ge V(T_1) + \cdots + V(T_M) = V(A, \mathcal{PQ}). \end{aligned}$$

With the other inequalities established, it is clear that

$$v^*(A) \equiv \sup v(A, \mathcal{P}) \le \inf V(A, \mathcal{Q}) \equiv V^*(A). \qquad \square$$

Definition. The bounded set A is **Archimedean** if $v^*(A) = V^*(A)$, in which event we define the **volume** of A to be their common value.

Example 2. Every box is Archimedean, and its volume under the new definition is, as before, the product of its dimensions.

Write $B \equiv [a, b]$ and $V \equiv (b_1 - a_1) \cdots (b_n - a_n)$. Let $m \ge 3$ be an integer, and look at the cross-partition $\mathcal{R}$ whose markers are

$$\left\{a_j, \left(1 - \frac{1}{m}\right) a_j + \left(\frac{1}{m}\right) b_j, \left(\frac{1}{m}\right) a_j + \left(1 - \frac{1}{m}\right) b_j, b_j\right\}$$

along the x_j-axis. The big box C (dashed border in Figure 4.6) in the middle of

B is the only interior subinterval. It is $\left(1-\frac{2}{m}\right)$ times the length of B in every direction, making its volume (by the old rules) $(1-\frac{2}{m})^n V$. Hence

$$v^*(B) \geq v(B, \mathcal{R}) = \left(1 - \frac{2}{m}\right)^n V.$$

This being true for arbitrary m, we conclude that $v^*(B) \geq V$.

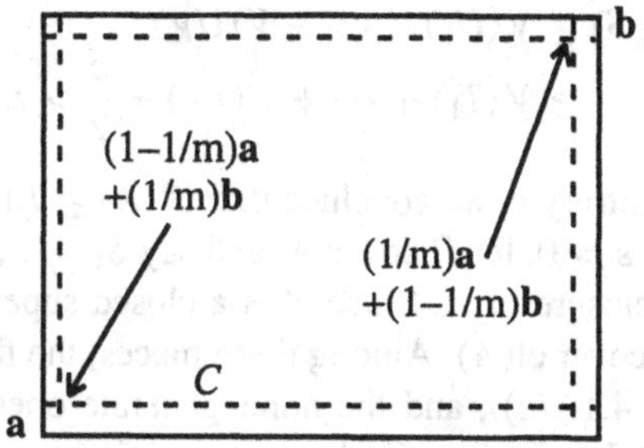

Figure 4.6.

At the same time, $V^*(B) \leq V(B, \mathcal{R}) = V$. Theorem 2 forces $v^*(B) = V^*(B) = V$, so that the box is Archimedean and has volume $(b_1 - a_1)\cdots(b_n - a_n)$.

The "big-box technique" just used—filling a given box with a concentric one entirely interior to it and using up most of the oxygen—is a terrific weapon. Given any $\varepsilon > 0$, we can make the interior box fill up all but ε of the volume of B by taking m sufficiently large. We will call several times on this construction.

In view of Example 2, there is no clash between the new and the familiar notions of volume for a box, and we may use the notation $V(A)$ for the volume of any Archimedean set A.

Theorem 3. *Let B be any box containing A.*

(a) *$v^*(A)$ is the lower integral of χ_A on B.*

(b) *$V^*(A)$ is the upper integral of χ_A on B.*

(c) *A is Archimedean iff $g(\mathbf{x}) \equiv 1$ is integrable on A, in which case $\int_A g = V(A)$.*

Proof. (a) Let $\mathcal{P}$ be a partition (not necessarily of B) whose subintervals cover A, and suppose $S_1, \ldots, S_J$ are the interior subintervals. Since $A \subseteq B$, we also have each $S_j \subseteq B$. Let $\mathcal{Q}$ be some partition of B that includes $S_1, \ldots, S_J$ among its subintervals. Then χ_A is identically 1 on each S_j, so that

$$l(\chi_A, \mathcal{Q}) \geq 1V(S_1) + \cdots + 1V(S_J) = v(A, \mathcal{P}).$$

Hence $L(\chi_A, B) \geq v(A, \mathcal{P})$ independent of $\mathcal{P}$, and so $L(\chi_A, B) \geq v^*(A)$.

For the reverse, let $\varepsilon > 0$ be named. By definition of lower integral, there is a partition $\mathcal{R}$ of B with $l(\chi_A, \mathcal{R}) > L(\chi_A, B) - \frac{\varepsilon}{2}$. This lower sum is simply $1V(T_1) + \cdots + 1V(T_K)$, where $T_1, \ldots, T_K$ are the subintervals of $\mathcal{R}$ on which χ_A is identically 1. Of necessity, each T_k is a subset of A. Let U_k be a box interior to T_k having $V(U_k) \geq V(T_k) - \frac{\varepsilon}{2K}$. Then some partition $\mathcal{S}$ includes the U_k, and perhaps other boxes, among its subintervals interior to A. Hence

$$\begin{aligned} v^*(A) \geq v(A, \mathcal{S}) &\geq V(U_1) + \cdots + V(U_K) \\ &\geq V(T_1) + \cdots + V(T_K) - \frac{\varepsilon}{2} > L(\chi_A, B) - \varepsilon. \end{aligned}$$

This being true for arbitrary ε, we conclude that $v^*(A) \geq L(\chi_A, B)$.

(b) Beginning with $\varepsilon > 0$, let $\mathcal{P}$ cover A, and say $S_1, \ldots, S_J$ are the subintervals intersecting the closure of A. Since B is a closed superset of A, the pieces $B \cap S_1, \ldots, B \cap S_J$ cover $\mathrm{cl}(A)$. Among these pieces, the flat ones lie along the walls of B (Exercise 4.1:2(c)), and the nondegenerate ones are subintervals of some partition $\mathcal{Q}$ of B. Let $T_1, \ldots, T_K$ be the remaining subintervals of $\mathcal{Q}$. These can have points of A on their boundaries only. Build boxes U_k interior to T_k having $V(U_k) \geq V(T_k) - \frac{\varepsilon}{K}$. Then every U_k is disjoint from (in fact, exterior to) A. If now $\mathcal{R}$ is a partition of B with the U_k as subintervals, then

$$\begin{aligned} V(A, \mathcal{P}) \equiv V(S_1) + \cdots + V(S_J) &\geq V(B \cap S_1) + \cdots + V(B \cap S_J) \\ &= V(B) - [V(T_1) + \cdots + V(T_K)] \\ &\geq V(B) - [V(U_1) + \cdots + V(U_K) + \varepsilon] \\ &\geq u(A, \mathcal{R}) - \varepsilon \geq U(\chi_A, B) - \varepsilon. \end{aligned}$$

It follows that $V^*(A) \geq U(\chi_A, B)$.

Start over with $\varepsilon > 0$ specified. Let $\mathcal{S} \equiv \{W_1, \ldots, W_L\}$ partition B such that

$$u(\chi_A, \mathcal{S}) < U(\chi_A, B) + \frac{\varepsilon}{2}.$$

If $W_1, \ldots, W_M$ are the subintervals of $\mathcal{S}$ in which there is at least one point of A, then

$$u(\chi_A, \mathcal{S}) = 1V(W_1) + \cdots + 1V(W_M).$$

In each of W_{M+1} to W_L, put a big box X_m such that $V(X_m) \geq V(W_m) - \frac{\varepsilon}{2L}$. Then the X_m are exterior to A. If $\mathcal{T}$ is a partition of B comprising the X_m, then

$$\begin{aligned} V(W_1) + \cdots + V(W_M) &= V(B) - \left[V(W_{M+1}) + \cdots + V(W_L)\right] \\ &\geq V(B) - \left[V(X_{M+1}) + \cdots + V(X_L) + \frac{\varepsilon}{2}\right] \\ &\geq V(A, \mathcal{T}) - \frac{\varepsilon}{2} \geq V^*(A) - \frac{\varepsilon}{2}. \end{aligned}$$

Hence $U(\chi_A, B) + \frac{\varepsilon}{2} > V^*(A) - \frac{\varepsilon}{2}$. It follows that $U(\chi_A, B) \geq V^*(A)$.

(c) is trivial from (a), (b), and the definitions. □

Exercises

1. Show that if B and C are boxes and $B \cap C$ has points from two different walls of B, that is, $[B \cap C] \cap \text{bd}(B)$ is not a subset of one wall of B, then B and C overlap.

2. Suppose B is a box, and the function f is zero at every interior point of B and bounded on the walls. Show that f is integrable on B and $\int_B f = 0$.

3. Use the definition to prove these sets Archimedean, and find their volumes:

   (a) the finite set $\{\mathbf{x}_1, \ldots, \mathbf{x}_m\} \subseteq \mathbf{R}^n$.

   (b) the hyperrectangle $\{\mathbf{x}\colon x_1 = a_1, a_2 \le x_2 \le b_2, \ldots, a_n \le x_n \le b_n\}$ in $\mathbf{R}^n$.

   (c) the right triangle with vertices at $(0, 0)$, $(a, 0)$, and $(0, b)$ in $\mathbf{R}^2$.

4. Show that if the inscribed volume of A is zero, then $\text{int}(A)$ is empty.

5. (a) Give an example of a bounded set with empty interior but not zero volume.

   (b) Show that if a set is Archimedean and has empty interior, then it has zero volume.

6. Prove that if A has zero volume, then an arbitrary function f that is bounded on A is integrable over A, and $\int_A f = 0$.

7. Show that characteristic functions satisfy:

   (a) $\chi_{A \cap B} = \chi_A \chi_B$.

   (b) $\chi_{A \cup B} = \chi_A + \chi_B - \chi_A \chi_B$.

   (c) A is bounded iff $\chi_A(\mathbf{x}) \to 0$ as $\mathbf{x} \to \infty$.

8. For an arbitrary set S:

   (a) At what points does χ_S have a limit?

   (b) At what points is χ_S continuous?

4.4 Integrability and Sets of Zero Volume

We have decided that Archimedean sets are the domains where integrability is a worthwhile notion. Right now, our supply of examples is limited: We know that boxes are Archimedean, with continuous functions integrable there; and Archimedean sets of zero volume have all bounded functions integrable on them. In this section we characterize Archimedean sets and show that the category includes familiar regions. We also show that familiar functions are integrable, and discuss the question of characterizing integrable functions.

The discussion leading to definitions of inscribed volume (hereafter **involume**) and circumscribed volume (**circumvolume**) makes clear that the difference between them has to do with how much space is occupied by the boundary of the set. Our first theorem formalizes the idea that "Archimedean" amounts to "thin-skinned."

Theorem 1. *A bounded set is Archimedean iff its boundary has zero circumvolume.*

Proof. $\Rightarrow$ Assume that A is Archimedean, with volume V. Name $\varepsilon > 0$. Some partition gives interior volume above $V - \frac{\varepsilon}{2}$; a second gives closure volume under $V + \frac{\varepsilon}{2}$; and by Theorem 4.3:2 their intersection-partition $\mathcal{P}$ has

$$V - \frac{\varepsilon}{2} < v(A, \mathcal{P}) \leq V(A, \mathcal{P}) < V + \frac{\varepsilon}{2}.$$

Suppose $S_1, \ldots, S_J$ are the subintervals of $\mathcal{P}$ interior to A, and $S_{J+1}, \ldots, S_K$ are the additional ones intersecting $\mathrm{cl}(A)$. In other words, $S_{J+1}, \ldots, S_K$ are the only subintervals of $\mathcal{P}$ touching $\mathrm{bd}(A)$, which is its own closure (Why?). By definition,

$$V\big(\mathrm{bd}(A), \mathcal{P}\big) = V(S_{J+1}) + \cdots + V(S_K) = V(A, \mathcal{P}) - v(A, \mathcal{P}) < \varepsilon.$$

Consequently, $V^*\big(\mathrm{bd}(A)\big) < \varepsilon$, leading to $V^*\big(\mathrm{bd}(A)\big) = 0$.

$\Leftarrow$ Assume that $D \equiv \mathrm{bd}(A)$ has zero circumvolume. Take some box B containing A. Necessarily, $D \subseteq \mathrm{cl}(A) \subseteq B$.

Name $\varepsilon > 0$. By Theorem 4.3:3(b), χ_D has zero upper integral on B. Hence there is a partition $\mathcal{Q}$ of B in which $u(\chi_D, \mathcal{Q}) < \varepsilon$. Each subinterval of $\mathcal{Q}$ either has or lacks points of D. A subinterval that does not intersect D must be all interior or all exterior to A (second paragraph following Example 4.3:1). Hence the subintervals fall into three groups: $S_1, \ldots, S_J$, made up entirely of points from $\mathrm{int}(A)$; $T_1, \ldots, T_K$, which are subsets of $\mathrm{ext}(A)$; and $U_1, \ldots, U_M$, in which there are boundary points. We see that

$$\begin{aligned} u(\chi_D, \mathcal{Q}) &= 1V(U_1) + \cdots + 1V(U_M), \\ V(A, \mathcal{Q}) &= V(S_1) + \cdots + V(S_J) + V(U_1) + \cdots + V(U_M), \\ v(A, \mathcal{Q}) &= V(S_1) + \cdots + V(S_J). \end{aligned}$$

Thus, $V(A, \mathcal{Q}) - v(A, \mathcal{Q}) = u(\chi_D, \mathcal{Q}) < \varepsilon$. We conclude that $V^*(A) \leq v^*(A)$, and A is Archimedean. □

Notice that zero circumvolume forces zero involume. Any set of zero circumvolume is therefore Archimedean and has zero volume. Hence if A is Archimedean, then so is its boundary. Is the converse true? (Exercise 1)

We will deal enough with sets of zero volume to make it worthwhile to give the property a name. There is no standard name for it, so we choose among appropriate synonyms: We say that D is **meager** if $V^*(D) = 0$.

Theorem 2.

(a) *A meager set is one that can be packed into small boxes. In symbols: The bounded set D is meager iff for each* $\varepsilon > 0$, *there is a class* $\{C_1, \ldots, C_J\}$ *of boxes that covers D and has* $V(C_1) + \cdots + V(C_J) < \varepsilon$.

(b) *A finite union of meager sets is meager.*

Proof. (a) $\Rightarrow$ Assume that D has zero volume. For $\varepsilon > 0$, there exists a partition $\mathcal{P}$ with $V(D, \mathcal{P}) < \varepsilon$, and $V(D, \mathcal{P})$ is precisely a sum of volumes of boxes that cover D.

$\Leftarrow$ Suppose, conversely, that each $\varepsilon > 0$ yields a cover $\{C_1, \ldots, C_J\}$ with volume-sum under $\frac{\varepsilon}{2}$. Let $\mathcal{Q}$ be the cross-partition created by $C_1, \ldots, C_J$ on a box B. By our familiar argument, the subintervals $T_1, \ldots, T_K$ of $\mathcal{Q}$ contained in $C_1 \cup \ldots \cup C_J$ have

$$V(T_1) + \cdots + V(T_K) \leq V(C_1) + \cdots + V(C_J) < \frac{\varepsilon}{2};$$

the remaining subintervals can be filled with big boxes exterior to D and having volume-sum at least

$$V(B) - [V(T_1) + \cdots + V(T_K)] - \frac{\varepsilon}{2}\,.$$

Any partition $\mathcal{R}$ comprising the big boxes has $V(D, \mathcal{R}) = V(T_1) + \cdots + V(T_K) + \frac{\varepsilon}{2} < \varepsilon$. It follows that $V^*(D) = 0$.

(b) Exercise 2.

Example 1. Every smooth hypersurface is meager (has meager range).

Recall that by smooth hypersurface, we mean a mapping $\mathbf{F}$ into $\mathbf{R}^n$ from box $B \equiv [\mathbf{r}, \mathbf{s}] \subseteq \mathbf{R}^{n-1}$ such that the Jacobian matrix $\frac{\partial \mathbf{F}}{\partial \mathbf{t}} = \mathbf{F}'$ is continuous and has rank $n-1$ at every $\mathbf{t}$. In this argument the rank is inessential. Only the continuity is needed; it guarantees that $\mathbf{F}'$ is bounded, say $\|\mathbf{F}'(\mathbf{t})\| = M$ for $\mathbf{t} \in B$.

Let k be a natural number, and partition B into k^{n-1} congruent subintervals. Suppose $\mathbf{p}$ is the center of a typical subinterval S, and $\mathbf{q} \in S$. By the mean value theorem, each component F_j satisfies

$$\begin{aligned} |F_j(\mathbf{q}) - F_j(\mathbf{p})| &= \left|\frac{\partial F_j}{\partial \mathbf{t}}(\mathbf{q}_j)\langle \mathbf{q} - \mathbf{p}\rangle\right| \qquad (\text{for some } \mathbf{q}_j \in S) \\ &\leq \left\|\frac{\partial \mathbf{F}}{\partial \mathbf{t}}(\mathbf{q}_j)\right\| \|\mathbf{q} - \mathbf{p}\| \leq \frac{M\|\mathbf{s} - \mathbf{r}\|}{2k}\,. \end{aligned}$$

That is, $\mathbf{F}(\mathbf{q})$ lies within the box in $\mathbf{R}^n$ centered at $\big(F_1(\mathbf{p}), \ldots, F_n(\mathbf{p})\big)$ having equal sides $\frac{M\|\mathbf{s}-\mathbf{r}\|}{k}$. Hence the image of S is contained in a box of volume

$\left(\frac{M\|\mathbf{s}-\mathbf{r}\|}{k}\right)^n$. Taking all the subintervals into account, we find that the entire hypersurface is covered by a union of boxes with volume-sum

$$k^{n-1}\left(\frac{M\|\mathbf{s}-\mathbf{r}\|}{k}\right)^n = O\left(\frac{1}{k}\right).$$

Since k is arbitrary, Theorem 2(a) tells us that the hypersurface is meager.

Now we see that every neighborhood is Archimedean. For a neighborhood, the boundary sphere can be given by a single continuously differentiable parametrization with singularities (Exercise 4) or as the union of smooth parametrizations (based, for example, on Theorem 2.6:2). For most regions we want to study, the boundaries are locally smooth hypersurfaces. That is, they are finite unions of smooth hypersurfaces, which unions are meager sets by Theorem 2(b) and Example 1.

Since neighborhoods are Archimedean, we might hope that all open sets would be.

Example 2. Let $(\mathbf{x}_i) \equiv ((r_i, s_i))$ be an enumeration of the members of $\mathbf{Q}^2$ interior to the unit square B. We will define a non-Archimedean open set by recursion.

Let a_1, b_1, c_1, d_1 be irrational numbers such that

$$0 < a_1 < r_1 < c_1 < 1, \qquad 0 < b_1 < s_1 < d_1 < 1,$$

and

$$(c_1 - a_1)(d_1 - b_1) < \frac{1}{4}.$$

In words: Pick points (a_1, b_1) to the lower left of $\mathbf{x}_1$ and (c_1, d_1) to the upper right, both having two irrational coordinates, so close to $\mathbf{x}_1$ that they are still in B and that their box $[(a_1, b_1), (c_1, d_1)]$ has area under $\frac{1}{4}$. Let O_1 be the interior of that box.

Next, $\mathbf{x}_2$ is not on $\text{bd}(O_1)$, because on that boundary every point has at least one irrational coordinate. Hence $\mathbf{x}_2 \in \text{int}(O_1)$ or $\mathbf{x}_2 \in \text{ext}(O_1)$. If it is interior, then we can fit a box $[(a_2, b_2), (c_2, d_2)]$ around it, entirely interior to the first one, having area under $\frac{1}{8}$, the corner coordinates irrational. If exterior, then we can fit $[(a_2, b_2), (c_2, d_2)]$ so that it is exterior to the first box, still contained in B, having area under $\frac{1}{8}$. Then we make O_2 the interior of the second box.

Assume that $O_1, \ldots, O_k$ have been similarly picked. The next point $\mathbf{x}_{k+1}$ is not on the boundary of any of them, because every boundary point has at least one irrational coordinate. Therefore, $\mathbf{x}_{k+1}$ is at positive distance ε_{k+1} from the union of the boundaries (Reason?). Take $a_{k+1}, b_{k+1}, c_{k+1}, d_{k+1}$ irrational numbers between 0 and 1 such that

$$r_{k+1} - \frac{\varepsilon_{k+1}}{2} < a_{k+1} < r_{k+1} < c_{k+1} < r_{k+1} + \frac{\varepsilon_{k+1}}{2},$$

$$s_{k+1} - \frac{\varepsilon_{k+1}}{2} < b_{k+1} < s_{k+1} < d_{k+1} < s_{k+1} + \frac{\varepsilon_{k+1}}{2},$$

and

$$(c_{k+1} - a_{k+1})(d_{k+1} - b_{k+1}) < \frac{1}{2^{k+2}}.$$

Set $O_{k+1} \equiv \big((a_{k+1}, b_{k+1}), (c_{k+1}, d_{k+1})\big)$. Note that $\mathbf{x}_{k+1} \in O_{k+1}$, which has volume under $\frac{1}{2^{k+2}}$. Also, O_{k+1} does not reach the boundaries of $O_1, \dots, O_k$. If we let $O \equiv O_1 \cup O_2 \cup \cdots$, then it follows (Exercise 5) that the components (maximal connected subsets) of O constitute a subsequence $O_{k(1)}, O_{k(2)}, \dots$ of these same open boxes.

Now let $\mathcal{P}$ be a partition that covers O. First, the subintervals that intersect $\text{cl}(O)$ cover all of B (Why?). Hence $V(O, \mathcal{P}) \geq 1$, forcing $V^*(O) \geq 1$. (Does it equal 1?) Next, suppose $S_1, \dots, S_J$ are the subintervals interior to O. Each S_j is contained in a unique component $O_{m(j)}$ of O (Why?). We can rearrange $V(S_1) + \cdots + V(S_J)$ by components to write

$$\begin{aligned} V(S_1) + \cdots + V(S_J) &\leq V(O_{m(1)}) + \cdots + V(O_{m(K)}) \qquad \text{(for some } K \leq J) \\ &< V(O_1) + V(O_2) + \cdots < \frac{1}{4} + \frac{1}{8} + \cdots. \end{aligned}$$

Consequently, $v(O, \mathcal{P}) < \frac{1}{2}$, and $v^*(O) \leq \frac{1}{2}$. It follows that O is not Archimedean.

The existence of non-Archimedean open sets means that there are non-Archimedean closed sets (Exercise 8b). Thus, some fundamental sets are not amenable to integration.

We have just two more results involving only sets, without reference to functions.

Theorem 3. *Assume that A and B are Archimedean. Then:*

(a) *$A \cup B$, $A \cap B$, and $A - B$ are Archimedean.*

(b) *A and B overlap iff $V(A \cap B) > 0$.*

Proof. (a) By Theorem 1, we need to prove that these sets have meager boundaries. For the union and difference, we will show that the Venn diagram (see Figure 4.7) does not lie: The boundary of the union (solid border, half of it thick in Figure 4.7) is a subset of the union of the boundaries, and so is $\text{bd}(A - B)$ (solid thin border plus dashed thick). We leave $A \cap B$ to Exercise 6.

Assume that $\mathbf{x} \in \text{bd}(A \cup B)$. Then there is a sequence $(\mathbf{x}_i)$ converging to $\mathbf{x}$ from $A \cup B$ and a second one $(\mathbf{y}_i) \to \mathbf{x}$ from outside $A \cup B$. (Reason?) Either A or B contributes infinitely many of the terms $\mathbf{x}_i$; assume that it is A. Thus, some subsequence $\big(\mathbf{x}_{j(i)}\big)$ comes from A. Since every $\mathbf{y}_{j(i)}$ is outside A, we conclude that $\mathbf{x}$ is on $\text{bd}(A)$. Similarly, if $\big(\mathbf{x}_{j(i)}\big)$ comes from B, then $\mathbf{x} \in \text{bd}(B)$. Hence $\text{bd}(A \cup B) \subseteq \text{bd}(A) \cup \text{bd}(B)$. Because the individual boundaries must be meager, it follows (Theorem 2(b) and Exercise 3) that $\text{bd}(A \cup B)$ is likewise, and $A \cup B$ is Archimedean.

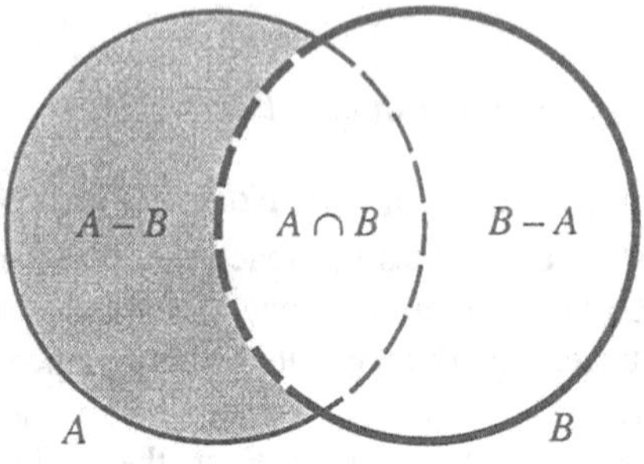

Figure 4.7.

Assume now that $\mathbf{y} \in \mathrm{bd}(A - B)$. There is one sequence $(\mathbf{v}_i)$ from $A - B$ and one $(\mathbf{w}_i)$ from outside $A - B$ converging to $\mathbf{y}$. A given $\mathbf{w}_i$ is outside $A - B$ for one of two reasons: it is not in A, or it is in A but also B. One of these two excuses must account for a subsequence $\left(\mathbf{w}_{k(i)}\right)$. If $\left(\mathbf{w}_{k(i)}\right)$ lies outside A, then $(\mathbf{v}_i) \to \mathbf{y}$ from A and $\left(\mathbf{w}_{k(i)}\right) \to \mathbf{y}$ from outside, meaning that $\mathbf{y} \in \mathrm{bd}(A)$. If instead $\left(\mathbf{w}_{k(i)}\right)$ lies within B, then $(\mathbf{v}_i) \to \mathbf{y}$ from outside B and $\left(\mathbf{w}_{k(i)}\right) \to \mathbf{y}$ from inside, meaning $\mathbf{y} \in \mathrm{bd}(B)$. We conclude that $\mathbf{y} \in \mathrm{bd}(A) \cup \mathrm{bd}(B)$, and, as with the union, $A - B$ is Archimedean.

(b) Exercise 7.

Creation of the concept of Archimedean set would be too much labor if the reward were limited to integrability of constants. It should come as no surprise that we also earn information about the functions that keep us employed.

Theorem 4. *Assume that f is bounded on the Archimedean set A. If the points of A where f is discontinuous (the "discontinuities" of f) constitute a meager set, then f is integrable on A.*

Proof. Let B be some box superset of A. Say $|f| \le M$ on A. With χ_A zero outside A, we have $|f\chi_A| \le M$ throughout B. We need to prove that $f\chi_A$ is integrable on B. To do so, we specify $\varepsilon > 0$ and find a lower sum within ε of an upper sum.

Let D comprise $\mathrm{bd}(A)$, together with the set of discontinuities. Being a union of two meager sets, D is itself meager. Hence there is a partition $\mathcal{P}$, whose box we may assume to be B, such that $V^*(D, \mathcal{P}) < \frac{\varepsilon}{4M}$. Let $S_1, \ldots, S_I$ be the subintervals of $\mathcal{P}$ that intersect D. These have $V^*(D, \mathcal{P}) = V(S_1) + \cdots + V(S_I)$. Any remaining subinterval, having no points from the boundary, cannot have points from both $\mathrm{int}(A)$ and $\mathrm{ext}(A)$. Separate these remaining subintervals into S_{I+1} to S_J, in which all the points are from $\mathrm{ext}(A)$, and T_1 to T_K, which have points from $\mathrm{int}(A)$ only.

On $T_1 \cup \cdots \cup T_K$, f is continuous. Hence f is uniformly continuous, guaranteeing that there exists δ such that $|f(\mathbf{x}) - f(\mathbf{y})| < \frac{\varepsilon}{2V(B)}$ whenever $\mathbf{x}$ and $\mathbf{y}$ are in this union and $\|\mathbf{x} - \mathbf{y}\| < \delta$. Break up every T_k into boxes with diagonal less than

δ, and label the boxes so produced S_{J+1} to S_L. The partition $\mathcal{Q} \equiv \{S_1, \ldots, S_L\}$ has the sums we need.

The difference $u(f\chi_A, \mathcal{Q}) - l(f\chi_A, \mathcal{Q})$ is a sum of terms of the form $[M_l - m_l]V(S_l)$, with M_l and m_l the supremum and infimum of $f\chi_A$ on S_l. On the subintervals S_1 to S_I, $|f\chi_A| \le M$ forces $M_l - m_l \le 2M$. Hence

$$\begin{aligned}[M_1 - m_1]V(S_1) + \cdots + [M_I - m_I]V(S_I) &\le 2M[V(S_1) + \cdots + V(S_I)] \\ &< 2M\left(\frac{\varepsilon}{4M}\right) = \frac{\varepsilon}{2}.\end{aligned}$$

On S_{I+1} to S_J, $f\chi_A$ is identically zero, and there is no contribution to the sum. Finally, on each of S_{J+1} to S_L, f is continuous. Hence M_l is the maximum value $f(\mathbf{c}_l)$; and similarly $m_l = f(\mathbf{d}_l)$. The diagonal of S_l is less than δ, so $\|\mathbf{c}_l - \mathbf{d}_l\| < \delta$. Therefore $|f(\mathbf{c}_l) - f(\mathbf{d}_l)| < \frac{\varepsilon}{2V(B)}$, and

$$\begin{aligned}&[M_{J+1} - m_{J+1}]V(S_{J+1}) + \cdots + [M_L - m_L]V(S_L) \\ &\qquad \le \left[\frac{\varepsilon}{2V(B)}\right][V(S_{J+1}) + \cdots + V(S_L)] \le \frac{\varepsilon}{2}.\end{aligned}$$

We have shown that $u(f\chi_A, \mathcal{Q}) - l(f\chi_A, \mathcal{Q}) < \varepsilon$, and the proof is complete.

There is some ambiguity in the phrase "points of A where f is discontinuous," but it does not harm the argument in Theorem 4. In our context, f is defined on $\mathbf{R}^n$, and it is natural to interpret the phrase to mean the points, among those where f is discontinuous, that happen to lie in A. Those points are described as follows:

$$\mathbf{x} \in A, \text{ and some sequence } (\mathbf{x}_i) \text{ has } (\mathbf{x}_i) \to \mathbf{x} \text{ but not } f(\mathbf{x}_i) \to f(\mathbf{x}).$$

The narrower interpretation would refer to the points at which the restriction of f to A is discontinuous. Its test would be more demanding:

$$\mathbf{x} \in A, \text{ and some sequence } (\mathbf{x}_i) \textit{ from } A \text{ has } (\mathbf{x}_i) \to \mathbf{x} \text{ but not } f(\mathbf{x}_i) \to f(\mathbf{x}).$$

Our usual understanding is the former; if not inventing counterexamples, we do not routinely deal with functions tailored to a particular set.

It would be nice if the criterion in Theorem 4 were necessary, as well as sufficient, for Riemann integrability. No such luck.

Example 3. We have given the uninspiring name **Modified Dirichlet function** to

$$G(\mathbf{x}) \equiv \begin{cases} \frac{1}{j_1 j_2 \cdots j_n} & \text{if } x = \left(\frac{i_1}{j_1}, \ldots, \frac{i_n}{j_n}\right) \in \mathbf{Q}^n, \\ 0 & \text{if } \mathbf{x} \notin \mathbf{Q}^n, \end{cases}$$

the rational numbers in the upper line understood to be in lowest terms. We will show that G is integrable on the unit square $B \subseteq \mathbf{R}^2$, much the way we argued that its discontinuities are precisely $\mathbf{Q}^n$ in $\mathbf{R}^n$ [Guzman, Example 3.4:3].

Fix an integer k. The finite array

$$\begin{array}{ccc} & \frac{0}{1} \quad \frac{1}{1} & \\ & \frac{0}{2} \quad \frac{1}{2} \quad \frac{2}{2} & \\ & \cdots & \\ \frac{0}{k}\,\frac{1}{k} & \cdots & \frac{(k-1)}{k}\,\frac{k}{k} \end{array}$$

has $2 + 3 + \cdots + (k+1) = \frac{(k+1)(k+2)}{2} - 1 < 3k^2$ entries. It lists every rational number from 0 to 1 having denominator k or less, obviously with repetitions. For each $\frac{i}{j}$ on the list, draw the two vertical lines at $x = \frac{i}{j} \pm \frac{1}{k^3}$, and paint the strip of B on and between the lines. Similarly, for $\frac{l}{m}$ on the list, paint the strip bordered by the horizontal lines $y = \frac{l}{m} \pm \frac{1}{k^3}$. The process produces fewer than $2\left(3k^2\right)$ strips, each of area $\frac{2}{k^3}$, some duplicated.

Evidently, vertical strips cross horizontal ones. However, no two vertical strips overlap, nor do two horizontal ones, if $k \geq 3$. The reason is that two unequal fractions $\frac{i}{j}$ and $\frac{p}{q}$ are separated by

$$\left|\frac{iq - jp}{jq}\right| \geq \frac{1}{jq} \geq \frac{1}{k^2} > \frac{2}{k^3},$$

so the strip holding $\frac{i}{j}$ cannot reach the one holding $\frac{p}{q}$. Consequently, the process results in alternating painted (equally wide) and unpainted (varying widths) strips horizontally, and the same thing vertically.

The lines create a cross-partition $\mathcal{P}$. On any subinterval lying within a painted band, $\min G = 0$ and $\max G \leq 1$; and the area-sum of these subintervals is less than $(6k^2)\left(\frac{2}{k^3}\right)$. On a subinterval between painted bands, there are no members of $\mathbf{Q}^2$ with denominators k or less, not even on the boundary, because those members are interior to the painted region. Hence on such subintervals, $\min G = 0$ and

$$\max G \leq \max\left\{\frac{1}{jm} : j \text{ and } m > k\right\} < \frac{1}{k^2};$$

and the area-sum for these is less than 1. Hence

$$0 = l(G, \mathcal{P}) < u(G, \mathcal{P}) < 1\left(\frac{12}{k}\right) + \frac{1}{k^2}(1).$$

It follows that G is integrable, and $\int_B G = 0$, even though the discontinuities of G do not constitute a meager set.

The example shows that continuity on all but a subset of volume zero does not characterize integrable functions. Indeed, that the set of discontinuities—even for an integrable function on an Archimedean set—may fail to be Archimedean, is a sign that our tools do not have the power to provide such characterization.

The complete description was given in, remarkably, a doctoral thesis. Our notions of "Archimedean" and "volume" approximate definitions given by Peano around 1890. In his 1903 dissertation, Henri Lebesgue introduced a finer instrument ("finer" in the sense of resolving smaller detail) along with other of the most important ideas of twentieth-century mathematics. We will not even describe Lebesgue measure, except to hint at it in Exercise 14; its development has to wait until the next course in analysis.

[As usual, Kline gives an illuminating account of both how and why the theory progressed. Pages 1040–1042 cover the development from Riemann to Peano, 1043–1046 from Peano to Lebesgue. The latter segment is comprehensible at our level, but pertains to the next one.]

We end the section with some information about the nature of integration.

Theorem 5. *Assume that A is Archimedean.*

(a) *Suppose $A \subseteq S$ and S is Archimedean. Then f is integrable on S iff it is integrable on both pieces A and $S - A$.*

(b) *Suppose* $\text{int}(A) \subseteq S \subseteq \text{cl}(A)$; *in words, S contains the interior and perhaps some of* $\text{bd}(A)$. *Then f is integrable on S iff it is integrable on A.*

Proof. (a) $\Rightarrow$ Assume that f is integrable on S, with $|f| \leq M$. Fix a box B containing S and let $\varepsilon > 0$. There must exist a partition $\mathcal{P}$ of B with $u(f\chi_S, \mathcal{P}) - l(f\chi_S, \mathcal{P}) < \frac{\varepsilon}{2}$. Because χ_A is also integrable on B, there is a partition $\mathcal{Q}$ of B with $u(\chi_A, \mathcal{Q}) - l(\chi_A, \mathcal{Q}) < \frac{\varepsilon}{4M}$.

We examine the difference $u(f\chi_A, \mathcal{PQ}) - l(f\chi_A, \mathcal{PQ})$. On the subintervals of $\mathcal{PQ}$ contained in subintervals of $\mathcal{Q}$ disjoint from A, $\sup f\chi_A = \inf f\chi_A = 0$; these contribute nothing to the difference. On a subinterval T of $\mathcal{PQ}$ contained in a subinterval of $\mathcal{Q}$ that is in turn a subset of A,

$$\sup f\chi_A = \sup f = \sup f\chi_S$$

and

$$\inf f\chi_A = \inf f = \inf f\chi_S.$$

Therefore, the term $(\sup f\chi_A)V(T) - (\inf f\chi_A)V(T)$ is also one of the constituents of $u(f\chi_S, \mathcal{PQ}) - l(f\chi_S, \mathcal{PQ})$, and the sum of those terms is no more than

$$u(f\chi_S, \mathcal{PQ}) - l(f\chi_S, \mathcal{PQ}) \leq u(f\chi_S, \mathcal{P}) - l(f\chi_S, \mathcal{P}) < \frac{\varepsilon}{2}.$$

Finally, if T is a subinterval of $\mathcal{PQ}$ within a subinterval of $\mathcal{Q}$ that meets both A and its complement, then $\sup f\chi_A - \inf f\chi_A = 2M$. The volume sum of those T's is the volume-sum of the subintervals of $\mathcal{Q}$ that house points from A and A^*. This sum is precisely $u(\chi_A, \mathcal{Q}) - l(\chi_A, \mathcal{Q})$. Hence this last group of T's contributes less than $2M\frac{\varepsilon}{4M} = \frac{\varepsilon}{2}$ to the difference. We conclude that $u(f\chi_A, \mathcal{PQ}) - l(f\chi_A, \mathcal{PQ}) < \varepsilon$, and f is integrable on A.

Since $S - A$ is just another Archimedean subset of S (Theorem 3), the same argument can be worked on it, showing that f is integrable on $S - A$.

$\Leftarrow$ Assume that f is integrable on A and $S - A$. Let box B cover S. By definition, $f\chi_A$ and $f\chi_{S-A}$ are integrable on B. By linearity (Exercise 4.2:4), $f\chi_A + f\chi_{S-A}$ is integrable on B. Since $\chi_A + \chi_{S-A} = \chi_S$ (Verify!), $f\chi_S$ is integrable on B.

(b) We need two topological facts (Exercise 15): The boundary of int(A) is a subset of bd(A); cl(A) is the (disjoint) union of int(A) and bd(A).

Given that A is Archimedean, the first fact tells us that the boundary of int(A) is a subset of a meager set, is therefore meager, makes therefore int(A) an Archimedean set. Given that int$(A) \subseteq S \subseteq$ cl(A), the second tells us that S is the union of int(A) and part of bd(A), so that S is the union of Archimedean sets, is therefore Archimedean. The part $S - A \subseteq$ bd(A) is meager. Hence f is trivially integrable on $S - A$. In view of part (a), f is integrable on S iff f is integrable on A. □

Statements like Theorems 4 and 5 above and Exercise 4.3:6 (whose converse is Exercise 12 here) indicate that Riemann integration and integrals are blind to what happens inside meager sets. As a mathematical matter, this is to be expected. We have described integration as an averaging process. In fact, it produces weighted averages, weighting function values by the volumes of the regions in which they prevail. Values that occur in meager sets are given arbitrarily small weights in the approximating (upper, lower, and Riemann) sums, and therefore affect neither integrability nor integrals. As a practical matter, this property means that for integration on an Archimedean set, we can keep or throw away any or all of the boundary of the set (provided the function is bounded on the boundary), at our convenience.

Exercises

1. Give an example of a non-Archimedean set whose boundary is Archimedean.

2. Prove that if $V(D_1) = \cdots = V(D_J) = 0$, then $V(D_1 \cup \cdots \cup D_J) = 0$.

3. Prove that every subset of a meager set is meager.

4. Show that the sphere $S(\mathbf{O}, a) \equiv \{\mathbf{x}\colon \|\mathbf{x}\| = a\}$ is given by

$$\begin{aligned} x_1 &= a\cos t_1, \quad x_2 = a\sin t_1 \cos t_2, \quad x_3 = a \sin t_1 \sin t_2 \cos t_3, \ldots, \\ x_{n-1} &= a \sin t_1 \sin t_2 \cdots \sin t_{n-2} \cos t_{n-1}, \\ x_n &= a \sin t_1 \sin t_2 \cdots \sin t_{n-2} \sin t_{n-1}, \end{aligned}$$

over the intervals $0 \le t_1 \le \pi, \ldots, 0 \le t_{n-2} \le \pi, 0 \le t_{n-1} \le 2\pi$.

Where is this parametrization smooth?

5. (a) In Example 2, show that if $i < j$ and O_i intersects O_j, then O_j is a proper subset of O_i.

   (b) Use (a) to show that for a given i, the union of all the O_j that intersect O_i is itself one of the open boxes $O_{k(i)}$, $k(i) \leq i$.

   (c) Show that each $O_{k(i)}$ is a component of O.

6. (a) Prove that for arbitrary sets A and B, $\text{bd}(A \cap B) \subseteq \text{bd}(A) \cup \text{bd}(B)$.

   (b) Prove that if A and B are Archimedean, then so is $A \cap B$.

7. Prove that Archimedean sets A and B overlap iff $V(A \cap B) > 0$.

8. Give examples of:

   (a) an infinite union of Archimedean sets that is not Archimedean.

   (b) a closed set that is not Archimedean.

   (c) a set whose boundary has empty interior but is not meager.

9. Which, if any, of these functions are integrable on the given set:

   (a) $f(x, y, z) \equiv 1$ if x, y, z are rational, $\equiv 0$ otherwise, on $A \equiv \{(x, y, z): x = 0, y^2 + z^2 \leq 1\}$.

   (b) $g(x, y, z) \equiv 1$ if x, y, z are rational numbers between $-\frac{1}{2}$ and $\frac{1}{2}$ inclusive, $\equiv 0$ otherwise, on $S \equiv \{(x, y, z): x^2 + y^2 + z^2 \leq 1\}$.

   (c) $h(x, y, z) \equiv \frac{1}{rst}$ if $x = \frac{i}{r}$, $y = \frac{j}{s}$, $z = \frac{k}{t}$ are rational numbers in lowest terms between $-\frac{1}{2}$ and $\frac{1}{2}$ inclusive, $\equiv 0$ otherwise, on $S \equiv \{(x, y, z): x^2 + y^2 + z^2 \leq 1\}$.

10. Show that if a function is bounded on an Archimedean set and continuous at all but finitely many points of the set, then it is integrable.

11. Assume that f is continuous on $\mathbf{R}^n$. Show that f is integrable on every Archimedean set.

12. Suppose every bounded function on S is integrable there. Show that S is meager. (Note that it is not given that S is Archimedean.)

13. Show that the Cantor set is Archimedean. [Refer to the following in Guzman: Section 5.3, for definition of the Cantor set; Exercises 5.3:5 and 5.4:6 for topological information; Exercises 5.2:5, 6 for sets to which a similar analysis applies.]

14. A set S is said to have **zero (Lebesgue) measure** if for each $\varepsilon > 0$, there is a sequence $B_1, B_2, \ldots$ of boxes that covers S and has $V(B_1)+V(B_2)+\cdots < \varepsilon$.

    (a) Show that every set of zero volume has zero measure.

(b) Give an example of a set with zero measure but not zero volume.

(c) Give an example of an uncountable set with zero measure.

(d) Show that every bounded set of zero measure has zero involume. (Hint: You need the Heine–Borel theorem.)

(e) Show that if a set of zero measure is Archimedean, then it has zero volume.

(f) Show that if a set of zero measure is closed and bounded, then it is Archimedean and has zero volume.

15. (a) Prove that for an arbitrary set S, the boundary of $\text{int}(S)$ is a subset of $\text{bd}(S)$.

(b) Show that the boundary of $\text{int}(S)$ may be a proper subset of $\text{bd}(S)$.

(c) Show that $\text{cl}(S) = \text{int}(S) \cup \text{bd}(S)$.

[Compare Guzman: Exercises 5.4:5b, 9a.]

5
Integrals of Scalar Functions

Having defined integrability and integrals, we turn next to their properties and to extensions. Some of the properties are general, and reflect the way we expect integrals to behave. Two are specific, and address a need: we defined integrals, but did not say how a typical integral might be evaluated. We will show that two methods familiar from elementary calculus apply to the objects we have defined. More important, we extend the notion of integral to a whole new variety of functions and domains.

5.1 Fubini's Theorem

In elementary calculus, the things that look like volume integrals are evaluated as iterated integrals. Here we discuss why that reduction to one-variable integrals works.

We will deal with a function f bounded on the box $B \equiv [\mathbf{a}, \mathbf{b}] \subseteq \mathbf{R}^n$. It is helpful to adapt notation we used for the inverse function and implicit function theorems. Thus, pick a fixed m, $1 \leq m < n$. Let $\mathbf{x} = (x_1, \ldots, x_n) \in B$. We write

$$\mathbf{v} = \Pi_1(\mathbf{x}) \equiv (x_1, \ldots, x_m) \in \mathbf{R}^m,$$
$$\mathbf{w} = \Pi_2(\mathbf{x}) \equiv (x_{m+1}, \ldots, x_n) \in \mathbf{R}^{n-m}.$$

We identify $\mathbf{x}$ with $(\mathbf{v}, \mathbf{w})$, so that $f(\mathbf{v}, \mathbf{w}) = f(\mathbf{x})$. Note that

$$\|\mathbf{x}\|^2 = \|\mathbf{v}\|^2 + \|\mathbf{w}\|^2.$$

For a fixed $\mathbf{w}$,

$$f_{\mathbf{w}}(\mathbf{v}) \equiv f(\mathbf{v}, \mathbf{w}), \qquad (a_1, \ldots, a_m) \le \mathbf{v} \le (b_1, \ldots, b_m)$$

defines a bounded function on the (fixed) box

$$C \equiv [\Pi_1(\mathbf{a}), \Pi_1(\mathbf{b})] \subseteq \mathbf{R}^m.$$

If $f_{\mathbf{w}}$ is integrable on C, then

$$\phi(\mathbf{w}) \equiv \int_C f_{\mathbf{w}}(\mathbf{v})\, d\mathbf{v}$$

defines a scalar function of $\mathbf{w}$. We have put in the "differential" $d\mathbf{v}$ just to remind us what the variable of integration is. We next want to integrate ϕ as a function of $\mathbf{w}$ on the box $D \equiv \left[\Pi_2(\mathbf{a}), \Pi_2(\mathbf{b})\right]$ in $\mathbf{R}^{n-m}$. We are entitled to try, because by the operator-boundedness of integrals (Exercise 4.2:4(e)),

$$|\phi(\mathbf{w})| \le \sup |f_{\mathbf{w}}(\mathbf{v})| V(C) \le (\sup |f|) V(C)$$

is a bounded function on D.

Theorem 1 (Fubini's Theorem). *Assume that f is integrable on B. Assume further that for each $\mathbf{w} \in D$, $f_{\mathbf{w}}(\mathbf{v})$ is integrable on C. Then $\phi(\mathbf{w})$ is integrable on D, and the integral of f is the "iterated integral"*

$$\int_B f = \int_D \phi(\mathbf{w})\, d\mathbf{w} = \int_D \left(\int_C f_{\mathbf{w}}(\mathbf{v})\, d\mathbf{v} \right) d\mathbf{w}.$$

Proof. We need to prove that $\phi(\mathbf{w})$ is integrable and has integral $\int_B f$. By Theorem 4.2:4, we must demonstrate that $\int_B f$ is the limit of the Riemann sums for ϕ. Accordingly, we name $\varepsilon > 0$ and find a fineness below which all the Riemann sums (hereinafter "R-sums") for ϕ fall within ε of $\int_B f$.

Because f is integrable, there exists Δ such that over any partition of B with norm less than Δ, all the R-sums for f are within $\frac{\varepsilon}{2}$ of $\int_B f$. We set $\delta \equiv \frac{\Delta}{2}$.

Suppose that $\mathcal{P} \equiv \{S_1, \ldots, S_K\}$ is any partition of D with $\|\mathcal{P}\| < \delta$. Any R-sum for ϕ on $\mathcal{P}$ looks like

$$\sigma \equiv \phi(\mathbf{w}_1)V(S_1) + \cdots + \phi(\mathbf{w}_K)V(S_K).$$

Here every $\mathbf{w}_k \in S_k$ and every value $\phi(\mathbf{w}_k)$ is an integral of the form

$$\phi(\mathbf{w}_k) = \int_C f(\mathbf{v}, \mathbf{w}_k)\, d\mathbf{v}.$$

Each such integral is the limit of its R-sums. Hence there is δ_k for which any partition of C with norm less than δ_k yields R-sums for $f(\mathbf{v}, \mathbf{w}_k)$ within $\frac{\varepsilon}{2V(D)}$ of $\phi(\mathbf{w}_k)$.

Next let $\mathcal{Q} \equiv \{T_1, \ldots, T_J\}$ be some partition of C with $\|\mathcal{Q}\| < \min\{\delta_1, \ldots, \delta_K, \delta\}$, and let $\mathbf{v}_1 \in T_1, \ldots, \mathbf{v}_J \in T_J$ be some sampling from the subintervals of $\mathcal{Q}$. Then for any k,

$$\phi(\mathbf{w}_k) - \frac{\varepsilon}{2V(D)} < f(\mathbf{v}_1, \mathbf{w}_k)V(T_1) + \cdots + f(\mathbf{v}_J, \mathbf{w}_k)V(T_J) < \phi(\mathbf{w}_k) + \frac{\varepsilon}{2V(D)}.$$

Hence

$$\begin{aligned}
\sigma - \frac{\varepsilon}{2} &= \sum_{1 \le k \le K} \left[\phi(\mathbf{w}_k) - \frac{\varepsilon}{2V(D)}\right] V(S_k) \\
&< \sum_{1 \le k \le K} \left[\sum_{1 \le j \le J} f(\mathbf{v}_j, \mathbf{w}_k)V(T_j)\right] V(S_k) \\
&< \sum_{1 \le k \le K} \left[\phi(\mathbf{w}_k) + \frac{\varepsilon}{2V(D)}\right] V(S_k) = \sigma + \frac{\varepsilon}{2}.
\end{aligned}$$

The expression in the middle line is an R-sum for f. It uses the values $f(\mathbf{v}_j, \mathbf{w}_k)$ at points in the Cartesian products

$$T_j \times S_k \equiv \{(\mathbf{v}, \mathbf{w}) : \mathbf{v} \in T_j, \mathbf{w} \in S_k\}.$$

These products are boxes that (Exercise 1) make up a partition of B, have volumes $V(T_j \times S_k) = V(T_j)V(S_k)$, and have diagonals

$$\text{diag}(T_j \times S_k) = \left[\text{diag}(T_j)^2 + \text{diag}(S_k)^2\right]^{1/2} < \left(\delta^2 + \|\mathcal{P}\|^2\right)^{1/2}.$$

Since $(\delta^2 + \|\mathcal{P}\|^2)^{1/2} < \Delta$, the R-sum for f has to lie within $\frac{\varepsilon}{2}$ of $\int_B f$. We conclude that

$$\int_B f - \varepsilon < \sigma < \int_B f + \varepsilon. \qquad \square$$

The idea of iterated integration is about as old as calculus. The name "Fubini's theorem" actually belongs to a twentieth-century extension of the idea to Lebesgue's integrals, but it is a good name for the principle.

Example 1. The hypotheses we list for Fubini's theorem are essential.

(a) From integrability of f, we cannot conclude that each $f(\mathbf{v}, \mathbf{w})$ is $\mathbf{v}$-integrable.

On the unit square, let $f(x, y) \equiv \frac{1}{k}$ if $y = \frac{j}{k}$ (lowest terms) and x are both rational, $\equiv 0$ otherwise. This function is integrable, but for any fixed rational r, $f_r(x) \equiv f(x, r)$ is an unintegrable function of x (Exercise 2). Thus, existence of the original integral in $\mathbf{R}^n$ does not guarantee existence of the iterated integral.

(b) From integrability of every $f_{\mathbf{w}}(\mathbf{v})$, we cannot infer that $\phi(\mathbf{w}) \equiv \int_C f_{\mathbf{w}}(\mathbf{v})\, d\mathbf{v}$ is $\mathbf{w}$-integrable.

Given

$$g(x, y) \equiv \begin{cases} 1 & \text{if } y \text{ is rational,} \\ 0 & \text{otherwise,} \end{cases}$$

we find that each $g_y(x)$ is integrable, but $\phi(y) \equiv \int_0^1 g_y(x)\,dx$ is not (Exercise 3).

(c) Worse than (b): Knowing that $f_{\mathbf{w}}(\mathbf{v})$ is integrable and $\phi(\mathbf{w}) \equiv \int_C f_{\mathbf{w}}(\mathbf{v})\,d\mathbf{v}$ happens also to be integrable, we still may not conclude that f is.

That is the situation with

$$h(x, y) \equiv \begin{cases} 1 & \text{if } y \text{ is rational XOR } 0 \le x \le \frac{1}{2}, \\ -1 & \text{otherwise} \end{cases}$$

(Exercise 4); here XOR represents "exclusive or" (one or the other *and* not both). This example turns (a) around; it shows that existence of the iterated integral does not guarantee existence of the main integral.

(d) If the hypotheses are satisfied, so that $\int_B f$ matches the iterated integral, it may not be possible to change the order of integration.

For the function in (a), if x is irrational, then $f_x(y) \equiv 0$ and $\phi(x) = 0$. If instead x is rational, then

$$f_x(y) = \begin{cases} \frac{1}{k} & \text{if } y = \frac{j}{k}, \\ 0 & \text{for other } y. \end{cases}$$

This is the single-variable modified Dirichlet function. It is integrable, with zero integral; again $\phi(x) \equiv \int_0^1 f_x(y)\,dy = 0$ for all x. By the theorem,

$$0 = \int f = \int_0^1 \left[\int_0^1 f(x, y)\,dy \right] dx,$$

even though $\int_0^1 \left[\int_0^1 f(x, y)\,dx \right] dy$ is meaningless.

The example informs us that we have to exercise caution in our applications of Fubini's theorem. Nevertheless, where integration is concerned, continuity washes away many sins. It certainly simplifies integrals on boxes.

Theorem 2. *Suppose f is continuous on $[\mathbf{a}, \mathbf{b}]$. Then*

$$\int_{[\mathbf{a},\mathbf{b}]} f = \int_{a_j}^{b_j} \left[\cdots \left[\int_{a_k}^{b_k} f(x_1, \ldots, x_n)\,dx_k \right] \cdots \right] dx_j,$$

and the order of the integrations is immaterial.

Proof. Exercise 5.

There are many ways to extend Fubini's theorem to regions more general than boxes. We settle for discussion of the most familiar kind, the region between two graphs. Let us agree to say that $A \subseteq \mathbf{R}^n$ **is a region between graphs** if there is a set $C \subseteq \mathbf{R}^{n-1}$ such that A consists of those $\mathbf{x} = \left(x_1, \mathbf{x}^\#\right) = (x_1, \ldots, x_n)$ for which

$$\mathbf{x}^\# \in C \quad \text{and} \quad g\left(\mathbf{x}^\#\right) \le x_1 \le G\left(\mathbf{x}^\#\right).$$

We will assume that g and G are continuous on the closure $\mathrm{cl}(C)$. We would like to find that

$$\int_A f = \int_C \left[\int_{g(\mathbf{x}^\#)}^{G(\mathbf{x}^\#)} f\left(x_1, \mathbf{x}^\#\right) dx_1\right] d\mathbf{x}^\#.$$

Before even considering the integrals, we need to know that A and C are Archimedean. Since the graphs themselves are reasonable objects, it is not surprising that the key is C, specifically, the assumption that C is Archimedean. (Compare Exercise 6.)

Theorem 3. *Assume that $C \subseteq \mathbf{R}^{n-1}$ is Archimedean, and g and G are continuous on $\mathrm{cl}(C)$. Let*

$$A \equiv \left\{\left(x_1, \mathbf{x}^\#\right) : \mathbf{x}^\# \in C \text{ and } g\left(\mathbf{x}^\#\right) \le x_1 \le G\left(\mathbf{x}^\#\right)\right\}.$$

(a) *A is an Archimedean subset of $\mathbf{R}^n$.*

(b) *Suppose that f is integrable on A, and that for each $\mathbf{x}^\# \in C$, $f\left(x_1, \mathbf{x}^\#\right)$ is integrable over the closed interval $g\left(\mathbf{x}^\#\right) \le x_1 \le G\left(\mathbf{x}^\#\right)$. Then*

$$\phi(\mathbf{x}^\#) \equiv \int_{g(\mathbf{x}^\#)}^{G(\mathbf{x}^\#)} f\left(x_1, \mathbf{x}^\#\right) dx_1$$

is integrable over C, and

$$\int_A f = \int_C \phi = \int_C \left[\int_{g(\mathbf{x}^\#)}^{G(\mathbf{x}^\#)} f\left(x_1, \mathbf{x}^\#\right) dx_1\right] d\mathbf{x}^\#.$$

Proof. (a) We must prove that the boundary of A has zero volume. With respect to that boundary, Figure 5.1 is clear: $\mathrm{bd}(A)$ consists of the surface on top, the one on the bottom, and the vertical sides. We will make this description precise, then show that these points can be packed into a small volume.

Suppose $\mathbf{x} = \left(x_1, \mathbf{x}^\#\right) \in A$ has $\mathbf{x}^\#$ interior to C and

$$g\left(\mathbf{x}^\#\right) < x_1 < G\left(\mathbf{x}^\#\right).$$

By continuity, there exists a neighborhood $N(\mathbf{x}^\#, \delta) \subseteq C$ in which g and G are bounded away from x_1. That is, there exist δ and Δ such that

$$\mathbf{y}^\# \in N\left(\mathbf{x}^\#, \delta\right) \Rightarrow g\left(\mathbf{y}^\#\right) < x_1 - \Delta < x_1 + \Delta < G\left(\mathbf{y}^\#\right).$$

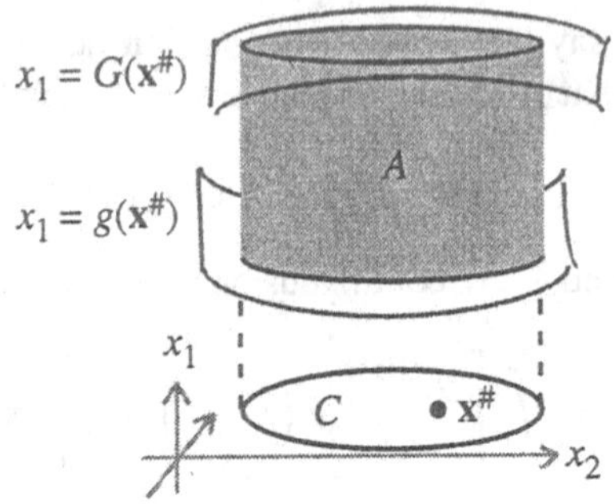

Figure 5.1.

Setting $r \equiv \min\{\Delta, \delta\}$, we have

$$\begin{aligned}\|\mathbf{y}-\mathbf{x}\| < r &\Rightarrow \left\|\mathbf{y}^\# - \mathbf{x}^\#\right\| < \delta \text{ and } \quad \|y_1 - x_1\| < \Delta \\ &\Rightarrow \mathbf{y}^\# \in C \text{ and } \quad g\left(\mathbf{y}^\#\right) < y_1 < G\left(\mathbf{y}^\#\right) \\ &\Rightarrow \mathbf{y} \in A.\end{aligned}$$

In other words, $N(\mathbf{x}, r) \subseteq A$, showing that $\mathbf{x} \in \operatorname{int}(A)$. By contraposition, if $\mathbf{x}$ is on $\operatorname{bd}(A)$, then either $\mathbf{x}^\# \in \operatorname{bd}(C)$ or $x_1 = g\left(\mathbf{x}^\#\right)$ or $x_1 = G\left(\mathbf{x}^\#\right)$; the picture rules.

To evaluate the volume of the vertical wall $\{\mathbf{x} \in A\colon \mathbf{x}^\# \in \operatorname{bd}(C)\}$ of A, let $\varepsilon > 0$ be specified. Write $M \equiv \max\left\{G\left(\mathbf{x}^\#\right) : \mathbf{x}^\# \in \operatorname{cl}(C), m \equiv \min\left\{g\left(\mathbf{x}^\#\right) : \mathbf{x}^\# \in \operatorname{cl}(C)\right\}\right.$. (Why do those extremes exist?) Because C is Archimedean, there exists a partition $\mathcal{P}$ covering C in $\mathbf{R}^{n-1}$ for which the subintervals $S_1, \dots, S_J$ intersecting $\operatorname{bd}(C)$ have

$$V(S_1) + \cdots + V(S_J) < \frac{\varepsilon}{M-m}.$$

Above these, the boxes $S_1 \times [m, M], \dots, S_J \times [m, M]$ in $\mathbf{R}^n$ cover the wall, together with the parts of top and bottom over $\operatorname{bd}(C)$. The boxes have volume-sum $V(S_1)(M-m) + \cdots + V(S_J)(M-m) < \varepsilon$. We conclude that the wall is meager.

Look next at the part of the top graph over the interior subintervals $T_1, \dots, T_K$ in $\mathcal{P}$. Since G is uniformly continuous on $T_1 \cup \cdots \cup T_K$, there exists δ^* such that

$$\left\|\mathbf{y}^\# - \mathbf{x}^\#\right\| < \delta^* \text{ in } T_1 \cup \cdots \cup T_K \Rightarrow \left|G(\mathbf{y}^\#) - G\left(\mathbf{x}^\#\right)\right| < \frac{\varepsilon}{V(C)}.$$

Subdivide each T_k into boxes with diagonal δ^* or less. Call the boxes so produced $U_1, \dots, U_L$, and write m_l and M_l for the minimum and maximum, respectively, of G on U_l. If $\mathbf{x}^\# \in U_l$, then

$$\left(G\left(\mathbf{x}^\#\right), \mathbf{x}^\#\right) \in [m_l, M_l] \times U_l.$$

Hence this part of the upper graph can be encapsulated in a union of boxes with volume

$$(M_1 - m_1)V(U_1) + \cdots + (M_L - m_L)V(U_L) \\ \leq [V(U_1) + \cdots + V(U_L)]\frac{\varepsilon}{V(C)} < \varepsilon.$$

We conclude that the upper skin is meager; similarly for the lower.

(b) In $\mathbf{R}^n$, let $B \equiv [\mathbf{a}, \mathbf{b}]$ be some box containing A. We need to prove $\phi(\mathbf{x}^\#)$ integrable on C in $\mathbf{R}^{n-1}$. This requires showing that $\phi\chi_C$ is integrable on $B^\# \equiv [\mathbf{a}^\#, \mathbf{b}^\#] \subseteq \mathbf{R}^{n-1}$. We will do that by applying Theorem 1 to the function $f\chi_A$ on B.

We know that $f\chi_A$ is integrable on B; this follows from the assumption that f is integrable on A. Also, for a fixed $\mathbf{x}^\# \in B^\#$,

$$(f\chi_A)_{\mathbf{x}^\#}(x_1) \equiv (f\chi_A)\left(x_1, \mathbf{x}^\#\right) = f\left(x_1, \mathbf{x}^\#\right)\chi_A\left(x_1, \mathbf{x}^\#\right)$$

is integrable on $[a_1, b_1]$. To see this, examine two cases. If $\mathbf{x}^\# \notin C$, then $(x_1, \mathbf{x}^\#) \notin A$, and $(f\chi_A)_{\mathbf{x}^\#}$ is identically zero. If instead $\mathbf{x}^\# \in C$, then

$$(f\chi_A)_{\mathbf{x}^\#}(x_1) = \begin{cases} f\left(x_1, \mathbf{x}^\#\right) & \text{if } g\left(\mathbf{x}^\#\right) \leq x_1 \leq G\left(\mathbf{x}^\#\right), \\ 0 & \text{otherwise} \end{cases} \\ = f\left(x_1, \mathbf{x}^\#\right)\chi_{[g(\mathbf{x}^\#), G(\mathbf{x}^\#)]}(x_1);$$

this last is integrable on $[a_1, b_1]$, owing to the assumption that $f\left(x_1, \mathbf{x}^\#\right)$ is integrable over the interval $g\left(\mathbf{x}^\#\right) \leq x_1 \leq G\left(\mathbf{x}^\#\right)$. Hence $f\chi_A$ satisfies the hypothesis of Fubini.

By the theorem,

$$F\left(\mathbf{x}^\#\right) \equiv \int_{a_1}^{b_1} f\left(x_1, \mathbf{x}^\#\right)\chi_A\left(x_1, \mathbf{x}^\#\right) dx_1$$

is integrable over $B^\#$. But F is just $\phi\chi_C$:

$$F(\mathbf{x}^\#) = \begin{cases} 0 & \text{if } \mathbf{x}^\# \notin C, \\ \int_{g(\mathbf{x}^\#)}^{G(\mathbf{x}^\#)} f\left(x_1, \mathbf{x}^\#\right) dx_1 & \text{if } \mathbf{x}^\# \in C \end{cases} \\ = \phi\left(\mathbf{x}^\#\right)\chi_C\left(\mathbf{x}^\#\right).$$

Since $\phi\chi_C$ is integrable on $B^\#$, we conclude that ϕ is integrable on C. Moreover, the equality from Theorem 1 says that

$$\int_B f\chi_A = \int_{B^\#} F = \int_{B^\#} \phi\chi_C,$$

or

$$\int_A f = \int_C \phi = \int_C \left[\int_{g(\mathbf{x}^\#)}^{G(\mathbf{x}^\#)} f\left(x_1, \mathbf{x}^\#\right) dx_1 \right] d\mathbf{x}^\#. \qquad \square$$

Example 2. What is the volume of a ball in $\mathbf{R}^4$?

Writing $A \equiv \{(x, y, z, w) : x^2 + y^2 + z^2 + w^2 \le a^2\}$, we see that A is characterized by

$$|w| \le \left(a^2 - x^2 - y^2 - z^2\right)^{1/2}$$

for

$$(x, y, z) \in C \equiv \{(x, y, z) : x^2 + y^2 + z^2 \le a^2\}.$$

The hypotheses of Theorem 3 are satisfied, and we may write the volume as

$$V = \int_C \left[\int_{-\sqrt{a^2-x^2-y^2-z^2}}^{\sqrt{a^2-x^2-y^2-z^2}} 1\, dw \right] d(x, y, z).$$

Now, the bracketed integral is a continuous function of its limits (fundamental theorem of calculus), which are continuous functions of (x, y, z). Hence Theorem 3 applies again, and we have

$$V = \int_D \left[\int_{-\sqrt{a^2-x^2-y^2}}^{\sqrt{a^2-x^2-y^2}} \left[\int_{-\sqrt{a^2-x^2-y^2-z^2}}^{\sqrt{a^2-x^2-y^2-z^2}} 1\, dw \right] dz \right] d(x, y),$$

D being the origin-centered disk of radius a in $\mathbf{R}^2$. This is the procedure we always want to follow: Apply Theorem 3 repeatedly to reduce an $\mathbf{R}^n$ integral to the iterated integral. Here the result is

$$V = \int_{-a}^{a} \left[\int_{-\sqrt{a^2-x^2}}^{\sqrt{a^2-x^2}} \left[\int_{-\sqrt{a^2-x^2-y^2}}^{\sqrt{a^2-x^2-y^2}} \left[\int_{-\sqrt{a^2-x^2-y^2-z^2}}^{\sqrt{a^2-x^2-y^2-z^2}} 1\, dw \right] dz \right] dy \right] dx.$$

The actual evaluation is rendered easy if we recognize that the inner integrals represent the quantity one level down, namely, the volume of the ball in $\mathbf{R}^3$ of radius $\sqrt{a^2 - x^2}$. Accordingly,

$$V = \int_{-a}^{a} \left(\frac{4\pi}{3}\right) \left(a^2 - x^2\right)^{3/2} dx = \frac{\pi^2 a^4}{2}.$$

In Theorem 3, suppose we have $g\left(\mathbf{x}^\#\right) \equiv 0$ and $f(\mathbf{x}) \equiv 1$. Then

$$A = \{(x_1, \mathbf{x}^\#) : 0 \le x_1 \le G\left(\mathbf{x}^\#\right)\}$$

is the region "under the graph of G." Part (a) says that this region is Archimedean. Also,

$$\int_g^G f\left(x_1, \mathbf{x}^\#\right) dx_1 = G\left(\mathbf{x}^\#\right),$$

which is certainly integrable. Then (b) tells us that

$$\int_C G\left(\mathbf{x}^\#\right) d\mathbf{x}^\# = \int_A f = V(A).$$

Just as in elementary calculus, the integral of a continuous multivariable function is the volume under its graph.

Exercises

1. Suppose $B = [\mathbf{a}, \mathbf{b}]$ projects onto $C \subseteq \mathbf{R}^m$ and $D \subseteq \mathbf{R}^{n-m}$, as described at the beginning of the section. Show that:

    (a) If $\{T_1, \dots, T_K\}$ is a partition of C and $\{S_1, \dots, S_J\}$ is a partition of D, then $\{T_k \times S_j\}$ is a partition of B.

    (b) The subinterval $T_k \times S_j$ has volume $V(T_k)V(S_j)$.

    (c) The diagonals satisfy $\text{diag}(T_k \times S_j) = [\text{diag}(T_k)^2 + \text{diag}(S_j)^2]^{1/2}$.

2. Show that in Example 1(a):

    (a) $f(x, y)$ is integrable on the unit square. (Hint: Adapt Example 4.4:3).

    (b) For a rational number $\frac{j}{k}$, $f\left(x, \frac{j}{k}\right)$ is not integrable over $0 \leq x \leq 1$.

3. Show that in Example 1(b):

    (a) For any fixed y, $g(x, y)$ is integrable with respect to x.

    (b) $\phi(y) \equiv \int_0^1 g(x, y)\, dx$ is not integrable.

4. Show that in Example 1(c):

    (a) For any fixed y, $h(x, y)$ is integrable with respect to x.

    (b) $\phi(y) \equiv \int_0^1 h(x, y)\, dx$ is integrable, $0 \leq y \leq 1$.

    (c) $h(x, y)$ is not integrable on the unit square.

5. Prove Theorem 2. (Hint: Prove the "reduction formula"

$$\int_{[\mathbf{a},\mathbf{b}]} f = \int_{a_1}^{b_1} \left[\int_C f_{x_1}(x_2, \dots, x_n)\right] dx_1,$$

C being the box from $(a_2, \dots, a_n)$ to $(b_2, \dots, b_n)$, and make the necessary extensions.)

6. Give an example of sets $A \subseteq \mathbf{R}^2$ and $C \subseteq \mathbf{R}$ exhibiting the property

$$A = \{(x, y) : x \in C \text{ and } g(x) \le y \le G(x)\}$$

for continuous functions g and G, and A is Archimedean but C is not.

7. Evaluate $\int e^{x+y+z}$ on $\{(x, y, z) : |x| \le 1, |y| \le 1, |z| \le 1\}$.

8. (a) Calculate the average value of $x + y$ in the unit square $0 \le x \le 1$, $0 \le y \le 1$.

   (b) Calculate the same average on the triangle bordered by the two axes and the line $x + y = 1$. Is this result surprising?

9. Find the volume of the region in $\mathbf{R}^3$ that is above the xy-plane, within the cylinder $x^2 + y^2 = 1$, and below the paraboloid $z = 3 - x^2 - 3y^2$.

10. Find the volume of a ball of radius a in $\mathbf{R}^5$.

5.2 Properties of Integrals

There are certain properties that anything called an integral should have, like linearity. We develop them for our integrals in this section. Throughout the section, any set referred to is assumed to be Archimedean.

Theorem 1. *Assume that A is a fixed subset of $\mathbf{R}^n$ and f and g are integrable on A. Then:*

(a) (Linearity) *Each linear combination $\alpha f + \beta g$ is integrable, with*

$$\int_A (\alpha f + \beta g) = \alpha \int_A f + \beta \int_A g.$$

(b) (Products) *The product fg is integrable on A.*

(c) (Function Monotonicity) *If $f \ge g$ throughout A, then $\int_A f = \int_A g$. (Note that "throughout A" can be replaced by "except possibly on a meager subset of A.")*

(d) (Triangle Inequality) *The absolute value of f is integrable on A, and*

$$\left|\int_A f\right| \le \int_A |f|.$$

Proof. Let B be some box containing A.

(a) This is immediate from linearity on boxes (Exercise 4.2:4(a)). By hypothesis, $f\chi_A$ and $g\chi_A$ are integrable on B. Hence $(\alpha f+\beta g)\chi_A = \alpha(f\chi_A)+\beta(g\chi_A)$ is integrable on B, with

$$\int_B (\alpha f + \beta g)\chi_A = \alpha \int_B f\chi_A + \beta \int_B g\chi_A.$$

(b) The standard way to handle products is to deal first with squares. Let $M \equiv \sup_A |f| = \sup_B |f|\chi_A$. If S is a subinterval of any partition $\mathcal{P}$ of B and $\mathbf{x}, \mathbf{y} \in S$, then

$$\begin{aligned} f^2(\mathbf{x})\chi_A(\mathbf{x}) - f^2(\mathbf{y})\chi_A(\mathbf{y}) &= [f(\mathbf{x})\chi_A(\mathbf{x}) + f(\mathbf{y})\chi_A(\mathbf{y})] \\ &\quad \times [f(\mathbf{x})\chi_A(\mathbf{x}) - f(\mathbf{y})\chi_A(\mathbf{y})] \\ &\le 2M\left[\sup_S f\chi_A - \inf_S f\chi_A\right]. \end{aligned}$$

Hence

$$\sup_S f^2\chi_A - \inf_S f^2\chi_A \le 2M\left[\sup_S f\chi_A - \inf_S f\chi_A\right].$$

It follows that the upper and lower sums of $f\chi_A$ and $f^2\chi_A$ are related by

$$u\left(f^2\chi_A, \mathcal{P}\right) - l\left(f^2\chi_A, \mathcal{P}\right) \le 2M[u(f\chi_A, \mathcal{P}) - l(f\chi_A, \mathcal{P})].$$

Given that f is integrable, we can make the difference on the right arbitrarily small. We conclude that f^2 is integrable on A.

The preceding paragraph tells us that the square of an integrable function is integrable. By hypothesis, f and g are integrable. By part (a), so is

$$\frac{(f+g)^2 - (f-g)^2}{4} = fg.$$

(c) has the same sort of proof as (a).

(d) If S is a subinterval of any partition $\mathcal{P}$ of B and $\mathbf{x}, \mathbf{y} \in S$, then

$$\begin{aligned} |f(\mathbf{x})\chi_A(\mathbf{x})| - |f(\mathbf{y})\chi_A(\mathbf{y})| &\le |f(\mathbf{x})\chi_A(\mathbf{x}) - f(\mathbf{y})\chi_A(\mathbf{y})| \\ &\le \sup_S f\chi_A - \inf_S f\chi_A. \end{aligned}$$

It follows (compare part (b)) that

$$u(|f|\chi_A, \mathcal{P}) - l(|f|\chi_A, \mathcal{P}) \le u(f\chi_A, \mathcal{P}) - l(f\chi_A, \mathcal{P}),$$

and $|f|$ is integrable on A. (Compare also Exercise 4.2:7.)

From $|f| \ge f \ge -|f|$, part (c), and part (a), we conclude that

$$\int_A |f| \ge \int_A f \ge \int_A -|f| = -\int_A |f|. \qquad \square$$

Theorem 1 addresses different functions on a single set. In the next family of results, the function is fixed but we use different sets. In effect, we treat the integral as a function of sets.

Theorem 2. *Assume that f is defined on $S \cup T$. Then:*

(a) *f is integrable on $S \cup T$ iff f is integrable on each of S and T.*

(b) (Set Additivity) *If f is integrable on S and T, then*

$$\int_{S\cup T} f = \int_S f + \int_T f - \int_{S\cap T} f.$$

In particular, if S and T do not overlap, then

$$\int_{S\cup T} f = \int_S f + \int_T f.$$

(c) (Set Monotonicity) *If f is nonnegative and integrable on T and $S \subseteq T$, then f is integrable on S and $\int_S f \le \int_T f$.*

(d) (Set Continuity) *If f is integrable on T, then*

$$\lim_{V(S)\to 0} \int_S f = \lim_{V(S)\to 0} \int_S |f| = 0, \qquad S \subseteq T;$$

that is, for any $\varepsilon > 0$, there exists $\delta > 0$ such that every Archimedean subset S of T with $V(S) < \delta$ has $\int_S |f| < \varepsilon$.

Proof. (a) All sets S and T have $S \cup T = (S - T) \cup (S \cap T) \cup (T - S)$, the three pieces on the right being disjoint. If S and T are Archimedean, then so are the pieces, by Theorem 4.4:3(a). By extension of Theorem 4.4:5(a), f is integrable on $S \cup T$ $\Leftrightarrow$ it is integrable on the pieces $\Leftrightarrow$ it is integrable on S and on T.

(b) We have $f\chi_{S\cup T} = f\chi_S + f\chi_T - f\chi_{S\cap T}$ (Exercise 4.3:7). Applying this relation (and linearity) in any box containing $S \cup T$, we get

$$\int_{S\cup T} f = \int_S f + \int_T f - \int_{S\cap T} f.$$

If S and T do not overlap, then Theorem 4.4:3(b) tells us that $V(S \cap T) = 0$. By Exercise 4.3:6, $\int_{S\cap T} f = 0$.

(c) Exercise 4.

(d) Letting $M \equiv \sup_T |f|$, we have $|f| \le M$ on any Archimedean subset S. By (function) monotonicity, $\int_S |f| \le \int_S M = MV(S)$, and the conclusion follows. □

In the last of these results there is interaction between the function and the set.

Theorem 3. *Assume that f is integrable on A.*

(a) (Operator Boundedness) *If $k \le f(\mathbf{x}) \le K$ throughout A, then*

$$kV(A) \le \int_A f \le KV(A).$$

(b) (Average Value Theorem) *If f is continuous on A and A is connected, then there exists $\mathbf{c} \in A$ with*

$$\int_A f = f(\mathbf{c})V(A).$$

Proof. (a) Exercise 5.

(b) The equality is trivial if A is meager, so we may assume $V(A) > 0$. Consequently, the interior of A is nonempty.

Suppose there is some $\mathbf{b} \in \operatorname{int}(A)$ where $f(\mathbf{b}) > m \equiv \inf_A f$. Then in some neighborhood of $\mathbf{b}$, f is bounded away from m. That is, there is an $\varepsilon > 0$ and some neighborhood $T \subseteq \operatorname{int}(A)$ of $\mathbf{b}$ in which $f(\mathbf{x}) \ge m + \varepsilon$. Hence

$$\begin{aligned}\int_A f = \int_T f + \int_{A-T} f &\ge \int_T (m+\varepsilon) + \int_{A-T} m \\ &= (m+\varepsilon)V(T) + mV(A-T) > mV(A).\end{aligned}$$

(Reasons? The last step needs Exercise 3.) By contraposition, if $\int_A f = mV(A)$, then $f(\mathbf{x}) \le m$ for all $\mathbf{x} \in \operatorname{int}(A)$. In this case $f(\mathbf{x}) = m$ throughout the interior, and we may take for $\mathbf{c}$ any vector there.

Similarly, if $\frac{\int_A f}{V(A)} = M \equiv \sup_A f$, then $f(\mathbf{x}) = M$ on $\operatorname{int}(A)$, and we take $\mathbf{c}$ there.

Part (a) says that $m \le \frac{\int_A f}{V(A)} \le M$. We have accounted for the two extremes. The only possibility left is that both inequalities are strict. In that case, the quantity $\int_A \frac{f}{V(A)}$ lies strictly between $\inf_A f$ and $\sup_A f$. It follows (by corollaries to the intermediate value theorem) that some value of f matches it. □

It is worth repeating that integrals cannot resolve details within meager sets. As a result, we can often soften an assumption that something hold "throughout A" to the condition that it hold "on most of A," meaning on all but a subset of zero volume. Thus, in Theorem 3(a), if $k \le f \le K$ is only "mostly" true—if $\{\mathbf{x}\colon f(\mathbf{x}) < k \text{ or } f(\mathbf{x}) > K\}$ is nonempty, but has zero volume—then we still have $k\,V(A) \le \int_A f \le K\,V(A)$.

Exercises

1. Find examples of:

   (a) functions f and g such that $f > g$ throughout some set A, but $\int_A f = \int_A g$.

   (b) a function $h \geq 0$ and a set A with $\int_A h = 0$, $V(A) > 0$, and $h \neq 0$ on some part of A.

   (c) Is it possible to find a *continuous* h that has all the properties in (b)?

2. Prove that integrals of continuous functions are "positive semidefinite": If f is continuous and $f \geq 0$ throughout A, then $\int_A f > 0$, unless the only places where $f > 0$ are on the boundary of A. (Equivalently, if $f \geq g$ are continuous, then $\int_A f > \int_A g$, unless f and g match throughout the interior of A.)

3. Prove that if A and B do not overlap, then $V(A \cup B) = V(A) + V(B)$.

4. Prove Theorem 2(c): If $f \geq 0$ on T and $S \subseteq T$, then $\int_S f \leq \int_T f$.

5. Prove Theorem 3(a).

6. Prove that root-mean-square gives a bigger mean than average-value does: If f is integrable on A and $V(A) > 0$, then

$$\frac{1}{V(A)} \int_A |f| \leq \left[\frac{1}{V(A)} \int_A f^2\right]^{1/2}.$$

   (Hint: Write the Riemann sum $|f(\mathbf{x}_1)|V(S_1) + \cdots + |f(\mathbf{x}_K)|V(S_K)$ as

$$|f(\mathbf{x}_1)|\sqrt{V(S_1)}\sqrt{V(S_1)} + \cdots + |f(\mathbf{x}_K)|\sqrt{V(S_K)}\sqrt{V(S_K)}$$

   and apply Cauchy's inequality.)

5.3 Change of Variable

Substitution, or change of variable, is a powerful elementary tool in integration. We will show that it can be applied to multivariable functions. To get the analytic result, we will do some geometric thinking related to a linear-algebra theorem.

Let $\mathbf{v}_1, \ldots, \mathbf{v}_n$ be vectors in $\mathbf{R}^n$. Given numbers $a_1 \leq b_1, \ldots, a_n \leq b_n$, consider the set

$$S \equiv \{\alpha_1\mathbf{v}_1 + \cdots + \alpha_n\mathbf{v}_n : a_1 \leq \alpha_1 \leq b_1, \ldots, a_n \leq \alpha_n \leq b_n\}.$$

We call S a **parallelepiped**, and refer to each $(b_j - a_j)\mathbf{v}_j$ as the **edge** of S **along** $\mathbf{v}_j$. Write

$$\begin{aligned}
\mathbf{w}_1 &\equiv (b_1 - a_1)\mathbf{v}_1, \\
\mathbf{w}_2 &\equiv (b_2 - a_2)\mathbf{v}_2 - [(b_2 - a_2)\mathbf{v}_2 \bullet \mathbf{w}_1]\frac{\mathbf{w}_1}{\|\mathbf{w}_1\|^2}, \\
&\cdots \\
\mathbf{w}_n &\equiv (b_n - a_n)\mathbf{v}_n - [(b_n - a_n)\mathbf{v}_n \bullet \mathbf{w}_1]\frac{\mathbf{w}_1}{\|\mathbf{w}_1\|^2} \\
&\quad - \cdots - [(b_n - an)\mathbf{v}_n \bullet \mathbf{w}_{n-1}]\frac{\mathbf{w}_{n-1}}{\|\mathbf{w}_{n-1}\|^2},
\end{aligned}$$

with $\mathbf{w}_{j+1} \equiv \mathbf{O}$ if $\mathbf{w}_j = \mathbf{O}$. In words, each $\mathbf{w}_{j+1}$ is the vector component of $(b_{j+1} - a_{j+1})\mathbf{v}_{j+1}$ perpendicular to the subspace spanned by $\mathbf{v}_1, \ldots, \mathbf{v}_j$. Except for not being normalized (not being unit vectors), the $\mathbf{w}_j$ are what the Gram–Schmidt process would produce from the edges $(b_1 - a_1)\mathbf{v}_1, \ldots, (b_n - a_n)\mathbf{v}_n$ of S.

Our geometric experience would lead us to call $\mathbf{w}_2$ the "altitude" and

$$\|\mathbf{v}_1\|\,\|\mathbf{w}_2\| = \|\mathbf{w}_1\|\,\|\mathbf{w}_2\|$$

the "area" of the parallelogram determined by $(b_1 - a_1)\mathbf{v}_1$ and $(b_2 - a_2)\mathbf{v}_2$. Similarly, $\mathbf{w}_3$ would be the "altitude" and $\|\mathbf{w}_1\|\,\|\mathbf{w}_2\|\,\|\mathbf{w}_3\|$ the "volume" of the prism having edges $(b_1 - a_1)\mathbf{v}_1$, $(b_2 - a_2)\mathbf{v}_2$, and $(b_3 - a_3)\mathbf{v}_3$. Our visualization ends there, but it is natural to extend to n dimensions and call $\|\mathbf{w}_1\| \cdots \|\mathbf{w}_n\|$ the "(n-dimensional) volume" of the parallelepiped. Unfortunately, of course, this proposed name is taken; we must seek permission to reuse it.

Theorem 1. *With the notation introduced, the following quantities are equal:*

(a) *the product* $\|\mathbf{w}_1\|\,\|\mathbf{w}_2\| \cdots \|\mathbf{w}_n\|$ *of the "altitudes"*

(b) *the absolute value* $|\det[\mathbf{w}_1 \cdots \mathbf{w}_n]|$ *of the determinant that has the* $\mathbf{w}_j$ *for columns*

(c) $|\det[(b_1 - a_1)\mathbf{v}_1 \cdots (b_n - a_n)\mathbf{v}_n]|$, *the "absolute determinant of the edges"*

(d) *the product* $(b_1 - a_1) \cdots (b_n - a_n)\big(\det[\mathbf{v}_j \bullet \mathbf{v}_k]\big)^{1/2}$ *of the edge markers and the square root of the determinant whose* jk*-entry is the dot product of* $\mathbf{v}_j$ *and* $\mathbf{v}_k$.

Proof. If $\mathbf{L}$ is any square matrix, then

$$(\det \mathbf{L})^2 = (\det \mathbf{L}^t)(\det \mathbf{L}) = \det(\mathbf{L}^t\mathbf{L}).$$

In the matrix product on the right, the jk-entry is the dot product of row number j from $\mathbf{L}^t$, which is column number j from $\mathbf{L}$, with column number k from $\mathbf{L}$.

Factoring the differences in (c), we quickly see that (c) and (d) match. If the columns of $\mathbf{L}$ are orthogonal, then the dot products are 0 if $j \neq k$ and $\|\mathbf{w}_j\|^2$ if $j = k$. In this case, the matrix on the right is diagonal, with $\|\mathbf{w}_j\|^2$ at position j. That gives us (a) = (b).

Finally, $|\det[(b_1 - a_1)\mathbf{v}_1 \cdots (b_n - a_n)\mathbf{v}_n]|$ matches (b), because each column in (b) is obtained from the same column in this last determinant by subtracting multiples of the previous columns.

It is standard to abbreviate det(L) by det L and the matrix whose jk-entry is a_{jk} by $[a_{jk}]$. Unfortunately, it is also standard to signify $\det \mathbf{L}$ by $|\mathbf{L}|$, making the absolute value sign represent unlike things. Because we need to talk so much about the absolute value of a determinant, we will adopt the notation $\operatorname{absdet} \mathbf{L} \equiv |\det \mathbf{L}|$.

In Theorem 1, part (c) is surprising in that it gives a simple spatial interpretation to determinants, which are defined by numerical processes divorced from geometry. Note how it fits with the principle that the determinant is zero iff the vectors are dependent; in that case their "unit parallelepiped"

$$\{\alpha_1\mathbf{v}_1 + \cdots + \alpha_n\mathbf{v}_n : 0 \le \alpha_1 \le 1, \ldots, 0 \le \alpha_n \le 1\}$$

is flat, and the vectors are all in some hyperplane.

Part (d) is a nice formula, and we will have important use for it later.

We still have to prove that the product of the altitudes matches our Archimedean definition of volume.

Theorem 2. *The volume of a parallelepiped is the absolute value of the determinant of its edges.*

Proof. Let $\mathbf{L} \equiv [\mathbf{v}_1 \cdots \mathbf{v}_n]$ be the matrix with columns $\mathbf{v}_1, \ldots, \mathbf{v}_n$. We have

$$\alpha_1\mathbf{v}_1 + \cdots + \alpha_n\mathbf{v}_n = \mathbf{L}\begin{bmatrix}\alpha_1\\ \vdots \\ \alpha_n\end{bmatrix} = \mathbf{L}\langle\alpha_1\mathbf{e}_1 + \cdots + \alpha_n\mathbf{e}_n\rangle.$$

Hence S is the image of the box $[\mathbf{a}, \mathbf{b}] \equiv \big[(a_1, \ldots, a_n), (b_1, \ldots, b_n)\big]$ under the map that takes $\mathbf{x}$ to $\mathbf{L}\langle\mathbf{x}\rangle$. If $\mathbf{v}_1, \ldots, \mathbf{v}_n$ are dependent, then S is a bounded subset of some hyperplane; by (among other results) Example 4.4:1, $V(S) = 0 = \det \mathbf{L}$. We may therefore limit our attention to the case of independent vectors.

With $\mathbf{v}_1, \ldots, \mathbf{v}_n$ independent, $\mathbf{L}$ is an invertible matrix, which can be written as the product $\mathbf{L} = E_1 \cdots E_J$ of elementary matrices. Since the linear image of one parallelepiped is another one, we may find the volume of

$$\mathbf{L}([\mathbf{a}, \mathbf{b}]) = E_1(\ldots(E_J([\mathbf{a}, \mathbf{b}]))\ldots)$$

by tracking the effects of the E_j on the volumes of the parallelepipeds on which they operate. There being three types of elementary matrix, there are three types of effect.

(a)

(b)

(c)

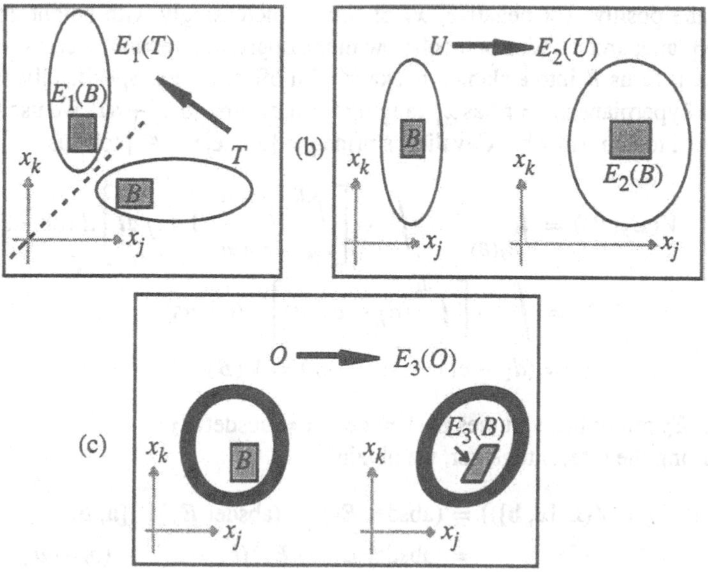

Figure 5.2.

(1) In part (a) of Figure 5.2, we label E_1 the type of elementary matrix made by switching rows j and k in the identity. The corresponding linear transformation reflects $\mathbf{R}^n$ about the hyperplane $x_j = x_k$. It turns the region T into its mirror image $E_1(T)$. It also turns the shaded box B into its reflection $E_1(B)$. Suppose

$$B \equiv \left[(\ldots, c_j, \ldots, c_k, \ldots), (\ldots, d_j, \ldots, d_k, \ldots)\right]$$

is one of the subintervals used to measure either the interior or the boundary of T. Then its image

$$E_1(B) = \left[(\ldots, c_k, \ldots, c_j, \ldots), (\ldots, d_k, \ldots, d_j, \ldots)\right]$$

obviously has the same volume, and would serve to tile the interior or cover the boundary of $E_1(T)$. We reason that E_1 preserves meagerness of boundary, and therefore Archimedeanness, as well as volume of the interior. Consequently, E_1 multiplies volumes by $1 = -(\det E_1) = \text{absdet } E_1$.

(2) In part (b) of the figure, we use E_2 to denote the elementary matrix resulting from multiplying row j by constant α. Its linear transformation stretches $\mathbf{R}^n$ by a factor of $|\alpha|$ in the x_j-direction, reflects about the hyperplane $x_j = 0$ if $\alpha < 0$, and leaves locations and sizes in the other coordinate directions unchanged. The effect on the measuring box B is to multiply one of its widths by $|\alpha|$ and, perhaps, turn it left-for-right. This time the transformation multiplies the volume of the tiling box by $|\alpha| = \text{absdet } E_2$, and we conclude that it has the same effect on any Archimedean volume.

(3) In part (c) of Figure 5.2, E_3 comes from adding the multiple (α times row k) to row j. The ("shear") transformation slides points above the hyperplane $x_k =$

0 in the positive (or negative) x_j-direction, increasingly with height above the hyperplane; and it slides points below in the opposite direction, increasingly with depth. It turns B into a slanted package with offset layers. Specifically, the layer in the hyperplane $x_k = t$ has x_j ranging from $c_j + \alpha t$ to $d_j + \alpha t$. We use Fubini's theorem to confirm what Cavalieri's principle [Kline, p. 349] tells us:

$$\begin{aligned} V(E_3(B)) = \int_{E_3(B)} 1 &= \int \cdots \left[\int_{c_k}^{d_k} \int_{c_j+\alpha t}^{d_j+\alpha t} 1\, dx_j\, dt \right] d\,(\text{others}) \\ &= \int \cdots \left[\int_{c_k}^{d_k} (d_j - c_j)\, dt \right] d\,(\text{others}) \\ &= (d_1 - c_1) \cdots (d_n - c_n) = V(B). \end{aligned}$$

Thus, E_3 multiplies volumes by $1 = \det E_3 = \text{absdet } E_3$.

Putting the effects together, we obtain

$$\begin{aligned} V(S) = V(\mathbf{L}([\mathbf{a}, \mathbf{b}])) &= (\text{absdet } E_1) \cdots (\text{absdet } E_J) V([\mathbf{a}, \mathbf{b}]) \\ &= (\text{absdet } E_1 \cdots E_J)(b_1 - a_1) \cdots (b_n - a_n) \\ &= \text{absdet}[\mathbf{v}_1 \cdots \mathbf{v}_n](b_1 - a_1) \cdots (b_n - a_n) \\ &= \text{absdet}[(b_1 - a_1)\mathbf{v}_1 \cdots (b_n - a_n)\mathbf{v}_n]. \qquad \square \end{aligned}$$

[I am grateful to my colleague Ethan Akin for pointing out the "elementary" approach in the proof of Theorem 2. Mathematicians are fond of methods that combine simplicity and power. "Ask Ethan" is a prime example of such techniques.]

The analytic result we want is the "substitution rule." In one variable, the rule looks like

$$\int f(u)\, du = \int f(u(x)) \left(\frac{du}{dx} \right) dx.$$

We have to make some adjustments, like making x and $u(x)$ vectors. Leaving that aside, examine the pieces here. Assume that there is no problem with f. For the right side to make sense, u has to be differentiable, and $\frac{du}{dx}$ has to be integrable over its domain. For the left side in turn to make sense, the range $\{u(x)\}$ must allow integration; it must be Archimedean. These considerations inform the assumptions we are about to make.

Assume that $\Phi = (\phi_1, \ldots, \phi_n)\colon \mathbf{R}^n \to \mathbf{R}^n$ is a continuously differentiable function on the open set O. Near any point where the Jacobian $\det\left(\frac{\partial \Phi}{\partial \mathbf{x}}\right)$ is nonzero, the inverse function theorem tells us that Φ maps open sets to open sets (Exercise 6(a) here). If $\det\left(\frac{\partial \Phi}{\partial \mathbf{x}}\right) \neq 0$ throughout O, then $\Phi(P)$ is open whenever $P \subseteq O$ is open (Exercise 6(b)); in particular, $\Phi(O)$ is an open set. Assume, therefore, that $\det\left(\frac{\partial \Phi}{\partial \mathbf{x}}\right) \neq 0$ on O. What we have said about open sets ensures that if $A \subseteq O$, then Φ maps the interior of A into the interior of $\Phi(A)$. The same cannot be said about boundary; it is possible for $\Phi(\text{bd}(A))$ to have points outside $\text{bd}(\Phi(A))$ (Exercise 6(d)). We need to track boundaries, because they decide whether sets are

Archimedean. To guarantee that boundary points map to boundary points—and exterior points to exterior points—we need to assume that Φ is (globally) one-to-one on O (Exercise 6(c)).

For the rest of this section, then, we study a continuously differentiable function Φ, with nonzero Jacobian, mapping the open set O one-to-one onto the open set $P \equiv \Phi(O)$. We call Φ a **transformation of coordinates** or **change of variable.** By the inverse function theorem, Φ has a differentiable inverse $\Psi = \Phi^{-1}$, and Ψ is likewise a transformation of coordinates. We write $J(\mathbf{x})$ for $\operatorname{absdet}(\frac{\partial \Phi}{\partial \mathbf{x}})$.

Theorem 3. *Suppose* $\mathrm{cl}(A) \subseteq O$. *(In other words, A is a subset of O, and the boundary of A is where* Φ *can apply to it.) Then A is Archimedean iff* $\Phi(A)$ *is Archimedean.*

Proof. The closure of a bounded set in $\mathbf{R}^n$ is compact, so there exists $2r > 0$ such that the distance from $\mathrm{cl}(A)$ to the complement O^* exceeds $2r$. The points at distance r or less from $\mathrm{cl}(A)$ form a closed, bounded set S with $\mathrm{cl}(A) \subseteq S \subseteq O$. Hence $\Phi'(\mathbf{x})$ is bounded on S; there exists a real M such that $\|\Phi'(\mathbf{x})\| \leq M$ for every $\mathbf{x} \in S$. Since J is continuous on O, we may assume that M is large enough to make $J(\mathbf{x}) \leq M$ on S, as well.

$\Rightarrow$ Assume that A is Archimedean, and let $\varepsilon > 0$. The boundary of A is meager, so we can find a partition $\mathcal{Q}$ with these properties: Its subintervals are cubes; its fineness satisfies $\|\mathcal{Q}\| < r$; and the subintervals $B_1, \ldots, B_K$ that intersect $\mathrm{bd}(A)$ have

$$V(B_1) + \cdots + V(B_K) < \frac{\varepsilon}{(2Mn)^n}.$$

These subintervals, each having points from $\mathrm{cl}(A)$ and diameter (diagonal) less than r, cannot reach outside S. Thus, $\mathrm{bd}(A) \subseteq B_1 \cup \cdots \cup B_K \subseteq S$.

Let $B \equiv [\mathbf{a}, \mathbf{b}]$ be one of these boundary subintervals. We want to measure the image $\Phi(B)$. For any $\mathbf{x} \in B$, the component ϕ_m satisfies

$$\begin{aligned} |\phi_m(\mathbf{x}) - \phi_m(\mathbf{a})| &= |\phi_m{}'(\mathbf{d}_m)\langle \mathbf{x} - \mathbf{a}\rangle| \quad (\text{for some } \mathbf{d}_m \in B) \\ &\leq \|\Phi'(\mathbf{d}_m)\| \, \|\mathbf{x} - \mathbf{a}\| \leq M\|\mathbf{b} - \mathbf{a}\|. \end{aligned}$$

Hence

$$\|\Phi(\mathbf{x}) - \Phi(\mathbf{a})\| \leq \sqrt{n}\, M\|\mathbf{b} - \mathbf{a}\|.$$

This tells us that $\Phi(\mathbf{x})$ lies within a box of sides $2\sqrt{n}\, M\|\mathbf{b} - \mathbf{a}\|$ centered at $\Phi(\mathbf{a})$. We conclude that $\Phi(B)$ is contained in a box of volume $(2\sqrt{n}\, M\|\mathbf{b} - \mathbf{a}\|)^n$. Since B is a cube, its diagonal is $\sqrt{n}$ times any side, so that its volume is

$$V(B) = \left(\frac{\|\mathbf{b} - \mathbf{a}\|}{\sqrt{n}}\right)^n.$$

Therefore, $\Phi(B)$ is covered by a box of volume $(2\sqrt{n}\, M)^n(\sqrt{n})^n V(B) = (2Mn)^n V(B)$.

With $\text{bd}(A) \subseteq B_1 \cup \cdots \cup B_K$, we have

$$\text{bd}(\Phi(A)) = \Phi(\text{bd}(A)) \subseteq \Phi(B_1) \cup \cdots \cup \Phi(B_K)$$

(the assumption that Φ is one-to-one being extremely important). From what we found about the $\Phi(B_k)$, we conclude that $\text{bd}(\Phi(A))$ is contained in an Archimedean set of volume no more than $[V(B_1) + \cdots + V(B_K)](2nM)^n \leq \varepsilon$. Therefore, the *circumvolume* of $\text{bd}(\Phi(A))$ is ε or less. It follows that $\text{bd}(\Phi(A))$ is meager; $\Phi(A)$ is Archimedean.

$\Leftarrow$ We have proved that if A is Archimedean, then $\Phi(A)$ is. The symmetry between O and Φ on the one hand and P and Ψ on the other tells us that the converse also holds. □

The next theorem is a lemma about the volume of the image of a cube. Suppose $B \subseteq O$ is cubic. By Theorem 3, $\Phi(B)$ is Archimedean, so we may speak of its volume. Within the proof of Theorem 3, we used a bound on $\|\Phi'\|$ to show that the volume is not too big. Now we sharpen our estimate to get an important approximation: If the cube is small, then the volume of its image is roughly the (absolute) Jacobian times the volume of the cube.

Theorem 4. *Assume that* $B \equiv [\mathbf{a}, \mathbf{b}] \subseteq O$ *is a cube and* $\mathbf{c}$ *is any point in* B. *Let* $K \equiv \max \|\Psi'(\Phi(\mathbf{x}))\|$ *and* $\Delta \equiv \max \|\Phi'(\mathbf{x}) - \Phi'(\mathbf{y})\|$ *over all* $\mathbf{x}, \mathbf{y} \in B$. *Then*

$$V(\Phi(B)) \leq V(B)(1 + 2K\Delta n)^n J(\mathbf{c}).$$

Proof. Note first that the two maxima exist: B is compact and Φ' is continuous in it; the continuous image $\Phi(B)$ must be compact, and Ψ' is continuous there.

Put the components

$$\phi_m(\mathbf{x}) - \phi_m(\mathbf{c}) = \phi_m'(\mathbf{d}_m)\langle \mathbf{x} - \mathbf{c}\rangle \text{ for some points } \mathbf{d}_1, \ldots, \mathbf{d}_n \in B$$

into

$$\Phi(\mathbf{x}) - \Phi(\mathbf{c}) = \begin{bmatrix} \phi_1'(\mathbf{d}_1) \\ \ddots \\ \phi_n'(\mathbf{d}_n) \end{bmatrix} \langle \mathbf{x} - \mathbf{c}\rangle.$$

Recalling that $\Phi'(\mathbf{x}) = \Psi'(\Phi(\mathbf{x}))^{-1}$, write

$$\Phi(\mathbf{x}) - \Phi(\mathbf{c}) = \Phi'(\mathbf{c})\Psi'(\Phi(\mathbf{c})) \begin{bmatrix} \phi_1'(\mathbf{d}_1) \\ \ddots \\ \phi_n'(\mathbf{d}_n) \end{bmatrix} \langle \mathbf{x} - \mathbf{c}\rangle.$$

On the right, the last three factors have dimensions $n \times n$, $n \times n$, and $n \times 1$, so their product is a column $\boldsymbol{\alpha}(\mathbf{x}) \equiv [\alpha_1(\mathbf{x}) \cdots \alpha_n(\mathbf{x})]^t$. The equation

$$\Phi(\mathbf{x}) - \Phi(\mathbf{c}) = \Phi'(\mathbf{c})\boldsymbol{\alpha}(\mathbf{x}) = \alpha_1(\mathbf{x}) \begin{bmatrix} \frac{\partial \phi_1}{\partial x_1}(\mathbf{c}) \\ \vdots \\ \frac{\partial \phi_n}{\partial x_1}(\mathbf{c}) \end{bmatrix} + \cdots + \alpha_n(\mathbf{x}) \begin{bmatrix} \frac{\partial \phi_1}{\partial x_n}(\mathbf{c}) \\ \vdots \\ \frac{\partial \phi_n}{\partial x_n}(\mathbf{c}) \end{bmatrix}$$

says that $\Phi(\mathbf{x})$ is in the $\Phi(\mathbf{c})$-translate of a parallelepiped determined by the vectors $\frac{\partial\Phi}{\partial x_k}(\mathbf{c})$, with the multipliers ranging from $\inf\{\alpha_k(\mathbf{x})\colon \mathbf{x} \in B\}$ to $\sup\{\alpha_k(\mathbf{x})\colon \mathbf{x} \in B\}$.

How long are those ranges? Write

$$\alpha(\mathbf{x}) = \Psi'(\Phi(\mathbf{c})) \begin{bmatrix} \phi_1'(\mathbf{c}) \\ \ddots \\ \phi_n'(\mathbf{c}) \end{bmatrix} \langle \mathbf{x} - \mathbf{c} \rangle + \Psi'(\Phi(\mathbf{c})) \begin{bmatrix} \phi_1'(\mathbf{d}_1) - \phi_1'(\mathbf{c}) \\ \ddots \\ \phi_n'(\mathbf{d}_n) - \phi_n'(\mathbf{c}) \end{bmatrix} \langle \mathbf{x} - \mathbf{c} \rangle.$$

The first product on the right is $\mathbf{x} - \mathbf{c}$. The other product has norm no more than

$$K(\sqrt{n}\Delta)\|\mathbf{x} - \mathbf{c}\| \le K(\sqrt{n}\Delta)\|\mathbf{b} - \mathbf{a}\| = K(\sqrt{n}\Delta)(b_k - a_k)\sqrt{n}.$$

Therefore,

$$\sup \alpha_k(\mathbf{x}) \le b_k - c_k + Kn\Delta(b_k - a_k),$$
$$\inf \alpha_k(\mathbf{x}) \ge a_k - c_k - Kn\Delta(b_k - a_k),$$

and

$$\sup \alpha_k(\mathbf{x}) - \inf \alpha_k(\mathbf{x}) \le b_k - a_k + 2Kn\Delta(b_k - a_k).$$

Hence $\Phi(\mathbf{x})$ is in a parallelepiped of volume no more than

$$(b_1 - a_1)(1 + 2Kn\Delta) \cdots (b_n - a_n)(1 + 2Kn\Delta) \operatorname{absdet} \left[\frac{\partial\Phi}{\partial x_1}(\mathbf{c}) \cdots \frac{\partial\Phi}{\partial x_n}(\mathbf{c}) \right]$$
$$= V(B)(1 + 2Kn\Delta)^n J(\mathbf{c}).$$

It follows that

$$V(\Phi(B)) \le V(B)(1 + 2K\Delta n)^n J(\mathbf{c}).$$ □

Theorem 5 (The Change of Variable Theorem, or Substitution Rule). *Assume that $\Phi(A)$ is Archimedean and f is continuous and bounded there. Then*

$$\int_{\Phi(A)} f(\mathbf{u})\, d\mathbf{u} = \int_A f\big(\Phi(\mathbf{x})\big) J(\mathbf{x})\, d\mathbf{x}.$$

In particular, if A is Archimedean, then $V\big(\Phi(A)\big) = \int_A J(\mathbf{x})$.

Proof. Suppose $\Phi(A)$ is Archimedean. Theorem 3 tells us that A is Archimedean. Also, the first paragraph of its proof establishes a compact set $S \subseteq O$ containing cl(A) and every vector within distance r of cl(A). Likewise, $\Phi(S)$ is compact. Therefore, Φ' is bounded on S, and Ψ' is bounded on $\Phi(S)$. With f bounded on A, let us assume that M is a bound for all three.

Let $\varepsilon > 0$ be specified. First, Φ' is uniformly continuous on S. Hence there is δ for which

$$\|\mathbf{x} - \mathbf{y}\| < \delta \text{ in } S \Rightarrow \|\Phi'(\mathbf{x}) - \Phi'(\mathbf{y})\| = \frac{\varepsilon}{2nM}.$$

Second, write BOX1 for some cube (probably not contained in O) containing A, BOX2 for one containing $\Phi(A)$. Let $\mathcal{Q}$ be any partition of BOX1 with five properties: Its subintervals are cubes; its norm is smaller than both r and δ; the subintervals $B_1, \dots, B_I$ that intersect bd(A) have volume-sum less than $\frac{\varepsilon}{M^2}$; the remaining subintervals are $B_{I+1}, \dots, B_K$ interior to A and $B_{K+1}, \dots, B_L$ exterior; the upper sum over $\mathcal{Q}$ for the function $f(\Phi(\mathbf{x}))J(\mathbf{x})$ is within ε of the function's integral on BOX1.

Next, $A \subseteq B_1 \cup \cdots \cup B_K$, giving $\Phi(A) \subseteq T \equiv \Phi(B_1) \cup \cdots \cup \Phi(B_K)$. The B_k do not overlap, so neither do their images. Further, the images of $B_{K+1}, \dots, B_L$ are exterior to $\Phi(A)$, forcing $\chi_{\Phi(A)} = 0$ outside T. Therefore,

$$\begin{aligned}\int_{\Phi(A)} f \equiv \int_{\text{BOX2}} f\chi_{\Phi(A)} &= \int_T f\chi_{\Phi(A)}\\ &= \int_{\Phi(B_1)} f\chi_{\Phi(A)} + \cdots + \int_{\Phi(B_K)} f\chi_{\Phi(A)}.\end{aligned}$$

On the boundary images $\Phi(B_1), \dots, \Phi(B_I)$, $f\chi_{\Phi(A)}$ matches either f or 0, so $|f\chi_{\Phi(A)}| \le M$. By Theorem 4, the volumes of those images satisfy

$$\begin{aligned}V(\Phi(B_i)) &\le V(B_i)\big[1 + 2n \max\|\Psi'\| \max\|\Phi'(\mathbf{x}) - \Phi'(\mathbf{y})\|\big]^n \max J(\mathbf{c})\\ &\le V(B_i)\left(1 + 2nM\left[\frac{\varepsilon}{2nM}\right]\right)^n M = MV(B_i)(1+\varepsilon)^n.\end{aligned}$$

(The maxima of Ψ' and J are M or less, because $B_i \subseteq S$.) Hence

$$\begin{aligned}&\int_{\Phi(B_1)} f\chi_{\Phi(A)} + \cdots + \int_{\Phi(B_I)} f\chi_{\Phi(A)}\\ &\qquad \le M^2[1+\varepsilon]^n\big(V(B_1) + \cdots + V(B_I)\big) < \varepsilon[1+\varepsilon]^n.\end{aligned}$$

On the interior images $\Phi(B_{I+1}), \dots, \Phi(B_K)$, $\chi_{\Phi(A)} = 1$ and f is continuous. By the average value theorem and Theorem 4,

$$\begin{aligned}\int_{\Phi(B_k)} f\chi_{\Phi(A)} = \int_{\Phi(B_k)} f &= f(\mathbf{d}_k)V\big(\Phi(B_k)\big) \quad \text{for some } \mathbf{d}_k = \Phi(\mathbf{c}_k),\\ &\le f\big(\Phi(\mathbf{c}_k)\big)V(B_k)J(\mathbf{c}_k)[1+\varepsilon]^n.\end{aligned}$$

Therefore,

$$\int_{\Phi(A)} f < \varepsilon[1+\varepsilon]^n + [1+\varepsilon]^n \sum_{k=I+1}^{K} f\big(\Phi(\mathbf{c}_k)\big)J(\mathbf{c}_k)V(B_k).$$

Now consider $\int_A f(\Phi(\mathbf{x}))J(\mathbf{x})\,d\mathbf{x}$. By construction of $\mathcal{Q}$, the upper sum for $f(\Phi)J$ has

$$\begin{aligned} u(f(\Phi)J, \mathcal{Q}) &= \left[\sup_{B_1} f(\mathbf{x})J(\mathbf{x})\chi_A(\mathbf{x})\right] V(B_1) \\ &\quad + \cdots + \left[\sup_{B_K} f(\mathbf{x})J(\mathbf{x})\chi_A(\mathbf{x})\right] V(B_K) \\ &\le \int_A f(\Phi(\mathbf{x}))J(\mathbf{x})\,d\mathbf{x} + \varepsilon. \end{aligned}$$

In the middle expression, the first I terms are small:

$$\begin{aligned} &\left|\left[\sup_{B_1} f(\mathbf{x})J(\mathbf{x})\chi_A(\mathbf{x})\right] V(B_1) + \cdots + \left[\sup_{B_I} f(\mathbf{x})J(\mathbf{x})\chi_A(\mathbf{x})\right] V(B_I)\right| \\ &\qquad \le MM\big(V(B_1) + \cdots + V(B_I)\big) \le \frac{MM\varepsilon}{M^2} < \varepsilon. \end{aligned}$$

Hence

$$\begin{aligned} &\sum_{k=I+1}^{K} f(\Phi(\mathbf{c}_k))J(\mathbf{c}_k)V(B_k) \\ &\qquad \le \left[\sup_{B_{I+1}} f(\Phi(\mathbf{x}))J(\mathbf{x})\right] V(B_{I+1}) + \cdots + \left[\sup_{B_K} f(\Phi(\mathbf{x}))\right] V(B_K) \\ &\qquad \le \int_A f(\Phi(\mathbf{x}))J(\mathbf{x})\,d\mathbf{x} + \varepsilon + \varepsilon. \end{aligned}$$

We have arrived at

$$\int_{\Phi(A)} f < \varepsilon[1+\varepsilon]^n + [1+\varepsilon]^n \left(\int_A f(\Phi(\mathbf{x}))J(\mathbf{x})\,d\mathbf{x} + 2\varepsilon\right).$$

This being true for arbitrary ε, we conclude that

$$\int_{\Phi(A)} f(\mathbf{u})\,d\mathbf{u} \le \int_A f(\Phi(\mathbf{x}))J(\mathbf{x})\,d\mathbf{x}.$$

We have shown that the transformed integral is smaller. But transformations work both ways. We invoke symmetry again. Write $F(\mathbf{x}) \equiv f(\Phi(\mathbf{x}))J(\mathbf{x})$. Because F is a continuous function on $A = \Psi(\Phi(A))$, the argument above tells us that

$$\int_{\Psi(\Phi(A))} F(\mathbf{x})\,d\mathbf{x} \le \int_{\Phi(A)} F(\Psi(\mathbf{u}))\left[\operatorname{absdet}\frac{\partial\Psi}{\partial\mathbf{u}}(\mathbf{u})\right] d\mathbf{u}.$$

The value $F(\Psi(\mathbf{u}))$ is $f(\Phi(\Psi(\mathbf{u})))J(\Psi(\mathbf{u})) = f(\mathbf{u})J(\Psi(\mathbf{u}))$. By the inverse function theorem, $\left[\frac{\partial\Psi}{\partial\mathbf{u}}(\mathbf{u})\right]$ is $\left[\frac{\partial\Phi}{\partial\mathbf{x}}(\Psi(\mathbf{u}))\right]^{-1}$. Hence its absolute determinant is

$\frac{1}{J(\Psi(\mathbf{u}))}$. The transformation inequality therefore translates as

$$\int_A f(\Phi(\mathbf{x}))J(\mathbf{x})\,d\mathbf{x} \le \int_{\Phi(A)} f(\mathbf{u})J(\Psi(\mathbf{u}))\frac{1}{J(\Psi(\mathbf{u}))}\,d\mathbf{u} = \int_{\Phi(A)} f(\mathbf{u})\,d\mathbf{u}. \qquad \square$$

Example 1. On average, how far from the origin are the points in the unit disk?

Write $D \equiv \{(x, y) : x^2+y^2 \le 1\}$, $F(x, y) \equiv (x^2+y^2)^{1/2}$. We need to calculate

$$\frac{\int_D F(x, y)d(x, y)}{V(D)} = \frac{\int_D F}{\int_D 1}.$$

(a) The most elementary of transformations is in $\mathbf{R}^2$ between rectangular and polar coordinates, which are suggested by the radial symmetry in the question. We know that the transformation Φ from polar to rectangular coordinates is given by

$$(x, y) = \Phi(r, \theta) \equiv (r\cos\theta, r\sin\theta), \qquad 0 \le r,\ 0 \le \theta \le 2\pi.$$

This description is problematic for two reasons: The stated domain is not an open subset of the $r\theta$-plane, and Φ is not one-to-one there. Those objections are unavoidable whenever the nonnegative x-axis is in the domain of integration.

Nevertheless, such integrals can be transformed in a standard way. Figure 5.3 shows $D(\varepsilon)$, the image of

$$A(\varepsilon) \equiv \{(r, \theta) : \varepsilon \le r \le 1, \varepsilon \le \theta \le 2\pi - \varepsilon\}.$$

The last is a closed subset of the open set O defined by $0 < r, 0 < \theta < 2\pi$. On O, Φ is one-to-one, and therefore Theorem 5 applies:

$$\int_{D(\varepsilon)} F(x, y)\,d(x, y) = \int_{A(\varepsilon)} F(\Phi(r, \theta)) \left|\frac{\partial(x, y)}{\partial(r, \theta)}\right| d(r, \theta).$$

By continuity of integrals (Theorem 5.2:2(d)), $\int_{D(\varepsilon)} F(\mathbf{x})$ tends to $\int_D F$ as $\varepsilon \to 0$, because the volume $V(D - D(\varepsilon)) \le 4\tan\varepsilon \to 0$. Similarly, $\int_{A(\varepsilon)} F(\Phi(\mathbf{r}))J(\mathbf{r})$

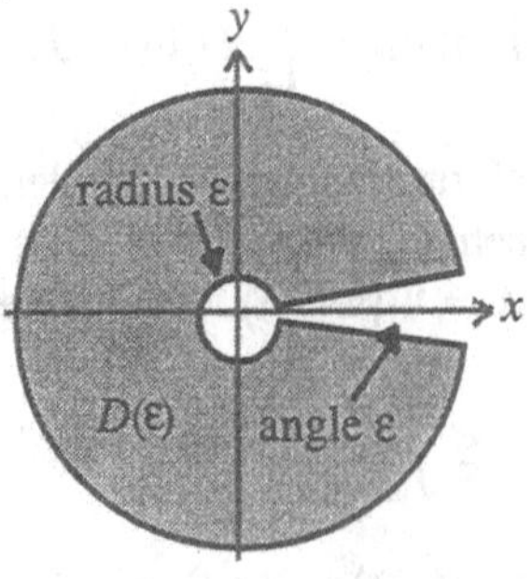

Figure 5.3.

tends to $\int_A F(\Phi(\mathbf{r}))J(\mathbf{r})$, where now A is the rectangle $0 \le r \le 1, 0 \le \theta \le 2\pi$. Therefore, we may treat the original integral as though Theorem 5 applied to it directly.

(b) The calculation is easy. We have

$$\det \frac{\partial \Phi}{\partial(r,\theta)} = \begin{vmatrix} \cos\theta & -r\sin\theta \\ \sin\theta & r\cos\theta \end{vmatrix} = r,$$

the last being positive in O. The volume part of the theorem gives us

$$V(D) = \int_A 1\, r\, d(r,\theta) = \int_0^{2\pi}\int_0^1 r\, dr\, d\theta = \pi.$$

(Compare the hint in Exercise 1.) By the transformation part,

$$\int_D F(x,y)d(x,y) = \int_A F(\Phi(r,\theta))\, r\, d(r,\theta) = \int_0^{2\pi}\int_0^1 r\, r\, dr\, d\theta = \frac{2\pi}{3}.$$

Hence the average distance is $\frac{2}{3}$.

(c) Why is this average more than half the radius? Compare Exercise 5.1:8.

Observe that part (b) of the change of variables theorem attaches geometric significance to the Jacobian. If $\det\left(\frac{\partial\Phi}{\partial\mathbf{x}}\right)$ is nonzero at $\mathbf{x} = \mathbf{a}$ and $\frac{\partial\Phi}{\partial\mathbf{x}}$ is continuous nearby—so that Φ is one-to-one nearby—then $J(\mathbf{a})$ is the instantaneous rate at which Φ causes volume to expand near $\mathbf{a}$. We could write $J(\mathbf{a}) = \frac{dV(\Phi(A))}{dV(A)}$, and the transformation equation would take the form

$$\int_{\Phi(A)} f(\mathbf{u})\, dV(\Phi(A)) = \int_A f(\Phi(\mathbf{x}))\left[\frac{dV(\Phi(A))}{dV(A)}\right] dV(A).$$

Alternatively, we may think of the Jacobian as

$$J(\mathbf{a}) = \lim_{A\to\{\mathbf{a}\}} \frac{V(\Phi(A))}{V(A)}, \qquad A \text{ near } \mathbf{a},$$

where the precise meaning is that for any $\varepsilon > 0$, there is $\delta > 0$ with the property that if A is an Archimedean subset (not necessarily including $\mathbf{a}$) of $N(\mathbf{a},\delta)$, then

$$J(\mathbf{a}) - \varepsilon < \frac{V(\Phi(A))}{V(A)} < J(\mathbf{a}) + \varepsilon.$$

Notice also that this principle goes back to our discussion of differentiability and its relation to approximate linearity. If Φ is differentiable at $\mathbf{x} = \mathbf{a}$, then its action near $\mathbf{a}$ resembles that of a linear mapping. Accordingly, it transforms small boxes near $\mathbf{a}$ into almost-parallelepipeds. Because they approximate parallelepipeds, the image volumes are given roughly by determinants.

Finally, consider the meaning of the absolute-value sign. Recall that the sign of a determinant has to do with reflections. Reflections preserve volumes, so whether the Jacobian is 4 or -4, its effect at small scale is to multiply volumes by 4. (Compare Exercise 5.)

Exercises

1. Use the transformation $x = au$, $y = bv$ to find the area of the elliptical region $\frac{x^2}{a^2} + \frac{y^2}{b^2} \leq 1$. (Hint: The area of a unit disk is π.)

2. In the triangle with vertices at $(0, 0)$, $(3, 0)$, and $(3, 4)$, use polar coordinates to find the average values of $f(x, y) \equiv x$ and $g(x, y) \equiv y$. (For $A \subseteq \mathbf{R}^n$, the averages
$$\frac{\int_A x_1}{V(A)}, \ldots, \frac{\int_A x_n}{V(A)}$$
are the coordinates of the **centroid** of A.)

3. For the cone $\sqrt{(x^2 + y^2)} \leq z \leq 4$, use spherical coordinates (in which the determinant of $\frac{\partial(x,y,z)}{\partial(\rho,\theta,\phi)}$ is $\rho^2 \sin\phi$) to find:

   (a) the volume,

   (b) the (volume-)average distance from the origin,

   (c) the centroid.

4. In $\mathbf{R}^4$, spherical coordinates (r, s, t, u) are given (compare Exercise 4.4:4) by $x_1 = r\cos s$, $x_2 = r\sin s\cos t$, $x_3 = r\sin s\sin t\cos u$, $x_4 = r\sin s\sin t\sin u$, $0 \leq r$, $0 \leq s \leq \pi$, $0 \leq t \leq \pi$, $0 \leq u \leq 2\pi$. Use these to find the volume of the ball of radius a. (Compare the answer with Example 5.1:2.)

5. Assume that $u = \phi(x)$ is continuously differentiable on the real interval $[a, b]$ and beyond, with $\phi'(x) \neq 0$ throughout. (The "beyond" is not essential; it saves worry about boundaries.) Let $c = \phi(a)$, $d = \phi(b)$. Use Theorem 5 to show that if $f(u)$ is continuous on $[c, d]$, then (the one-variable substitution rule)
$$\int_c^d f(u)\,du = \int_a^b f(\phi(x))\left[\frac{d\phi}{dx}\right]dx.$$
Explain the missing absolute-value sign.

6. Given $\Phi\colon \mathbf{R}^n \to \mathbf{R}^n$, continuously differentiable on the open set O:

   (a) Assume that $\mathbf{a} \in O$ and $J(\mathbf{a}) \neq 0$ (equivalently, $\frac{\partial\Phi}{\partial\mathbf{x}}(\mathbf{a})$ is invertible). Use Theorems 3.2:2 and 3.2:3 to show that in some open set surrounding $\mathbf{a}$, Φ maps open sets to open sets. (A function with this property is called an **open mapping**.)

   (b) Show that if $J(\mathbf{x}) \neq 0$ throughout O, then Φ maps open subsets of O into open sets.

(c) Assume that Φ is one-to-one on O. Show that if A and its boundary are contained in O, then Φ maps the interior of A onto the interior of $\Phi(A)$, the boundary of A onto the boundary of $\Phi(A)$, and the remaining $O - \mathrm{cl}(A)$ onto the remainder $\Phi(O) - \mathrm{cl}(\Phi(A))$.

(d) Adapt Example 3.2:1 to show that if Φ is not globally one-to-one, then Φ can map points from both $\mathrm{bd}(A)$ and $\mathrm{ext}(A)$ into the interior of $\Phi(A)$.

7. Prove the assertion, in the paragraph right after Example 1 of this section, that

$$J(\mathbf{a}) = \lim_{A \to \{\mathbf{a}\}} \frac{V(\Phi(A))}{V(A)}.$$

5.4 Generalized Integrals

Our definitions related to Riemann integrals restrict us to bounded regions and bounded functions. We know, however, that it is worthwhile to attach meanings to integrals with arbitrarily large domains or integrands. In this section we extend our notions to allow regions and functions that reach toward infinity.

Handling the regions is direct. Given $S \subseteq \mathbf{R}^n$, set

$$S_k \equiv S \cap B(\mathbf{O}, k) = \{\mathbf{x} \in S : \|\mathbf{x}\| \leq k\}.$$

In terms of set containment, (S_k) is an increasing sequence of sets, whose supremum, and therefore limit, is S. It would be handy to think of integrals on S as limits of integrals on S_k. The latter integrals require S_k to be Archimedean. If each S_k is Archimedean, then the part of S in any fixed neighborhood is Archimedean.

Definition. A set is **locally Archimedean** (we abbreviate it to LA) if its intersection with any neighborhood is Archimedean. If every such intersection is meager, then we say that the set is **locally meager**.

Notice that this really is a generalization: Every Archimedean set is LA (Exercise 6). To develop a wide class of LA sets, we extend Theorems 4.4:1, 3.

Theorem 1. *Assume $S, T \subseteq \mathbf{R}^n$.*

(a) *S is* LA *iff its boundary is locally meager.*

(b) *If S and T are* LA, *then so are S^*, $S \cup T$, $S \cap T$, and $S - T$.*

Proof. (a) $\Rightarrow$ Suppose S is LA. Let N be a neighborhood. If $\mathbf{x} \in \mathrm{bd}(S) \cap N$, then every small-enough neighborhood of $\mathbf{x}$ has points entirely from N, some from S, and some from S^*. Thus, $\mathbf{x} \in \mathrm{bd}(S \cap N)$. Hence $\mathrm{bd}(S) \cap N \subseteq \mathrm{bd}(S \cap N)$. Since

$S \cap N$ is Archimedean, its boundary is meager. We conclude that $\text{bd}(S) \cap N$ is meager.

$\Leftarrow$ Suppose instead that every intersection $\text{bd}(S) \cap N$ is meager. If $\mathbf{y} \in \text{bd}(S \cap N)$, then every neighborhood of $\mathbf{y}$ has points from N, so $\mathbf{y}$ cannot be exterior to N. Thus, either $\mathbf{y}$ is in N, in which case it must also be on $\text{bd}(S)$, or $\mathbf{y} \in \text{bd}(N)$. Consequently,

$$\text{bd}(S \cap N) \subseteq \left[\text{bd}(S) \cap N\right] \cup \text{bd}(N),$$

so that $\text{bd}(S \cap N)$ is meager. Hence $S \cap N$ is Archimedean, and S is LA.

(b) Exercise 5. □

Next we treat the functions. We want to allow unbounded functions, but we cannot afford widespread unboundedness. Specifically, we cannot allow the function to be unbounded near every point of an entire neighborhood. Such a function would have to be discontinuous at every point of the neighborhood, and we have already admitted (Section 4.4) that we are unable to deal with such mischief. Accordingly, we confine the unboundedness to the vicinity of a locally meager set. Thus, $f(x_1, \ldots, x_n) \equiv \frac{1}{x_1}$ is unbounded in any neighborhood of a point on the hyperplane $x_1 = 0$, but near any other point, it is bounded. In fact, in any neighborhood whose closure does not intersect the hyperplane, f is integrable. That "local integrability" is the kind of behavior we require.

Definition. The function f is **locally integrable** on the LA set S if there is a locally meager set U such that f is integrable on every Archimedean set A with $\text{cl}(A) \subseteq S - U$.

The definition requires f to be defined on S; as before, we extend a given function to $\mathbf{R}^n$ by setting it to zero wherever it is not defined. Observe that in the definition, we may replace "set A with $\text{cl}(A) \subseteq S - U$" with "closed subset of $S - U$". Further, it suffices for f to be integrable on closed sets interior to S and exterior to U, because a function satisfying this requirement also satisfies the definition, with U replaced by $U \cup \text{bd}(U) \cup \text{bd}(S)$. Finally, local integrability is not too expensive an assumption. We know that, for instance, a function continuous on an LA domain T (and zero outside T) fits this description, with $U \equiv \text{bd}(T)$.

Now to the integrals. Our preference would be to extend directly the definition of improper integral on the real line by, for example, setting

$$\int_{\mathbf{R}^n} f = \lim_{M \to \infty} \int_{\|\mathbf{x}\| \le M} f.$$

This provision would be troublesome, because the elementary definition of improper integral has a number of weaknesses.

Example 1. Consider $\int_0^\infty \frac{\sin x}{x}\, dx$. The impropriety is due only to the unbounded domain, since the integrand has a finite limit at $x = 0$. If $0 < 2k\pi \le M <$

$(2k+2)\pi$, then

$$\left|\int_0^M \frac{\sin x}{x}\,dx - \sum_{i=0}^{2k-1}\left[\int_{i\pi}^{(i+1)\pi} \frac{\sin x}{x}\,dx\right]\right| = \int_{2k\pi}^M \frac{\sin x}{x}\,dx$$
$$\leq \left(\frac{1}{2k\pi}\right)\int_0^{2\pi} |\sin x|\,dx = \frac{4}{2k\pi}.$$

Hence

$$\lim_{M\to\infty}\int_0^M \frac{\sin x}{x}\,dx = \sum_{i=0}^{\infty}\int_{i\pi}^{(i+1)\pi} \frac{\sin x}{x}\,dx.$$

From interval to interval, the integrands on the right switch signs and are squeezed toward zero. Therefore, the series falls under the alternating series test, and the improper integral converges.

This convergence-by-cancellation means that $f(x) \equiv \frac{\sin x}{x}$ has an integral on $[0, \infty)$, but not on the subset $[0, \pi] \cup [2\pi, 3\pi] \cup \cdots$ (Exercise 4). Similarly, f has an integral, but $|f|$ does not. These typify the kind of bad feature we wish to eliminate.

To avoid the situation where an integral exists because an infinity of negative contributions cancels an infinity of positives, we will take an approach that accounts for the two separately.

Definition. Assume that S is LA and f is locally integrable on S. If

$$\mathrm{Itg}(f, S) \equiv \left\{\int_A f : A \subseteq S \text{ is closed and Archimedean and } f \text{ is integrable on } A\right\}$$

is a bounded set of real numbers, then we say that f **is integrable on** S, and call $\sup\ \mathrm{Itg}(f, S) + \inf\ \mathrm{Itg}(f, S)$ the **integral of** f **over** S.

Notice that $\inf \mathrm{Itg}(f, S) \leq 0 \leq \sup\ \mathrm{Itg}(f, S)$ is trivial.

Example 2. Roughly speaking, a function is integrable near infinity if it is (absolutely) small enough out there, and integrable near $\mathbf{x} = \mathbf{a}$ if it is not too big nearby.

(a) Let $f(\mathbf{x}) \equiv (1 + \|\mathbf{x}\|)^{-n-1}$. Since f is continuous, it is locally integrable on $\mathbf{R}^n$. Given any closed Archimedean set A, let K be an integer such that $A \subseteq B(\mathbf{O}, K)$. First, $\int_A f \geq 0$, so $\inf\ \mathrm{Itg}(f, \mathbf{R}^n) = 0$. Second,

$$\int_A f \leq \int_{\|\mathbf{x}\|\leq K} f = \sum_{k=1}^{K}\int_{k-1\leq\|\mathbf{x}\|\leq k} f$$
$$\leq \sum k^{-n-1} V\big(B(\mathbf{O}, k) - N(\mathbf{O}, k-1)\big)$$
$$= \sum k^{-n-1}[k^n - (k-1)^n] V\,(\text{unit ball}).$$

(Why does multiplying the radius by k increase the volume to the tune of k^n?) Since the series converges, we have $\sup \mathrm{Itg}(f, \mathbf{R}^n) < \infty$. Hence f is integrable on $\mathbf{R}^n$. (Compare Exercise 1.)

(b) Let $g(x, y) \equiv \frac{1}{xy}$ and D be the unit disk in $\mathbf{R}^2$. It is clear from the symmetry that we need only study g in quadrant I. However, even in the absence of symmetry, we may deal separately with the sets on which f has positive and negative values; see Exercise 10.

Let $k > 0$ be an integer. Consider, from the band bordered by the hyperbolas $xy = 2^{-k-3}$ and $xy = 2^{-k-2}$, the part A_k between $x = 2^{-k-1}$ and $x = 2^{-k}$. (Verify that for this set of x, the higher of the hyperbolas is below the unit circle.) We have $g \geq 2^{k+2}$ on A_k and

$$V(A_k) = \int_{2^{-k-1}}^{2^{-k}} \frac{(2^{-k-2} - 2^{-k-3})}{x}\, dx = 2^{-k-3} \ln 2.$$

(Why can we find the volume with a one-variable integral?) Hence

$$\int_{A_k} g \geq \frac{\ln 2}{2}.$$

Since the regions A_k are nonoverlapping subsets of D, we conclude that $\sup \int_A g = \infty$, and g is not integrable on D. (Compare Exercise 2.)

(c) The same g, which is too big at $\mathbf{O}$, is insufficiently small at ∞. In the infinite box $S \equiv \big[(1, 1), \infty\big]$, let $T \equiv \big[(1, 1), (M, M)\big]$. Then

$$\int_T f = \int_1^M \int_1^M \frac{1}{xy}\, dx\, dy = (\ln M)^2.$$

It follows that g is not integrable on S.

The extended definitions allow us to extend theorems about properties of integrals.

Theorem 2. *Assume that S and T are LA, and f is locally integrable on S.*

(a) *If S has empty interior, then f is integrable on S, with zero integral.*

(b) *Suppose $T \subseteq S$. Then f is integrable on S iff it is integrable on both T and $S - T$, in which case*

$$\int_S f = \int_T f + \int_{S-T} f.$$

(c) (Set Additivity) *Suppose f is also locally integrable on T. Then f is integrable on $S \cup T$ iff it is integrable on each of S and T, in which case*

$$\int_{S\cup T} f = \int_S f + \int_T f - \int_{S\cap T} f.$$

In particular, if S and T do not overlap, then $\int_{S\cup T} f = \int_S f + \int_T f$.

(d) (Function and Set Monotonicity) *If $f \geq 0$ is integrable on S and $T \subseteq S$, then*

$$0 \leq \int_T f \leq \int_S f.$$

(e) (Set Continuity) *Suppose f is integrable on S. Then the integral is concentrated in a bounded part of S: For any $\varepsilon > 0$, there is an Archimedean $A \subseteq S$ with the property that*

$$T \subseteq S - A \Rightarrow \left|\int_T f\right| < \varepsilon.$$

(f) (Linearity) *If f and g are integrable on S, then so is $\alpha f + \beta g$, and*

$$\int_S \alpha f + \beta g = \alpha \int_S f + \beta \int_S g.$$

(g) (The Triangle Inequality) *Assume that $T^+ \equiv \{\mathbf{x} \in S\colon f(\mathbf{x}) > 0\}$ and $T^- \equiv \{\mathbf{x} \in S\colon f(\mathbf{x}) < 0\}$ are LA. Then f is integrable on S iff $|f|$ is, in which case $\left|\int_S f\right| \leq \int_S |f|$. (Recall that if $f > 0$ on a non-Archimedean subset of its domain, then it is possible to have $|f|$ integrable and f not. Check Exercise 10d.)*

Proof. In these arguments, $A \subseteq S$ is a closed Archimedean set on which f is integrable.

(a) If S has empty interior, then so does A. Hence f is integrable with $\int_f = 0$ for every A, and the conclusion is immediate.

(b) $\Rightarrow$ First observe that f must be locally integrable on T. Suppose now f is integrable on S and $A \subseteq T$. Then also $A \subseteq S$, so

$$-\infty < \inf \operatorname{Itg}(f, S) \leq \int_A f \leq \sup \operatorname{Itg}(f, S) < \infty.$$

Hence f is integrable on T, and similarly for $S - T$, which is LA by Theorem 1.

$\Leftarrow$ Suppose, conversely, f is integrable on T and $S - T$, and let $A \subseteq S$. Then

$$\int_A f = \int_{A \cap T} f + \int_{A \cap (S-T)} f,$$

because the two intersections are Archimedean (Exercise 6) and nonoverlapping. On the right, $\int_{A\cap T} f$ is a member of $\operatorname{Itg}(f, T)$. Similarly, $\int_{A\cap(S-T)} f \in \operatorname{Itg}(f, S - T)$. Hence $\int_A f$ is a sum of numbers from bounded sets. We conclude that f is integrable on S.

From the last equality, we also have

$$\sup \operatorname{Itg}(f, S) \equiv \sup_{A \subseteq S} \int_A f \leq \sup \operatorname{Itg}(f, T) + \sup \operatorname{Itg}(f, S - T).$$

On the other hand, for any $\varepsilon > 0$, there exist Archimedean sets $C \subseteq T$ and $D \subseteq S-T$ such that $\int_C f$ and $\int_D f$ are within $\frac{\varepsilon}{2}$ of $\sup \mathrm{Itg}(f, T)$ and $\sup \mathrm{Itg}(f, S-T)$, respectively. Hence

$$\sup \mathrm{Itg}(f, T) + \sup \mathrm{Itg}(f, S-T) - \varepsilon < \int_C f + \int_D f$$
$$= \int_{C \cup D} f \leq \sup \mathrm{Itg}(f, S).$$

Necessarily,

$$\sup \mathrm{Itg}(f, T) + \sup \mathrm{Itg}(f, S-T) \leq \sup \ \mathrm{Itg}(f, S).$$

It follows that these quantities are equal, and similarly for the infimums. We conclude that $\int_S f = \int_T f + \int_{S-T} f$.

(c) and (d) are Exercise 7.

(e) Let f be integrable and $\varepsilon > 0$. We will invent a set B to take care of the positive values of f and a set C to take in the negatives.

By definition, there exists B with

$$\int_B f > \sup \mathrm{Itg}(f, S) - \frac{\varepsilon}{2}.$$

By (b),

$$\int_{S-B} f = \int_S f - \int_B f = \sup \ \mathrm{Itg}(f, S) + \inf \ \mathrm{Itg}(f, S) - \int_B f$$
$$< \inf \ \mathrm{Itg}(f, S) + \frac{\varepsilon}{2}.$$

Now,

$$\inf \ \mathrm{Itg}(f, S-B) = \int_{S-B} f - \sup \ \mathrm{Itg}(f, S-B) < \inf \ \mathrm{Itg}(f, S) + \frac{\varepsilon}{2} - 0,$$

so there must be a closed Archimedean subset C of $S - B$ for which

$$\int_C f < \inf \ \mathrm{Itg}(f, S) + \frac{\varepsilon}{2}.$$

The concentration is in $A \equiv B \cup C$.

Let T be any LA subset of $S-A$. Every closed, Archimedean $D \subseteq T$ is disjoint from B, so that

$$\int_D f = \int_{B \cup D} f - \int_B f \leq \sup \mathrm{Itg}(f, S) - \int_B f < \frac{\varepsilon}{2},$$

and is likewise disjoint from C, so that

$$\int_D f = \int_{C \cup D} f - \int_C f \geq \inf \ \mathrm{Itg}(f, S) - \int_C f > -\frac{\varepsilon}{2}.$$

Hence

$$\left|\int_T f\right| \le \left|\sup \int_D f\right| + \left|\inf \int_D f\right| \le \varepsilon.$$

(f) By hypothesis, there are meager sets U and V such that f is integrable on every $A \subseteq S - U$, and g is integrable on $A \subseteq S - V$. Let $A \subseteq S - (U \cup V)$. Then f and g are integrable on A. By Theorem 5.2:1(a), $\alpha f + \beta g$ is also integrable on A, and

$$\int_A \alpha f + \beta g = \alpha \int_A f + \beta \int_A g.$$

Hence $\alpha f + \beta g$ is locally integrable, and its local integrals are sums from the bounded sets α Itg(f, S) and β Itg(g, S). We infer that $\alpha f + \beta g$ is integrable on S.

We may therefore use part (e) on f, g, and $\alpha f + \beta g$. For any $\varepsilon > 0$, there are Archimedean sets B, C, D such that integrals of f, g, and $\alpha f + \beta g$ on $S - B$, $S - C$, $S - D$, respectively, are dominated by ε. Let $A \equiv B \cup C \cup D$. Then

$$\begin{aligned}
&\left|\int_S (\alpha f + \beta g) - \left(\alpha \int_S f + \beta \int_S g\right)\right| \\
&\quad \le \left|\int_S (\alpha f + \beta g) - \int_A (\alpha f + \beta g)\right| + \left|\int_A (\alpha f + \beta g) - \left(\alpha \int_S f + \beta \int_S g\right)\right| \\
&\quad = \left|\int_{S-A} \alpha f + \beta g\right| + \left|-\alpha \int_{S-A} f - \beta \int_{S-A} g\right| \quad \text{(by part (c))} \\
&\quad \le \varepsilon + |\alpha|\varepsilon + |\beta|\varepsilon.
\end{aligned}$$

We conclude that

$$\int_S \alpha f + \beta g = \alpha \int_S f + \beta \int_S g.$$

(g) First observe that the local integrability of f guarantees that $|f|$ is locally integrable; the opposite implication is false.

Let T^0 be the subset of S on which $f = 0$. Assume that T^+ and T^- are LA. Then $T^0 = S - T^+ - T^-$ is LA. From (b), (f), and (d), we obtain

$$\begin{aligned}
f \text{ is integrable on } S &\Leftrightarrow f \text{ is integrable on } T^+, T^0 \text{ (automatic), and } T^- \\
&\Leftrightarrow f \text{ is integrable on } T^+ \text{ and } -f \text{ is integrable on } T^- \\
&\Leftrightarrow |f| \text{ is integrable on } T^+ \text{ and } T^- \\
&\Leftrightarrow |f| \text{ is integrable on } S,
\end{aligned}$$

and

$$\left|\int_S f\right| = \left|\int_{T+} f + \int_{T-} f\right| \le \left|\int_{T+} f\right| + \left|\int_{T-} f\right| = \int_{T+} f + \int_{T-} -f = \int_S |f|.$$

□

Example 3. One standard, but interesting, use of our integrals is the evaluation of $\int_{-\infty}^{\infty} \exp\left(-x^2\right) dx$. It cannot be performed by the fundamental theorem, because the integrand does not have an elementary antiderivative. Still, its value is essential in probability, so somebody has to do it.

Let $f(x, y) \equiv \exp\left(-x^2 - y^2\right)$. This function is extremely small at infinity, so it is integrable (compare Exercise 1(a)), and it has one sign. For such a function, we have

$$\begin{aligned}\int_{\mathbf{R}^2} f &= \lim_{M\to\infty} \int_{|x|\le M, |y|\le M} e^{-x^2-y^2} \\ &= \lim_{M\to\infty} \int_{-M}^{M} e^{-x^2}\,dx \int_{-M}^{M} e^{-y^2}\,dy \qquad \text{(Reason?)} \\ &= \left(\int_{-\infty}^{\infty} \exp\left(-x^2\right)\,dx\right)^2.\end{aligned}$$

On the other hand, we also have

$$\begin{aligned}\int_{\mathbf{R}^2} f &= \lim_{M\to\infty} \int_{x^2+y^2\le M^2} e^{-x^2-y^2} \\ &= \lim_{M\to\infty} \int_{0\le r\le M, 0\le\theta\le 2\pi} \exp\left(-r^2\right) r\,dr\,d\theta \qquad \text{(Reason?)} \\ &= \lim_{M\to\infty} \left[2\pi \frac{\exp(-r^2)}{-2}\right]_0^M = \pi.\end{aligned}$$

Hence $\int_{-\infty}^{\infty} \exp\left(-x^2\right)\,dx = \sqrt{\pi}$.

Exercises

1. (a) Show that if $f(\mathbf{x}) = O(\|\mathbf{x}\|^{-p})$, for a fixed $p > n$, then f is integrable near ∞.

   (b) Find the integral of $\left(1 + \sqrt{x^2 + y^2}\right)^{-p}$, $p > 2$ fixed, on $\mathbf{R}^2$.

   (c) Show that $(1 + \|\mathbf{x}\|)^{-p}$ is not integrable on $\mathbf{R}^n$, $p \le n$ fixed.

2. (a) Show that if $f(\mathbf{x}) = O\left(\|\mathbf{x}\|^{-q}\right)$, for a fixed $q < n$, then f is integrable in the unit ball.

   (b) Find the integral of $\left(x^2 + y^2\right)^{-q/2}$, $q < 2$ fixed, on the unit disk in $\mathbf{R}^2$.

   (c) Show that $\|\mathbf{x}\|^{-q}$ is not integrable on $\mathbf{R}^n$, $q \ge n$ fixed.

3. Decide whether the given function is integrable on the unit disk D in $\mathbf{R}^2$, and separately on D^*.

   (a) $\frac{xy}{(x^2+y^2)^2}$

(b) $\frac{\sin xy}{xy}$

(c) $\sin\left(x^2+y^2\right)^{-2}$

(d) $\left(x^4+y^4\right)^{-1}$.

4. Show that $\frac{\sin x}{x}$ is not integrable on $[0,\pi]\cup[2\pi,3\pi]\cup\cdots$.

5. Prove that the complement, union, intersection, and difference of LA sets are LA.

6. (a) Show that a set is Archimedean iff it is LA and bounded.

   (b) Show that if S is LA and A is Archimedean, then $S\cap A$ is Archimedean.

7. Prove parts (c) and (d) of Theorem 2.

8. (The Comparison Test) Given locally integrable functions f and g on an LA set S, prove that if $|g|\le f$ and f is integrable on S, then g is integrable on S.

9. Is the product of integrable functions on an LA set necessarily integrable?

10. A standard approach to generalized integrals in $\mathbf{R}^n$ is to separate the positive and negative values of f. The function $f^+(\mathbf{x})\equiv\max\{f(\mathbf{x}),0\}$ is called the **positive part** of f, $f^-(\mathbf{x})\equiv\max\{-f(\mathbf{x}),0\}$ the **negative part** of f. (These definitions apply to any real number, not just function values. Notice that just as the imaginary part of a complex number is not imaginary, the negative part of a real number is not negative.)

    (a) Show that $f=f^+-f^-$ and $|f|=f^++f^-$.

    (b) Show that on an LA set, f is locally integrable iff each of f^+ and f^- is.

    (c) Assume that S, $T^+\equiv\{\mathbf{x}\in S\colon f>0\}$, and $T^-\equiv\{\mathbf{x}\in S\colon f<0\}$ are LA. Show that f is integrable on S iff f^+ and f^- are, in which case $\int_S f=\int_S f^+-\int_S f^-$.

    (d) Give an example to show that $\{\mathbf{x}\in S\colon f(\mathbf{x})>0\}$ may fail to be LA.

5.5 Line Integrals

The mathematical objects called "integrals" always involve limits of weighted sums, in which function values are weighted by some measure of size of the region in which they prevail. Our notions so far have used volume in $\mathbf{R}^n$ for the weight. In the next two sections we relate integrals to lower-dimensional indicators of size.

We begin with what we call "one-dimensional" sets. Suppose we have an arc in $\mathbf{R}^n$ and a scalar function f defined on its range C. We would like to define the integral $\int_C f\,ds$ of f along C in terms of sums like $y_1 \Delta s_1 + \cdots + y_J \Delta s_J$, each y_j representing a value—or the infimum, or supremum—of f along a piece of the range and Δs_j the length of that piece. To do so, we must address such questions as what is meant by "length" of a nonstraight point set and whether the pieces add up to something that can be called the length of the arc (so that, for example, $\int_C 1\,ds$ matches the length of C). As usual, our attack is based on the geometry of the Greeks: We approximate the curve by polygonal paths.

Definition. Let $\mathbf{g}\colon [a, b] \to \mathbf{R}^n$ be continuous and $T \equiv \{a = t_0 < t_1 < \cdots < t_J = b\}$ a partition of $[a, b]$. The sum

$$\text{len}(\mathbf{g}, T) \equiv \|\mathbf{g}(t_1) - \mathbf{g}(t_0)\| + \cdots + \|\mathbf{g}(t_J) - \mathbf{g}(t_{J-1})\|$$

is the length of a polygonal path anchored along the **contour** $C \equiv \{\mathbf{g}(t)\colon a \le t \le b\}$, with the same endpoints. We call $\text{len}(\mathbf{g}) \equiv \sup\{\text{len}(\mathbf{g}, T)\colon T \text{ partitions } [a, b]\}$ the **length** of $\mathbf{g}$; if $\text{len}(\mathbf{g})$ is finite, then we say that $\mathbf{g}$ is **rectifiable**.

Notice that it is essential to assume continuity. The function

$$\mathbf{g}(t) \equiv \begin{cases} (2t, 0) & \text{if } 0 \le t \le \frac{1}{2}, \\ (2t + 1, 0) & \text{if } \frac{1}{2} < t \le 1, \end{cases}$$

maps the unit interval into the union of two disjoint pieces of the x-axis in $\mathbf{R}^2$. It should be clear that $\text{len}(\mathbf{g}, T) = 3$ for any partition of $[0, 1]$. Thus, $\mathbf{g}$ has polygons of bounded length. However, they all ascribe to $\mathbf{g}$ the length of the gap between the pieces, a set of points that is not in the range of $\mathbf{g}$.

We have defined length and rectifiability in terms of the parametrization. There is no doubt that they are representation-dependent (Exercise 2). However, it is an important fact that rectifiability and arclength are, with certain provisos, *intrinsic to the contour*. That is, as long as a parametrization does not cheat by retracing parts of the contour—if it does not turn and go backward or go around the contour more than once—then the length it ascribes to the contour matches what any other honest parametrization would.

Theorem 1. *Suppose* $\mathbf{g}\colon [a, b] \to \mathbf{R}^n$ *and* $\mathbf{h}\colon [c, d] \to \mathbf{R}^n$ *have equal range, starts, and ends, and both are one-to-one. Then* $\text{len}(\mathbf{g}) = \text{len}(\mathbf{h})$. (See also Exercise 8.)

Proof. First we prove that $\mathbf{h}^{-1}$ is continuous. (There is a theorem in topology that guarantees this, but we are analysts.) Let $t \in [c, d]$ and $\mathbf{x} = \mathbf{h}(t)$. Name a small $\varepsilon > 0$, and consider the image $S \equiv \mathbf{h}\big([c, t-\varepsilon] \cup [t+\varepsilon, d]\big)$. This is a compact set, to which $\mathbf{x}$ does not belong. Hence $\mathbf{x}$ has positive distance 2δ from S. If $\mathbf{y}$ is on

C (the range of g) and within δ of $\mathbf{x}$, then $y = \mathbf{h}(t^*)$ for a t^* that can be nowhere except $(t - \varepsilon, t + \varepsilon)$. Thus,

$$\|\mathbf{x} - \mathbf{y}\| < \delta \Rightarrow |\mathbf{h}^{-1}(\mathbf{x}) - \mathbf{h}^{-1}(\mathbf{y})| = |t - t^*| < \varepsilon,$$

proving that $\mathbf{h}^{-1}$ is continuous.

Now let $T \equiv \{t_0 < t_1 < \cdots < t_J\}$ partition $[a, b]$. We know that $\mathbf{h}^{-1}(\mathbf{g})$: $[a, b] \to [c, d]$ is continuous and one-to-one. By a theorem from the real case [Ross, Theorem 18.6], $\mathbf{h}^{-1}(\mathbf{g})$ must be strictly monotonic. Hence

$$c = \mathbf{h}^{-1}(\mathbf{g}(t_0)) < \mathbf{h}^{-1}(\mathbf{g}(t_1)) < \cdots < \mathbf{h}^{-1}(\mathbf{g}(t_J)) = d$$

is a partition of $[c, d]$. Write $u_j \equiv \mathbf{h}^{-1}(\mathbf{g}(t_j))$, $j = 0, \ldots, J$. Since

$$\begin{aligned}&\|\mathbf{g}(t_1) - \mathbf{g}(t_0)\| + \cdots + \|\mathbf{g}(t_J) - \mathbf{g}(t_{J-1})\| \\ &\quad = \|\mathbf{h}(u_1) - \mathbf{h}(u_0)\| + \cdots + \|\mathbf{h}(u_J) - \mathbf{h}(u_{J-1})\|,\end{aligned}$$

we conclude that len(**g**) $\le$ len(**h**). By symmetry, the two lengths are equal. □

We do not want to restrict ourselves to one-to-one functions, so the question of parametrization-dependency will always be with us. Nevertheless, we will allow ourselves to abuse the language with such phrases as "length of C" and "integral on C."

Rectifiability is characterized by "bounded variation" (see Exercises 6 and 7). We must leave detailed study of that concept to the next course. However, we do want to show that rectifiability is a property of the arcs that interest us most.

Example 1. (a) Every smooth curve is rectifiable.

Assume that **g** is continuously differentiable on $[a, b]$. Then $\mathbf{g}'$ is bounded, say $\|\mathbf{g}'\| \le M$, on the interval. Applying the mean-value theorem to the components of **g**, we have

$$\begin{aligned}&\|\mathbf{g}(t_1) - \mathbf{g}(t_0)\| + \cdots + \|\mathbf{g}(t_J) - \mathbf{g}(t_{J-1})\| \\ &= \left\| \begin{bmatrix} g_1'(t_{11}^*) \\ \ddots \\ g_n'(t_{1n}^*) \end{bmatrix} (t_1 - t_0) \right\| + \cdots + \left\| \begin{bmatrix} g_1'(t_{J1}^*) \\ \ddots \\ g_n'(t_{Jn}^*) \end{bmatrix} (t_j - t_{J-1}) \right\| \\ &\le \sqrt{n}M(t_1 - t_0 + \cdots + t_{J-1} - t_{J-1}) = \sqrt{n}M(b - a).\end{aligned}$$

(b) An arc need not be rectifiable.

Examine the graph of $y = x \sin \frac{1}{x}$, described by

$$\mathbf{h}(t) \equiv \left(t, t \sin\left(\frac{1}{t}\right)\right), \qquad 0 < t \le \frac{2}{\pi}, \quad \mathbf{h}(0) = (0, 0).$$

Fix J odd, and let T have $t_0 = 0$, $t_1 = \frac{1}{J\pi/2}$, $t_2 = \frac{1}{(J-1)\pi/2}, \ldots, t_J = \frac{1}{\pi/2}$.

For each odd j,

$$\mathbf{h}(t_j) = \left(\frac{2}{[J-j+1]\pi}, \pm\frac{2}{[J-j+1]\pi}\right),$$

while $\mathbf{h}(t_{j\pm 1})$ are both on the x-axis. The distance from $\mathbf{h}(t_j)$ to its predecessor or follower exceeds $\frac{2}{(J-j+1)\pi}$. Hence

$$\operatorname{len}(\mathbf{h}, T) > \frac{4}{J\pi} + \cdots + \frac{4}{3\pi} + \frac{2}{\pi}.$$

The last is a partial sum from a divergent series of positive terms. We conclude that $\mathbf{h}$ is unrectifiable.

Notice that in part (a) above, we did not need the smoothness condition $\mathbf{g}'(t) \neq \mathbf{O}$. (Actually, this particular argument did not even use continuous differentiability, just boundedness of the derivative; compare Exercise 6(b).) The situation will recur often enough to reward an abbreviation. We will say that a function **is of class** (or **belongs to class** or simply **is**) C^1 if it is differentiable and the derivative is continuous. More generally, $\mathbf{f} \in C^j$ if $\mathbf{f}$ is j-times differentiable and the jth derivative is continuous.

Example 2. (Space-Filling Arc) Example 1(b) shows that mere continuity does not guarantee rectifiability. Still, the arc there is recognizably one-dimensional. In this example we show that the behavior of an arc can be much worse: (The range of) an arc may have not just infinite length but positive volume.

We work from a sequence of arcs in the unit square of $\mathbf{R}^2$. Consider the path in Figure 5.4(a). We see that it has piecewise-linear parametrization $\mathbf{F}_1(t) =$

(a)

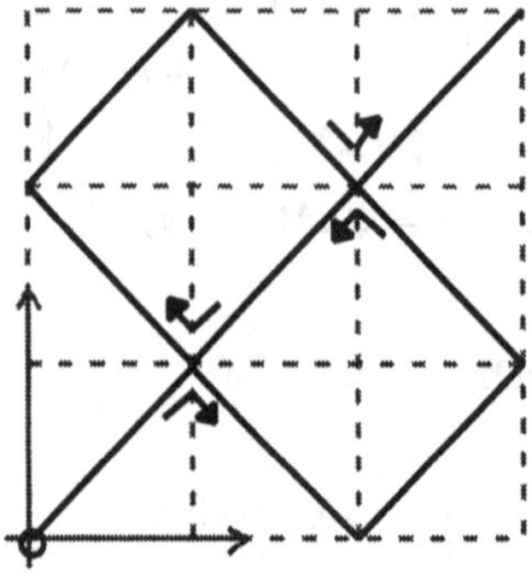

(b)

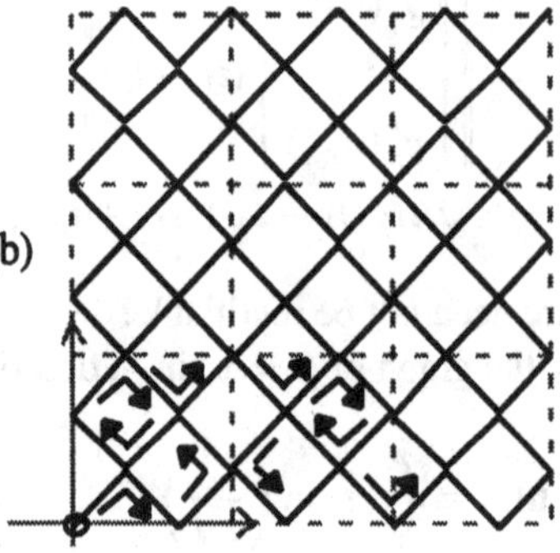

Figure 5.4.

$\left(F_{1x}(t), F_{1y}(t)\right)$ given by

$$F_{1x}(t) = \begin{cases} 3t, & 0 \le t \le \frac{1}{3}, \\ 1 - 3\left(t - \frac{1}{3}\right), & \frac{1}{3} \le t \le \frac{2}{3}, \\ 3\left(t - \frac{2}{3}\right), & \frac{2}{3} \le t \le 1, \end{cases}$$

$$F_{1y}(t) = \begin{cases} 3t, & 0 \le t \le \frac{1}{9} \\ \frac{1}{3} - 3\left(t - \frac{1}{9}\right), & \frac{1}{9} \le t \le \frac{2}{9} \\ \cdots & \text{(seven others).} \end{cases}$$

In Figure 5.4(b) we replace each straight part of the first arc by a one-third-size model of the whole arc, oriented so as to have the same start and end as the replaced part. Thus, we define $\mathbf{F}_2 = (F_{2x}, F_{2y})$ in the interval $\frac{j}{9} \le t \le \frac{(j+1)}{9}$ by

$$F_{2x}(t) \equiv F_{1x}\left(\frac{j}{9}\right) + F_{1x}\left(9\left[t - \frac{j}{9}\right]\right)\left[F_{1x}\left(\frac{j+1}{9}\right) - F_{1x}\left(\frac{j}{9}\right)\right],$$

$$F_{2y}(t) \equiv F_{1y}\left(\frac{j}{9}\right) + F_{1y}\left(9\left[t - \frac{j}{9}\right]\right)\left[F_{1y}\left(\frac{j+1}{9}\right) - F_{1y}\left(\frac{j}{9}\right)\right].$$

Notice that the figure is accurate in suggesting that the image of the interval $\left[\frac{j}{9}, \frac{j+1}{9}\right]$ under $\mathbf{F}_2$ is contained in the box whose diagonal ends at $\mathbf{F}_1\left(\frac{j}{9}\right)$ and $\mathbf{F}_1\left(\frac{j+1}{9}\right)$.

We continue this process. In general, $\mathbf{F}_k$ is linear in each closed interval $\left[\frac{j}{9^k}, \frac{j+1}{9^k}\right]$, the two endpoint images determining one diagonal of a box $\frac{1}{3^k}$ on each side. In this box $\mathbf{F}_{k+1}$ replaces the diagonal, substituting instead the zigzag defined by

$$F_{(k+1)x}(t) \equiv F_{kx}\left(\frac{j}{9^k}\right) + F_{1x}\left(9^k\left[t - \frac{j}{9^k}\right]\right)\left[F_{kx}\left(\frac{j+1}{9^k}\right) - F_{kx}\left(\frac{j}{9^k}\right)\right],$$

$$F_{(k+1)y}(t) \equiv F_{ky}\left(\frac{j}{9^k}\right) + F_{1y}\left(9^k\left[t - \frac{j}{9^k}\right]\right)\left[F_{ky}\left(\frac{j+1}{9^k}\right) - F_{ky}\left(\frac{j}{9^k}\right)\right].$$

We observe that for $\frac{j}{9^k} \le t \le \frac{j+1}{9^k}$, $\mathbf{F}_{k+1}(t)$ starts at $\mathbf{F}_k\left(\frac{j}{9^k}\right)$, ends at $\mathbf{F}_k\left(\frac{j+1}{9^k}\right)$, and stays in the box they determine. It easily follows that the images of this interval under $\mathbf{F}_{k+2}, \mathbf{F}_{k+3}, \ldots$ also travel within the box from $\mathbf{F}_k\left(\frac{j}{9^k}\right)$ to $\mathbf{F}_k\left(\frac{j+1}{9^k}\right)$.

The last remark implies that $(\mathbf{F}_k)$ is a uniformly Cauchy sequence. Hence it converges uniformly. Since each $\mathbf{F}_k$ is continuous, the limit function $\mathbf{F}$ is continuous.

Now let $\mathbf{p}$ be a point in the unit square. The range of $\mathbf{F}_1$ bisects every one-third square, so some point $\mathbf{F}_1(t_1)$ is within $\frac{1}{3}$ of $\mathbf{p}$. Similarly, some point $\mathbf{F}_2(t_2)$ is within $\frac{1}{9}$ of $\mathbf{p}, \ldots, \mathbf{F}_i(t_i)$ is within $\frac{1}{3^i}$ of $\mathbf{p}, \ldots$. The real-number sequence (t_i) must have a subsequence $(t_{m(i)})$ convergent to a real t. Then for any i,

$$\|\mathbf{p} - \mathbf{F}(t)\| \le \|\mathbf{p} - \mathbf{F}_{m(i)}(t_{m(i)})\| + \|\mathbf{F}_{m(i)}(t_{m(i)}) - \mathbf{F}(t_{m(i)})\| + \|\mathbf{F}(t_{m(i)}) - \mathbf{F}(t)\|.$$

The three terms on the right can be squeezed to zero, owing respectively to convergence to $\mathbf{p}$, uniform convergence, and uniform continuity. We conclude that $\mathbf{p} = \mathbf{F}(t)$. Thus, the range of $\mathbf{F}$ is the whole square; $\mathbf{F}$ is a two-dimensional arc.

Rectifiable arcs will be our domains of length-integration. We will define these integrals after we exhibit some properties of arclength. (In the exercises for this section we make a similar split: arc length problems, then integral problems.) Because these properties are shared by the integrals, we will attach to them the names they bear in the context of regular (volume-) integrals.

Theorem 2. *Assume* $C = \mathbf{g}([a, b])$ *is an arc in* $\mathbf{R}^n$.

(a) (Partition Additivity) *Let* $a < c < b$ *and* $\mathbf{h}$ *and* $\mathbf{H}$ *be the restrictions of* $\mathbf{g}$ *to* $[a, c]$ *and* $[c, b]$, *respectively. Then* $\text{len}(\mathbf{g}) = \text{len}(\mathbf{h}) + \text{len}(\mathbf{H})$.

(b) (Darboux's Lemma) *The length of C is the limiting length of the polygons, as the norm of the partition approaches zero.*

(c) (Continuity) *Suppose C is rectifiable. Then for any $\varepsilon > 0$, there is $\delta > 0$ such that*

$$0 < u - t < \delta \text{ in } [a, b] \Rightarrow \text{len}\big(\mathbf{g}([t, u])\big) < \varepsilon.$$

Equivalently, the function $s(t) \equiv \text{len}\big(\mathbf{g}([a, t])\big)$ *is continuous.*

Proof. It is handy to remember that refinement of a partition increases the length of the polygon. Thus, let $T \equiv \{t_0 < t_1 < \cdots < t_J\}$, and suppose U refines T by adding the single point u between t_j and t_{j+1}. Then $\text{len}(\mathbf{g}, U)$ differs from $\text{len}(\mathbf{g}, T)$ by the detour $\|\mathbf{g}(u) - \mathbf{g}(t_j)\| + \|\mathbf{g}(t_{j+1}) - \mathbf{g}(u)\| - \|\mathbf{g}(t_{j+1}) - \mathbf{g}(t_j)\|$, which is nonnegative. (Reason?)

(a) Suppose $L \equiv \{a < t_1 < \cdots < t_K = c\}$ and $R \equiv \{c < t_{K+1} < \cdots < t_J = b\}$ are partitions of their respective intervals. Then $L \cup R$ partitions $[a, b]$, and clearly,

$$\text{len}(\mathbf{h}, L) + \text{len}(\mathbf{H}, R) = \text{len}(\mathbf{g}, L \cup R) \leq \text{len}(\mathbf{g}).$$

Since L and R are arbitrary and independent, we conclude that

$$\text{len}(\mathbf{h}) + \text{len}(\mathbf{H}) \leq \text{len}(\mathbf{g}).$$

On the other hand, let $T \equiv \{t_0 < t_1 < \cdots < t_J\}$ partition $[a, b]$. Assuming $t_K \leq c < t_{K+1}$, let $L \equiv \{a < t_1 < \cdots < t_K \leq c\}$ and $R \equiv \{c < t_{K+1} < \cdots < t_J = b\}$. Then $L \cup R$ is a refinement of T, so

$$\text{len}(\mathbf{g}, T) \leq \text{len}(\mathbf{g}, L \cup R) = \text{len}(\mathbf{h}, L) + \text{len}(\mathbf{H}, R) \leq \text{len}(\mathbf{h}) + \text{len}(\mathbf{H}).$$

Hence

$$\text{len}(\mathbf{g}) \leq \text{len}(\mathbf{h}) + \text{len}(\mathbf{H}).$$

(b) The argument in (a) does not require **g**, **h**, or **H** to be rectifiable. The one we give here—a variation on our usual ε–δ argument—has the same property.

Assume $M < \text{len}(\mathbf{g})$. By definition, there exists a partition $T = \{t_0 < t_1 < \cdots < t_J\}$ such that $\text{len}(\mathbf{g}, T) > M$. Set $\varepsilon \equiv \frac{\text{len}(\mathbf{g},T)-M}{3}$. Because **g** is uniformly continuous, there is δ with $|u - t| < \delta$ in $[a, b] \Rightarrow \|\mathbf{g}(u) - \mathbf{g}(t)\| < \frac{\varepsilon}{J}$.

Now let U be a partition with $\|U\| < \delta$. A given subinterval $[u_{k-1}, u_k]$ of U may have points from T interior to it; write, for example,

$$u_{k-1} < t_{j+1} < \cdots < t_{j+m(k)} < u_k.$$

If there are no such points—if $m(k) = 0$—then $\Delta_k \equiv \|\mathbf{g}(u_k) - \mathbf{g}(u_{k-1})\|$ is one of the terms that make up $\text{len}(\mathbf{g}, T \cup U)$. If there are such points, then

$$\Delta_k \equiv \|\mathbf{g}(t_{j+1}) - \mathbf{g}(u_{k-1})\| + \cdots + \|\mathbf{g}(u_k) - \mathbf{g}(t_{j+m(k)})\|$$

is the contribution to $\text{len}(\mathbf{g}, T \cup U)$ from $[u_{k-1}, u_k]$. Each term is less than $\frac{\varepsilon}{J}$, so their sum is at most $[m(k) + 1]\frac{\varepsilon}{J} \le 2m(k)\frac{\varepsilon}{J}$. Either way,

$$\Delta_k \le \|\mathbf{g}(u_k) - \mathbf{g}(u_{k-1})\| + 2m(k)\frac{e}{J}.$$

Summing the Δ_k, we obtain

$$\text{len}(\mathbf{g}, T \cup U) \le \text{len}(\mathbf{g}, U) + 2\varepsilon.$$

Thus,

$$\text{len}(\mathbf{g}, U) \ge \text{len}(\mathbf{g}, T \cup U) - 2\varepsilon \ge \text{len}(\mathbf{g}, T) - 2\varepsilon > M.$$

Beginning with $M < \text{len}(\mathbf{g})$, we have found δ such that

$$\|U\| < \delta \Rightarrow M < \text{len}(\mathbf{g}, U) \le \text{len}(\mathbf{g}).$$

This proves that

$$\text{len}(\mathbf{g}) = \lim_{\|U\| \to 0} \text{len}(\mathbf{g}, U).$$

(c) Assume that C is rectifiable and $\varepsilon > 0$. Pick δ so small that

$$\|T\| < \Delta \Rightarrow \text{len}(\mathbf{g}, T) > \text{len}(\mathbf{g}) - \frac{\varepsilon}{2} \qquad \text{(part(b))}$$

and

$$|u - t| < \delta \text{ in } [a, b] \Rightarrow \|\mathbf{g}(u) - \mathbf{g}(t)\| < \frac{\varepsilon}{2} \qquad \text{(uniform continuity).}$$

Pick u and t with $0 < u - t < \delta$. Let L be a δ-fine partition of $[a, t]$ and **h** the restriction of **g** there; similarly with R and **H** on $[u, b]$. Then $T \equiv L \cup R$ is a δ-fine partition of $[a, b]$, so

$$\begin{aligned} \text{len}(\mathbf{g}) - \frac{\varepsilon}{2} &< \text{len}(\mathbf{g}, T) \\ &= \text{len}(\mathbf{h}, L) + \|\mathbf{g}(u) - \mathbf{g}(t)\| + \text{len}(\mathbf{H}, R) \\ &< \text{len}(\mathbf{h}) + \frac{\varepsilon}{2} + \text{len}(\mathbf{H}). \end{aligned}$$

Hence

$$\varepsilon > \text{len}(\mathbf{g}) - \text{len}(\mathbf{h}) - \text{len}(\mathbf{H}),$$

and by additivity, the last is the length of $\mathbf{g}([t, u])$.

The equivalence of the two conclusions is also a consequence of additivity. □

Definition. Assume that $\mathbf{g}$ is rectifiable and f is bounded along the range C of $\mathbf{g}$. Let $T \equiv \{a = t_0 < t_1 < \cdots < t_J = b\}$ be a partition of the domain of $\mathbf{g}$. Write $C_j \equiv \mathbf{g}([t_{j-1}, t_j])$ for the "jth piece of the curve," Δs_j for $\text{len}(C_j)$, $M_j \equiv \sup\{f(\mathbf{x})\colon \mathbf{x} \in C_j\}$ for the supremum of f along that part of the curve, $m_j \equiv \inf\{f(\mathbf{x})\colon \mathbf{x} \in C_j\}$ for the corresponding infimum. It is natural to call $u(f, T) \equiv M_1\Delta s_1 + \cdots + M_J \Delta s_J$ the **upper sum** for f on T, $l(f, T) \equiv m_1\Delta s_1 + \cdots + m_J\Delta s_J$ the **lower sum**. If

$$\inf\{u(f, T)\colon T \text{ partitions } [a, b]\} = \sup\{l(f, T)\colon T \text{ partitions } [a, b]\},$$

then we say that f **is integrable over** C and call the common value the **integral of** f **on** C.

We next exhibit some expected results, together with the main method for evaluating these **line** (or **contour**) **integrals**, denoted by $\int_C f\, ds$.

Theorem 3. *Let C be rectifiable and f bounded on C.*

(a) *If the restriction of f to C is continuous, then f is integrable over C. In particular, 1 is integrable over C, and $\int_C 1\, ds = \text{len}(C)$.*

(b) *f is integrable over C iff the "Riemann sums" $\sum f(\mathbf{x}_j)\Delta s_j$, $\mathbf{x}_j \in C_j$, have a limit as the norm of T vanishes, in which case the integral is the limit of the Riemann sums.*

(c) *f is integrable over C iff the "chord-length Riemann sums" $\sum f(\mathbf{x}_j)\Delta L_j$, with $\Delta L_j \equiv \|\mathbf{g}(t_j) - \mathbf{g}(t_{j-1})\|$, have a limit; and the integral is the limit of these sums.*

(d) *If $\mathbf{g}$ is C^1 and f is integrable on C, then*

$$\int_C f\, ds = \int_a^b f(\mathbf{g}(t))\|\mathbf{g}'(t)\|\, dt.$$

In particular, the length of C is $\int_a^b \|\mathbf{g}'(t)\|\, dt$.

Proof. (a) is sufficiently like Theorem 4.2:2(b) that we may simply sketch its proof. We have observed that partitioning squeezes sums together. Hence, as in Theorem 4.2:1(d), no lower sum exceeds any upper sum. Consequently, a function is integrable iff it has upper sums close to lower sums (like 4.2:2(a)). If f is continuous on the compact set C, then uniform continuity guarantees this closeness.

The statement about the integral of 1 is trivial.

(b) is like Theorem 4.2:4. By the argument of Darboux's lemma (Theorem 4.2:3 and Theorem 2(b) here), the infimum of the upper sums and the supremum of the lower sums are actually their limits. If f is integrable, then its Riemann sums are sandwiched; if, conversely, the Riemann sums have a limit, then upper and lower sums, which can be approximated by Riemann sums, must be close together.

(c) Keep in mind that ΔL_j is the straight distance between the ends of C_j, and $\Delta s_j = \text{len}(C_j)$. Consequently, $\Delta L_j \le \Delta s_j$. Also, by extension of Theorem 2(a), $\sum \Delta s_j = \text{len}(C)$.

Let $\varepsilon > 0$ and $|f| \le M$. By Theorem 2(b), for fine-enough partitions we have

$$0 \le \text{len}(C) - \sum \|\mathbf{g}(t_j) - \mathbf{g}(t_{j-1})\| < \frac{\varepsilon}{M}.$$

For such partitions,

$$\begin{aligned}\left|\sum f(\mathbf{x}_j)\Delta s_j - \sum f(\mathbf{x}_j)\Delta L_j\right| &\le M \sum |\Delta s_j - \Delta L_j| \\ &= M\left(\sum \Delta s_j - \sum \Delta L_j\right) \\ &= M\left(\text{len}(C) - \sum \Delta L_j\right) < \varepsilon.\end{aligned}$$

Hence if either type of sum has a limit, then the other has the same limit, and the conclusion follows by (b).

(d) Assume that C is C^1 and f is integrable on C. Let $\varepsilon > 0$. By part (c), there exists $\delta > 0$ such that if $\|T\| < \delta$, then every chord-length sum for f over T lies within ε of $\int_C f\,ds$. By taking δ small enough, we may assume also that

$$|u - t| < \delta \Rightarrow \|\mathbf{g}'(u) - \mathbf{g}'(t)\| < \frac{\varepsilon}{M\sqrt{n}[b-a]}. \qquad \text{(Reason?)}$$

Let $\|T\| < \delta$. For any Riemann sum

$$R \equiv \sum f(\mathbf{g}(t_j{}^*))\|\mathbf{g}'(t_j{}^*)\|(t_j - t_{j-1})$$

built from $f(\mathbf{g}(t))\|\mathbf{g}'(t)\|$ and T, we have

$$\begin{aligned}\left|\int_C f\,ds - R\right| &\le \left|\int_C f\,ds - \sum f(\mathbf{g}(t_j{}^*))\Delta L_j\right| \\ &\quad + \left|\sum f(\mathbf{g}(t_j{}^*))\left[\Delta L_j - \|\mathbf{g}'(t_j{}^*)\|(t_j - t_{j-1})\right]\right|.\end{aligned}$$

The first absolute value is ε or less. For the second, by the mean value theorem,

$$\Delta L_j = \|\mathbf{g}(t_j) - \mathbf{g}(t_{j-1})\| = \left\|\begin{bmatrix} g_1'(u_{j1}) \\ \cdots \\ g_n'(u_{jn}) \end{bmatrix}(t_j - t_{j-i})\right\|.$$

Hence

$$\left|\Delta L_j - \|\mathbf{g}'(t_j{}^*)\|(t_j - t_{j-1})\right| \le \left\| \begin{bmatrix} g_1'(u_{j1}) - g_1'(t_{j1}{}^*) \\ \cdots \\ g_n'(u_{jn}) - g_n'(t_{jn}{}^*) \end{bmatrix} \right\| (t_j - t_{j-1})$$

(the difference of norms being less than the norm of the difference), and

$$\left| \int_C f\,ds - R \right| \le \varepsilon + \sum M\sqrt{n} \frac{\varepsilon}{M\sqrt{n}[b-a]}(t_j - t_{j-1}) = 2\varepsilon.$$

We conclude that the Riemann sums for $f\big(\mathbf{g}(t)\big)\|\mathbf{g}'(t)\|$ have a limit, so that this function is integrable, and

$$\int_C f\,ds = \int_a^b f\big(\mathbf{g}(t)\big)\|\mathbf{g}'(t)\|\,dt. \qquad \square$$

In view of the last relation, it is natural to define the "element of arc length" by $ds = \|\mathbf{g}'(t)\|\,dt$. Since

$$\mathbf{g}'(t) = \left(\frac{dx}{dt}, \frac{dy}{dt}, \ldots, \frac{dz}{dt}\right),$$

we may write

$$ds = \sqrt{(dx/dt)^2 + (dy/dt)^2 + \cdots + (dz/dt)^2}\,dt,$$

or

$$ds = \sqrt{dx^2 + dy^2 + \cdots + dz^2}.$$

Note that in (a), we require f to be a continuous function of position along C. Continuity of $f(\mathbf{x})$ with respect to $\mathbf{x} \in \mathbf{R}^n$ is sufficient, but not necessary, for the restriction to C to be continuous. For example, Dirichlet's function is discontinuous everywhere, but it is constant along the line $x = \sqrt{2}$ in $\mathbf{R}^2$.

Example 3. What is the length of the graph of $x^2 + y^2 = a^2$.

Having claimed to be analysts, we take a strictly analytic approach. We know that the two series

$$s(t) \equiv t - \frac{t^3}{3!} + \frac{t^5}{5!} - \frac{t^7}{7!} + \cdots$$

and

$$c(t) \equiv 1 - \frac{t^2}{2!} + \frac{t^4}{4!} - \frac{t^6}{6!} + \cdots$$

converge for all real t, and define functions with $s' = c$, $c' = -s$, and $s^2 + c^2 = 1$. With some work [see Beardon, Theorem 6.3.1], we may prove that the first positive t for which $s(t) = 0$—let us give it a name, like "Pi"—satisfies

$c(t) = s\left(\frac{\text{Pi}}{2}+t\right)$ and $s(\text{Pi}+t) = -s(t)$, and that both functions range from -1 to 1. Hence, given $y \in [-1, 1]$, there exist two places $t \in [0, 2(\text{Pi})]$ where $s(t) = y$ (or one place, if $y = \pm 1$), at which $c(t)$ takes the two (or one) values $\pm\sqrt{1-y^2}$. In other words,

$$x = ac(t), \quad y = as(t), \qquad 0 \le t \le 2(\text{Pi}),$$

is a one-to-one parametrization of the graph.

Theorem 3(d) now says that

$$\begin{aligned}\text{len}(C) &= \int_0^{2\,\text{Pi}} \sqrt{(dx/dt)^2 + (dy/dt)^2}\,dt \\ &= \int_0^{2\text{Pi}} \left[a^2s^2(t) + a^2c^2(t)\right]^{1/2} dt = 2a\,\text{Pi}.\end{aligned}$$

In words: The length of every circle bears to its radius a ratio equal to the period of s and c. This result allows us to connect the analytic quantity Pi to those parts of geometry related to circles.

Example 4. (a) What is the length of the helix C given by $x = a\cos t$, $y = a\sin t$, $z = bt$, a and b both positive constants, $0 \le t \le 2\pi$?

By Theorem 3(d),

$$\begin{aligned}\text{len}(C) &= \int_0^{2\pi} \|(-a\sin t, a\cos t, b)\|\,dt \\ &= \int_0^{2\pi} \left(a^2\sin^2 t + a^2 cos^2 t + b^2\right)^{1/2} dt = 2\pi\left(a^2+b^2\right)^{1/2}.\end{aligned}$$

(b) Suppose the helix is a wire, whose mass is so distributed that at the point (x, y, z), its linear density is e^{-z} grams per meter. What is the mass of the wire?

The "linear density" description is a way of saying that at small ("micro") scale, a piece of the wire Δs meters long near the point (x, y, z) has mass $\Delta m \approx e^{-z}\Delta s$ grams. Summing and passing to the limit, we have the macro-scale relation

$$\begin{aligned}m &= \int_C e^{-z}\,ds = \int_0^{2\pi} e^{-bt}\sqrt{(dx/dt)^2 + (dy/dt)^2 + (dz/dt)^2}\,dt \\ &= \left(a^2+b^2\right)^{1/2}\left[\frac{e^{-bt}}{b}\right]_{2\pi}^{0} = \left(a^2+b^2\right)^{1/2}\frac{\left(1-e^{-2\pi b}\right)}{b}.\end{aligned}$$

Notice that if we extend this wire to infinity ($0 \le t < \infty$), its length necessarily becomes infinite, but its mass stays bounded.

Line integrals have other familiar properties of integrals. The proofs are also familiar, and we leave them to Exercise 11.

Theorem 4. *Assume that C is rectifiable and f, g are integrable on C.*

(a) (Linearity) *$\alpha f + \beta g$ is integrable, and*

$$\int_C (\alpha f + \beta g)\, ds = \alpha \int_C f\, ds + \beta \int_C g\, ds.$$

(b) (Partition Additivity) *If $a < c < b$, $D = \mathbf{g}([a, c])$, $E = \mathbf{g}([c, b])$, then*

$$\int_C f\, ds = \int_D f\, ds + \int_E f\, ds.$$

(c) (Function Monotonicity) *If $f \geq g$ on C, then $\int_C f\, ds = \int_C g\, ds$.*

(d) (Operator Boundedness) *If $m \leq f \leq M$ on C, then*

$$m \operatorname{len}(C) \leq \int_C f\, ds \leq M \operatorname{len}(C).$$

(e) (Triangle Inequality) *$|f|$ is integrable on C, and*

$$\left|\int_C f\, ds\right| \leq \int_C |f|\, ds.$$

(f) (Arc Continuity) *Given $\varepsilon > 0$, there exists $\delta > 0$ such that whenever $0 < u - t < \delta$,*

$$\int_{\mathbf{g}([t,u])} |f|\, ds < \varepsilon.$$

(g) (Average Value Theorem) *If the restriction of f to C is continuous, then there exists $\mathbf{c} \in C$ with*

$$\int_C f\, ds = f(\mathbf{c}) \operatorname{len}(C).$$

Partition additivity is especially important for line integrals, because it allows us to employ Theorem 3(d) on the union of C^1 curves. An arc is **piecewise smooth** (or **piecewise** C^1) if there is a partition of its domain for which the arc is smooth (respectively C^1) on each subinterval. The most elementary arcs for our work are the boundaries of rectangles; these, just like any polygon, are piecewise smooth curves.

There is another class of line integral we want to introduce. An alternative weight to the arc length Δs_j is the increment $\Delta(x_k)_j \equiv g_k(t_j) - g_k(t_{j-1})$ in the kth component of $\mathbf{g}$.

Definition. Assume that $\mathbf{g} = (g_1, \ldots, g_n)\colon [a, b] \to C \subseteq \mathbf{R}^n$ is rectifiable, and f is bounded along C. If the Riemann sums

$$\sum_{1 \leq j \leq J} f(\mathbf{x}_j)\Delta(x_k)_j = \sum_{1 \leq j \leq J} f\big(\mathbf{g}(t_j{}^*)\big)\big[g_k(t_j) - g_k(t_{j-1})\big]$$

have a limit as $\|T\| \to 0$, then we say that f is **integrable with respect to x_k along C**, and write $\int_C f\,dx_k$ for the limit. We refer to the result as a **coordinate integral.**

This is the first integral in which we allow the weights $\Delta(x_k)_j$ to be negative. We will see that the signed weights do not pose a problem with our usual functions, but they do have implications.

First, we have to use the *limit*-of-sums definition. Our usual approach does not work, because, for example, the "upper sum" $\sum M_j \Delta(x_k)_j$ may be smaller than the "lower" $\sum m_j \Delta(x_k)_j$ and may increase with refinement of partition.

Second, a new element comes into these integrals: **orientation** of the contour. Consider $C \subseteq \mathbf{R}^2$ given by $x = t$, $y = 0$, $0 \le t \le 1$. Clearly,

$$\begin{aligned}\int_C 1\,dx &= \lim_{\|T\|\to 0} \sum_{1\le j\le J} 1[x(t_j) - x(t_{j-1})] \\ &= \lim_{\|T\|\to 0} \sum_{1\le j\le J} 1[t_j - t_{j-1}] = t_J - t_0 = 1.\end{aligned}$$

But D defined by $x = 1 - t$, $y = 0$, $0 \le t \le 1$, is the same set of points, whereas

$$\int_D 1\,dx = \lim_{\|T\|\to 0} \sum_{1\le j\le J} 1[(1 - t_j) - (1 - t_{j-1})] = t_0 - t_J = -1.$$

Thus, although the integral over C would be the same for (say) any two one-to-one parametrizations that cover the curve in the same **sense**, a parametrization that travels in the opposite sense gives values of the opposite sign. With this in mind, we will write $D = -C$ if the end of C is the start of D and vice versa. We may then write

$$\int_C f\,dx = -\int_{-C} f\,dx.$$

Notice, in contrast, that distance integrals are unaffected by orientation. Here, we have

$$\begin{aligned}\int_D 1\,ds &= \lim_{\|T\|\to 0} \sum_{1\le j\le J} 1\Delta s_j \\ &= \lim \sum 1\sqrt{[x(t_j) - x(t_{j-1})]^2 + [y(t_j) - y(t_{j-1})]^2} \\ &= \lim \sum 1\sqrt{[(1 - t_j) - (1 - t_{j-1})]^2} \\ &= t_J - t_0 = \int_C 1\,ds.\end{aligned}$$

(See also Exercises 8a and 12.)

Finally, observe that the value of $\int_C 1\,dx_k$ is always $g_k(b) - g_k(a)$. This suggests that we may discard the assumption of rectifiability in the definition. However, the assumption is essential to avoid the problems that accompany convergence by cancellation. View

$$F(x,y) \equiv \begin{cases} 1, & x \in \left[\frac{2}{3\pi}, \frac{2}{\pi}\right] \cup \left[\frac{2}{7\pi}, \frac{2}{5\pi}\right] \cup \cdots, \\ -1, & \text{otherwise in } \mathbf{R}^2, \end{cases}$$

along the contour H of Example 1(b). This function is 1 on (roughly speaking) the rising parts of H, -1 on the descents. The Riemann sums for $\int_H F\,dy$ include sums like

$$(-1)\left[\frac{-2}{99\pi} - 0\right] + (1)\left[\frac{2}{97\pi} - \frac{-2}{99\pi}\right] + (-1)\left[\frac{-2}{95\pi} - \frac{2}{97\pi}\right]$$
$$+ \cdots + (1)\left[\frac{2}{1\pi} - \frac{-2}{3\pi}\right] = \frac{4}{99\pi} + \frac{4}{97\pi} + \cdots + \frac{4}{3\pi} + \frac{2}{\pi}.$$

It follows that the sums are unbounded, and F cannot be y-integrable. This is an unpleasant surprise, since F is reasonably continuous.

Nevertheless, some expected results hold. We write three important ones in the next theorem, and note another, as well as two exceptions, in Exercise 13.

Theorem 5. *Assume that* $\mathbf{g}$: $[a, b] \to C$ *is rectifiable and* f *is bounded on* C.

(a) *f is (length-)integrable along C iff f is integrable along C with respect to every x_k.*

(b) *If the restriction of f to C is continuous, then f is x_k-integrable over C.*

(c) *If $\mathbf{g}$ is C^1 and f is integrable on C with respect to x_k, then*

$$\int_C f\,dx_k = \int_a^b f(\mathbf{g}(t))g_k'(t)\,dt.$$

Proof. (a) $\Rightarrow$ Assume that f is integrable. Let $\varepsilon > 0$. By extension of Darboux's lemma, there is δ such that partitions finer than δ have their upper sums (for $\int_C f\,ds$) within $\frac{\varepsilon}{2}$ of their lower sums. Let T and U be two such partitions. We inspect Riemann x_k-sums.

First, let W be a refinement of T. Suppose $t_0 = w_0 < \cdots < w_I = t_1$ is how W breaks up the first subinterval of T. Where an x_k-sum for T has a single term $f(\mathbf{x}_1)\left[g_k(t_1) - g_k(t_0)\right]$, an x_k-sum for W has

$$f(\mathbf{y}_1)\left[g_k(w_1) - g_k(w_0)\right] + \cdots + f(\mathbf{y}_I)\left[g_k(w_I) - g_k(w_{I-1})\right].$$

Rewrite

$$f(\mathbf{x}_1)\left[g_k(t_1) - g_k(t_0)\right] = f(\mathbf{x}_1)\left[g_k(w_1) - g_k(w_0)\right]$$
$$+ \cdots + f(\mathbf{x}_1)\left[g_k(w_I) - g_k(w_{I-1})\right].$$

Since $|f(\mathbf{x}_1) - f(\mathbf{y}_i)| \le M_1 - m_1$ (M_1 and m_1 being the supremum and infimum of f on that first subinterval) and

$$|g_k(w_i) - g_k(w_{i-1})| \le \operatorname{len}\big(\mathbf{g}([w_{i-1}, w_i])\big) \qquad \text{for each } i,$$

the difference between the one term for T and the terms for W is at most

$$(M_1 - m_1)\big[\operatorname{len}\big(\mathbf{g}([w_0, w_1])\big) + \cdots + \operatorname{len}\big(\mathbf{g}([w_{k-1}, w_k])\big)\big] = (M_1 - m_1)\Delta s_1;$$

similarly for the other subintervals of T. Summing over the subintervals, we find that a Riemann x_k-sum for T and one for W are no further apart than the upper and lower sums for T. Thus, an x_k-sum for T and an x_k-sum for any refinement are separated by no more than $\frac{\varepsilon}{2}$.

Next, look at an x_k-sum R_T for T and an x_k-sum R_U for U. Since $T \cup U$ refines T and U, these two sums are within $\frac{\varepsilon}{2}$ of any one sum from $T \cup U$. Hence $|R_T - R_U| < \varepsilon$. We have shown that the Riemann x_k-sums for f form a Cauchy family as the norm vanishes. We conclude that they have a limit.

$\Leftarrow$ Assume that f is integrable along each x_k. Let $\varepsilon > 0$. Pick δ such that, independent of k, if $\|T\| < \delta$, then any two Riemann x_k-sums for T differ by less than $\frac{\varepsilon}{n}$.

We examine $R \equiv \sum f(\mathbf{x}_j)\Delta L_j$ and $R^* \equiv \sum f(\mathbf{y}_j)\Delta L_j$, two chord-length sums for such a T. The chords have

$$\begin{aligned}\Delta L_j &= \sqrt{[g_1(t_j) - g_1(t_{j-1})]^2 + \cdots + [g_n(t_j) - g_n(t_{j-1})]^2}\\ &\le |g_1(t_j) - g_1(t_{j-1})| + \cdots + |g_n(t_j) - g_n(t_{j-1})|.\end{aligned}$$

Reorganizing the sums, we get

$$\begin{aligned}|R - R^*| \le &\sum_{j=1}^{J} \big|f(\mathbf{x}_j) - f(\mathbf{y}_j)\big|\,\big|g_1(t_j) - g_1(t_{j-1})\big|\\ &+ \cdots + \sum_{j=1}^{J} \big|f(\mathbf{x}_j) - f(\mathbf{y}_j)\big|\,\big|g_n(t_j) - g_n(t_{j-1})\big|.\end{aligned}$$

The first summation on the right is the difference of two x_1-sums. That is,

$$\begin{aligned}&\sum_{j=1}^{J} \big|f(\mathbf{x}_j) - f(\mathbf{y}_j)\big|\,\big|g_1(t_j) - g_1(t_{j-1})\big|\\ &\quad = \sum_{j=1}^{J} f(\mathbf{X}_j)\big[g_1(t_j) - g_1(t_{j-1})\big] - \sum_{j=1}^{J} f(\mathbf{Y}_j)\big[g_1(t_j) - g_1(t_{j-1})\big],\end{aligned}$$

where $f(\mathbf{X}_j)$ and $f(\mathbf{Y}_j)$ are the bigger and smaller of $f(\mathbf{x}_j)$ and $f(\mathbf{y}_j)$ (or the other way around) when $g_1(t_j)$ is more than (respectively, less than) $g_1(t_{j-1})$. Hence this summation totals $\frac{\varepsilon}{n}$ or less. Then $|R - R^*| \le \varepsilon$, the chord-length sums for f are close together, and it follows (Theorem 3(c)) that f is integrable.

(b) follows from (a) and Theorem 3(a).

(c) follows, like Theorem 3(d), from the mean value theorem and the uniform continuity of g_k'. □

Exercises

1. Find the length of the curve given by

$$x_1 = (1-t)a_1 + tb_1, \ldots, x_n = (1-t)a_n + tb_n, \qquad 0 \le t \le 1.$$

(Hint: If your answer does not match the distance formula, then "length" is the wrong name for what we have defined.)

2. (a) Find a continuous mapping of $[0, 1]$ onto $\{(x, y): 0 \le x \le 1, y = 0\}$ with length 2.
   (b) Find a continuous map, same domain and range, with length $= \infty$.
   (c) Show that there cannot exist such a continuous map with length < 1.

3. Assume that h is C^1 on $[a, b]$. Show that the length of the graph $y = h(x)$, $a \le x \le b$, matches $\int_a^b (1 + h'(x)^2)^{1/2}\, dx$. (Why is the graph even rectifiable?)

4. Which of these infinite spirals is rectifiable? (In which is the length of the portion of the spiral connecting any two of its points bounded?)
   (a) $r = \frac{1}{\theta}, \theta \ge 2\pi$.
   (b) $r = e^{-\theta}, \theta \ge 0$.

5. Suppose an object's position as a function of time is given by

$$x_1 = g_1(t), \ldots, x_n = g_n(t) \qquad 0 \le t,$$

with each $g_j \in C^1$. By time $t = b$, the object has described a curve whose length is the distance the object has traveled, possibly back and forth.

   (a) Prove that the rate of change of this distance (the "speed") is the magnitude of the derivative $\frac{d(g_1,\ldots,g_n)}{dt}$ of position (the magnitude of the "velocity").
   (b) Prove that if the speed is constant, then the derivative $\frac{d^2(g_1,\ldots,g_n)}{dt^2}$ of the velocity (the "acceleration") is orthogonal to the velocity.

6. Suppose h is defined on a closed interval $[a, b]$. For a partition T of the interval, the quantity

$$\text{var}(h, T) \equiv |h(t_1) - h(t_0)| + \cdots + |h(t_J) - h(t_{J-1})|$$

is called the **variation of** h **over** T. We say that h is of **bounded variation** ("$h \in$ BV") if $\sup\{\text{var}(h, T)\text{: all } T\} < \infty$.

   (a) Show that $\mathbf{g}$ is rectifiable iff each component of $\mathbf{g}$ is of bounded variation.

(b) Show that if h is Lipschitz ($|h(x) - h(y)| < K|x - y|$ throughout), then $h \in \mathrm{BV}$. In particular, if h is differentiable and h' bounded, then $h \in \mathrm{BV}$.

(c) Show that $H(x) \equiv x^{\frac{3}{2}} \sin\left(\frac{1}{x}\right)$, $H(0) \equiv 0$, has unbounded derivative on $[0, 1]$ (such a function cannot be Lipschitz), but still has bounded variation. (Hint: Given a partition, refine it by adding the points between t_1 and 1 where the function realizes local extremes; then the changes in H are all positive on the intervals of increase and all negative on the way down.)

(d) Give an example of a differentiable function on $[0, 1]$ that has unbounded derivative and unbounded variation.

(e) Show that every monotone function is of bounded variation.

(f) Show that a function of bounded variation need not be continuous.

(g) Show that a continuous function need not be of bounded variation.

7. (a) Is the graph of $y = x^2 \sin\left(\frac{1}{x}\right)$, $0 < x \leq 1$, together with the origin, rectifiable?

(b) Is the graph of $y = x^{3/2} \sin\left(\frac{1}{x}\right)$, $0 < x \leq 1$, together with the origin, rectifiable?

8. (a) Assume that $\mathbf{g}$ and $\mathbf{h}$ are arcs with the same range C, and both are one-to-one. Suppose $\mathbf{g}$ starts where $\mathbf{h}$ ends, and ends where $\mathbf{h}$ starts. Prove that $\mathrm{len}(\mathbf{g}) = \mathrm{len}(\mathbf{h})$.

(b) Assume that $\mathbf{g}$ and $\mathbf{h}$ are arcs with the same range, same start, same end. Suppose $\mathbf{g}$: $[a, b] \to \mathbf{R}^n$ is one-to-one on $[a, b)$, but $\mathbf{g}(a) = \mathbf{g}(b)$; similarly, h: $[c, d] \to \mathbf{R}^n$ is one-to-one on $[c, d)$, but $\mathbf{h}(c) = \mathbf{h}(d)$. Prove that $\mathrm{len}(\mathbf{g}) = \mathrm{len}(\mathbf{h})$.

9. (a) Find the centroid of the helix in Example 4a. (The x_k-coordinate of the centroid is the length-average $\frac{\int_C x_k\, ds}{\mathrm{len}(C)}$ of x_k along the curve. Compare Exercise 5.3:2.)

(b) Suppose the helix carries a linear density e^{-z}gm/m. Find its center of mass. (The x_k-coordinate of the center of mass is the mass-average

$$\frac{\int_C x_k dm}{\mathrm{mass}(C)} = \frac{\int_C x_k e^{-z}\, ds}{\int_C e^{-z}\, ds}$$

of x_k along the curve.)

10. Prove that in polar coordinates, the "element of arc length" is $ds = \left(dr^2 + r^2 d\theta^2\right)^{1/2}$; that is:

(a) If the curve C is given by $r = g_1(t), \theta = g_2(t), a \le t \le b$, then

$$\int_C f(r, \theta)\, ds = \int_a^b f(g_1(t), g_2(t))\big[g_1'(t)^2 + g_1(t)^2 g_2'(t)^2\big]^{1/2}\, dt.$$

(b) If C is given by $r = g(\theta), \alpha \le \theta \le \beta$, then

$$\int_C f\, ds = \int_\alpha^\beta f(g(\theta), \theta)\big[g'(\theta)^2 + g(\theta)^2\big]^{1/2}\, d\theta.$$

11. (a)–(g) Prove Theorem 4.

12. Evaluate $\int_C x^2 + y^2\, ds$, $\int_C x^2 + y^2\, dx$, and $\int_C x^2 + y^2\, dy$, where C is:

    (a) the line segment from (0, 0) to (2, 1).

    (b) the line segment from (2, 1) to (0, 0).

    (c) the semicircle centered at the origin, from (2, 0) to (−2, 0).

    (d) the circle of radius 2 centered at the origin, oriented counterclockwise.

13. (a) (Triangle Inequality, Set Continuity) Show that if f is integrable on C, then
$$\left|\int_C f\, dx_k\right| \le \int_C |f|\, ds.$$

    (b) Show that x_k-integrals are not monotonic: $f \ge 0$ does not force $\int_C f\, dx_k \ge 0$.

    (c) Show that x_k-integrals do not have the average value property: Give an example in which f is continuous, the x-component $g_1(t)$ has $g_1(b) \ne g_1(a)$, but
$$\frac{\int_C f\, dx}{\int_C 1\, dx} = \frac{\int_C f\, dx}{g_1(b) - g_1(a)}$$
    is not a value of f.

5.6 Surface Integrals

In our world, length measures one-dimensional extent and area measures two-dimensional extent. We next introduce area of a surface and its higher-dimensional analogues.

Recall (Section 2.6) that by "(2-dimensional) surface," we mean a differentiable mapping $\mathbf{f}(t, u)$ from a rectangle $R \subseteq \mathbf{R}^2$ into $\mathbf{R}^n$. To get a close-up look at the surface, partition R, and examine the image (shaded, in the lower part of Figure 5.5) of any subinterval $B \equiv \big[(a, b), (c, d)\big]$ (upper part). Just as we expect a small

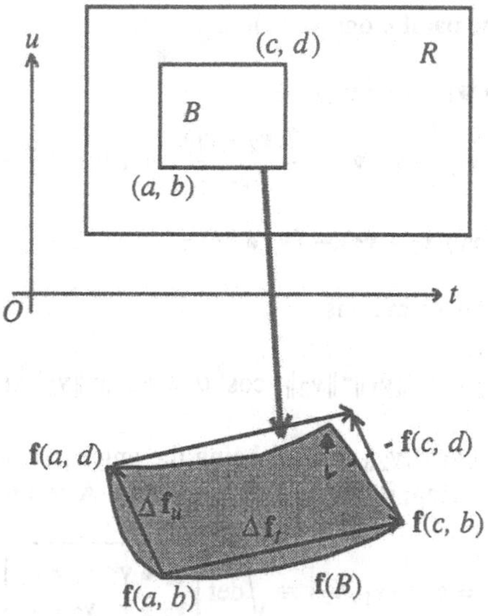

Figure 5.5.

piece of a curve to be almost straight, we figure this part of the surface to be roughly flat, with roughly straight sides. We try to approximate its area by using a quadrilateral.

The images of the four corners of B will not work, because they will in general not be coplanar. However, the vectors

$$\Delta\mathbf{f}_t \equiv \mathbf{f}(c, b) - \mathbf{f}(a, b),$$

from the image of (a, b) to the image of the lower-right corner (c, b), and

$$\Delta\mathbf{f}_u \equiv \mathbf{f}(a, d) - \mathbf{f}(a, b),$$

from the image of (a, b) to the image of the upper-left corner (a, d), determine a parallelogram (bounded by arrows in the figure) whose area looks like a decent approximation to the area of the piece of surface.

Let us, therefore, discuss the area σ of the parallelogram determined by two vectors $\mathbf{v}_1$, $\mathbf{v}_2$. Write (compare the start of Section 5.3)

$$\mathbf{w}_1 \equiv \mathbf{v}_1,$$
$$\mathbf{w}_2 \equiv \mathbf{v}_2 - \frac{[\mathbf{v}_2 \bullet \mathbf{w}_1]\mathbf{w}_1}{\|\mathbf{w}_1\|^2}.$$

We have described $\mathbf{w}_2$ as the "altitude" of the parallelogram, so that $\sigma = \|\mathbf{w}_1\|\,\|\mathbf{w}_2\|$. (In different words, $\mathbf{w}_1$ and $\mathbf{w}_2$ determine a rectangle of the same base and height as the parallelogram.) Hence

$$\begin{aligned}\sigma^2 &= (\mathbf{w}_1 \bullet \mathbf{w}_1)(\mathbf{w}_2 \bullet \mathbf{w}_2)\\ &= (\mathbf{v}_1 \bullet \mathbf{v}_1)\left[\mathbf{v}_2 \bullet \mathbf{v}_2 - \frac{2(\mathbf{v}_2 \bullet \mathbf{v}_1)^2}{\|\mathbf{v}_1\|^2} + (\mathbf{v}_2 \bullet \mathbf{v}_1)^2 \frac{(\mathbf{v}_1 \bullet \mathbf{v}_1)}{\|\mathbf{v}_1\|^4}\right]\\ &= (\mathbf{v}_1 \bullet \mathbf{v}_1)(\mathbf{v}_2 \bullet \mathbf{v}_2) - (\mathbf{v}_2 \bullet \mathbf{v}_1)^2.\end{aligned}$$

One way to look at this formula is

$$\sigma^2 = \|\mathbf{v}_1\|^2\|\mathbf{v}_2\|^2 - \|\mathbf{v}_1\|^2\|\mathbf{v}_2\|^2\cos^2\theta = \|\mathbf{v}_1\|^2\|\mathbf{v}_2\|^2\left(1-\cos^2\theta\right).$$

That says that $\sigma = \|\mathbf{v}_1\|\,\|\mathbf{v}_2\|\sin\theta$, θ being the angle between $\mathbf{v}_1$ and $\mathbf{v}_2$. This interpretation is reassuring; it matches our geometry. A second way is

$$\sigma = \operatorname{area}(\mathbf{v}_1, \mathbf{v}_2) \equiv \sqrt{\det\begin{bmatrix}\mathbf{v}_1 \bullet \mathbf{v}_1 & \mathbf{v}_1 \bullet \mathbf{v}_2\\ \mathbf{v}_2 \bullet \mathbf{v}_1 & \mathbf{v}_2 \bullet \mathbf{v}_2\end{bmatrix}}.$$

This one is powerful. It is a size-2 version of Theorem 5.3:1(d), and therefore suggests that we will be able to generalize.

Returning to our task, we have the two vectors $\Delta\mathbf{f}_t$ and $\Delta\mathbf{f}_u$, which determine a parallelogram, whose area approximates the contribution to the surface coming from the subinterval B. We sum up the parallelograms corresponding to all the subintervals, do a limiting process, and the limit is then our definition of area.

Definition. Let $\mathbf{f}$ be a surface and $\mathcal{P} \equiv \{B_1, \ldots, B_J\}$ a partition of its domain rectangle. For $B_j \equiv \big[(a_j, b_j), (c_j, d_j)\big]$, let

$$\begin{aligned}\mathbf{p}_j &\equiv \mathbf{f}(c_j, b_j) - \mathbf{f}(a_j, b_j),\\ \mathbf{q}_j &\equiv \mathbf{f}(a_j, d_j) - \mathbf{f}(a_j, b_j).\end{aligned}$$

If the sums $\operatorname{area}(\mathbf{f}, \mathcal{P}) \equiv \sum_{1\le j\le J} \operatorname{area}(\mathbf{p}_j, \mathbf{q}_j)$ have a limit as $\|\mathcal{P}\| \to 0$, then we say that **f is rectifiable** and call the limit the **(surface) area** of $\mathbf{f}$, written area($\mathbf{f}$).

Theorem 1. *Every C^1 surface is rectifiable.*

Proof. Assume that $\mathbf{f}$ is C^1. Applying the mean value theorem to the notation of the definition, we have

$$\mathbf{p}_j = \begin{bmatrix}\frac{\partial f_1}{\partial t}(\mathbf{r}_{j1})\\ \vdots\\ \frac{\partial f_n}{\partial t}(\mathbf{r}_{jn})\end{bmatrix}(c_j - a_j), \quad \mathbf{q}_j = \begin{bmatrix}\frac{\partial f_1}{\partial u}(\mathbf{s}_{j1})\\ \vdots\\ \frac{\partial f_n}{\partial u}(\mathbf{s}_{jn})\end{bmatrix}(d_j - b_j).$$

Abbreviate the first column by $\mathbf{f}_{j1}$, the second by $\mathbf{f}_{j2}$. Also observe that $(c_j - a_j)(d_j - b_j)$, which factors out of the determinant, is just the area of B_j, for which we use the volume notation $V(B_j)$. We then have

$$\begin{aligned}\operatorname{area}(\mathbf{f}, \mathcal{P}) &= \sum_{1 \le j \le J} \operatorname{area}(\mathbf{p}_j, \mathbf{q}_j)\\ &= \sum_{1 \le j \le J} V(B_j)\sqrt{\det\begin{bmatrix}\mathbf{f}_{j1}\bullet\mathbf{f}_{j1} & \mathbf{f}_{j1}\bullet\mathbf{f}_{j2}\\ \mathbf{f}_{j2}\bullet\mathbf{f}_{j1} & \mathbf{f}_{j2}\bullet\mathbf{f}_{j2}\end{bmatrix}}.\end{aligned}$$

By a now-familiar argument based on the uniform continuity of $\frac{\partial \mathbf{f}}{\partial \mathbf{r}}$ (see, for example, the proof of Theorem 5.5:3(d); this argument is sometimes called **Duhamel's principle**), area($\mathbf{f}$, $\mathcal{P}$) has the same limit as

$$\sum_{1 \le j \le J} \operatorname{area}\left(\frac{\partial \mathbf{f}}{\partial t}(a_j, b_j), \frac{\partial \mathbf{f}}{\partial u}(a_j, b_j)\right) V(B_j).$$

This last is a Riemann sum over R for the function

$$F(t, u) \equiv \operatorname{area}\left(\frac{\partial \mathbf{f}}{\partial t}(t, u), \frac{\partial \mathbf{f}}{\partial u}(t, u)\right) = \sqrt{\det\begin{bmatrix}\frac{\partial \mathbf{f}}{\partial t}\bullet\frac{\partial \mathbf{f}}{\partial t} & \frac{\partial \mathbf{f}}{\partial t}\bullet\frac{\partial \mathbf{f}}{\partial u}\\ \frac{\partial \mathbf{f}}{\partial u}\bullet\frac{\partial \mathbf{f}}{\partial t} & \frac{\partial \mathbf{f}}{\partial u}\bullet\frac{\partial \mathbf{f}}{\partial u}\end{bmatrix}}.$$

Since F is continuous, it is integrable. Consequently, $\lim_{\|\mathcal{P}\|\to 0} \operatorname{area}(\mathbf{f}, \mathcal{P})$ exists, and

$$\operatorname{area}(\mathbf{f}) = \int_A F = \int_A \operatorname{area}\left(\frac{\partial \mathbf{f}}{\partial t}, \frac{\partial \mathbf{f}}{\partial u}\right). \qquad \square$$

In view of Theorem 1, we may as well restrict our attention to piecewise-C^1 surfaces and *define* the area of a C^1 piece by the integral

$$\int_B \operatorname{area}\left(\frac{\partial \mathbf{f}}{\partial t}, \frac{\partial \mathbf{f}}{\partial u}\right).$$

This seeming restriction actually expands our reach, because we may now define "C^1 surface" as a continuously differentiable map $\mathbf{f}$ from *any closed, connected Archimedean* $A \subseteq \mathbf{R}^2$ to $\mathbf{R}^n$. (We should remember, in this context, that what is needed is $\mathbf{f}'$ uniformly continuous on the interior of A.) Then the area of the surface is

$$\int_A \operatorname{area}\left(\frac{\partial \mathbf{f}}{\partial t}, \frac{\partial \mathbf{f}}{\partial u}\right).$$

Example 1. What is the area of a sphere of radius a in $\mathbf{R}^3$?

(a) We have parametrized the sphere (Example 2.6:3) by

$$x = a\cos\theta\sin\phi,\ y = a\sin\theta\sin\phi,\ z = a\cos\phi, \quad 0 \le \theta \le 2\pi,\ 0 \le \phi \le \pi,$$

wherein

$$\frac{\partial(x,y,z)}{\partial\theta} = \begin{bmatrix} -a\sin\theta\sin\phi \\ a\cos\theta\sin\phi \\ 0 \end{bmatrix}, \quad \frac{\partial(x,y,z)}{\partial\phi} = \begin{bmatrix} a\cos\theta\cos\phi \\ a\sin\theta\cos\phi \\ -a\sin\phi \end{bmatrix}.$$

We could calculate area $\left(\frac{\partial \mathbf{x}}{\partial\theta}, \frac{\partial \mathbf{x}}{\partial\phi}\right)$ directly, doing the three needed dot products. But that is unnecessary work. Recall that latitude and longitude curves meet at right angles, a manifestation of the orthogonality of $\frac{\partial \mathbf{x}}{\partial\theta}$ and $\frac{\partial \mathbf{x}}{\partial\phi}$. Their parallelogram is a rectangle, with area

$$\left\|\frac{\partial \mathbf{x}}{\partial\theta}\right\| \left\|\frac{\partial \mathbf{x}}{\partial\phi}\right\| = (a\sin\phi)a.$$

Consequently,

$$\sigma = \int_0^{2\pi}\int_0^{\pi} a^2\sin\phi\,d\phi\,d\theta = 4\pi a^2.$$

(b) The sphere is also half-given by $z = \left(a^2 - x^2 - y^2\right)^{1/2}, 0 \le x^2 + y^2 \le a^2$. We have

$$\frac{\partial(x,y,z)}{\partial x} = \left[1\ 0\ \frac{\partial z}{\partial x}\right]^t, \quad \frac{\partial(x,y,z)}{\partial y} = \left[0\ 1\ \frac{\partial z}{\partial y}\right]^t,$$

and

$$\begin{aligned} \text{area} &= \int_{\text{disk}} \sqrt{\left(1+\left[\frac{\partial z}{\partial x}\right]^2\right)\left(1+\left[\frac{\partial z}{\partial y}\right]^2\right) - \left(\left[\frac{\partial z}{\partial x}\right]\left[\frac{\partial z}{\partial y}\right]\right)^2} \\ &= \int_{\text{disk}} \sqrt{1+\left[\frac{\partial z}{\partial x}\right]^2 + \left[\frac{\partial z}{\partial y}\right]^2}. \end{aligned}$$

We will come back to this formula in a more general setting. Here, we observe that the integral is improper:

$$\begin{aligned} 1+\left(\frac{\partial z}{\partial x}\right)^2 + \left(\frac{\partial z}{\partial y}\right)^2 &= 1 + \frac{(-x)^2}{a^2-x^2-y^2} + \frac{(-y)^2}{a^2-x^2-y^2} \\ &= \frac{a^2}{a^2-x^2-y^2} \end{aligned}$$

is unbounded in the disk. Nevertheless, it is easy to show that the hemisphere has

$$\text{area} = \lim_{t\to a}\int_0^{2\pi}\int_0^{t} \frac{a}{\sqrt{a^2-r^2}}\,r\,dr\,d\theta = 2\pi a^2.$$

Example 2. Our first definition of surface area went straight to the limit of sums. Why not upper and lower sums?

Let

$$\mathbf{f}(x, y) \equiv \left(x, \left[1 - \frac{x}{2}\right] y, 1\right),$$
$$\mathbf{g}(x, y) \equiv \left(x, \left[1 + \frac{x}{2}\right] y, 1\right), \qquad 0 \le x \le 1,\ 0 \le y \le 1.$$

Each maps the unit square S in $\mathbf{R}^2$ to a trapezoid at the $z = 1$ level in $\mathbf{R}^3$. With the partition $\{S\}$, we have

$$\text{area}(\mathbf{f}, \{S\}) = \text{area}(\mathbf{g}, \{S\}) = \text{area}(\mathbf{i}, \mathbf{j}) = 1.$$

If we split S into left- and right-hand halves S_l and S_r, then

$$\text{area}(\mathbf{f}, \{S_l, S_r\}) = \text{area}(0.5\mathbf{i}, \mathbf{j}) + \text{area}(0.5\mathbf{i}, 0.75\mathbf{j}) = \frac{7}{8},$$
$$\text{area}(\mathbf{g}, \{S_l, S_r\}) = \text{area}(0.5\mathbf{i}, \mathbf{j}) + \text{area}(0.5\mathbf{i}, 1.25\mathbf{j}) = \frac{9}{8}.$$

With refinement, the estimate of area may increase or decrease. Therefore, we cannot approach surface area the way we introduced arc length; we have to retreat to the path we followed to define coordinate (x_k)-integrals.

(Our approach to arc length used the increase of polygon length with refinement of partition, which results entirely from the triangle inequality. Clearly, the trouble here is that there is no corresponding inequality for areas. Buck [opening paragraph of Section 6.3 and near Figure 6-24] comments on this difficulty, and says that it cannot be resolved in an elementary way.)

We now have two-dimensional area under control. The matter of whether we can generalize to higher dimension comes down to one question: Can we show that the prism spanned by three vectors $\mathbf{v}_1, \mathbf{v}_2, \mathbf{v}_3$ in $\mathbf{R}^n$ is measured by the determinant $\det(\mathbf{v}_j \bullet \mathbf{v}_k)$? That is, our natural measure of the prism's size is the product of the altitudes $\mathbf{w}_1, \mathbf{w}_2, \mathbf{w}_3$ produced by the Gram–Schmidt process. If we can show that

$$\|\mathbf{w}_1\|^2 \|\mathbf{w}_2\|^2 \|\mathbf{w}_3\|^2 = \det(\mathbf{v}_j \bullet \mathbf{v}_k),$$

then the recursion needed to define higher-dimensional size will be evident.

To establish the result, we may assume that the $\mathbf{v}_j$ are nonzero. Begin with $\mathbf{w}_1 = \mathbf{v}_1$, so that

$$\det(\mathbf{v}_j \bullet \mathbf{v}_k) = \det \begin{bmatrix} \mathbf{w}_1 \bullet \mathbf{w}_1 & \mathbf{w}_1 \bullet \mathbf{v}_2 & \mathbf{w}_1 \bullet \mathbf{v}_3 \\ \mathbf{v}_2 \bullet \mathbf{w}_1 & \mathbf{v}_2 \bullet \mathbf{v}_2 & \mathbf{v}_2 \bullet \mathbf{v}_3 \\ \mathbf{v}_3 \bullet \mathbf{w}_1 & \mathbf{v}_3 \bullet \mathbf{v}_2 & \mathbf{v}_3 \bullet \mathbf{v}_3 \end{bmatrix}.$$

Multiply row 1 by $-\frac{\mathbf{v}_2 \bullet \mathbf{w}_1}{\|\mathbf{w}_1\|^2}$ and add the result to row 2. The second row becomes

$$\left[\mathbf{v}_2 \bullet \mathbf{w}_1 - \frac{\mathbf{v}_2 \bullet \mathbf{w}_1}{\|\mathbf{w}_1\|^2} \mathbf{w}_1 \bullet \mathbf{w}_1 \quad \mathbf{v}_2 \bullet \mathbf{v}_2 - \frac{\mathbf{v}_2 \bullet \mathbf{w}_1}{\|\mathbf{w}_1\|^2} \mathbf{w}_1 \bullet \mathbf{v}_2 \right.$$
$$\left. \mathbf{v}_2 \bullet \mathbf{v}_3 - \frac{\mathbf{v}_2 \bullet \mathbf{w}_1}{\|\mathbf{w}_1\|^2} \mathbf{w}_1 \bullet \mathbf{v}_3 \right]$$
$$= [\mathbf{w}_2 \bullet \mathbf{w}_1 \quad \mathbf{w}_2 \bullet \mathbf{v}_2 \quad \mathbf{w}_2 \bullet \mathbf{v}_3].$$

We know that $\mathbf{w}_2 \bullet \mathbf{w}_1 = 0$ and $\mathbf{w}_2 \bullet \mathbf{v}_2 = \mathbf{w}_2 \bullet (\mathbf{w}_2 + \alpha \mathbf{w}_1) = \mathbf{w}_2 \bullet \mathbf{w}_2$. That makes the second row

$$[0 \quad \mathbf{w}_2 \bullet \mathbf{w}_2 \quad \mathbf{w}_2 \bullet \mathbf{v}_3].$$

(It also makes the upper left subdeterminant

$$\det \begin{bmatrix} \mathbf{v}_1 \bullet \mathbf{v}_2 & \mathbf{v}_1 \bullet \mathbf{v}_2 \\ \mathbf{v}_2 \bullet \mathbf{v}_1 & \mathbf{v}_2 \bullet \mathbf{v}_2 \end{bmatrix} = \det \begin{bmatrix} \mathbf{w}_1 \bullet \mathbf{w}_1 & \mathbf{w}_1 \bullet \mathbf{v}_2 \\ 0 & \mathbf{w}_2 \bullet \mathbf{w}_2 \end{bmatrix} = \|\mathbf{w}_1\|^2 \|\mathbf{w}_2\|^2;$$

our dimension-2 result is part of a pattern.) Now add to row 3, $-\frac{\mathbf{v}_3 \bullet \mathbf{w}_1}{\|\mathbf{w}_1\|^2}$ times row 1 and $-\frac{\mathbf{v}_3 \bullet \mathbf{v}_2}{\|\mathbf{w}_2\|^2}$ times the current row 2. Row 3 becomes

$$\left[\mathbf{v}_3 \bullet \mathbf{w}_1 - \frac{\mathbf{v}_3 \bullet \mathbf{w}_1}{\|\mathbf{w}_1\|^2} \mathbf{w}_1 \bullet \mathbf{w}_1 \quad \mathbf{v}_3 \bullet \mathbf{v}_2 - \frac{\mathbf{v}_3 \bullet \mathbf{w}_1}{\|\mathbf{w}_1\|^2} \mathbf{w}_1 \bullet \mathbf{v}_2 - \frac{\mathbf{v}_3 \bullet \mathbf{w}_2}{\|\mathbf{w}_2\|^2} \mathbf{w}_2 \bullet \mathbf{w}_2 \right.$$
$$\left. \mathbf{v}_3 \bullet \mathbf{v}_3 - \frac{\mathbf{v}_3 \bullet \mathbf{w}_1}{\|\mathbf{w}_1\|^2} \mathbf{w}_1 \bullet \mathbf{v}_3 - \frac{\mathbf{v}_3 \bullet \mathbf{w}_2}{\|\mathbf{w}_2\|^2} \mathbf{w}_2 \bullet \mathbf{v}_3 \right].$$

The first entry is clearly 0. Using $\mathbf{w}_2 \bullet \mathbf{v}_2 = \mathbf{w}_2 \bullet \mathbf{w}_2$, we turn the second entry into

$$\left(\mathbf{v}_3 - \frac{\mathbf{v}_3 \bullet \mathbf{w}_1}{\|\mathbf{w}_1\|^2} \mathbf{w}_1 - \frac{\mathbf{v}_3 \bullet \mathbf{w}_2}{\|\mathbf{w}_2\|^2}\right) \bullet \mathbf{v}_2 = \mathbf{w}_3 \bullet \mathbf{v}_2 = 0.$$

The third is

$$\left(\mathbf{v}_3 - \frac{\mathbf{v}_3 \bullet \mathbf{w}_1}{\|\mathbf{w}_1\|^2} \mathbf{w}_1 - \frac{\mathbf{v}_3 \bullet \mathbf{w}_2}{\|\mathbf{w}_2\|^2} \mathbf{w}_2\right) \bullet \mathbf{v}_3 = \mathbf{w}_3 \bullet \mathbf{v}_3$$
$$= \mathbf{w}_3 \bullet (\mathbf{w}_3 + \beta \mathbf{v}_2 + \gamma \mathbf{v}_1) = \mathbf{w}_3 \bullet \mathbf{w}_3.$$

We have arrived at the relation we wanted:

$$\det(\mathbf{v}_j \bullet \mathbf{v}_k) = \det \begin{bmatrix} \mathbf{w}_1 \bullet \mathbf{w}_1 & \mathbf{w}_1 \bullet \mathbf{v}_2 & \mathbf{w}_1 \bullet \mathbf{v}_3 \\ 0 & \mathbf{w}_2 \bullet \mathbf{w}_2 & \mathbf{w}_2 \bullet \mathbf{v}_3 \\ 0 & 0 & \mathbf{w}_3 \bullet \mathbf{w}_3 \end{bmatrix} = \|\mathbf{w}_1\|^2 \|\mathbf{w}_2\|^2 \|\mathbf{w}_3\|^2.$$

We now have the tool for measurement in any dimension, so we proceed to generalized definitions.

Definition. (a) A **surface of dimension** k is a C^1 mapping $\mathbf{g}(t)$ of a closed, connected Archimedean set $A \subset \mathbf{R}^k$ into $\mathbf{R}^n$. The surface is **smooth** if $\mathbf{g}'$ has rank k throughout A.

(b) The **area** of the surface is

$$\int_A \sqrt{\det\left(\frac{\partial \mathbf{g}}{\partial t_i} \bullet \frac{\partial \mathbf{g}}{\partial t_j}\right)}.$$

(c) If the restriction of function f is continuous on the range S of $\mathbf{g}$, then the **surface integral** $\int_S f\,ds$ of f on S is

$$\int_A f(\mathbf{g}(\mathbf{t})) \sqrt{\det\left(\frac{\partial \mathbf{g}}{\partial t_i}(\mathbf{t}) \bullet \frac{\partial \mathbf{g}}{\partial t_j}(\mathbf{t})\right)}.$$

In view of (c), it is usual to introduce the "element of surface area"

$$d\sigma \equiv \sqrt{\det\left(\frac{\partial \mathbf{g}}{\partial t_i}(\mathbf{t}) \bullet \frac{\partial \mathbf{g}}{\partial t_j}(\mathbf{t})\right)}\, dt_1 \cdots dt_k.$$

Also as usual, (c) reflects our willingness to call S "the surface," even though integrals, including the area of S, are dependent on parametrization.

The definition does not specify $k > 1$: If $k = 1$, then A is an interval $[a, b]$, the "surface" is a curve, and "area" is the arc length

$$\int_a^b \sqrt{\left(\frac{\partial \mathbf{g}}{\partial t} \bullet \frac{\partial \mathbf{g}}{\partial t}\right)}\, dt.$$

We employ both names, "surface" and "area," in all the cases $2 \le k < n - 1$. The case $k = n - 1 > 2$ is special; we use **hypersurface** and **hyperarea**, and will discuss the latter at length. The definition is also silent about $k < n$; $k = n$ is allowed, and we deal with that case in Exercise 5.

Example 3. What is the hyperarea of the sphere of radius a in $\mathbf{R}^4$?

We parametrize (Exercise 5.3:4) $\mathbf{g} = (x_1, x_2, x_3, x_4)$ by

$$x_1 = a\cos s, \quad x_2 = a\sin s\cos t,$$
$$x_3 = a\sin s\sin t\cos u, \quad x_4 = a\sin s\sin t\sin u,$$

$0 \le s \le \pi, 0 \le t \le \pi, 0 \le u \le 2\pi$. Then

$$\frac{\partial \mathbf{g}}{\partial s} = \begin{bmatrix} -a\sin s \\ a\cos s\cos t \\ a\cos s\sin t\cos u \\ a\cos s\sin t\sin u \end{bmatrix}, \quad \frac{\partial \mathbf{g}}{\partial t} = \begin{bmatrix} 0 \\ -a\sin s\sin t \\ a\sin s\cos t\cos u \\ a\sin s\cos t\sin u \end{bmatrix},$$

$$\frac{\partial \mathbf{g}}{\partial u} = \begin{bmatrix} 0 \\ 0 \\ -a\sin s\sin t\sin u \\ a\sin s\sin t\cos u \end{bmatrix}.$$

These are mutually orthogonal, so they determine a hyperrectangle of 3-area

$$\left\|\frac{\partial \mathbf{g}}{\partial s}\right\| \left\|\frac{\partial \mathbf{g}}{\partial t}\right\| \left\|\frac{\partial \mathbf{g}}{\partial u}\right\| = a(a\sin s)(a\sin s\sin t).$$

Therefore,

$$\sigma = \int_0^\pi \int_0^\pi \int_0^{2\pi} a^3 \sin^2 s\sin t\, du\, dt\, ds = 2\pi^2 a^3.$$

(Check also Exercise 7.)

Integrals on surfaces have the expected properties, imitating Theorem 5.5:4 with only small modifications.

Theorem 2. *Assume that $S \equiv \mathbf{h}(A)$ is a surface and f, g are continuous on S.*

(a) (Linearity) *$\alpha f + \beta g$ is integrable, and*

$$\int_S (\alpha f + \beta g)\, d\sigma = \alpha \int_S f\, d\sigma + \beta \int_S g\, d\sigma.$$

(b) (Partition Additivity) *If D, E, and $D \cup E$ are closed, connected Archimedean subsets of A, and D, E do not overlap, then*

$$\int_{\mathbf{g}(D\cup E)} f\, d\sigma = \int_{\mathbf{g}(D)} f\, d\sigma + \int_{\mathbf{g}(E)} f\, d\sigma.$$

(c) (Function Monotonicity) *If $f \geq g$ on S, then $\int_S f\, d\sigma = \int_S g\, d\sigma$.*

(d) (Operator Boundedness) *If $m \leq f \leq M$ on S, then*

$$m\ \text{area}(S) \leq \int_S f\, d\sigma \leq M\ \text{area}(S).$$

(e) (Triangle Inequality) *$|f|$ is integrable on S, and $\left|\int_S f\, d\sigma\right| \leq \int_S |f|\, d\sigma$.*

(f) (Continuity) *Given $\varepsilon > 0$, there exists $\delta > 0$ such that*

$$\int_{\mathbf{g}(D)} |f|\, d\sigma < \varepsilon$$

for every Archimedean $D \subseteq A$ having $V(D) < \delta$.

(g) (Average Value Theorem) *There exists $\mathbf{c} \in S$ with*

$$\int_S f\, d\sigma = f(\mathbf{c})\ \text{area}(S).$$

Proof. No separate proofs are needed. Because we defined surface integrals via integrals on A, these statements follow from theorems in Section 5.2.

We reiterate that partition additivity is important because it lets us work with piecewise-C^1 surfaces. It also reminds us that the notation $\int_{\mathbf{g}(D)} f\, d\sigma$ merely pretends to depend on just the range $\mathbf{g}(D)$. If $\mathbf{g}$ is not one-to-one, then it overcounts integrals on its range, to the extent that it repeats points there.

We have ascribed special importance to the case $k = n - 1$, and we gave some justification in Theorem 3.3:3. Let us look closely at hypersurfaces and hyperarea.

The most elementary hypersurface is the graph of a function, $x_n = f(x_1, \ldots, x_{n-1})$. Write $x_1 = t_1, \ldots, x_{n-1} = t_{n-1}, x_n = f(t_1, \ldots, t_{n-1})$. We saw (proof of Theorem 2.6:2) that $\mathbf{G} \equiv (x_1, \ldots, x_n)$ has

$$\frac{\partial \mathbf{G}}{\partial t_1} = \left[1\ 0 \cdots 0\ \frac{\partial f}{\partial t_1}\right]^t, \ldots, \frac{\partial \mathbf{G}}{\partial t_{n-1}} = \left[0 \cdots 0\ 1\ \frac{\partial f}{\partial t_{n-1}}\right]^t,$$

and

$$\det\left(\frac{\partial \mathbf{G}}{\partial t_j} \bullet \frac{\partial \mathbf{G}}{\partial t_k}\right) = \det \begin{bmatrix} 1+\left(\frac{\partial f}{\partial t_1}\right)^2 & \frac{\partial f}{\partial t_1}\frac{\partial f}{\partial t_2} & \cdots & \frac{\partial f}{\partial t_1}\frac{\partial f}{\partial t_{n-1}} \\ \frac{\partial f}{\partial t_2}\frac{\partial f}{\partial t_1} & 1+\left(\frac{\partial f}{\partial t_2}\right)^2 & \cdots & \frac{\partial f}{\partial t_2}\frac{\partial t_1}{\partial t_{n-1}} \\ & & \ddots & \\ \frac{\partial f}{\partial t_{n-1}}\frac{\partial f}{\partial t_1} & \cdots & \frac{\partial f}{\partial t_{n-1}}\frac{\partial f}{\partial t_{n-2}} & 1+\left(\frac{\partial f}{\partial t_{n-1}}\right)^2 \end{bmatrix}.$$

The determinant can be found recursively. Break it up into

$$\begin{vmatrix} 1\ \ 0\ \ \cdots\ \ 0 \\ \text{other rows unchanged} \end{vmatrix} + \begin{vmatrix} \left(\frac{\partial f}{\partial t_1}\right)^2 \left(\frac{\partial f}{\partial t_1}\right)\left(\frac{\partial f}{\partial t_2}\right) \cdots \left(\frac{\partial f}{\partial t_1}\right)\left(\frac{\partial f}{\partial t_2}\right) \\ \text{other rows unchanged} \end{vmatrix}.$$

The first of these is $1 + \left(\frac{\partial f}{\partial t_2}\right)^2 + \cdots + \left(\frac{\partial f}{\partial t_{n-1}}\right)^2$, by the induction hypothesis. In the second, we may factor $\frac{\partial f}{\partial t_1}$ from the first row, then use the resulting first row to reduce the determinant to

$$\frac{\partial f}{\partial t_1} \begin{vmatrix} \frac{\partial f}{\partial t_1} & \frac{\partial f}{\partial t_2} & \cdots & \frac{\partial f}{\partial t_{n-1}} \\ 0 & 1 & \cdots & 0 \\ & & \ddots & \\ 0 & \cdots & 0 & 1 \end{vmatrix} = \left(\frac{\partial f}{\partial t_1}\right)^2.$$

We end up with

$$\det\left(\frac{\partial \mathbf{G}}{\partial t_j} \bullet \frac{\partial \mathbf{G}}{\partial t_k}\right) = 1 + \left(\frac{\partial f}{\partial t_2}\right)^2 + \cdots + \left(\frac{\partial f}{\partial t_{n-1}}\right)^2 + \left(\frac{\partial f}{\partial t_1}\right)^2.$$

In agreement with Example 1,

$$\sigma = \int \sqrt{1 + \left(\frac{\partial f}{\partial x_1}\right)^2 + \cdots + \left(\frac{\partial f}{\partial x_{n-1}}\right)^2}\, dx_1 \cdots dx_{n-1}.$$

Our second characterization of hypersurface was as level surface, say $F(x) = 0$. Wherever F is C^1 and $\nabla F \neq \mathbf{O}$, the implicit function theorem guarantees that we can recast the equation as, say, $x_n = h(x_1, \ldots, x_{n-1})$. As above, we have

$$d\sigma = \sqrt{1 + \left(\frac{\partial h}{\partial x_1}\right)^2 + \cdots + \left(\frac{\partial h}{\partial x_{n-1}}\right)^2}\, dx_1 \cdots dx_{n-1}$$

for the element of surface area. To put this in terms of F, recall that

$$\frac{\partial x_n}{\partial x_j} = \frac{-\partial F/\partial x_j}{\partial F/\partial x_n}.$$

Hence

$$\frac{d\sigma}{dx_1 \cdots dx_{n-1}} = \sqrt{1 + \left(\frac{\partial F/\partial x_1}{\partial F/\partial x_n}\right)^2 + \cdots + \left(\frac{\partial F/\partial x_{n-1}}{\partial F/\partial x_n}\right)^2}$$

$$= \sqrt{\frac{(\partial F/\partial x_1)^2 + \cdots + (\partial F/\partial x_n)^2}{(\partial F/\partial x_n)^2}} = \sqrt{\frac{\|\nabla F\|^2}{(\nabla F \bullet \mathbf{e}_n)^2}},$$

or

$$\sigma = \int \frac{\|\nabla F\|}{|\nabla F \bullet \mathbf{e}_n|}\, dx_1 \cdots dx_{n-1}.$$

The formula $d\sigma = \frac{\|\nabla F\|}{|\nabla F \bullet \mathbf{e}_n|}\, dx_1 \cdots dx_{n-1}$ suggests that if we describe the hypersurface with a different function, then the numerator and denominator are rescaled by the same factor, and the element of area stays the same. Even better, it gives us a geometric way to think of the surface area. Write

$$\nabla F \bullet \mathbf{e}_n = \|\nabla F\|\, \|\mathbf{e}_n\| \cos\alpha,$$

in which α is the angle between ∇F and $\mathbf{e}_n$. Then

$$\frac{\|\nabla F\|}{|\nabla F \bullet \mathbf{e}_n|} = \frac{1}{|\cos\alpha|},$$

and

$$d\sigma = |\sec\alpha|\, dx_1 \cdots dx_{n-1}.$$

Now look at Figure 5.6. The small piece of hypersurface shown has hyperarea $d\sigma$. It casts onto the $x_1 \cdots x_{n-1}$-hyperplane a rectangular shadow of area $dA = dx_1 \cdots dx_{n-1}$. It is clear that if the normal $\mathbf{n}$ to the hypersurface is parallel to the x_n-axis, then the hypersurface is parallel to the $x_1 \cdots x_{n-1}$-hyperplane, and $dA = d\sigma$. If we incline the normal by the acute angle α, then the shadow is shortened by a factor of $\cos\alpha$. Thus, $dA = \cos\alpha\, d\sigma$, and

$$d\sigma = \pm(\sec\alpha)\, dx_1 \cdots dx_{n-1},$$

the minus sign applying if α is obtuse.

The figure makes clear the role of the normal, so we should wonder why the normal did not come up in the other formulation. We look back at the $x_n = f(x_1, \ldots, x_{n-1})$ model, and find that

$$\sqrt{1 + \left(\frac{\partial f}{\partial x_1}\right)^2 + \cdots + \left(\frac{\partial f}{\partial x_{n-1}}\right)^2} = \left\|\left(\frac{\partial f}{\partial x_1}, \ldots, \frac{\partial f}{\partial x_{n-1}}, -1\right)\right\|.$$

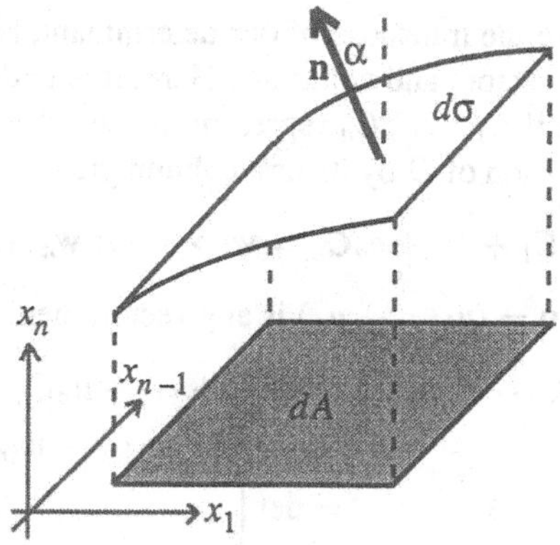

Figure 5.6.

The last vector is normal to the hypersurface: The tangent plane is spanned by

$$\left(1, 0, \ldots, 0, \frac{\partial f}{\partial x_1}\right), \ldots, \left(0, \ldots, 0, 1, \frac{\partial f}{\partial x_{n-1}}\right),$$

and we easily check that $\left(\frac{\partial f}{\partial x_1}, \ldots, \frac{\partial f}{\partial x_{n-1}}, -1\right)$ is orthogonal to all the tangent vectors. We therefore have

$$d\sigma = \|\mathbf{n}\|\, dx_1 \cdots dx_{n-1},$$

where $\mathbf{n}$ is either of the normals to the hypersurface that have x_n-component ± 1.

We will have good reason to return to the normals. Our last task here is to reconcile what we have found to the parametric definition of hypersurface. (See Exercises 3, 4.)

If $(x_1, \ldots, x_n) = \mathbf{g}(t_1, \ldots, t_{n-1})$, we have

$$d\sigma \equiv \sqrt{\det\left(\frac{\partial \mathbf{g}}{\partial t_i}(\mathbf{t}) \bullet \frac{\partial \mathbf{g}}{\partial t_j}(\mathbf{t})\right)}\, dt_1 \cdots dt_{n-1}.$$

To relate this determinant to a normal, write $\frac{\partial \mathbf{g}}{\partial t_j}(\mathbf{t})$ as $\mathbf{w}_j \equiv (w_{j1}, \ldots, w_{jn})$. We define the **vector product** or **cross product** of $w_1, \ldots, w_{n-1}$ by

$$\mathbf{w}_1 \times \cdots \times \mathbf{w}_{n-1} \equiv \mathbf{e}_1 \det\big(\mathbf{e}_1\ \mathbf{w}_1 \cdots \mathbf{w}_{n-1}\big) + \cdots + \mathbf{e}_n \det\big(\mathbf{e}_n\ \mathbf{w}_1 \cdots \mathbf{w}_{n-1}\big).$$

It is usual to denote this vector by the "formal" determinant

$$\mathbf{D} \equiv \det \begin{bmatrix} \mathbf{e}_1 & w_{11} & \cdots & w_{(n-1)1} \\ & & \ddots & \\ \mathbf{e}_n & w_{1n} & \cdots & w_{(n-1)n} \end{bmatrix}.$$

(It is also usual to write the transpose of our determinant, but we are being consistent about equating vectors and columns.) Here, it is understood that **D** is not a real determinant, but if $C_1, \dots, C_n$ represent the cofactors of the first column, then the cofactor expansion of D by its first column yields

$$\mathbf{e}_1 C_1 + \cdots + \mathbf{e}_n C_n = \mathbf{w}_1 \times \cdots \times \mathbf{w}_{n-1}.$$

It is easy to see that if $\mathbf{u} = (u_1, \dots, u_n)$ is any vector, then

$$\begin{aligned}\mathbf{u} \bullet (\mathbf{w}_1 \times \cdots \times \mathbf{w}_{n-1}) &= u_1 C_1 + \cdots + u_n C_n \\ &= \det \begin{bmatrix} u_1 & w_{11} & \cdots & w_{(n-1)1} \\ & & \ddots & \\ u_n & w_{1n} & \cdots & w_{(n-1)n} \end{bmatrix}.\end{aligned}$$

In particular, $\mathbf{w}_j \bullet (\mathbf{w}_1 \times \cdots \mathbf{w}_{n-1})$, being a determinant with two identical columns, is zero; the cross product is orthogonal to all the $\mathbf{w}_j$. Since the $\mathbf{w}_j = \frac{\partial \mathbf{g}}{\partial t_j}$ span the tangent hyperplane, we see that $\frac{\partial \mathbf{g}}{\partial t_1} \times \cdots \times \frac{\partial \mathbf{g}}{\partial t_{n-1}}$ is the normal.

How long is it? We have

$$\begin{aligned}\left(C_1^2 + \cdots + C_n^2\right)^2 &= \det{}^2 \begin{bmatrix} C_1 & w_{11} & \cdots & w_{(n-1)1} \\ & & \ddots & \\ C_n & w_{1n} & \cdots & w_{(n-1)n} \end{bmatrix} \\ &= \det \left(\begin{bmatrix} C_1 & \cdots & C_n \\ w_{11} & & w_{1n} \\ \vdots & \ddots & \vdots \\ w_{(n-1)1} & & w_{(n-1)n} \end{bmatrix} \begin{bmatrix} C_1 & w_{11} & \cdots & w_{(n-1)1} \\ & & \ddots & \\ C_n & w_{1n} & \cdots & w_{(n-1)n} \end{bmatrix} \right).\end{aligned}$$

In the matrix product on the right, the top left entry is $C_1^2 + \cdots + C_n^2$. For $i \neq 1$, the entry at position $1i$ or $i1$ is the dot product of column i of **D** with the column of cofactors from column 1. In every determinant, that mixed product is 0. If $i \neq 1$ and $j \neq 1$, then the entry at position ij is the dot product of column i in **D** with column j, namely, $\frac{\partial \mathbf{g}}{\partial t_i} \bullet \frac{\partial \mathbf{g}}{\partial t_j}$. Hence

$$\begin{aligned}\left(C_1^2 + \cdots + C_n^2\right)^2 &= \det \begin{bmatrix} C_1^2 + \cdots + C_n^2 & 0 & \cdots & 0 \\ 0 & \frac{\partial \mathbf{g}}{\partial t_1} \bullet \frac{\partial \mathbf{g}}{\partial t_1} & \cdots & \frac{\partial \mathbf{g}}{\partial t_1} \bullet \frac{\partial \mathbf{g}}{\partial t_{n-1}} \\ \vdots & & \ddots & \\ 0 & \frac{\partial \mathbf{g}}{\partial t_{n-1}} \bullet \frac{\partial \mathbf{g}}{\partial t_1} & & \frac{\partial \mathbf{g}}{\partial t_{n-1}} \bullet \frac{\partial \mathbf{g}}{\partial t_{n-1}} \end{bmatrix} \\ &= \left(C_1^2 + \cdots + C_n^2\right) \det \left(\frac{\partial \mathbf{g}}{\partial t_i} \bullet \frac{\partial \mathbf{g}}{\partial t_j} \right).\end{aligned}$$

This reduces to

$$\left\| \frac{\partial \mathbf{g}}{\partial t_1} \times \cdots \times \frac{\partial \mathbf{g}}{\partial t_{n-1}} \right\| = \left(C_1^2 + \cdots + C_n^2\right)^{1/2} = \sqrt{\det \left(\frac{\partial \mathbf{g}}{\partial t_i} \bullet \frac{\partial \mathbf{g}}{\partial t_j} \right)}.$$

We conclude that

$$d\sigma = \left\| \frac{\partial \mathbf{g}}{\partial t_1} \times \cdots \times \frac{\partial \mathbf{g}}{\partial t_{n-1}} \right\| dt_1 \cdots dt_{n-1},$$

in which the cross product is a vector whose direction is normal to the hyperplane spanned by $\frac{\partial \mathbf{g}}{\partial t_1}, \ldots, \frac{\partial \mathbf{g}}{\partial t_{n-1}}$, and whose magnitude is the hyperarea of their hyperparallelogram.

Exercises

1. Find the area of the part of the paraboloid $z = x^2 + y^2$ below the plane $z = 2$.

2. Use the formula area($\mathbf{f}$) $= \int_A \text{area}\left(\frac{\partial \mathbf{f}}{\partial r}, \frac{\partial \mathbf{f}}{\partial \theta}\right)$ to find the area of the surface parametrized by $x = r\cos\theta$, $y = r\sin\theta$, $z = r^2$, $0 \le r \le \sqrt{2}$, $0 \le \theta \le 2\pi$. (Your answer should match Exercise 1.)

3. For the hemisphere given by $z = \left(a^2 - x^2 - y^2\right)^{1/2}$, the radius from $(0, 0, 0)$ to (x, y, z) is normal to the surface. Find the secant of the angle α between this segment and the z-axis. Then show that it matches:

   (a) $\left(1 + \left[\frac{\partial z}{\partial x}\right]^2 + \left[\frac{\partial z}{\partial y}\right]^2\right)^{1/2}$;

   (b) $\frac{\|\nabla F\|}{|\nabla F \bullet \mathbf{e}_n|}$ for the function $F(x, y, z) = x^2 + y^2 + z^2$ at the level surface $F = a^2$, with $\mathbf{e}_n = \mathbf{k}$;

   (c) the secant of the angle between $\frac{\partial(x,y,z)}{\partial \phi} \times \frac{\partial(x,y,z)}{\partial \theta}$ and $\mathbf{k}$, for the parametrization $x = a\cos\theta\sin\phi$, $y = a\sin\theta\sin\phi$, $z = a\cos\phi$.

4. Find the area of the surface given by

   $$x = r\cos\theta,\ \ y = r\sin\theta,\ \ z = r\cos\theta, \qquad 0 \le r \le a,\ 0 \le \theta \le 2\pi:$$

   (a) directly from $d\sigma = \sec\alpha\, dA$, A representing area in the xy-plane;

   (b) employing the gradient of $G(x, y, z) \equiv x - z$, on the level surface $G = 0$.

   (c) as $\displaystyle\int \left\| \frac{\partial(x, y, z)}{\partial r} \times \frac{\partial(x, y, z)}{\partial \theta} \right\|$.

5. Show that, for a smooth n-dimensional surface, "area" is just volume. Thus, let $\mathbf{g}\colon A \subseteq \mathbf{R}^n \to \mathbf{R}^n$ be smooth and one-to-one. Then:

   (a) The area of $\mathbf{g}$ is the volume of $\mathbf{g}(A)$.

(b) If f is continuous, then $\int_{\mathbf{g}(A)} f \, d\sigma = \int_{\mathbf{g}(A)} f$.

Is the one-to-one condition essential?

6. Suppose the graph of $y = f(x)$, with $f(x) \geq 0$ for $a \leq x \leq b$, is revolved around the x-axis. Find the surface area of the curved part of the solid of revolution.

7. (a) In $\mathbf{R}^3$, the area $\sigma = 4\pi r^2$ of a sphere and the volume $V = \frac{4\pi r^3}{3}$ of the ball are related by $\sigma = \frac{dV}{dr}$. Give a geometric explanation for this relationship.

   (b) Show that the same applies to the hyperarea and volume of the sphere and ball in $\mathbf{R}^4$. (Consult Example 3 and Example 5.1:2.)

   (c) Find the hyperarea of the sphere of radius a in $\mathbf{R}^5$.

8. Show that $|\mathbf{v}_1 \bullet (\mathbf{v}_2 \times \cdots \times \mathbf{v}_n)|$ is the volume of the parallelepiped spanned by $\mathbf{v}_1, \ldots, \mathbf{v}_n$. (In $\mathbf{R}^3$, $\mathbf{u} \bullet (\mathbf{v} \times \mathbf{w})$ is called the **triple product** of **u**, **v**, **w**.)

6

Vector Integrals and the Vector-Field Theorems

In our final chapter we look at two kinds of integrals involving interaction between a vector function and the curve or surface on which we want to calculate an integral. This study will show us many connections among the topics that we have viewed separately. We end up with small versions of two results that helped complete the analytic description of classical physics.

That description began with Newton's invention of calculus and differential equations, for the purpose of explaining mechanics. By the time of Lagrange (second half of the 1700s), it had evolved into partial differential equations. The 1800s brought an enormous body of work on partial differential equations, roughly bracketed by Joseph Fourier's study of heat and James Maxwell's of electricity and magnetism. We will use our two results to express two of (the four) Maxwell's equations. The equations are so fundamental that modern physics was built on the assumption that they must be satisfied universally.

6.1 Integrals of the Tangential and Normal Components

Throughout this chapter we will deal with vector fields. A **vector field** is just a continuous function $\mathbf{F}$ mapping (some of) $\mathbf{R}^n$ to $\mathbf{R}^n$. We may think of $\mathbf{F}$ as attaching a vector $\mathbf{F}(\mathbf{x})$ to each point $\mathbf{x}$. Our archetype is gravitation. A fixed mass occupying the origin will pull on a unit mass at $\mathbf{x} \neq \mathbf{O}$ with a force pointing toward the origin and having magnitude inversely proportional to the square of the distance. Thus, we think of the fixed mass as creating the field $\mathbf{F}(\mathbf{x}) \equiv \frac{-k\mathbf{x}}{\|\mathbf{x}\|^3}$.

We have no reason to study integrals of vector functions with respect to length, area, or volume, because all of them reduce to the scalar case. Those integrals are defined by linear processes. It is easy to see that $\int_C \mathbf{F}\,ds$ would match

$$\left(\int_C F_1\,ds, \ldots, \int_C F_n\,ds\right),$$

and similarly for surface and volume integrals. The integrals we want are of the components of **F** tangent to a curve or normal to a hypersurface.

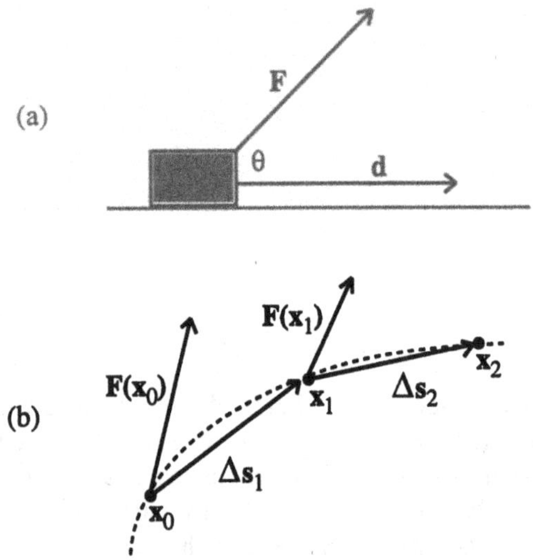

Figure 6.1.

The integral of the tangential component arises in the physical concept of work. The **work** done by a constant force F acting through a distance d is Fd. If the force does not act in the direction of the motion (Figure 6.1(a)), we turn the force and displacement into vectors and define the work as $(F\cos\theta)d = \mathbf{F}\bullet\mathbf{d}$. Finally, for a variable force field $\mathbf{F}(\mathbf{x})$ applied to an object moving along a curve (Figure 6.1(b)), we approximate the motion with a polygonal path, form sums like

$$\mathbf{F}_0\bullet\Delta\mathbf{s}_1+\cdots+\mathbf{F}_{J-1}\bullet\Delta\mathbf{s}_J \equiv \mathbf{F}(\mathbf{x}_0)\bullet(\mathbf{x}_1-\mathbf{x}_0)+\cdots+\mathbf{F}(\mathbf{x}_{J-1})\bullet(\mathbf{x}_J-\mathbf{x}_{J-1}),$$

and look for their limit.

Definition. Assume that $\mathbf{F}(\mathbf{x})$ is a vector field and C a rectifiable curve in the domain of $\mathbf{F}$. The **path integral of F along** C is

$$\int_C \mathbf{F}\bullet d\mathbf{s} = \int_C (F_1,\ldots,F_n)\bullet(dx_1,\ldots,dx_n) \equiv \int_C F_1\,dx_1+\cdots+\int_C F_n\,dx_n.$$

(The coordinate integrals on the right must exist, by Theorem 5.5:5(b).)

Theorem 1. *If* $\mathbf{g}$: $[a,b] \to C$ *is* C^1, *then the path integral of* $\mathbf{F}$ *along* C *is*

$$\int_C \mathbf{F} \bullet d\mathbf{s} = \int_a^b \left[F_1(\mathbf{g}(t))g_1'(t) + \cdots + F_n(\mathbf{g}(t))g_n'(t)\right] dt$$

$$= \int_a^b \mathbf{F}(\mathbf{g}(t)) \bullet \mathbf{g}'(t)\, dt.$$

Proof. The first equality follows from Theorem 5.5:5(c). The second is a restatement in dot-product form. □

It is usual to write $\int_C F_1\, dx_1 + \cdots + \int_C F_n\, dx_n$ with a single integral sign, as $\int_C F_1\, dx_1 + \cdots + F_n\, dx_n$.

As soon as we introduced x_k-integrals, we noted that they are directed. We will use the word **path** to indicate the combination of a piecewise-C^1 curve and orientation, or equivalently, a curve with specified start point and end point.

Example 1. How much work does the gravitational field $\mathbf{F}(x,y) \equiv \frac{-k(x,y)}{\|(x,y)\|^3}$ do on an object moving from (3, 4) to (5, 12)?

(Note that the object must be under the influence of another force as well; gravity, acting alone, would not push it away. It is nevertheless fair to ask what work the field contributes. Sometimes the question is put as, what work is done by the other force? That version pretends that a force $-\mathbf{F}$ at each place will achieve the displacement, so the answer will be the negative of the upcoming one.)

(a) The segment C can be parametrized as $x = 3 + 2t$, $y = 4 + 8t$, $0 \le t \le 1$. Consequently,

$$\int_C \mathbf{F} \bullet d\mathbf{s} = \int_0^1 \frac{-k(3+2t, 4+8t)}{\left[(3+2t)^2 + (4+8t)^2\right]^{3/2}} \bullet (2,8)\, dt$$

$$= -k \int_0^1 (25 + 76t + 68t^2)^{-3/2}(38 + 68t)\, dt$$

$$= k\left(\frac{1}{13} - \frac{1}{5}\right).$$

(b) Suppose we take D from (3, 4) to (5, 4) to (5, 12). Let us employ the definition directly. On the horizontal, $dy = 0$, and

$$\int F_1\, dx = \int_3^5 \frac{-kx}{(x^2 + 4^2)^{3/2}}\, dx = k\left(\frac{1}{\sqrt{41}} - \frac{1}{5}\right).$$

On the vertical, $dx = 0$, and

$$\int F_2\, dy = \int_4^{12} \frac{-ky}{(5^2 + y^2)^{3/2}}\, dy = k\left(\frac{1}{13} - \frac{1}{\sqrt{41}}\right).$$

Hence

$$\int_D \mathbf{F} \bullet d\mathbf{s} = k\left(\frac{1}{13} - \frac{1}{\sqrt{41}} + \frac{1}{\sqrt{41}} - \frac{1}{5}\right) = k\left(\frac{1}{13} - \frac{1}{5}\right).$$

Example 2. In Example 1(b) the answers fit together in a way suggesting that alternative paths always give the same integral. A different field might behave otherwise.

Consider $\mathbf{G}(x, y) \equiv y\mathbf{i} = y(1, 0)$. On the path C of Example 1,

$$\int_C \mathbf{G} \bullet d\mathbf{s} = \int_0^1 (y, 0) \bullet (2, 8)\, dt = \int_0^1 2(4 + 8t)\, dt = 16.$$

On D,

$$\int_D G_1\, dx + G_2\, dy = \int_3^5 4\, dx + \int_4^{12} 0\, dy = 8.$$

The quantity $\mathbf{F} \bullet \mathbf{g}'$ is not the tangential component of $\mathbf{F}$. Certainly $\mathbf{g}'$ is tangent to the path, but by "(scalar) component of $\mathbf{F}$ along the tangent line," we should mean

$$F\cos\theta = \frac{\mathbf{F} \bullet \mathbf{g}'}{\|\mathbf{g}'\|} = \mathbf{F} \bullet \frac{\mathbf{g}'}{\|\mathbf{g}'\|}.$$

That leads us to introduce the unit tangent.

Definition. Assume that $\mathbf{g}\colon [a, b] \to C$ is smooth. The **unit tangent (vector)** at $\mathbf{g}(t)$ is

$$\mathbf{T}(t) \equiv \frac{\mathbf{g}'(t)}{\|\mathbf{g}'(t)\|}.$$

Theorem 2. *The path integral is the line integral of the tangential component. In symbols: If C is smooth, then $\int_C \mathbf{F} \bullet d\mathbf{s} = \int_C \mathbf{F} \bullet \mathbf{T}\, ds$.*

Proof. Exercise 4.

Example 3. Theorem 2 gives us a third way of doing the path integral, with obvious advantage when the tangential component is simple. Let $\mathbf{G}(x, y) \equiv y\mathbf{i}$.

(a) On the path D from Example 1, $\mathbf{G}$ is constantly $4\mathbf{i}$ and $\mathbf{T} = \mathbf{i}$ along the horizontal part, contributing $\int_3^5 4\, ds = 8$. On the vertical, $\mathbf{G} = y\mathbf{i}$ and $\mathbf{T} = \mathbf{j}$, so $\mathbf{G} \bullet \mathbf{T} = 0$; we need not integrate. Hence $\int_D \mathbf{G} \bullet d\mathbf{s} = 8$.

(b) Let us try the path E from $(3, 4)$ to $(3, 12)$ to $(5, 12)$. On the vertical, $\mathbf{G}$ is again orthogonal to $\mathbf{T}$. On the horizontal, $\mathbf{G} = 12\mathbf{i}$ and $\mathbf{T} = \mathbf{i}$, giving $\int_3^5 12\, ds = 24$. We obtain $\int_E \mathbf{G} \bullet d\mathbf{s} = 24$.

Having identified the expressions

$$\int_C \mathbf{F} \bullet d\mathbf{s} = \int_C F_1 dx_1 + \cdots + \int_C F_n\, dx_n = \int_C F \bullet \mathbf{g}'\, dt = \int_C \mathbf{F} \bullet \mathbf{T}\, ds,$$

let us adopt notation that identifies the differentials. We may write the "vector element of arc length" (or "element of displacement" or "differential of displacement" or many other names)

$$\begin{aligned} d\mathbf{s} = \mathbf{g}'\, dt &= \left(\frac{dg_1}{dt}, \ldots, \frac{dg_n}{dt}\right) dt = (dg_1, \ldots, dg_n) \\ &= (dx_1, \ldots, dx_n) \text{ along } C. \end{aligned}$$

The last has the advantage of involving length and position along the curve, without reference to a parametrization. To use our earlier word, it is "intrinsic" to the curve. Taking magnitudes, we get

$$\|d\mathbf{s}\| = \|\mathbf{g}'\|\, dt = ds.$$

Since $d\mathbf{s}$ has length ds and the direction of $\mathbf{T}$, we identify $d\mathbf{s} = \mathbf{T}\, ds$. Then

$$ds = \|d\mathbf{s}\| = \sqrt{dx_1^2 + \cdots + dx_n^2}$$

and

$$\mathbf{T} = \frac{d\mathbf{s}}{ds} = \frac{(dx_1, \ldots, dx_n)}{\sqrt{dx_1^2 + \cdots + dx_n^2}}$$

are intrinsic descriptions of ds and $\mathbf{T}$ along C. (Keep in mind that C incorporates an orientation. See Exercises 9 and 10 for more related to intrinsic properties.)

Our other interest, the integral of the normal component, is best described in terms of flows. Imagine the vector field $\mathbf{V}(\mathbf{x})$, representing the velocity of air at the point $\mathbf{x}$, with the wind blowing toward two windows of area $\Delta\sigma$. In Figure 6.2(a), $\mathbf{V}(\mathbf{x})$ is normal to the window (shaded); it has the direction of the unit normal $\mathbf{N}$. A column of air of cross-sectional area $\Delta\sigma$ advances through the window at a rate $\|\mathbf{V}(\mathbf{x})\|$ m/sec, so that volume $\|\mathbf{V}(\mathbf{x})\|\Delta\sigma$ of air flows through per unit time. At the place $\mathbf{y}$ (Figure 6.2 (b)) where $\mathbf{V}(\mathbf{y})$ makes angle θ with the normal, the air advances $\|\mathbf{V}(\mathbf{y})\| \cos\theta$ m/sec, cutting the flow rate to

$$\|\mathbf{V}(\mathbf{y})\| \cos\theta\ \Delta\sigma = \mathbf{V}(\mathbf{y}) \bullet \mathbf{N}\ \Delta\sigma.$$

Given a surface, then, we break it up into pieces of areas $\Delta\sigma_j$, small enough to be nearly flat, allowing a normal $\mathbf{N}_j$ to be established; we sum contributions $\mathbf{V}(\mathbf{y}_j) \bullet \mathbf{N}_j \Delta\sigma_j$; we proceed to the limit, which leads us to an integral.

Definition. Let $\mathbf{V}(\mathbf{x})$ be a vector field defined along a (by definition C^1) hypersurface S. Assume that $\mathbf{N}(\mathbf{x})$ is a continuous function of unit magnitude, normal to the hypersurface at each $\mathbf{x} \in S$. The surface integral $\int_S \mathbf{V} \bullet \mathbf{N}\, d\sigma$ of the normal

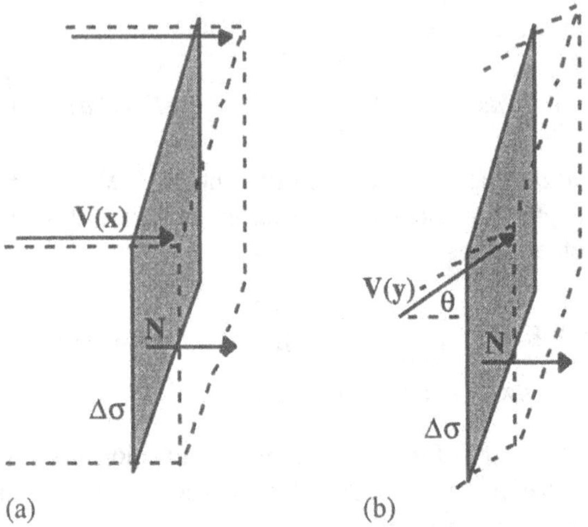

Figure 6.2.

component is called the **flux of V through** S (toward the side "determined" by the normals.)

Theorem 3. *The flux of* **V** *is given by:*

(a)

$$\int_A \mathbf{V}(x_1, \ldots, f(x_1, \ldots, x_{n-1})) \bullet \left(\frac{-\partial f}{\partial x_1}, \ldots, \frac{-\partial f}{\partial x_{n-1}}, 1\right) dx_1 \cdots dx_{n-1},$$

in the direction of increasing x_n, for the hypersurface defined by

$$x_n = f(x_1, \ldots, x_{n-1}) \quad \textit{for } (x_1, \ldots, x_{n-1}) \in A \subseteq \mathbf{R}^{n-1}.$$

(b)

$$\int_A \mathbf{V}(x_1, \ldots, x_n) \bullet \frac{\nabla F}{|\nabla F \bullet \mathbf{e}_n|} dx_1 \cdots dx_{n-1},$$

in the direction of increasing F, for the hypersurface defined by $F(x_1, \ldots, x_n) = 0$, assuming $\frac{\partial F}{\partial x_n} \neq 0$ along the hypersurface.

(c)

$$\int_A \mathbf{V}(\mathbf{g}(t)) \bullet \left(\frac{\partial \mathbf{g}}{\partial t_1} \times \cdots \times \frac{\partial \mathbf{g}}{\partial t_{n-1}}\right) dt_1 \cdots dt_{n-1},$$

in the direction of the cross product, for the hypersurface defined by

$$\mathbf{x} = \mathbf{g}(t_1, \ldots, t_{n-1}) = \mathbf{g}(\mathbf{t}), \qquad \mathbf{t} \in A.$$

Proof. All three formulations are consequences of the discussion of hyperarea in the last third of Section 5.6. In (c), for example, write

$$\mathbf{n}(\mathbf{t}) \equiv \frac{\partial \mathbf{g}}{\partial t_1}(\mathbf{t}) \times \cdots \times \frac{\partial \mathbf{g}}{\partial t_{n-1}}(\mathbf{t}).$$

We saw that $\mathbf{n}(\mathbf{t})$ is normal to the (tangent hyperplane of the) hypersurface, and $d\sigma = \|\mathbf{n}\|\, dt_1 \cdots dt_{n-1}$. Hence $\mathbf{N} \equiv \frac{\mathbf{n}}{\|\mathbf{n}\|}$ is a continuous unit normal, and

$$\begin{aligned}\int_S \mathbf{V} \bullet \mathbf{N}\, d\sigma &= \int_A \mathbf{V}(\mathbf{g}(t)) \bullet \left(\frac{\mathbf{n}}{\|\mathbf{n}\|}\right) \|\mathbf{n}\|\, dt_1 \cdots dt_{n-1} \\ &= \int_A \mathbf{V}(\mathbf{g}(t)) \bullet \mathbf{n}\, dt_1 \cdots dt_{n-1}.\end{aligned}$$

That proves (c), and we leave (a) and (b) to Exercise 8.

We see again a "vector differential." We define the **vector element of area** by

$$d\boldsymbol{\sigma} = \mathbf{N}\, d\sigma = \mathbf{n}\, dt_1 \cdots dt_{n-1},$$

where $\mathbf{n}$ is $\left(\frac{-\partial x_n}{\partial x_1}, \ldots, \frac{-\partial x_n}{\partial x_{n-1}}, 1\right)$ or $\frac{\nabla F}{|\nabla F \bullet \mathbf{e}_n|}$ or $\frac{\partial \mathbf{g}}{\partial t_1} \times \cdots \times \frac{\partial \mathbf{g}}{\partial t_{n-1}}$, corresponding to the parametrization of the hypersurface.

Example 4. Michael Faraday discovered that a changing magnetic flux ϕ "induces" an "electromotive force" (voltage) V given by **Faraday's law**: $V = -\frac{d\phi}{dt}$. (The sign has to do with the voltage's trying to oppose the change.) We will create a changing flux by rotating a loop of wire within a fixed magnetic field.

Suppose $\mathbf{R}^3$ carries the fixed magnetic field $\mathbf{B} = b\mathbf{k}$. Consider the disk described by

$$x = (r\cos\theta)\cos\omega t, \quad y = r\sin\theta, \quad z = (r\cos\theta)\sin\omega t,$$

$0 \le r \le a$, $0 \le \theta \le 2\pi$, whose edge (given by $r = a$) is a circle of radius a, rotating about the y-axis with angular speed ω. We can always find flux directly by recalling that the triple product $\mathbf{B} \bullet (\frac{\partial \mathbf{x}}{\partial r} \times \frac{\partial \mathbf{x}}{\partial \theta})$ is given by

$$\begin{aligned}\det\left[\mathbf{B}\ \frac{\partial \mathbf{x}}{\partial r}\ \frac{\partial \mathbf{x}}{\partial \theta}\right] &= \begin{vmatrix} 0 & \cos\theta\cos\omega t & -r\sin\theta\cos\omega t \\ 0 & \sin\theta & r\cos\theta \\ b & \cos\theta\sin\omega t & -r\sin\theta\cos\omega t \end{vmatrix} \\ &= b(r\cos^2\theta\cos\omega t + r\sin^2\theta\cos wt) = br\cos wt.\end{aligned}$$

Hence

$$\phi = \int_0^{2\pi}\int_0^a br\cos\omega t\, dr\, d\theta = b\cos\omega t\left(\frac{2\pi a^2}{2}\right),$$

and

$$V = -\frac{d\phi}{dt} = b(\pi a^2)\omega\sin\omega t.$$

The induced voltage is sinusoidal, with amplitude proportional to the strength of the field, the area of the wire loop, and the rotational speed.

Example 5. A charge q at the origin in $\mathbf{R}^3$ creates an **electrostatic field** $\mathbf{E}(\mathbf{x}) = Kq\frac{\mathbf{x}}{\|\mathbf{x}\|^3}$. (Notice that the field points outward—it is "repulsive"—if q is positive.) What is the outward flux through an origin-centered sphere of radius a?

Instead of calculating directly, we take advantage of the geometry. The radius is normal to the sphere, making the outward unit normal $\frac{\mathbf{x}}{\|\mathbf{x}\|}$. Hence

$$\mathbf{E} \bullet \mathbf{N} = Kq\mathbf{x} \bullet \frac{\mathbf{x}}{\|\mathbf{x}\|^4} = \frac{Kq}{\|\mathbf{x}\|^2} = \frac{Kq}{a^2}.$$

Since this is constant, we have

$$\int_S \mathbf{E} \bullet \mathbf{N}\, d\sigma = \left(\frac{Kq}{a^2}\right) \text{area} = \left(\frac{Kq}{a^2}\right) 4\pi a^2.$$

The flux is a multiple $4\pi K$ (characteristic of space) of the charge, independent of a.

There is a simple geometric argument that for such an inverse-square field, the flux will be the same as for the sphere on any closed surface surrounding the origin, as well as zero for a surface not surrounding it. [See Peck, Section 1.9, for illustration of the argument, and Exercise 7 for an example.] By that principle, *flux* $= (4\pi K)Q$ holds for any surface, where Q is the total charge surrounded by the surface. This flux relationship is called **Gauss's law**.

The direction referred to in Theorem 3(a), the positive x_n-direction, is obvious. The "direction of increasing F" in (b) is, if not geometrically obvious, simply the direction of ∇F. The line of

$$\mathbf{n} \equiv \frac{\partial \mathbf{g}}{\partial t_1} \times \cdots \times \frac{\partial \mathbf{g}}{\partial t_{n-1}}$$

in (c) is normal to the hypersurface, but which way does $\mathbf{n}$ point? The answer lies in the "right-hand rule."

In Section 5.6 we saw that the components of $\mathbf{n}$ are the cofactors $C_1, \ldots, C_n$ from the first column of the determinant $\det\left[\mathbf{e}_1 \ \frac{\partial \mathbf{g}}{\partial t_1} \ \cdots \ \frac{\partial \mathbf{g}}{\partial t_{n-1}}\right]$. For that reason,

$$\det\left[\mathbf{n} \ \frac{\partial \mathbf{g}}{\partial t_1} \ \cdots \ \frac{\partial \mathbf{g}}{\partial t_{n-1}}\right] = C_1{}^2 + \cdots + C_n{}^2 > 0.$$

The normal $-\mathbf{n}$ in the opposite direction has $\det\left[-\mathbf{n} \ \frac{\partial \mathbf{g}}{\partial t_1} \ \cdots \ \frac{\partial \mathbf{g}}{\partial t_{n-1}}\right] < 0$.

Definition. The vectors $\mathbf{v}_1, \ldots, \mathbf{v}_n$ **form a right-handed system** (or **follow the right-hand rule**) if $\det\left[\mathbf{v}_n \ \mathbf{v}_1 \ \cdots \mathbf{v}_{n-1}\right] > 0$, **left-handed** if < 0.

We previously interpreted the absolute value of a determinant as the volume of the parallelepiped spanned by the columns. Now we see that the sign of the determinant has to do with right-handedness. Notice that in $\mathbf{R}^3$, $\det\left[\mathbf{v}_1 \ \mathbf{v}_2 \ \mathbf{v}_3\right] = \det\left[\mathbf{v}_3 \ \mathbf{v}_1 \ \mathbf{v}_2\right]$, but moving $\mathbf{v}_n$ to the end changes the determinant sign if n is even.

Exercises

1. Suppose $\mathbf{p}$ and $\mathbf{q}$ are distinct points $\neq \mathbf{O}$ in $\mathbf{R}^2$. Show that there exist two paths from $\mathbf{p}$ to $\mathbf{q}$ such that the work done by the gravitational field is the same over both paths.

2. Prove the same as in Exercise 1 for the field $\mathbf{G}(x, y) \equiv y\mathbf{i}$.

3. Consider the "whirlpool" $\mathbf{H}(x, y) \equiv -y\mathbf{i} + x\mathbf{j}$. Make a sketch to illustrate the field vectors at numerous points. Then find the work done by $\mathbf{F}$ on an object moving:

   (a) along the segment from (a, b) to (c, d);

   (b) along the path that travels straight from $(2, 0)$ to $(4, 0)$, then counterclockwise around the circle $x^2 + y^2 = 16$ to $(0, 4)$;

   (c) counterclockwise once around $x^2 + y^2 = 16$, starting and ending at the same point.

4. Prove Theorem 2.

5. For $\mathbf{H}$ in Exercise 3:

   (a) Find the flux outward through the circle $x^2 + y^2 = a^2$.

   (b) Give a geometric argument for why the result should be the same for any loop (path with start = end).

6. Let $\mathbf{V}(\mathbf{x}) \equiv v\mathbf{k}$ (constant) in $\mathbf{R}^3$.

   (a) Find the flux of $\mathbf{V}$ upward through the hemisphere

   $$z = \sqrt{a^2 - x^2 - y^2}.$$

   (b) Give a physical argument for why the result is simply v times the area of the disk $x^2 + y^2 = a^2$, $z = 0$.

7. Show that the flux of the electrostatic field in Example 5, through the cube with vertices at $(\pm a, \pm a, \pm a)$, is also $(4\pi K)q$.

8. Prove Theorem 3(a) and (b).

9. Suppose $\mathbf{g}\colon [a, b] \to C \subseteq \mathbf{R}^n$ is a smooth curve.

   (a) Show that the arc length $s = L(t) \equiv \operatorname{len}\big(\mathbf{g}([a, t])\big)$ is differentiable, with $\frac{ds}{dt} > 0$.

   (b) Show that $\mathbf{G}(s) \equiv \mathbf{g}\big(L^{-1}(s)\big)$ is a smooth curve with the same start, end, and range as C. (Notice that the domain is $[0, \operatorname{len}(C)]$.)

   (c) Show that $\frac{d\mathbf{G}}{ds} = \frac{d\mathbf{g}/dt}{ds/dt}$.

(d) Show that $\mathbf{G}(s)$ has the same length as C.

(e) Show that $\mathbf{T} = \frac{d\mathbf{G}}{ds}$. This gives another intrinsic description of the unit tangent.

10. For $\mathbf{g}$ and $\mathbf{G}$ as in Exercise 9, assume that $\mathbf{T} = \frac{d\mathbf{G}}{ds}$ is differentiable. Its derivative $\frac{d\mathbf{T}}{ds} = \frac{d^2\mathbf{G}}{ds^2}$ indicates which way, and to what extent, the curve is changing direction. Its norm $K(s) \equiv \|\mathbf{G}''(s)\|$ is called the **curvature** of C at $\mathbf{G}(s)$.

(a) Assume that $\mathbf{g}(t)$ is twice differentiable. Show that

$$K\big(L(t)\big) = \frac{\sqrt{\|\mathbf{g}'(t)\|^2\|\mathbf{g}''(t)\|^2 - [\mathbf{g}'(t) \bullet \mathbf{g}''(t)]^2}}{\|\mathbf{g}'(t)\|^3}.$$

(b) Show that in $\mathbf{R}^3$, $K\big(L(t)\big) = \frac{\|\mathbf{g}'(t)\times\mathbf{g}''(t)\|}{\|\mathbf{g}'(t)\|^3}$.

(c) Find the curvature of the line $x_1 = a_1 + \alpha_1 t, \ldots, x_n = a_n + \alpha_n t$.

(d) If $K(s)$ is identically zero, does the curve have to be part of a line?

(e) Find the curvature at a point of the circle $x = a\cos\theta$, $y = a\sin\theta$. What if we change the parametrization to $x = a\cos e^t$, $y = a\sin e^t$?

(f) If $K(s)$ is constant, does the curve have to be a circle? Take a guess, then find the curvature of the helix $x = a\cos t$, $y = a\sin t$, $z = bt$.

(g) Assume that f is twice differentiable. Show that the curvature of the graph of $y = f(x)$ at the point (x, y) is $\frac{|f''(x)|}{[1+f'(x)]^{3/2}}$.

(h) Find the curvature at a point of the graph of $y = \left(a^2 - x^2\right)^{1/2}$, $x \neq \pm a$.

(i) Show that in $\mathbf{R}^3$, $K(s) = \left\|\mathbf{T} \times \frac{d\mathbf{T}}{ds}\right\|$. This gives an intrinsic description of curvature. Then show that this norm is the scalar component of $\frac{d\mathbf{T}}{ds}$ *perpendicular* to the line of $\mathbf{T}$.

6.2 Path-Independence

The path integral $\int_C \mathbf{F} \bullet d\mathbf{s}$ is a sort of directional integral. Accordingly, it should be related by something like the fundamental theorem of calculus to directional derivatives.

Theorem 1. *The path integral of a gradient is the change in its function. Thus, assume that f is C^1 and $\mathbf{F} = \nabla f$; then for a C^1 path C, in the domain of f, with initial point $\mathbf{p}$ and end point $\mathbf{q}$,*

$$\int_C \mathbf{F} \bullet d\mathbf{s} = f(\mathbf{q}) - f(\mathbf{p}).$$

Proof. Let $\mathbf{g}$: $[a, b] \to C$, and look at f along C. That is, define the scalar function of a scalar variable

$$h(t) \equiv f(\mathbf{g}(t)), \qquad a \leq t \leq b.$$

By the chain rule,

$$h'(t) = f'(\mathbf{g}(t))\mathbf{g}'(t) = \nabla f(\mathbf{g}(t)) \bullet \mathbf{g}'(t).$$

By the fundamental theorem,

$$h(b) - h(a) = \int_a^b \nabla f(\mathbf{g}(t)) \bullet \mathbf{g}'(t)\, dt.$$

In view of Theorem 6.1:1, the last says that

$$f(\mathbf{q}) - f(\mathbf{p}) = \int_C \mathbf{F} \bullet d\mathbf{s}. \qquad \square$$

Example 1. Let $f(\mathbf{x}) \equiv \|\mathbf{x}\| = (x_1^2 + \cdots + x_n^2)^{1/2}$. Then

$$\nabla f(\mathbf{x}) = \left(\frac{x_1}{\|\mathbf{x}\|}, \ldots, \frac{x_n}{\|\mathbf{x}\|}\right) = \frac{\mathbf{x}}{\|\mathbf{x}\|}.$$

If C goes from $\mathbf{p}$ to $\mathbf{q}$ without crossing the origin, then Theorem 1 says that

$$\int_C \nabla f \bullet d\mathbf{s} = \|\mathbf{q}\| - \|\mathbf{p}\|.$$

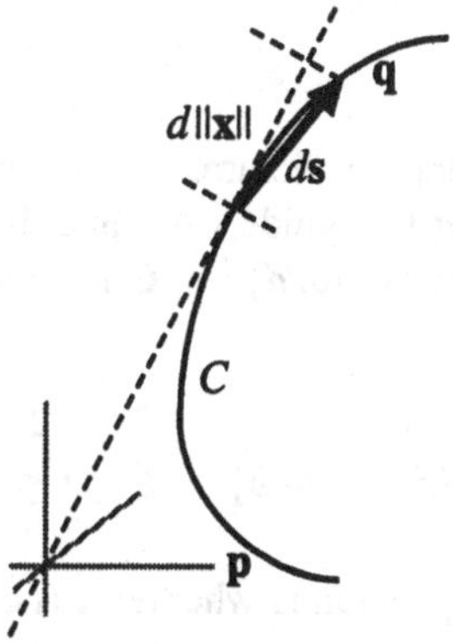

Figure 6.3.

Look at this relation geometrically. Along the curve in Figure 6.3, the displacement $d\mathbf{s}$ adds to $\|\mathbf{x}\|$ by exactly the component of $d\mathbf{s}$ in the direction away from the origin. That direction is given by the unit vector $\frac{\mathbf{x}}{\|\mathbf{x}\|}$. Therefore,

$$d\|\mathbf{x}\| = d\mathbf{s} \bullet \frac{\mathbf{x}}{\|\mathbf{x}\|} = \nabla f \bullet d\mathbf{s},$$

leading to the stated integral. (Compare Exercise 2.1:3.)

Theorem 1 introduces us to integrals for which it matters where you start and end, but not how you go. It also implies that for integrals of gradients, if the end point is the same as the start, then the integral is zero. These two properties characterize gradient fields, as we show in the next theorem.

Definition. (a) A vector field is **path-independent** (abbreviated PI) on O if $\int_C \mathbf{F} \bullet d\mathbf{s} = \int_D \mathbf{F} \bullet d\mathbf{s}$ for any two paths $C, D \subseteq O$ having equal starts and equal ends.
(b) A path $\mathbf{g}\colon [a, b] \to C$ is a **loop** if $\mathbf{g}(a) = \mathbf{g}(b)$.

It is almost universal to use "closed path" for what we call a "loop." We already have an important meaning for the word "closed." Since a path, being a compact image, is necessarily closed, we employ the new term.

Theorem 2. *Let O be a connected open set. For a vector field* $\mathbf{F}$ *defined on O, the following are equivalent*:

(a) $\mathbf{F}$ *is* PI.

(b) $\int_L \mathbf{F} \bullet d\mathbf{s} = 0$ *for every loop* $L \subseteq O$.

(c) $\mathbf{F}$ *is a gradient: There is a ("potential") function f defined on O for which* $\mathbf{F} = \nabla f$.

Proof. (b) $\Rightarrow$ (a) (If you draw a picture, the conclusion will be obvious. The picture is not analysis, but it is a guide.) Assume that $\int_L \mathbf{F} \bullet d\mathbf{s}$ is always zero. Suppose $\mathbf{g}\colon [a, b] \to C$ and $\mathbf{G}\colon [c, d] \to C$ have the same start and the same end. First,

$$\mathbf{h}(t) \equiv \begin{cases} \mathbf{g}(t), & a \le t \le b, \\ \mathbf{G}(d - [t - b]), & b \le t \le b + d - c, \end{cases}$$

is piecewise C^1. The only question is whether $\mathbf{h}$ is continuous at $t = b$, and we have

$$\lim_{t \to b^-} \mathbf{h}(t) = \mathbf{g}(b) = \mathbf{G}(d) = \lim_{t \to b^+} \mathbf{h}(t).$$

Second,

$$\mathbf{h}(a) = \mathbf{g}(a) = \mathbf{G}(c) = \mathbf{h}(b + d - c),$$

so $\mathbf{h}$ is a loop. By assumption,

$$\begin{aligned} 0 = \int_{\mathbf{h}} \mathbf{F} \bullet d\mathbf{s} &= \int_a^b \mathbf{F}(\mathbf{h}(t)) \bullet \mathbf{h}'(t)\,dt + \int_b^{b+d-c} \mathbf{F}(\mathbf{h}(t)) \bullet \mathbf{h}'(t)\,dt \\ &= \int_a^b \mathbf{F}(\mathbf{g}(t)) \bullet \mathbf{g}'(t)\,dt + \int_b^{b+d-c} \mathbf{F}(\mathbf{G}(d-t+b)) \\ &\quad \bullet \left(-\mathbf{G}'(d-t+b)\right) dt \\ &= \int_C \mathbf{F} \bullet d\mathbf{s} - \int_c^d \mathbf{F}(\mathbf{G}(u)) \bullet \mathbf{G}'(u)\,du \\ &= \int_C \mathbf{F} \bullet d\mathbf{s} - \int_D \mathbf{F} \bullet d\mathbf{s}. \end{aligned}$$

We conclude that $\mathbf{F}$ is PI.

(a) $\Rightarrow$ (c) Assume that $\mathbf{F}$ is PI, and fix one point $\mathbf{b} \in O$.

Suppose $\mathbf{x}$ is any point of O. By Theorem 2.2:2(a), $\mathbf{x}$ can be joined to $\mathbf{b}$ by a polygonal path $C(\mathbf{x})$ within O. Polygonal paths are piecewise smooth. Therefore,

$$f(\mathbf{x}) \equiv \int_{C(\mathbf{x})} \mathbf{F} \bullet d\mathbf{s}$$

defines a function of $\mathbf{x}$, independent (by hypothesis) of $C(\mathbf{x})$.

Now let $\varepsilon > 0$ be given. There exists a neighborhood $N \subseteq O$ of $\mathbf{x}$ in which $\|\mathbf{F}(\mathbf{y}) - \mathbf{F}(\mathbf{x})\| < \varepsilon$. Suppose $\mathbf{y} \in N$. Then $C(\mathbf{x}) \cup \mathbf{xy}$ is one path within O from $\mathbf{b}$ to $\mathbf{y}$, so $f(\mathbf{y}) = \int_{C(\mathbf{x})\cup\mathbf{xy}} \mathbf{F} \bullet d\mathbf{s}$. Using partition additivity, we obtain

$$f(\mathbf{y}) - f(\mathbf{x}) = \int_{\mathbf{xy}} \mathbf{F} \bullet d\mathbf{s}.$$

Parametrize $\mathbf{xy}$ by

$$\mathbf{u}(t) = \mathbf{x} + t(\mathbf{y} - \mathbf{x}), \qquad 0 \le t \le 1.$$

Then

$$\begin{aligned} &\left| f(\mathbf{y}) - f(\mathbf{x}) - \mathbf{F}(\mathbf{x}) \bullet (\mathbf{y} - \mathbf{x}) \right| \\ &\quad = \left| \int_0^1 \mathbf{F}(\mathbf{u}(t)) \bullet (\mathbf{y} - \mathbf{x})\,dt - \int_0^1 \mathbf{F}(\mathbf{x}) \bullet (\mathbf{y} - \mathbf{x})\,dt \right| \\ &\quad \le \max \|\mathbf{F}(\mathbf{u}(t)) - \mathbf{F}(\mathbf{x})\| \, \|\mathbf{y} - \mathbf{x}\| \le \varepsilon \|\mathbf{y} - \mathbf{x}\|. \end{aligned}$$

We have shown that $f(\mathbf{y}) - f(\mathbf{x})$ is approximated to within $\mathbf{o}(\mathbf{y} - \mathbf{x})$ by $\mathbf{F}(\mathbf{x}) \bullet (\mathbf{y} - \mathbf{x})$, proving simultaneously that f is differentiable and $\nabla f = \mathbf{F}$.

(c) $\Rightarrow$ (b) Assume $\mathbf{F} = \nabla f$. Then for a loop L from $\mathbf{p}$ to $\mathbf{p}$, Theorem 1 gives

$$\int_L \mathbf{F} \bullet d\mathbf{s} = f(\mathbf{p}) - f(\mathbf{p}) = 0. \qquad \square$$

Example 2. Is the gravitational field $\mathbf{F} = \frac{-k\mathbf{x}}{\|\mathbf{x}\|^3}$ PI?

The proof of (a) $\Rightarrow$ (c) above shows that if it is, then it is the gradient of its integral, and the path of integration does not matter. Accordingly, let C be the path from $(1, 0, \ldots, 0)$ to $\mathbf{x} \neq \mathbf{O}$ that follows the x_1-axis to $\mathbf{x}^* \equiv (\|\mathbf{x}\|, 0, \ldots, 0)$, then any path in the sphere of radius $\|\mathbf{x}\|$ from $\mathbf{x}^*$ to $\mathbf{x}$. (We leave it to the reader to show that such a path always exists and misses the origin.) Along the straight part of C, $\mathbf{T} = \pm\mathbf{e}_1$. On the sphere, $\mathbf{T}$ is orthogonal to the radius vector; that is a geometric principle, but Example 2.6:5 verifies it. Hence

$$\int_C \mathbf{F} \bullet \mathbf{T}\, ds = \int_1^{\|\mathbf{x}\|} \left(\frac{-k}{t^2}, 0, \ldots, 0\right) \bullet (\pm 1, 0, \ldots, 0)(\pm dt)$$

(same sign both times)

$$= \frac{k}{\|\mathbf{x}\|} - k.$$

To within a constant, this is the only possible potential. All we have to do is check:

$$\nabla\left(\frac{k}{\|\mathbf{x}\|} - k\right) = k\left(\frac{\partial}{\partial x_1}\left[\sum x_j^2\right]^{-1/2}, \ldots, \frac{\partial}{\partial x_n}\left[\sum x_j^2\right]^{-1/2}\right)$$

$$= -k\left(\left[\sum x_j^2\right]^{-3/2} x_1, \ldots, \left[\sum x_j^2\right]^{-3/2} x_n\right)$$

$$= \frac{-k\mathbf{x}}{\|\mathbf{x}\|^3};$$

$\mathbf{F}$ is a gradient.

How can this integration fail to produce a potential? See Exercise 1.

The test suggested by Example 2 always works: Integration produces the potential if there is one, or produces a candidate whose failure reveals that there is none. However, it is a clumsy tool if no path yields convenient integrals. The next result gives an easy necessary condition for a field to be a gradient.

Theorem 3. *Suppose* $\mathbf{F} = (F_1, \ldots, F_n)$ *is* C^1 *and PI on the open set* O. *Then on* O, $\frac{\partial F_k}{\partial x_j} = \frac{\partial F_j}{\partial x_k}$ *for each pair* j, k.

Proof. If $\mathbf{F}$ is PI, then there is f for which

$$(F_1, \ldots, F_n) = \nabla f = \left(\frac{\partial f}{\partial x_1}, \ldots, \frac{\partial f}{\partial x_n}\right).$$

Then

$$\frac{\partial F_j}{\partial x_k} = \left(\frac{\partial}{\partial x_k}\right)\frac{\partial f}{\partial x_j} = \frac{\partial^2 f}{\partial x_j \partial x_k},$$

and the conclusion follows from symmetry of mixed partials. □

Path-independence is an important concept in the study of force fields. We next give a fundamental principle in that area.

Theorem 4. *A gradient field is always* **conservative***: If* **F** *is a gradient in an open set, then for a given mass m moving in the set under the influence of total force* **F**, *there exists a function* $U(\mathbf{x})$ *such that the "kinetic energy"* $\frac{1}{2}mv^2$ *and "potential energy"* U *have constant sum.*

Proof. Interpreting **F** as force means that we invoke Newton's second law: $\mathbf{F} = m\mathbf{a}$. That is, the position $\mathbf{r}\colon [a, b] \to C$ of the given mass m satisfies $\mathbf{F} = m\frac{d^2\mathbf{r}}{dt^2}$. Consequently,

$$\int_C \mathbf{F} \bullet d\mathbf{s} = \int_a^b m\,\mathbf{r}''(t) \bullet \mathbf{r}'(t)\,dt.$$

By the dot-product rule (Exercise 1.3:6, adapted),

$$\mathbf{r}''(t) \bullet \mathbf{r}'(t) = \frac{1}{2}d\frac{\mathbf{r}'(t) \bullet \mathbf{r}'(t)}{dt} = \frac{1}{2}d\frac{\|\mathbf{r}'(t)\|}{dt}.$$

Write $v(t) \equiv \|\mathbf{r}'(t)\|$ for the **speed** of the object. (Compare Exercise 5.5:5a.) We now have

$$\int_C \mathbf{F} \bullet d\mathbf{s} = \int_a^b \frac{m}{2}d\frac{v(t)}{dt}dt = \frac{1}{2}m\,v(b)^2 - \frac{1}{2}m\,v(a)^2.$$

Notice that path-independence is not involved so far; if a field represents force, then the path integral = work *always* shows up as change in kinetic energy.

Now assume that **F** is the gradient of f. Then by Theorem 1,

$$\frac{1}{2}m\,v(b)^2 - \frac{1}{2}m\,v(a)^2 = f\big(\mathbf{r}(b)\big) - f\big(\mathbf{r}(a)\big),$$

or

$$\frac{1}{2}m\,v(b)^2 - f\big(\mathbf{r}(b)\big) = \frac{1}{2}m\,v(a)^2 - f\big(\mathbf{r}(a)\big).$$

We set $U \equiv -f$, and find that $\frac{1}{2}m\,v(t)^2 + U\big(\mathbf{r}(t)\big)$ is constant. □

Given our statement that path-independence is important, we should wonder why it did not come up before. Can line integrals, including coordinate integrals, have the property? Our final result shows that these integrals are PI only in the trivial cases.

Theorem 5. *Let* f *be continuous in a connected open set.*

(a) $\int_C f\,ds$ *is path-independent iff* $f \equiv 0$.

(b) $\int_C f\,dx_1$ *is path-independent iff* $f(\mathbf{x})$ *is actually a function of just* x_1.

Proof. (a) $\Leftarrow$ is trivial, so we focus on $\Rightarrow$. Suppose $f \neq 0$ at one point $\mathbf{b}$, say $f(\mathbf{b}) = \varepsilon > 0$. In some neighborhood $N(\mathbf{b}, 2\delta)$, $f(\mathbf{x})$ is confined to

$$0.9\varepsilon < f(\mathbf{x}) < 1.1\varepsilon.$$

On the segment C from $\mathbf{b}$ to $\mathbf{b} + \delta\mathbf{e}_1 + \delta\mathbf{e}_2$,

$$\int_C f\,ds < (1.1\varepsilon)\sqrt{2}\delta < 1.6\varepsilon\delta.$$

On the path D from $\mathbf{b}$ to $\mathbf{b} + \delta\mathbf{e}_1$ to $\mathbf{b} + \delta\mathbf{e}_1 + \delta\mathbf{e}_2$,

$$\int_D f\,ds > (0.9\varepsilon)\delta + (0.9\varepsilon)\delta = 1.8\varepsilon\delta.$$

The line integral depends on path.

(b) $\Leftarrow$ Suppose $f(x_1, \ldots, x_n) = g(x_1)$ throughout the set, call it O, for a function g, which must then be defined and continuous on the projection of O onto the x_1-axis. Let $\mathbf{p}, \mathbf{q} \in O$, with $\mathbf{h}: [a, b] \to C$ a path from $\mathbf{p}$ to $\mathbf{q}$. For a partition of $[a, b]$, the Riemann sums for $\int_C f\,dx_1$ look like

$$R = \sum f(\mathbf{h}(t_j{}^*))[h_1(t_j) - h_1(t_{j-1})] = \sum g(h_1(t_j{}^*))[h_1(t_j) - h_1(t_{j-1})].$$

By the average value theorem, among the values of x_1 between $h_1(t_{j-1})$ and $h_1(t_j)$, and irrespective of their order, there exists $x_{1j} \equiv h_1(t_j^{**})$ such that

$$g(h_1(t_j^{**}))[h_1(t_j) - h_1(t_{j-1})] = \int_{h_1(t_{j-1})}^{h_1(t_j)} g(h_1(t))\,dt.$$

(Why does the domain of g contain the interval from the lower of $h_1(t_{j-1})$ and $h_1(t_j)$ to the higher?) Take the sample point t_j^* to be any point where $\mathbf{h}([t_{j-1}, t_j])$ crosses the hyperplane $x_1 = h_1(t_j^{**})$. (Why must there be such a point?) That sampling makes

$$R = \sum g(h_1(t_j^{**}))[h_1(t_j) - h_1(t_{j-1})] = \sum \int_{h_1(t_{j-1})}^{h_1(t_j)} g(h_1(t))\,dt$$

$$= \int_{h_1(a)}^{h_1(b)} g(h_1(t))\,dt = \int_{p_1}^{q_1} g(h_1(t))\,dt.$$

The last is independent of partition. It follows that

$$\int_C f\,dx_1 = \int_{p_1}^{q_1} g(h_1(t))\,dt,$$

a quantity that depends on $\mathbf{p}$ and $\mathbf{q}$, but not on C. Therefore, the integral is PI.

$\Rightarrow$ Assume that $\int_C f\,dx_1$ is PI, and let B be a box contained in O. Suppose there existed two places $\mathbf{u} \equiv (x, u_2, u_3, \ldots, u_n)$ and $\mathbf{v} \equiv (x, v_2, v_3, \ldots, v_n)$

in B where f took different values, say w and $w + 3\varepsilon$, respectively. Along the segment from $\mathbf{u}$ to $\mathbf{v}$, the integral $\int f\,dx_1$ would be zero. Assume for definiteness that $u_2 \neq v_2$ (some u must differ from the corresponding v) and that $\mathbf{u}$ and $\mathbf{v}$ are interior to B (we could switch to a slightly bigger box). For small-enough δ, the path from $\mathbf{u}$ to $(x+\delta, u_2, u_3, \ldots, u_n)$ to $(x+\delta, v_2, v_3, \ldots, v_n)$ to $(x, v_2, v_3, \ldots, v_n) = \mathbf{v}$ would lie within B and give the integral

$$\int f\,dx_1 < (w+\varepsilon)\delta + 0 + (w+2\varepsilon)(-\delta) < 0,$$

making the integral path-dependent. We conclude that for any points $\mathbf{u}$, $\mathbf{v}$ in B, with $u_1 = v_1$, we must have $f(\mathbf{u}) = f(\mathbf{v})$.

Therefore,

$$g_B(x_1) \equiv f(x_1, \ldots, x_n)$$

defines a function of x_1 throughout B. Clearly, g_B is continuous in B. If B and D are intersecting boxes in O, then on their intersection, $g_B(x_1) = g_D(x_1)$. Therefore the union of the functions (view the functions as sets of ordered pairs) is a function $g(x_1)$, continuous at every x_1 for which there exist points $(x_1, \ldots, x_n) \in O$, and having $f(x_1, \ldots, x_n) = g(x_1)$ throughout. □

Exercises

1. Pick an easy path on which to integrate the field, then show that the field is not the gradient of the integral.

   (a) $\mathbf{G}(x, y) = y\mathbf{i}$,

   (b) $\mathbf{H}(x, y) = -y\mathbf{i} + x\mathbf{j}$.

2. For each field in Exercise 1, show that it fails the test of Theorem 3.

3. Is $\mathbf{K}(x, y) \equiv 2xy\mathbf{i} + x^2\mathbf{j}$ a gradient?

4. A field is **central** if its direction is always along the line to the origin and its magnitude depends only on distance to the origin; thus, $\mathbf{F}(\mathbf{x}) = f(\|\mathbf{x}\|)\mathbf{x}$, where $f(r)$ is continuous at least for $r > 0$. Prove that every central field is PI.

5. Prove that in a central field, the path integral is not only PI, it actually depends only on the *norms* of the start and end points.

6. Show that a field of the form $\mathbf{F}(x_1, \ldots, x_n) \equiv f_1(x_1)\mathbf{e}_1 + \cdots + f_n(x_n)\mathbf{e}_n$ is PI.

6.3 On the Edge: The Theorems of Green and Stokes

We move next to a family of results that relate path integrals along the edge of a two-dimensional region to flux through the region.

We begin with a lemma involving differentiation of integrals.

Theorem 1. *Suppose $f(x, y)$ and $\frac{\partial f}{\partial x}$ are continuous on the rectangle $\big[(a, c), (b, d)\big]$ in $\mathbf{R}^2$. Assume that $g(x)$ and $G(x)$ are differentiable, and valued in $[c, d]$, for $a \le x \le b$. Then*

$$h(x) \equiv \int_{g(x)}^{G(x)} f(x, y)\, dy$$

is differentiable for $a \le x \le b$, and

$$h'(x) = f(x, G(x))G'(x) - f(x, g(x))g'(x) + \int_{g(x)}^{G(x)} \frac{\partial f(x, y)}{\partial x}\, dy.$$

Proof. Write

$$H(u, v, w) \equiv \int_u^v f(w, y)\, dy, \qquad w \in [a, b],\ u \text{ and } v \in [c, d].$$

By the fundamental theorem, $\frac{\partial H}{\partial u} = -f(w, u)$ and $\frac{\partial H}{\partial v} = f(w, v)$. To find $\frac{\partial H}{\partial w}$, fix u and v, and suppose $w + t \in [a, b]$. By the mean value theorem,

$$\begin{aligned}\frac{H(u, v, w + t) - H(u, v, w)}{t} &= \int_u^v \frac{[f(w + t, y) - f(w, y)]}{t}\, dy \\ &= \int_u^v \frac{\partial f}{\partial x}(w + t^*, y)\, dy.\end{aligned}$$

Given $\varepsilon > 0$, there exists δ such that

$$\|\mathbf{x} - \mathbf{z}\| < \delta \Rightarrow \left|\frac{\partial f}{\partial x}(\mathbf{x}) - \frac{\partial f}{\partial x}(\mathbf{y})\right| < \frac{\varepsilon}{d - c}.$$

Then for $|t| < \delta$,

$$\begin{aligned}&\left|\frac{H(u, v, w + t) - H(u, v, w)}{t} - \int_u^v \frac{\partial f}{\partial x}(w, y)\, dy\right| \\ &= \left|\int_u^v \left[\frac{\partial f}{\partial x}(w + t^*, y) - \frac{\partial f}{\partial x}(x, y)\right] dy\right| \le |v - u| \frac{\varepsilon}{d - c} = \varepsilon.\end{aligned}$$

We conclude that

$$\frac{\partial H}{\partial w} = \int_u^v \frac{\partial f}{\partial x}(w, y)\, dy.$$

(Compare Exercise 2.2:10.) It is clear that $\frac{\partial H}{\partial u}$ and $\frac{\partial H}{\partial v}$ are continuous functions of u, v, w. The continuity of $\frac{\partial H}{\partial w}$ results from boundedness and uniform continuity of $\frac{\partial f}{\partial x}$:

$$\begin{aligned}
&\left|\int_{u+\Delta u}^{v+\Delta v} \frac{\partial f}{\partial x}(w+\Delta w, y)\,dy - \int_u^v \frac{\partial f}{\partial x}(w, y)\,dy\right| \\
&\leq \left|\int_u^v \left[\frac{\partial f}{\partial x}(w+\Delta w, y) - \frac{\partial f}{\partial x}(w, y)\right] dy\right| \\
&\quad + \left|\int_v^{v+\Delta v} \frac{\partial f}{\partial x}(w+\Delta w, y)\,dy\right| + \left|\int_{u+\Delta u}^{u} \frac{\partial f}{\partial x}(w+\Delta w, y)\,dy\right| \\
&\leq (d-c)\max\left\{\frac{\partial f}{\partial x}(w+\Delta w, y) - \frac{\partial f}{\partial x}(w, y)\right\} \\
&\quad + |\Delta v|\max\left\{\frac{\partial f}{\partial x}\right\} + |\Delta u|\max\left\{\frac{\partial f}{\partial x}\right\},
\end{aligned}$$

all of which tend to zero with Δu, Δv, Δw.

Since H has continuous partials, it is differentiable, and we may apply the chain rule. Set $u = g(x)$, $v = G(x)$, $w = x$. Then

$$\begin{aligned}
\frac{dh}{dx} &= \frac{\partial H}{\partial u}\frac{du}{dx} + \frac{\partial H}{\partial v}\frac{dv}{dx} + \frac{\partial H}{\partial w}\frac{dw}{dx} \\
&= -f(x, g(x))g'(x) + f(x, G(x))G'(x) + \int_{g(x)}^{G(x)} \frac{\partial f(x, y)}{\partial x}\,dy. \qquad \square
\end{aligned}$$

Theorem 2 (Green's Theorem). *Let $g \leq G$ be two C^1 functions on $[a, b]$. For the region $A \equiv \{(x, y) : a \leq x \leq b, g(x) \leq y \leq G(x)\}$ between the graphs, let C be the loop rightward along the lower graph, up (if necessary) from $(b, g(b))$ to $(b, G(b))$, leftward along the upper graph, down from $(a, G(a))$ to $(a, g(a))$. ("C is oriented counterclockwise.") Assume that the field $\mathbf{F}(x, y) \equiv (P(x, y), Q(x, y))$ is C^1 on an open set O containing A. Then*

$$\int_C \mathbf{F} \bullet d\mathbf{s} = \int_A \frac{\partial Q}{\partial x} - \frac{\partial P}{\partial y}.$$

Proof. We will first simplify our task, by reducing it to working inside a rectangle.

Theorem 5.1:3 tells us that A is Archimedean, and the proof there tells us that C is the boundary of A. Since $C \subseteq A$, we deduce that A is compact. Therefore, there is a positive ε such that no point of A lies within 3ε of the complement O^*. Since G and g are uniformly continuous, there exists δ such that $|s - t| < \delta \Rightarrow |g(s) - g(t)| < \varepsilon$ and $|G(s) - G(t)| < \varepsilon$.

Let $\{a = x_0 < \cdots < x_J = b\}$ be a partition with fineness smaller that δ and ε. Write A_j for the part of A between $x = x_{j-1}$ and $x = x_j$, with its boundary C_j oriented counterclockwise, as depicted for A_2 and A_3 in Figure 6.4. Clearly, the

A_j are nonoverlapping Archimedean sets whose union is A, so

$$\int_A \frac{\partial Q}{\partial x} - \frac{\partial P}{\partial y} = \int_{A_1} \frac{\partial Q}{\partial x} - \frac{\partial P}{\partial y} + \cdots + \int_{A_J} \frac{\partial Q}{\partial x} - \frac{\partial P}{\partial y}.$$

Also, the work done by $\mathbf{F}$ up the right edge of A_j cancels the work done down the left edge of A_{j+1}, $j = 1, \ldots, J-1$, so that by partition additivity,

$$\int_C \mathbf{F} \bullet ds = \int_{C_1} \mathbf{F} \bullet ds + \cdots + \int_{C_J} \mathbf{F} \bullet ds.$$

Therefore, it suffices to prove Green's theorem for each A_j.

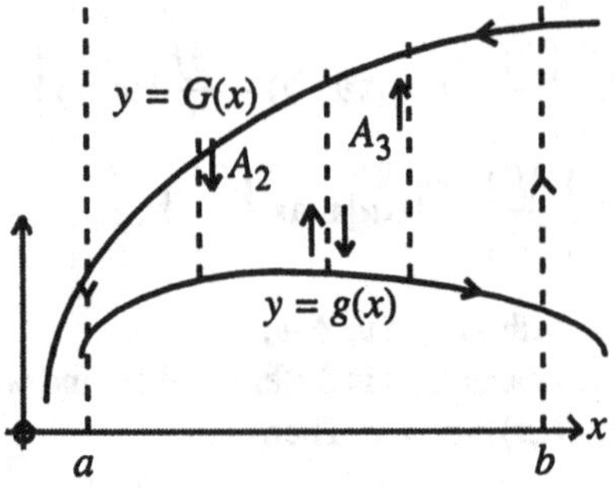

Figure 6.4.

Now look at the rectangle $R_j \equiv \{(x, y) \colon x_{j-1} - \varepsilon \le x \le x_j + \varepsilon, g(x_j) - \varepsilon \le y \le G(x_j) + \varepsilon\}$. Every point in R_j is within ε horizontally and 2ε vertically of points in A_j, so $R_j \subseteq O$. By construction, $A_j \subseteq \operatorname{int}(R_j)$. Thus have we reduced the problem to examining a region interior to a rectangle in which $\mathbf{F}$ is C^1. Accordingly, we drop the subscript, and pretend that A was surrounded by such a rectangle in the first place.

We calculate the area integral by linearity and Fubini's theorem. Thus,

$$\int_A \frac{\partial Q}{\partial x} - \frac{\partial P}{\partial y} = \int_a^b \int_{g(x)}^{G(x)} \frac{\partial Q(x,y)}{\partial x}\, dy\, dx - \int_a^b \int_{g(x)}^{G(x)} \frac{\partial P(x,y)}{\partial y}\, dy\, dx.$$

To calculate the second integral, fix x. By the fundamental theorem,

$$\int_{g(x)}^{G(x)} \frac{\partial P(x,y)}{\partial y}\, dy = P(x, G(x)) - P(x, g(x)).$$

Hence

$$\begin{aligned}
-\int_a^b \int_{g(x)}^{G(x)} \frac{\partial P(x,y)}{\partial y}\, dy\, dx &= \int_a^b \big[P(x, g(x)) - P(x, G(x))\big]\, dx \\
&= \int_a^b P(x, g(x))\, dx + \int_b^a P(x, G(x))\, dx \\
&= \int_{\text{bottom}} P\, dx + \int_{-\text{top}} P\, dx,
\end{aligned}$$

where we have written "−top" to indicate the orientation. To calculate the first integral, consider

$$h(x) \equiv \int_{g(x)}^{G(x)} Q(x,y)\,dy.$$

By Theorem 1,

$$h'(x) = G'(x)Q(x,G(x)) - g'(x)Q(x,g(x)) + \int_{g(x)}^{G(x)} \frac{\partial Q(x,y)}{\partial x}\,dy.$$

The integral we are seeking is then

$$\begin{aligned}
\int_a^b \int_{g(x)}^{G(x)} \frac{\partial Q(x,y)}{\partial x}\,dy\,dx
&= \int_a^b \left[h'(x) - G'(x)Q(x,G(x)) + g'(x)Q(x,g(x))\right] dx \\
&= h(b) - h(a) + \int_b^a Q(x,G(x))G'(x)\,dx \\
&\quad + \int_a^b Q(x,g(x))g'(x)\,dx \\
&= \int_{\text{right}} Q\,dy + \int_{-\text{left}} Q\,dy + \int_{-\text{top}} Q\,dy + \int_{\text{bottom}} Q\,dy.
\end{aligned}$$

Adding the six pieces we have evaluated to $\int_{-\text{left}} P\,dx = 0$ and $\int_{\text{right}} P\,dx = 0$, we obtain

$$\int_A \frac{\partial Q}{\partial x} - \frac{\partial P}{\partial y} = \int_C P\,dx + Q\,dy = \int_C \mathbf{F} \bullet d\mathbf{s}. \qquad \square$$

Suppose we place a small circular "pinwheel" into the force field $\mathbf{F}$ (see Figure 6.5). At a point $\mathbf{x}$ on the edge of the wheel, the tangential component $\mathbf{F}(\mathbf{x}) \bullet \mathbf{T}$ is also the component of $\mathbf{F}(\mathbf{x})$ normal to the radius. The quantity $r\mathbf{F}(\mathbf{x}) \bullet \mathbf{T}$, called the **torque** (or **moment**) of the force, measures tendency to make the pinwheel rotate. If $\int \mathbf{F} \bullet \mathbf{T}\,ds$ around the edge is nonzero, then the field exerts a net torque, and makes the wheel spin. Suppose now $\frac{\partial Q}{\partial x} - \frac{\partial P}{\partial y}$ is nonzero at some point $\mathbf{y}$. Then it maintains its sign in some neighborhood of $\mathbf{y}$. By Green's theorem, $\int_C \mathbf{F} \bullet d\mathbf{s}$ will be nonzero around any small-enough circle C surrounding $\mathbf{y}$; the field is **rotational**. If instead $\frac{\partial Q}{\partial x} - \frac{\partial P}{\partial y}$ is identically zero, then the field exerts no torque on small-enough pinwheels; it is **irrotational**. With this connection, we will call $\frac{\partial Q}{\partial x} - \frac{\partial P}{\partial y}$ the **curl** of the field.

Example 1. We already have two path-dependent examples. Let us see them in terms of rotation, and relate the curl.

(a) Let $\mathbf{G}(x,y) \equiv y\mathbf{i}$. Clearly, a pinwheel above the x-axis has its upper part in a stronger rightward wind than the lower. Therefore, it should rotate clockwise, the negative sense. We obtain $P = y$, $Q = 0$, and $\frac{\partial Q}{\partial x} - \frac{\partial P}{\partial y} = -1$.

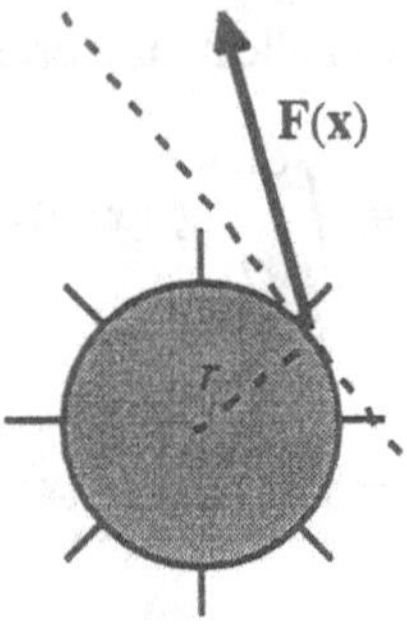

Figure 6.5.

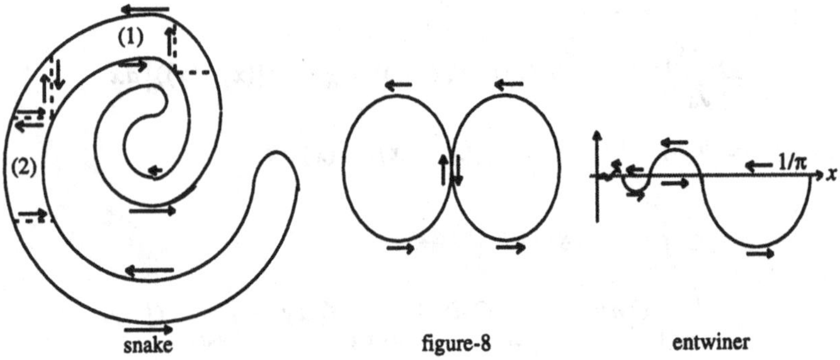

Figure 6.6.

Notice that below the x-axis, the stronger wind is on the bottom, but points leftward; the rotation is the same.

(b) The whirlpool $\mathbf{H}(x, y) \equiv -y\mathbf{i}+x\mathbf{j}$ certainly looks rotational. The rotation is obvious at the origin, but applies everywhere. Our pinwheel would have clockwise torque on the side near the origin, counterclockwise on the far side, and the latter would be greater. The wheel should turn counterclockwise. We obtain $P = -y$, $Q = x$, $\frac{\partial Q}{\partial x} - \frac{\partial P}{\partial y} = 2$. (See also Exercise 5.)

Our modest version of the region of applicability for Green's theorem generalizes to more complex shapes. We will adopt an informal name: A closed Archimedean set A bounded by a piecewise-C^1 path is a **Green region** if for every C^1 field on A, the counterclockwise path integral over bd(A) equals the integral on A of the curl.

Example 2. Figure 6.6 shows some Green regions.

(a) The "snake" in the figure has only a finite number of places where the tangent is vertical or horizontal. Using horizontal and vertical lines to isolate those places, we can break the region into a finite number of pieces between graphs. Some, like (1), have graphs of the form $y = h(x)$; for some, like (2), $x = k(y)$.

Each of the pieces is a Green region. Then the area integrals sum to the snake integral, while the line integrals over the interior walls cancel, making the separate path integrals add up to the integral on the boundary.

(b) Self-intersection is allowed. In the figure-8, the area integral over the enclosed region is the sum of the integrals over the two lobes, and each lobe integral is the work around its edge. Notice that to keep the signs of the area integrals, it is essential to orient the figure-8 as shown, with the boundary traversed counterclockwise around each lobe.

(c) Even an infinity of self-intersections is allowed. In "entwiner," we suggest the graph of $y = x^3 \sin(\frac{1}{x})$, $0 \le x \le \frac{1}{\pi}$, and the segment from $(\frac{1}{\pi}, 0)$ to $(0, 0)$. Between two intersections $x_{j+1} = \frac{1}{(j+1)\pi}$ and $x_j = \frac{1}{j\pi}$, orient the upper graph to the left and the lower graph to the right, and name the enclosed region A_j. The region $A \equiv A_1 \cup A_2 \cup \cdots$ has meager boundary, so $\int_A \frac{\partial Q}{\partial x} - \frac{\partial P}{\partial y}$ is defined. The boundary D is not a path, because it is not piecewise C^1 (Why?). Nevertheless, we may define

$$\int_D \mathbf{F} \bullet d\mathbf{s} = \lim_{j\to\infty} \int_{D(j)} \mathbf{F} \bullet d\mathbf{s}.$$

Here $D(j)$ is the part of D with $x \ge x_{j+1}$. The limit must exist, because both graphs being rectifiable, the arc length between the origin and $x = x_{j+1}$ vanishes as $j \to \infty$ (Theorem 5.5:4(f)). A similar statement holds for volumes: $V(A_{j+1} \cup A_{j+2} \cup \cdots) \to 0$ as $j \to \infty$. The volume statement implies that

$$\int_A \left(\frac{\partial Q}{\partial x} - \frac{\partial P}{\partial y}\right) - \int_{A_1 \cup \cdots \cup A_j} \left(\frac{\partial Q}{\partial x} - \frac{\partial P}{\partial y}\right)$$
$$= \int_{A-(A_1 \cup \cdots \cup A_j)} \left(\frac{\partial Q}{\partial x} - \frac{\partial P}{\partial y}\right) \to 0.$$

It follows that

$$\int_A \frac{\partial Q}{\partial x} - \frac{\partial P}{\partial y} = \lim_{j\to\infty} \int_{A_1 \cup \cdots \cup A_j} \left(\frac{\partial Q}{\partial x} - \frac{\partial P}{\partial y}\right)$$
$$= \lim_{j\to\infty} \int_{D(j)} \mathbf{F} \bullet d\mathbf{s} = \int_D \mathbf{F} \bullet d\mathbf{s}.$$

We should attempt to make "counterclockwise" more precise. Suppose we are at a point $\mathbf{x}$ through which the boundary $C = \text{bd}(A)$ is a smooth curve; refer to Figure 6.7. The nearby points of C must lie roughly in the line of the tangent $\mathbf{T}$; that is, they lie within the shaded cone. The remaining nearby points split like $\mathbf{y}$ and $\mathbf{z}$, to the left and right, respectively, of the cone. We can tell that $\mathbf{y}$ is to the left, because the vectors $\mathbf{y} - \mathbf{x}$ and $\mathbf{T}$ make a right-handed system: $\det[\mathbf{T}\ \mathbf{y} - \mathbf{z}] > 0$. Similarly, $\mathbf{z}$ is to the right, because $\mathbf{z} - \mathbf{x}$ and $\mathbf{T}$ make a left-handed system. If at each point of C, the nearby points of A lie to the left of the tangent vector, then the orientation of C is **counterclockwise**.

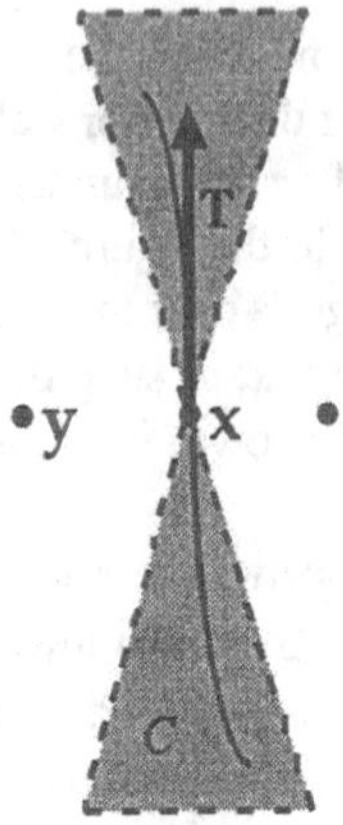

Figure 6.7.

The most important adjustment we have to make with Green's theorem is in the case where there are holes in the domain of (continuous differentiability of) the field.

Example 3. Look at a new whirlpool: $\mathbf{F}(\mathbf{x}) \equiv \frac{-y\mathbf{i}+x\mathbf{j}}{\|(x,y)\|^2}$. It is C^1 on $\mathbf{R}^2 - \{\mathbf{O}\}$.

(a) Let C be the circle given by $x = a\cos\theta$, $y = a\sin\theta$, $0 \le \theta \le 2\pi$, and D the enclosed disk. For the circle,

$$\int_C \mathbf{F} \bullet d\mathbf{s} = \int_0^{2\pi} \frac{(-a\sin\theta, a\cos\theta)}{a^2} \bullet (-a\sin\theta, a\cos\theta)\, d\theta = 2\pi.$$

On D,

$$\frac{\partial Q}{\partial x} - \frac{\partial P}{\partial y} = \frac{(x^2+y^2)\,1 - x(2x)}{\left(x^2+y^2\right)^2} - \frac{(x^2+y^2)\,(-1) - (-y)(2y)}{\left(x^2+y^2\right)^2} = 0,$$

excluding $\mathbf{O}$. Hence the curl is integrable, and its integral is $\int_D \frac{\partial Q}{\partial x} - \frac{\partial P}{\partial y} = 0$, not matching the path integral.

(b) Since the curl, where defined, is zero, Green's theorem should guarantee zero work over loops that do not surround the origin.

Instead of calculation, we give a geometric argument. In Figure 6.8, a (solid) curve in the first quadrant bounds a shaded area away from the origin. Think of the curve as travelling in stages: radially away from $\mathbf{O}$, then circling counterclockwise at distance R, then radially toward the origin, then clockwise at distance r; the dotted path does that directly. Zero work is done on the radials, positive work $Fd = \left(\frac{1}{R}\right)(R\Delta\theta)$ on the outside, negative work $-(\frac{1}{r})(r\Delta\theta)$ on the inside. The inverse-distance variation of the field accounts for the zero path integrals.

(c) The zero curl and integrals create a question: Does it follow that $\mathbf{F}$ is a gradient?

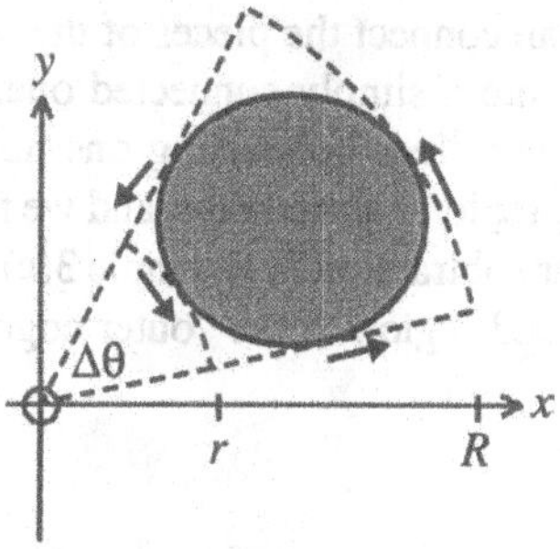

Figure 6.8.

The answer is yes, locally. The work from $(1, 0)$ to $\left([x^2+y^2]^{1/2}, 0\right)$ (then circling) to (x, y) is due just to the curved part. There, $\mathbf{F} \bullet d\mathbf{s} = d\theta$, and

$$\int \mathbf{F} \bullet d\mathbf{s} = \int_0^{\tan^{-1}(y/x)} d\theta = \tan^{-1}\left(\frac{y}{x}\right).$$

We check:

$$\frac{\partial}{\partial x} \tan^{-1}\left(\frac{y}{x}\right) = \frac{1}{1+(y/x)^2} \frac{-y}{x^2} = \frac{-y}{x^2+y^2},$$
$$\frac{\partial}{\partial y} \tan^{-1}\left(\frac{y}{x}\right) = \frac{1}{1+(y/x)^2} \frac{1}{x} = \frac{x}{x^2+y^2}.$$

The candidate $\tan^{-1}\left(\frac{y}{x}\right)$ has the right partials, but is undefined on the y-axis. Set

$$f(x, y) \equiv \begin{cases} \tan^{-1}\left(\frac{y}{x}\right), & x > 0, \\ \tan^{-1}\left(\frac{y}{x}\right) + \pi, & x < 0, \\ \frac{\pi}{2}, & x = 0 \text{ and } y > 0. \end{cases}$$

This one is continuously differentiable except on the nonpositive y-axis, and we may check that $\nabla f = \mathbf{F}$. We thereby see $\mathbf{F}$ as a gradient on $\mathbf{R}^2$ without the nonpositive y-axis, an open set with no holes. (We could choose to throw away any ray beginning at the origin; we just have to make an adjustment of the form $\tan^{-1}\left(\frac{y}{x}\right) + \theta_0$.)

Before going further, let us sharpen our notion of "holes."

Definition. A bounded set is **simply connected** if its boundary is connected.

In Figure 6.9(a), O is an open set with (shaded) holes. The boundary of O therefore consists of the surrounding oval and the edges of the holes, and is disconnected. When we speak of a bounded set with no holes, we mean that it is simply connected. If we can connect the pieces of the boundary, then we turn the "multiply connected" set into a simply connected one. This is visible in Figure 6.9(b), where we have drawn lines connecting one hole to the other and to the outer edge. The remaining region has no holes, and we may employ Green's theorem there. We did a similar subtraction in Example 3(c), connecting the boundary (the origin) of an unbounded region to the "outer edge" at infinity, by means of the negative y-axis.

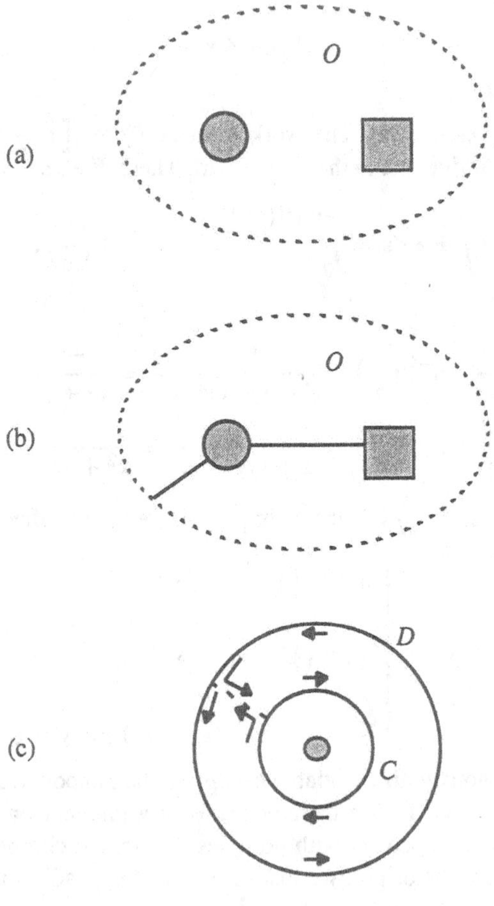

Figure 6.9.

Another feature of Example 3 that is typical comes from part (a). Suppose paths C and D, shown as circles in Figure 6.9(c), both go counterclockwise once around the hole in the domain of **F**. Define path E as shown by the arrows: once counterclockwise around D, inward to C, clockwise around C, outward along the same segment to the start on D. The work inward cancels the work outward, making

$$\int_E \mathbf{F} \bullet d\mathbf{s} = \int_D \mathbf{F} \bullet d\mathbf{s} + \int_{-C} \mathbf{F} \bullet d\mathbf{s}.$$

Still, E is a perfectly decent loop in the region where $\mathbf{F}$ is legal. Hence

$$\int_E \mathbf{F} \bullet d\mathbf{s} = \int_A \frac{\partial Q}{\partial x} - \frac{\partial P}{\partial y},$$

where A denotes the region between the circles. Therefore, we may write

$$\int_A \frac{\partial Q}{\partial x} - \frac{\partial P}{\partial y} = \int_{D \cup -C} \mathbf{F} \bullet d\mathbf{s}.$$

That is Green's theorem again, because $D \cup -C$ is a counterclockwise path around the boundary of A. Counterintuitively, $-C$ is the counterclockwise sense around C, because one keeps A on the left. Moreover, if $\mathbf{F}$ is irrotational, then

$$0 = \int_A \frac{\partial Q}{\partial x} - \frac{\partial P}{\partial y} = \int_D \mathbf{F} \bullet d\mathbf{s} - \int_C \mathbf{F} \bullet d\mathbf{s}.$$

We have $\int_D \mathbf{F} \bullet d\mathbf{s} = \int_C \mathbf{F} \bullet d\mathbf{s}$, which explains why the answer in Example 3(a) was independent of the circle.

We prove next that where there are no holes, *only* irrotational flows are PI.

Definition. A path $\mathbf{g}\colon [a, b] \to C$ is **simple** (or **non-self-intersecting**) if $\mathbf{g}$ is one-to-one on $[a, b)$.

Theorem 3. *Assume that O is a (bounded) simply connected open set in $\mathbf{R}^2$, and the field $\mathbf{F} \equiv (P, Q)$ is C^1 in O. Then $\mathbf{F}$ is PI iff $\frac{\partial Q}{\partial x} = \frac{\partial P}{\partial y}$.*

Proof. $\Rightarrow$ Theorem 6.2:3.

$\Leftarrow$ We will prove that $\int_L \mathbf{F} \bullet d\mathbf{s} = 0$ for every loop in O.

Suppose $\mathbf{g}\colon [a, b] \to L$ is a loop in O. Let $\varepsilon > 0$. Pick a partition $\{a = t_0 < \cdots < t_J = b\}$ that includes the singularities of $\mathbf{g}$, so that $\mathbf{g}$ is C^1 on each subinterval; is so fine that each $\|\mathbf{g}(t_j) - \mathbf{g}(t_{j-1})\|$ is less than half the distance from L to O^*; is so fine that P and Q vary by no more than ε over any subinterval. For the piece $D \equiv \mathbf{g}([t_{j-1}, t_j])$ of L,

$$\int_D \mathbf{F} \bullet d\mathbf{s} = \mathbf{F}(\mathbf{x}_j{}^*) \bullet \Delta\mathbf{s}_j = P(\mathbf{x}_j{}^*)\Delta x_j + Q(\mathbf{x}_j{}^*)\Delta y_j$$

for some $\mathbf{x}_j{}^* = \mathbf{g}(t_j{}^*)$ along the piece. The "step-path" E from $\mathbf{g}(t_{j-1}) = (x_{j-1}, y_{j-1})$ (horizontally) to (x_j, y_{j-1}) (then vertically) to $\mathbf{g}(t_j) = (x_j, y_j)$ is within $|\Delta x_j| + |\Delta y_j| \le 2\|\Delta\mathbf{s}_j\|$ of the loop, so it is contained in O. Along the step-path,

$$\int_E \mathbf{F} \bullet d\mathbf{s} = P(x_j{}^{**}, y_{j-1})\Delta x_j + Q(x_j, y_j{}^{**})\Delta y_j$$

for appropriate x_j^{**}, y_j^{**}. Link the step-paths to form a loop L^*. Then

$$\int_L \mathbf{F} \bullet ds - \int_{L^*} \mathbf{F} \bullet ds = \sum \left[P(\mathbf{x}_j{}^*) - P(x_j{}^{**}, y_{j-1})\right] \Delta x_j \\ + \left[Q(\mathbf{x}_j{}^*) - Q(x_j, y_j{}^{**})\right] \Delta y_j,$$

whose absolute value is no more than

$$\sum \varepsilon \|\Delta \mathbf{s}_j\| + \varepsilon \|\Delta \mathbf{s}_j\| = 2\varepsilon \sum \|\Delta \mathbf{s}_j\| = 2\varepsilon \text{ len}(L).$$

In words, we can approximate the work along L by the work along a "step-loop" L^* (a union of horizontal and vertical segments). Hence, we can accomplish our mission by proving that the work around every step-loop is zero.

Throw away the asterisk, and assume that L is itself a step-loop in O. We may assume that consecutive segments of L are noncollinear. The reason is that if $\mathbf{g}([t_j, t_{j+1}])$ goes back over $\mathbf{g}([t_{j-1}, t_j])$, then we can replace the two pieces by the segment from $\mathbf{g}(t_{j-1})$ to $\mathbf{g}(t_{j+1})$, parametrized from $t = t_{j-1}$ to $t = t_{j+1}$, with the integral over the replacement matching the sum of the integrals over the replaced pieces. Next, we may assume that L is simple. For suppose that $\mathbf{g}$ is one-to-one on $[a, u)$, for some $u < b$, but $\mathbf{g}(u)$ matches some earlier value $\mathbf{g}(t)$. (The repetition cannot be of the form $\mathbf{g}(u^+) = \mathbf{g}(u^-)$.) This must mean that the segment holding $\mathbf{g}(u)$ crosses the one holding $\mathbf{g}(t)$ at right angles, or approaches it along a horizontal or vertical line. In either case, $\mathbf{g}([t, u])$ is a simple step-loop (possibly with the wrong orientation), $\mathbf{g}([a, t] \cup [u, b])$ is a step-loop with fewer sides than L, and

$$\int_L \mathbf{F} \bullet ds = \int_{\mathbf{g}([t,u])} \mathbf{F} \bullet ds + \int_{\mathbf{g}([a,t]) \cup \mathbf{g}([u,b])} \mathbf{F} \bullet ds.$$

Continuing the isolation of subloops, we infer that $\int_L \mathbf{F} \bullet ds$ is the sum of integrals over simple loops, some perhaps traversed clockwise. The orientations do not matter to us; we have reduced the problem to proving that the work around a simple step-loop is always zero.

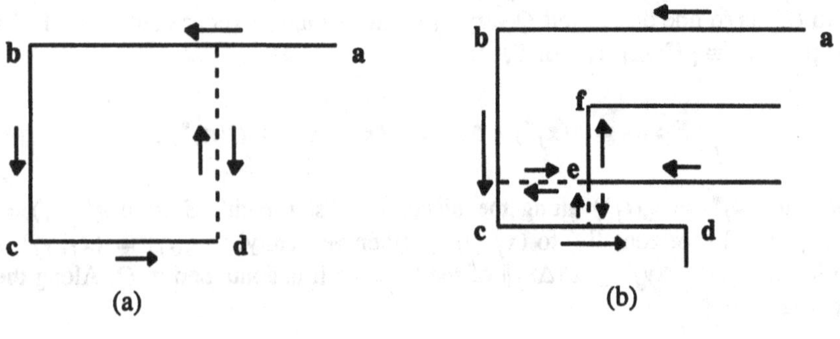

Figure 6.10.

Let therefore L be a simple step-loop, as in Figure 6.10. Take the leftmost segment **bc** of L; we show it oriented downward, but it could be upward. The rectangle determined by **bc** and the shorter of the adjoining segments **ab** and **cd**, which must extend to the right, might contain no other segments (part (a) of the figure). In that case, close the rectangle. The integral around L is the sum of the integrals around the rectangle and around the remainder of L, which is a simple step-loop with strictly fewer sides than L had. If instead the rectangle contains some other segments (part (b)), take the lowest, leftmost of those others. Call it **ef**, which must be oriented upward. Draw the perpendiculars from **e** to **bc** and **cd**. Then the integral around L is the sum of the integral around this new rectangle and around *two* simple step-loops. For those two, the total of sides is two more than L's; since each must have four or more sides, each must have strictly fewer sides than L. Continuing this reduction process, we eventually bring the question down to the sum of integrals around the boundaries of rectangles.

So, finally, we assume $L = \text{bd}(R)$, where R is a box. The interior $\text{int}(R)$ cannot contain any points of O^*. If it did, then it would also possess points from $\text{bd}(O)$; in that case, $\text{ext}(R)$ could not have points from $\text{bd}(O)$, because $\text{bd}(O) \cap \text{ext}(R)$ and $\text{bd}(O) \cap \text{int}(R)$ would disconnect $\text{bd}(O)$; and that would mean $\text{ext}(R) \subseteq O$, making O unbounded. Hence $R \subseteq O$. Therefore, we may apply Green's theorem, and

$$\int_L \mathbf{F} \bullet d\mathbf{s} = \int_R \frac{\partial Q}{\partial x} - \frac{\partial P}{\partial y} = 0. \qquad \square$$

We have striven to avoid making statements we cannot prove (see the remarks about Lebesgue in Section 4.4), but here we must make an exception. The key to Theorem 3 was in the last paragraph: the property of a rectangle that $\text{int}(R)$ and $\text{ext}(R)$ are connected open sets that make up the remainder $\mathbf{R}^2 - \text{bd}(R)$ of the plane. Any loop L is closed, so its complement is open, and L is bounded, implying that its complement has exactly one unbounded component. Camille Jordan discovered that if L is simple, then there is precisely one more component. In other words, the **Jordan curve theorem** says that every simple loop is the common boundary of an enclosed connected open region (a "Jordan region," which must, because its boundary is connected, be simply connected) and the region's unbounded, connected exterior (which must, naturally, have a hole). This elegantly elementary principle is beyond our means to prove. [It was beyond Jordan's, too; see Kline, p. 1017. Jordan was also part of the development "from Peano to Lebesgue"; Kline, pp. 1043–1046.]

We are not able to extend Green's theorem to Jordan regions, but we can move the region out into space.

Theorem 4. *Let A be a Green region with boundary C. Suppose $\mathbf{G}\colon A \to S \subseteq \mathbf{R}^n$ is a C^2 surface contained in an open set O, on which the field $\mathbf{F} = (F_1, \ldots, F_n)$ is C^1. Then*

$$\int_{\mathbf{G}(C)} \mathbf{F} \bullet d\mathbf{s} = \int_A \sum_{j=1}^{n} \sum_{k>j} \left(\frac{\partial F_k}{\partial x_j} - \frac{\partial F_j}{\partial x_k} \right) \left(\frac{\partial G_j}{\partial x} \frac{\partial G_k}{\partial y} - \frac{\partial G_j}{\partial y} \frac{\partial G_k}{\partial x} \right).$$

Proof. To start, note that $D \equiv \mathbf{g}(C)$ is piecewise C^1 and oriented. The path integral is

$$I = \int_D F_1\,dx_1 + \cdots + F_n\,dx_n$$
$$= \int_C F_1\big(\mathbf{G}(x,y)\big)dG_1(x,y) + \cdots + F_n\big(\mathbf{G}(x,y)\big)dG_n(x,y).$$

Along each C^1 part of $\mathbf{G}(C)$, we have

$$dG_k = \frac{\partial G_k}{\partial x}dx + \frac{\partial G_k}{\partial y}dy.$$

We substitute, then collect the dx and dy terms, to get

$$I = \int_C \left(\sum_{k=1}^n F_k\big(\mathbf{G}(x,y)\big)\frac{\partial G_k(x,y)}{\partial x}\right)dx + \left(\sum_{k=1}^n F_k\big(\mathbf{G}(x,y)\big)\frac{\partial G_k(x,y)}{\partial y}\right)dy.$$

For this last, Green's theorem and the product rule give

$$I = \int_A \sum_{k=1}^n \left[F_k(\mathbf{G})\frac{\partial^2 G_k}{\partial y \partial x} + \frac{\partial F_k(\mathbf{G})}{\partial x}\frac{\partial G_k}{\partial y}\right] - \sum_{k=1}^n \left[F_k(\mathbf{G})\frac{\partial^2 G_k}{\partial x \partial y} + \frac{\partial F_k(\mathbf{G})}{\partial y}\frac{\partial G_k}{\partial x}\right].$$

By symmetry of mixed partials, the terms with second-order derivatives offset each other, yielding

$$I = \int_A \sum_{k=1}^n \left[\frac{\partial F_k(\mathbf{G}(x,y))}{\partial x}\frac{\partial G_k(x,y)}{\partial y} - \frac{\partial F_k(\mathbf{G}(x,y))}{\partial y}\frac{\partial G_k(x,y)}{\partial x}\right].$$

(In our work, this is the only use of class C^2.) Finally, the chain rule

$$\frac{\partial F_k(\mathbf{G})}{\partial x} = \sum_{j=1}^n \frac{\partial F_k(\mathbf{G})}{\partial x_j}\frac{\partial G_j}{\partial x}, \quad \frac{\partial F_k(\mathbf{G})}{\partial y} = \sum_{j=1}^n \frac{\partial F_k(\mathbf{G})}{\partial x_j}\frac{\partial G_j}{\partial y}$$

implies

$$I = \sum_{k=1}^n \sum_{j=1}^n \int_A \left[\frac{\partial F_k}{\partial x_j}\frac{\partial G_j}{\partial x}\frac{\partial G_k}{\partial y} - \frac{\partial F_k}{\partial x_j}\frac{\partial G_j}{\partial y}\frac{\partial G_k}{\partial x}\right].$$

We will fold this square array along the diagonal, because there are zeros there. That is, observe that for an ordered pair (j,k), if $j = k$, then the summand is zero. Next, observe that if $j < k$, then the (j,k) and (k,j) summands are

$$\frac{\partial F_k}{\partial x_j}\frac{\partial G_j}{\partial x}\frac{\partial G_k}{\partial y} - \frac{\partial F_k}{\partial x_j}\frac{\partial G_j}{\partial y}\frac{\partial G_k}{\partial x} \quad\text{and}\quad \frac{\partial F_j}{\partial x_k}\frac{\partial G_k}{\partial x}\frac{\partial G_j}{\partial y} - \frac{\partial F_j}{\partial x_k}\frac{\partial G_k}{\partial y}\frac{\partial G_j}{\partial x},$$

which combine into

$$\left(\frac{\partial F_k}{\partial x_j} - \frac{\partial F_j}{\partial x_k}\right)\left(\frac{\partial G_j}{\partial x}\frac{\partial G_k}{\partial y} - \frac{\partial G_j}{\partial y}\frac{\partial G_k}{\partial x}\right).$$

(Notice the pattern: For the first factor, differentiate the later component of **F** by the earlier coordinate, then switch; second factor, differentiate the earlier component of **G** by the first parameter, the later by the second parameter, then switch.) The result is

$$\int_{\mathbf{G}(C)} \mathbf{F} \bullet d\mathbf{s} = \sum_{j=1}^{n} \sum_{k>j} \int_A \left(\frac{\partial F_k}{\partial x_j} - \frac{\partial F_j}{\partial x_k} \right) \left(\frac{\partial G_j}{\partial x} \frac{\partial G_k}{\partial y} - \frac{\partial G_j}{\partial y} \frac{\partial G_k}{\partial x} \right). \quad \square$$

The triangular sum in Theorem 4 has $(n-1)n/2$ terms, matching n iff $n = 3$. In that dimension, we can give the sum an interesting form. Write t, u (instead of x, y) for the parameters; $x = G_1, y = G_2, z = G_3$ (instead of x_1, x_2, x_3) for the coordinates; and D again for the edge $\mathbf{G}(C)$. The integral $\int_D \mathbf{F} \bullet d\mathbf{s}$ becomes

$$\int_A \left(\frac{\partial F_2}{\partial x} - \frac{\partial F_1}{\partial y} \right) \left(\frac{\partial x}{\partial t} \frac{\partial y}{\partial u} - \frac{\partial y}{\partial t} \frac{\partial x}{\partial u} \right) + \left(\frac{\partial F_3}{\partial x} - \frac{\partial F_1}{\partial z} \right) \left(\frac{\partial x}{\partial t} \frac{\partial z}{\partial u} - \frac{\partial z}{\partial t} \frac{\partial x}{\partial u} \right)$$
$$+ \left(\frac{\partial F_3}{\partial y} - \frac{\partial F_2}{\partial z} \right) \left(\frac{\partial y}{\partial t} \frac{\partial z}{\partial u} - \frac{\partial z}{\partial t} \frac{\partial y}{\partial u} \right).$$

That integrand is recognizably a dot product. Further, the three second-factor combinations come from

$$\frac{\partial(x,y,z)}{\partial t} \times \frac{\partial(x,y,z)}{\partial u} = \begin{vmatrix} \mathbf{i} & \frac{\partial x}{\partial t} & \frac{\partial x}{\partial u} \\ \mathbf{j} & \frac{\partial y}{\partial t} & \frac{\partial y}{\partial u} \\ \mathbf{k} & \frac{\partial z}{\partial t} & \frac{\partial z}{\partial u} \end{vmatrix} = \left(\frac{\partial y}{\partial t} \frac{\partial z}{\partial u} - \frac{\partial z}{\partial t} \frac{\partial y}{\partial u} \right) \mathbf{i}$$
$$+ \left(\frac{\partial z}{\partial t} \frac{\partial x}{\partial u} - \frac{\partial x}{\partial t} \frac{\partial z}{\partial u} \right) \mathbf{j} + \left(\frac{\partial x}{\partial t} \frac{\partial y}{\partial u} - \frac{\partial y}{\partial t} \frac{\partial x}{\partial u} \right) \mathbf{k},$$

except that the middle component has the wrong sign. We may adjust the factor in the middle term of the integral, to write

$$\int_D \mathbf{F} \bullet d\mathbf{s} = \int_A \left(\frac{\partial F_3}{\partial y} - \frac{\partial F_2}{\partial z}, \frac{\partial F_1}{\partial z} - \frac{\partial F_3}{\partial x}, \frac{\partial F_2}{\partial x} - \frac{\partial F_1}{\partial y} \right) \bullet \left(\frac{\partial \mathbf{x}}{\partial t} \times \frac{\partial \mathbf{x}}{\partial u} \right).$$

The last factor is what we identified in Section 6.2 as $d\boldsymbol{\sigma} = \mathbf{n} d\sigma$, the vector element of area on the surface $S = \mathbf{G}(A)$. Hence the last integral is the flux $\int_S \mathbf{H} \bullet d\boldsymbol{\sigma}$ of the quantity $\mathbf{H}$ with the partials of $\mathbf{F}$.

To give $\mathbf{H}$ an identity, we write

$$\mathbf{H} \equiv \left(\frac{\partial}{\partial y} F_3 - \frac{\partial}{\partial z} F_2 \right) \mathbf{i} + \left(\frac{\partial}{\partial z} F_1 - \frac{\partial}{\partial x} F_3 \right) \mathbf{j} + \left(\frac{\partial}{\partial x} F_2 - \frac{\partial}{\partial y} F_1 \right) \mathbf{k}$$
$$= \begin{vmatrix} \mathbf{i} & \frac{\partial}{\partial x} & F_1 \\ \mathbf{j} & \frac{\partial}{\partial y} & F_2 \\ \mathbf{k} & \frac{\partial}{\partial z} & F_3 \end{vmatrix}.$$

We have allowed not only vectors, but *operators*, into a determinant: Recall (Section 2.1) the **del operator** $\nabla \equiv \left(\frac{\partial}{\partial x}, \frac{\partial}{\partial y}, \frac{\partial}{\partial z}\right)$. Our quantity **H** looks like $\nabla \times \mathbf{F}$. We call $\nabla \times \mathbf{F}$ the **curl** of **F**.

Theorem 4 then turns into the following:

Theorem 5 (Stokes's Theorem). *In* $\mathbf{R}^3$, *under the hypothesis of Theorem 4, the path integral around the edge of the surface is the flux of the curl:*

$$\int_D \mathbf{F} \bullet d\mathbf{s} = \int_S (\nabla \times \mathbf{F}) \bullet d\boldsymbol{\sigma}.$$

Example 4. The electrostatic field from Example 6.1:5 can be defined in more general settings. In the setting of Example 6.1:4, its tangential integral gives the electromotive force: $V = \int_C \mathbf{E} \bullet d\mathbf{s}$. By Stokes's theorem, the path integral of **E** matches the flux of its curl through the disk bounded by the wire loop. Thus,

$$V = \int_C \mathbf{E} \bullet d\mathbf{s} = \int_D (\nabla \times \mathbf{E}) \bullet d\boldsymbol{\sigma}.$$

At the same time, Faraday's law says

$$V = -\frac{\partial \phi}{\partial t} = -\left(\frac{\partial}{\partial t}\right) \int_D \mathbf{B} \bullet d\boldsymbol{\sigma}.$$

By Theorem 1, if $\mathbf{B}(x, y, z, t)$ is continuously differentiable as a function of four variables—"in space and time"—then we can do the t-differentiation under the integral sign. Hence

$$\int_D (\nabla \times \mathbf{E}) \bullet d\boldsymbol{\sigma} = \int_D -\frac{\partial \mathbf{B}}{\partial t} \bullet d\boldsymbol{\sigma}.$$

Because D can be shrunk to arbitrarily small size and pointed in arbitrary directions, the last macroscale, or average-value, relation leads to

$$\nabla \times \mathbf{E} = -\frac{\partial \mathbf{B}}{\partial t}.$$

This one is microscale, or local. It appears as one of Maxwell's equations, under the same name "Faraday's law."

We referred to "surface" and "edge" in our statement of Stokes's theorem. There, they do not need definition. They refer specifically to the images of a special type of Archimedean set and its boundary. We can think of additional kinds of surfaces to which Stokes's theorem would extend, but will not seek systematic generalization.

Notice that our notation

$$\nabla \times \mathbf{F} = \begin{vmatrix} \mathbf{i} & \frac{\partial}{\partial x} & F_1 \\ \mathbf{j} & \frac{\partial}{\partial y} & F_2 \\ \mathbf{k} & \frac{\partial}{\partial z} & F_3 \end{vmatrix}$$

again transposes the usual. Given our penchant for such mirror-imaging, it should come as no surprise that the usual view is of Green's theorem as a special case of Stokes's. Suppose $\mathbf{F}(x, y, z)$ is confined to the xy-plane:

$$\mathbf{F}(x, y, z) = P(x, y)\mathbf{i} + Q(x, y)\mathbf{j}.$$

We verify that

$$\nabla \times \mathbf{F} \equiv \left(\frac{\partial 0}{\partial y} - \frac{\partial Q}{\partial z}\right)\mathbf{i} + \left(\frac{\partial P}{\partial z} - \frac{\partial 0}{\partial x}\right)\mathbf{j} + \left(\frac{\partial Q}{\partial x} - \frac{\partial P}{\partial y}\right)\mathbf{k} = \left(\frac{\partial Q}{\partial x} - \frac{\partial P}{\partial y}\right)\mathbf{k}.$$

If $A \subseteq xy$-plane is our usual region, bounded by C, then $\mathbf{n} = \mathbf{k}$ throughout, and Stokes's theorem reduces to

$$\int_C P\,dx + Q\,dy = \int_C \mathbf{F} \bullet d\mathbf{s} = \int_A (\nabla \times \mathbf{F}) \bullet \mathbf{N}\,d\sigma = \int_A \frac{\partial Q}{\partial x} - \frac{\partial P}{\partial y}.$$

The relation $\nabla \times \mathbf{F} = \left(\frac{\partial Q}{\partial x} - \frac{\partial P}{\partial y}\right)\mathbf{k}$ also reconciles our two uses of "curl." We had allowed ourselves a similar indulgence when we applied "component" and "projection" to both a vector quantity and its associated scalar. "Curl" is the name we want for $\nabla \times \mathbf{F}$, because where $\nabla \times \mathbf{F} \neq \mathbf{O}$, the field will make our pinwheel rotate within the plane perpendicular to the curl.

Finally, we use Stokes's theorem to extend Theorem 3 into $\mathbf{R}^3$.

Theorem 6. *Suppose O is a convex open set. Then $\mathbf{F}$ is PI iff $\nabla \times \mathbf{F} = \mathbf{O}$ throughout O.*

Proof. Let $\mathbf{g}: [a, b] \to C \subseteq O$ be C^1. Reparametrize C as C_{low}, given by

$$\mathbf{g}_{\text{low}}(t) \equiv \mathbf{g}\big(a + t[b - a]\big), \qquad 0 \le t \le 1.$$

Then

$$\begin{aligned} \int_{C_{\text{low}}} \mathbf{F} \bullet d\mathbf{s} &= \int_0^1 \mathbf{F}\big(\mathbf{g}_{\text{low}}(t)\big)\mathbf{g}'_{\text{low}}(t)\,dt \\ &= \int_a^b \mathbf{F}\big(\mathbf{g}(u)\big)\left[(b-a)\mathbf{g}'(u)\right]du/(b-a) \\ &= \int_C \mathbf{F} \bullet d\mathbf{s}. \end{aligned}$$

Next, suppose $\mathbf{h}\colon [c,d] \to D \subseteq O$ is another path with the same start as C and the same end. Do a similar reconstruction of D into D_{up}, given by

$$\mathbf{h}_{\text{up}}(t) \equiv \mathbf{h}\bigl(c + t[d-c]\bigr), \qquad 0 \le t \le 1.$$

We verify that

$$\begin{aligned}\int_{D_{\text{up}}} \mathbf{F} \bullet d\mathbf{s} &= \int_0^1 \mathbf{F}\bigl(\mathbf{h}_{\text{up}}(t)\bigr)\mathbf{h}'_{\text{up}}(t)\,dt \\ &= \int_c^d \mathbf{F}\bigl(\mathbf{h}(u)\bigr)\bigl[(d-c)\mathbf{h}'(u)\bigr]\,du/(d-c) \\ &= \int_D \mathbf{F} \bullet d\mathbf{s}.\end{aligned}$$

Now let A be the unit square in the tv-plane, and define the surface

$$\mathbf{G}(t,v) \equiv \mathbf{g}_{\text{low}}(t) + v\left[\mathbf{h}_{\text{up}}(t) - \mathbf{g}_{\text{low}}(t)\right], \qquad 0 \le t \le 1,\ 0 \le v \le 1.$$

(In space, $\mathbf{G}$ draws the segment from a point on C to the corresponding one on D. Such a construction produces a "ruled surface.") The convexity of O keeps $\mathbf{G}$ in O. Also, $\mathbf{G}$ has continuous partials; it is C^1. By Stokes's theorem,

$$\int_{\text{bd}(A)} \mathbf{F} \bullet d\mathbf{s} = \int_{\mathbf{G}(A)} (\nabla \times \mathbf{F}) \bullet d\boldsymbol{\sigma} = 0.$$

Along the left edge of A,

$$\mathbf{G}(t,v) \equiv \mathbf{g}_{\text{low}}(0) + v\left[\mathbf{h}_{\text{up}}(0) - \mathbf{g}_{\text{low}}(0)\right] = \mathbf{g}(a) + v\left[\mathbf{h}(c) - \mathbf{g}(a)\right] = \mathbf{g}(a),$$

because the start $\mathbf{g}(a)$ of C matches the start $\mathbf{h}(c)$ of D. Therefore, $\int_{\text{left}} \mathbf{F} \bullet d\mathbf{s} = 0$, and similarly on the right side. We arrive at

$$\begin{aligned}0 &= \int_{\text{bottom}} \mathbf{F} \bullet d\mathbf{s} + \int_{-\text{top}} \mathbf{F} \bullet d\mathbf{s} \\ &= \int_{C_{\text{low}}} \mathbf{F} \bullet d\mathbf{s} - \int_{D_{\text{up}}} \mathbf{F} \bullet d\mathbf{s} \\ &= \int_C \mathbf{F} \bullet d\mathbf{s} - \int_D \mathbf{F} \bullet d\mathbf{s}.\end{aligned}$$

That takes care of path-independence for C^1 paths.

If the paths are only piecewise C^1, then we simply split A into vertical strips, so that $\mathbf{g}_{\text{low}}$ is C^1 along the bottom and $\mathbf{h}_{\text{up}}$ C^1 along the top of each strip. Then the area integral over A is the sum of the separate integrals over the strips, all these area integrals being zero; and the path integral around A matches the sum of the path integrals around the strips, because the contribution from the right edge of one strip cancels the contribution from the left edge of the next. □

Exercises

1. Let A be a Green region. Show that

$$\text{area}(A) = \int_C -y\,dx = \int_C x\,dy = \frac{1}{2}\int_C x\,dy - y\,dx.$$

2. (a) By Exercise 1, area$(A) = \frac{1}{2}\int_C(-y\mathbf{i} + x\mathbf{j}) \bullet d\mathbf{s}$. Explain this relation geometrically.

   (b) Use the relation to show that if A is the region bounded by the polar graph $r = f(\theta)$, then

$$\text{area}(A) = \int_0^{2\pi} \frac{f^2(\theta)}{2}\,d\theta.$$

3. Verify Stokes's theorem for the field $\mathbf{F} \equiv -y\mathbf{i} + x\mathbf{j} + e^z \ln(1+z)\mathbf{k}$ and the surface $z = \sqrt{4 - x^2 + y^2}$. (Find separately the path integral around the edge and the flux of the curl. You must decide which sense of the normal to pick.)

4. Calculate the curl of a central field $\mathbf{F}(\mathbf{x}) = f(\|\mathbf{x}\|)\mathbf{x}$ in $\mathbf{R}^3$. Does it reconcile with Exercise 6.2:4 and Theorem 6? (Compare also Exercise 1.3:7.)

5. Give a geometric argument why a central field will not make a pinwheel rotate.

6. Show that $\nabla \times (f\mathbf{F}) = f(\nabla \times \mathbf{F}) + (\nabla f) \times \mathbf{F}$.

6.4 Gauss's Theorem

Our final section deals with a theorem about "outward flux," the integral of the component normal to a surface that bounds a region in $\mathbf{R}^n$, in the sense pointing toward the exterior of the enclosed region. We begin with a plane version of the theorem.

Theorem 1. *Suppose C is the boundary of a Green region $D \subseteq \mathbf{R}^2$ and $\mathbf{F} = \big(P(x, y), Q(x, y)\big)$ is a C^1 field on an open set O containing D. Then the outward flux is given by*

$$\int_C \mathbf{F} \bullet \mathbf{N}\,ds = \int_D \frac{\partial P}{\partial x} + \frac{\partial Q}{\partial y}.$$

Proof. Recall that where C is smooth, the tangent is given by $\mathbf{t} = \left(\frac{dx}{dt}, \frac{dy}{dt}\right)$. Since C is oriented counterclockwise, D is always on the left. Hence the outward normal points to the right of $\mathbf{t}$:

$$\mathbf{n} = \left(\frac{dy}{dt}, -\frac{dx}{dt}\right).$$

(Why is this normal pointing to the right?) Then

$$\int_C \mathbf{F} \bullet \mathbf{N}\,ds = \int_C (P, Q) \bullet \frac{(dy/dt, -dx/dt)}{\|(dy/dt, -dx/dt)\|} \left\| \left(\frac{dx}{dt}, \frac{dy}{dt} \right) \right\| dt$$
$$= \int_C -Q\,dx + P\,dy.$$

Green's theorem applies to $(-Q, P)$ as well as to $\mathbf{F}$, so we conclude that

$$\int_C \mathbf{F} \bullet \mathbf{N}\,ds = \int_D \frac{\partial P}{\partial x} - \frac{\partial(-Q)}{\partial y}. \qquad \square$$

Our attention moves immediately to the quantity $\frac{\partial P}{\partial x} + \frac{\partial Q}{\partial y}$. It is called the **divergence** of the field, a name clearly reflective of the view of $\mathbf{F}$ as a flow. If we divide both sides of

$$\int_C \mathbf{F} \bullet \mathbf{N}\,ds = \int_D \frac{\partial P}{\partial x} + \frac{\partial Q}{\partial y}$$

by the volume (area) of D, then we match the average divergence and the outward flux per unit volume. Passing to the limit as D shrinks, we see that divergence measures the pointwise tendency of the fluid to expand. In Figure 6.11(a) $\mathbf{F} \bullet \mathbf{N}$ is positive along C_1, and the picture clearly suggests expansion out of D_1. The suggestion is not as strong in part (b), but again divergence measures rate of flow out of the region. For the left-hand half of C_2, the outward flux is negative, precisely the negative of the rate of flow of fluid into D_2. Over the right-hand half, the outward flux matches the rate of flow out of D_2. Therefore, the total outward flux, determined by the divergence, is the net rate of flow out of D_2.

Example 1. (a) Recall $\mathbf{G}(x, y) \equiv y\mathbf{j}$. Referring to the region D_2 in Figure 6.11, we see that $\mathbf{G}$ is stronger in the upper part of D_2, but that it blows as much air into the left side of D_2 as it blows out along the right. Such a wind should have zero divergence. We verify: $\frac{\partial y}{\partial x} + \frac{\partial 0}{\partial y} = 0$.

(b) How about $\mathbf{H}(x, y) \equiv -y\mathbf{i} + x\mathbf{j}$? This wind hits D_2 from the southeast, and is stronger away from the origin. On the other hand, what it blows in along the lower right, it blows out along the upper left, or northwest, side. It appears to have zero divergence: $\frac{\partial(-y)}{\partial x} + \frac{\partial x}{\partial y} = 0$.

(c) Let us invent $\mathbf{H}(x, y) \equiv \mathbf{x} = x\mathbf{i} + y\mathbf{j}$. (This one, being central, has zero curl.) It hits D_2 from the southwest, and blows out through the northeast side. However, it has greater magnitude exiting than entering. This means that air is leaving D_2, and suggests positive divergence. We check: $\frac{\partial x}{\partial x} + \frac{\partial y}{\partial y} = 2$.

Example 2. To see that divergence has the same significance in $\mathbf{R}^n$, let us do some informal analysis of fluid motion there.

Suppose $\mathbf{R}^n$ is filled with a gas having density $\rho(\mathbf{x}, t)$ and velocity $\mathbf{v}(\mathbf{x}, t)$ at a given "point in space-time." At time t, the region A is occupied by mass

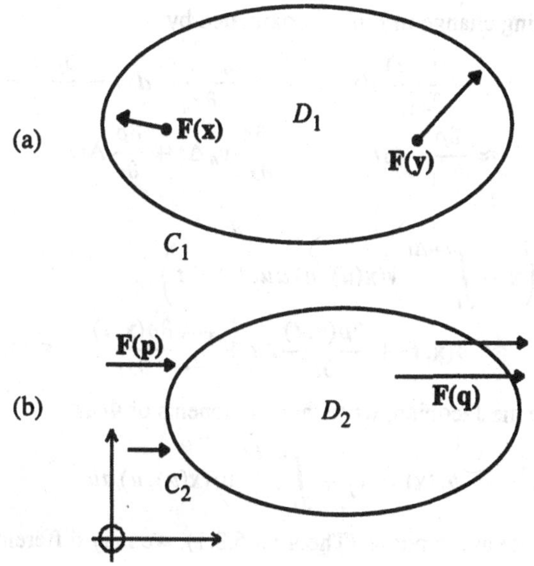

Figure 6.11.

$\int_A \rho(\mathbf{x}, t)\, d\mathbf{x}$ of gas. At time $t + \Delta t$, the occupied region has shifted to $A(\Delta t) \equiv \Phi(A)$, where the transformation Φ is given by

$$\mathbf{y} = \Phi(\mathbf{x}) \equiv \mathbf{x} + \int_t^{t+\Delta t} \mathbf{v}\big(\mathbf{x}(u), u\big)\, du.$$

The mass within $A(\Delta t)$ is

$$\int_{A(\Delta t)} \rho(\mathbf{y}, t + \Delta t)\, d\mathbf{y} = \int_A \rho(\Phi(\mathbf{x}), t + \Delta t) J(\mathbf{x}) d\mathbf{x},$$

in which $J(\mathbf{x})$ is the absolute Jacobian of $\Phi(\mathbf{x})$. Assuming that gas does not vanish into thin air—more scientifically, that mass is conserved—we demand

$$\int_A \rho(\mathbf{x}, t)\, d\mathbf{x} = \int_A \rho(\Phi(\mathbf{x}), t + \Delta t) J(\mathbf{x})\, d\mathbf{x}.$$

Matching average values and passing to the limit as A shrinks to a point, we get

$$\rho(\mathbf{x}, t) = \rho\left(\mathbf{x} + \int_t^{t+\Delta t} \mathbf{v}(\mathbf{x}(u), u)\, du, t + \Delta t\right) J(\mathbf{x}).$$

Let us do some linear approximation. Assume that Δt is small. The motion gives

$$\int_t^{t+\Delta t} \mathbf{v}(\mathbf{x}(u), u)\, du \approx \big(v_1(\mathbf{x}, t), \ldots, v_n(\mathbf{x}, t)\big)\, \Delta t.$$

The corresponding change in ρ is approximated by

$$\begin{aligned} d\rho(\mathbf{x},t) &= \frac{\partial\rho(\mathbf{x},t)}{\partial x_1}dx_1+\cdots+\frac{\partial\rho(\mathbf{x},t)}{\partial x_n}dx_n+\frac{\partial\rho(\mathbf{x},t)}{\partial t}dt \\ &\approx \frac{\partial\rho}{\partial x_1}v_1\Delta t+\cdots+\frac{\partial\rho}{\partial x_n}v_n\Delta t+\frac{\partial\rho}{\partial t}\Delta t. \end{aligned}$$

Hence

$$\begin{aligned} &\rho\left(\mathbf{x}+\int_t^{t+\Delta t}\mathbf{v}(\mathbf{x}(u),u)\,du,\,t+\Delta t\right) \\ &\quad\approx \rho(\mathbf{x},t)+\frac{\partial\rho(\mathbf{x},t)}{\partial t}\Delta t+\sum\frac{\partial\rho(\mathbf{x},t)}{\partial x_k}v_k(\mathbf{x},t)\Delta t. \end{aligned}$$

To approximate the Jacobian, write the components of Φ as

$$\phi_j(\mathbf{x})=x_j+\int_t^{t+\Delta t}v_j(\mathbf{x}(u),u)\,du.$$

Under appropriate assumptions (Theorem 6.3:1), we may differentiate under the integral sign. Thus,

$$\frac{\partial\phi_j}{\partial x_k}=\frac{\partial x_j}{\partial x_k}+\int_t^{t+\Delta t}\frac{\partial v_j}{\partial x_k}(\mathbf{x}(u),u)\,du\approx\frac{\partial x_j}{\partial x_k}+\frac{\partial v_j(\mathbf{x},t)}{\partial x_k}\Delta t.$$

Consequently,

$$J(\mathbf{x})\approx\operatorname{absdet}\begin{pmatrix} 1+\frac{\partial v_1}{\partial x_1}\Delta t & \frac{\partial v_1}{\partial x_2}\Delta t & \cdots & \frac{\partial v_1}{\partial x_n}\Delta t \\ \frac{\partial v_2}{\partial x_1}\Delta t & 1+\frac{\partial v_2}{\partial x_2}\Delta t & \cdots & \frac{\partial v_2}{\partial x_n}\Delta t \\ & & \ddots & \\ \frac{\partial v_n}{\partial x_1}\Delta t & \cdots & \frac{\partial v_n}{\partial x_{n-1}}\Delta t & 1+\frac{\partial v_n}{\partial x_n}\Delta t \end{pmatrix}.$$

This is a forbidding determinant, but we need only compute part of it. Observe that one product from the diagonal is 1^n; n others are

$$1^{n-1}\left(\frac{\partial v_1}{\partial x_1}\right)\Delta t,\ldots,1^{n-1}\left(\frac{\partial v_n}{\partial x_n}\right)\Delta t;$$

and all the others, from diagonal or otherwise, have $(\Delta t)^2$ or higher powers of Δt. For small enough Δt, the sum is close enough to 1 to be its own absolute value. Therefore,

$$J(\mathbf{x})=1+\frac{\partial v_1}{\partial x_1}\Delta t+\cdots+\frac{\partial v_n}{\partial x_n}\Delta t+(?)\Delta t^2.$$

The conservation equation becomes

$$\begin{aligned} \rho &= \left[\rho+\left(\frac{\partial\rho}{\partial t}+\frac{\partial\rho}{\partial x_1}v_1+\cdots+\frac{\partial\rho}{\partial x_n}v_n\right)\Delta t\right] \\ &\quad\times\left[1+\left(\frac{\partial v_1}{\partial x_1}+\cdots+\frac{\partial v_n}{\partial x_n}\right)\Delta t+(?)\Delta t^2\right]. \end{aligned}$$

We multiply out on the right, cancel ρ, divide by Δt, and let $\Delta t \to 0$. The result is

$$0 = \rho\left(\frac{\partial v_1}{\partial x_1} + \cdots + \frac{\partial v_n}{\partial x_n}\right) + \frac{\partial \rho}{\partial t} + \frac{\partial \rho}{\partial x_1} v_1 + \cdots + \frac{\partial \rho}{\partial x_n} v_n.$$

The expression $\frac{\partial \rho}{\partial x_1} v_1 + \cdots + \frac{\partial \rho}{\partial x_n} v_n$ is obviously a dot product. We turn

$$\frac{\partial v_1}{\partial x_1} + \cdots + \frac{\partial v_n}{\partial x_n}$$

into a dot product, and simultaneously produce a notation for divergence, by writing

$$\frac{\partial v_1}{\partial x_1} + \cdots + \frac{\partial v_n}{\partial x_n} = \left(\frac{\partial}{\partial x_1}, \ldots, \frac{\partial}{\partial x_n}\right) \bullet (v_1, \ldots, v_n) = \nabla \bullet \mathbf{v}.$$

That puts the conservation equation into the form

$$0 = \rho(\nabla \bullet \mathbf{v}) + \nabla\rho \bullet \mathbf{v} + \frac{\partial \rho}{\partial t}.$$

If we write

$$\frac{\partial \rho}{\partial t} = \mathbf{v} \bullet (-\nabla\rho) - \rho(\nabla \bullet \mathbf{v}),$$

we see that the thickness of the gas at a point is affected by two opposing factors: First, to the extent that $\mathbf{v}$ works opposite to $\nabla\rho$, it brings in denser gas, increasing the thickness; second, $\nabla \bullet \mathbf{v}$ represents an outflow of volume, which locally decreases mass per volume at the rate $\rho(\nabla \bullet \mathbf{v})$.

The relation

$$\rho(\nabla \bullet \mathbf{v}) + \nabla\rho \bullet \mathbf{v} + \frac{\partial \rho}{\partial t} = 0$$

is sometimes called **Euler's equation**. It is as fundamental in the study of fluid mechanics as is Newton's second law in particle mechanics. Observe that if the fluid is **incompressible**—if density is invariant in space and time, which is how we usually think of liquids—then the equation reduces to $\nabla \bullet \mathbf{v} = 0$.

We next extend Theorem 1 to simple regions in $\mathbf{R}^n$. Let us say that $A \subseteq \mathbf{R}^n$ is **simple in the x_k-direction** if it is described by $g_k(x^\#) \le x_k \le G_k(\mathbf{x}^\#)$, for C^1 functions g_k and G_k of the other variables, defined over a region $A_k \subseteq \mathbf{R}^{n-1}$ that is in turn described by $g_j(\mathbf{x}^{\#\#}) \le x_j \le G_j(\mathbf{x}^{\#\#}), \ldots$, all the way down to $g_i(x) \le y \le G_i(x)$ for the last two variables y and x. Over such a region, the outward flux of a field is the integral of its divergence.

Theorem 2 (The Divergence Theorem). *Assume that* $\mathbf{F}$ *is a* C^1 *field on the open set* $O \subseteq \mathbf{R}^n$. *Suppose* $A \subseteq O$ *is simple in one coordinate direction. Then*

$$\int_{\mathrm{bd}(A)} \mathbf{F} \bullet \mathbf{N}_{\mathrm{out}}\, d\sigma = \int_A \nabla \bullet \mathbf{F}.$$

Proof. By our line of argument in Green's theorem (Theorem 6.3:2), we may assume that A is a subset of a box contained in O. Here, as there, this reduction allows us to differentiate integrals by invoking Theorem 6.3:1.

To avoid a thicket of subscripts, we will treat the case $n = 3$, with variables x, y, z, assuming that A is simple in the z-direction. Our argument will establish a pattern clearly showing how we extend it to higher dimension.

We integrate $\nabla \bullet \mathbf{F} = \frac{\partial F_1}{\partial x} + \frac{\partial F_2}{\partial y} + \frac{\partial F_3}{\partial z}$ one term at a time. For the last term, we have

$$\int_A \frac{\partial F_3}{\partial z} = \int_a^b \int_{g(x)}^{G(x)} \int_{h(x,y)}^{H(x,y)} \frac{\partial F_3}{\partial z}\, dz\, dy\, dx$$
$$= \int_a^b \int_{g(x)}^{G(x)} \left[F_3(x, y, H(x, y)) - F_3(x, y, h(x, y))\right] dy\, dx.$$

The top of A is given by $z = H(x, y)$, so that

$$d\sigma = \left(-\frac{\partial H}{\partial x}, -\frac{\partial H}{\partial y}, 1\right) dy\, dx.$$

Therefore, the flux of $F_3\mathbf{k}$ out of the top is

$$\int_{\text{top}} (F_3\mathbf{k}) \bullet \left(-\frac{\partial H}{\partial x}, -\frac{\partial H}{\partial y}, 1\right) dy\, dx = \int_a^b \int_{g(x)}^{G(x)} F_3(x, y, H(x, y))\, dy\, dx.$$

Similarly, the flux out of the bottom is

$$\int_{\text{bottom}} (F_3\mathbf{k}) \bullet \left(\frac{\partial h}{\partial x}, \frac{\partial h}{\partial y}, -1\right) dy\, dx = \int_a^b \int_{g(x)}^{G(x)} -F_3(x, y, h(x, y))\, dy\, dx.$$

(Notice that we have to use the downward normal.) The remaining sides of bd(A) are "vertical." They are given by equations of the form $y = f(x)$ or $x = k$. The corresponding normals are of the form $\pm(-\frac{dy}{dx}, 1, 0)$, $\pm(1, 0, 0)$, with no component in the direction of $\mathbf{k}$. Thus, the top and bottom contribute the total flux for $F_3\mathbf{k}$, and

$$\int_A \frac{\partial F_3}{\partial z} = \int_{\text{bd}(A)} (F_3\mathbf{k}) \bullet d\sigma.$$

For the second term, we employ Theorem 6.3:1 to write

$$\int_A \frac{\partial F_2}{\partial y} = \int_a^b \int_{g(x)}^{G(x)} \int_{h(x,y)}^{H(x,y)} \frac{\partial F_2}{\partial y}\, dz\, dy\, dx$$

as

$$\int_a^b \int_{g(x)}^{G(x)} \left(\frac{\partial}{\partial y} \int_{h(x,y)}^{H(x,y)} F_2(x, y, z)\, dz \right.$$
$$\left. + \left[-F_2(x, y, H(x, y)) \frac{\partial H}{\partial y} + F_2(x, y, h(x, y)) \frac{\partial h}{\partial y} \right] dy\, dx \right.$$
$$= \left(\int_a^b \int_{h(x,G(x))}^{H(x,G(x))} F_2(x, G(x), z)\, dz\, dx \right.$$
$$\left. + \int_a^b \int_{h(x,g(x))}^{H(x,g(x))} -F_2(x, g(x), z)\, dz\, dx \right)$$
$$+ \left[\int_a^b \int_{g(x)}^{G(x)} -F_2(x, y, H(x, y)) \frac{\partial H}{\partial y}\, dy\, dx \right.$$
$$\left. + \int_a^b \int_{g(x)}^{G(x)} F_2(x, y, h(x, y)) \frac{\partial h}{\partial y}\, dy\, dx \right].$$

These four integrals are identifiable as fluxes of $F_2\mathbf{j}$. The last two belong to the top and bottom of A: Along the top, for example, $z = H(x, y)$, $g(x) \le y \le G(x)$, $a \le x \le b$, and

$$(F_2\mathbf{j}) \bullet d\sigma = \left[F_2(x, y, H(x, y))\mathbf{j} \right] \bullet \left(-\frac{\partial H}{\partial x}, -\frac{\partial H}{\partial y}, 1 \right) dy\, dx$$
$$= -F_2(x, y, H(x, y)) \left(\frac{\partial H}{\partial y} \right) dy\, dx;$$

that accounts for the third integral. The first pair of integrals belongs to the y-sides of A: Along the higher-y side, $y = G(x)$, $a \le x \le b$, $h(x, G(x)) \le z \le H(x, G(x))$, and

$$(F_2\mathbf{j}) \bullet d\sigma = \left[F_2(x, G(x), z)\mathbf{j} \right] \bullet \left(-\frac{dy}{dx}, 1, 0 \right) dz\, dx$$
$$= F_2(x, G(x), z)\, dz\, dx;$$

that accounts for the first integral. There remain two sides of A, given by $x = b$ and $x = a$. On those two, $d\sigma = (\pm 1, 0, 0)\, dz\, dy$, and $F_2\mathbf{j}$ has zero flux. We have obtained

$$\int_A \frac{\partial F_2}{\partial y} = \int_{\mathrm{bd}(A)} (F_2\mathbf{j}) \bullet d\sigma.$$

For the leading term of the divergence, we have

$$\begin{aligned}
\int_A \frac{\partial F_1}{\partial x} &= \int_a^b \int_{g(x)}^{G(x)} \int_{h(x,y)}^{H(x,y)} \frac{\partial F_1}{\partial x}\, dz\, dy\, dx \\
&= \left(\int_a^b \frac{d}{dx} \int_{g(x)}^{G(x)} \int_{h(x,y)}^{H(x,y)} F_1(x,y,z)\, dz\, dy\, dz \right) \\
&\quad + \left[\int_a^b \int_{h(x,G(x))}^{H(x,G(x))} -F_1(x,G(x),z) G'(x)\, dz\, dx \right. \\
&\qquad \left. + \int_a^b \int_{h(x,g(x))}^{H(x,g(x))} F_1(x,g(x),z) g'(x)\, dz\, dx \right] \\
&\quad + \left[\int_a^b \int_{g(x)}^{G(x)} -F_1(x,y,H(x,y)) \frac{\partial H}{\partial x}\, dy\, dx \right. \\
&\qquad \left. + \int_a^b \int_{g(x)}^{G(x)} F_1(x,y,h(x,y)) \frac{\partial h}{\partial x}\, dy\, dx \right]
\end{aligned}$$

by two applications of Theorem 6.3:1. Since we have established parametrizations of the sides of A, we leave to the reader to verify that the four integrals in brackets are the outward fluxes of $F_1\mathbf{i}$ on the upper-y, lower-y, upper-z, and lower-z sides. That is part of the pattern we sought: Each additional variable introduces fluxes across two more sides, as well as the integral in parentheses. That integral breaks into

$$\begin{aligned}
&\int_a^b \frac{d}{dx} \int_{g(x)}^{G(x)} \int_{h(x,y)}^{H(x,y)} F_1(x,y,z)\, dz\, dy\, dx \\
&\quad = \int_{g(b)}^{G(b)} \int_{h(b,y)}^{H(b,y)} F_1(b,y,z)\, dz\, dy - \int_{g(a)}^{G(a)} \int_{h(a,y)}^{H(a,y)} F_1(a,y,z)\, dz\, dy.
\end{aligned}$$

The integrals on the right are clearly the fluxes of $F_1\mathbf{i}$, in the direction of $(1,0,0)$ along the side $x = b$, and in the direction of $(-1,0,0)$ along $x = a$. We conclude that

$$\int_A \frac{\partial F_1}{\partial x} = \int_{\mathrm{bd}(A)} (F_1\mathbf{i}) \bullet d\sigma,$$

and

$$\int_{\mathrm{bd}(A)} (F_1\mathbf{i} + F_2\mathbf{j} + F_3\mathbf{k}) \bullet d\sigma = \int_A \frac{\partial F_1}{\partial x} + \frac{\partial F_2}{\partial y} + \frac{\partial F_3}{\partial z}. \qquad \square$$

Observe that the third paragraph of our proof, alone, would suffice for a region that is simple in every coordinate direction. Restricting ourselves to such regions would not be excessive. For example, a convex closed set is simple in every direction, provided its upper and lower boundaries in each direction (which are necessarily graphs) are C^1. We could then apply the divergence theorem to any region that can be viewed as the union of convex pieces (or as the difference; see Exercise 4).

Example 2. We said that Gauss's law relates the flux of the electric field **E** and the charge Q within A by

$$\int_{\mathrm{bd}(A)} \mathbf{E} \bullet d\sigma = 4\pi K Q.$$

If we use $\rho(\mathbf{x})$ to denote charge density at $\mathbf{x}$, then $Q = \int_A \rho(\mathbf{x})$. On the other side, the divergence theorem says that

$$\int_{\mathrm{bd}(A)} \mathbf{E} \bullet d\sigma = \int_A \nabla \bullet \mathbf{E}.$$

Again we reach a macroscale equality

$$4\pi K \int_A \rho(\mathbf{x}) = \int_A \nabla \bullet \mathbf{E}.$$

Its limiting, local, form

$$\nabla \bullet \left(\frac{\mathbf{E}}{4\pi K} \right) = \rho(\mathbf{x})$$

appears among Maxwell's equations, bearing still the name "Gauss's law."

The divergence theorem is also called "Gauss's theorem." The latter name, like "Euler's equation," is useless. (Similar statements apply wherever the names of Lagrange and Cauchy appear.) Euler and Gauss did not produce just solutions or theorems; they created worlds of mathematical inquiry, and approaches of enormous scope and power. They combined the elegance of Mozart with the inventiveness and productive longevity of Bach. (I am grateful to David Kramer for that description.) Still, it is fitting to attach the name, because the theorem has a wonderful grace of form. The form is symmetric with respect to the variables, and it is the same in all dimensions. That is not the case with the theorems of Green and Stokes. (All three theorems, we should note, can be cast as instances of a single principle, relating differential behavior in regions to changes around the edges.)

Typical examples for Gauss's theorem evaluate some surface integral as a volume integral, or vice versa; we give two of those as Exercises 1, 2. Here, we offer an example given by Jesse Douglas.

Example 3. What force does a liquid exert on a submerged object?

Figure 6.12 shows the $x_1 \ldots x_{n-1}$ hyperplane as the surface of the liquid (hereafter "water"), which occupies the lower half of $\mathbf{R}^n$. At each point of the hyperrectangle R, at depth $h \equiv -x_n$, the pressure p is the same, namely, enough to hold up the column of water above R and the column of air above that. The column of air weighs area$(R)p_0$, where standard force-per-area air pressure p_0 is

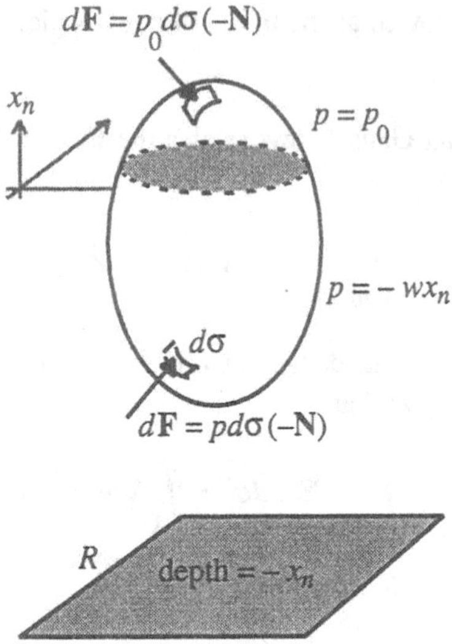

Figure 6.12.

14.7 lb/in^2 $=$ 2120 lb/ft^2. The column of water weighs [area$(R)h]w$, where the "weight density" w of water is 62.5 lb/ft^3. (Switching to metric does not improve the numbers: Pressure is a big number of newtons/m^2, and the weight density is 9800 n/m^3.) Hence

$$p \text{ area}(R) = [\text{area}(R)h]w + \text{area}(R)p_0,$$

and

$$p = p_0 - wx_n.$$

In the upper half of the figure we see an object of simple shape, partly submerged, partly exposed to the air. We make two simplifications. First, pretend that air pressure is constant. That is reasonably accurate for ordinary objects: At roughly $h = 34$ ft, p is already $2120 - 62.5(-34) \approx 4250$ ("two atmospheres"), a 100% increase in water pressure, but it takes thousands of feet to reduce air pressure by even 10%. Second, although water presses only on the outside of the submerged part, as air presses only the exposed part, we will pretend that the water presses down, and the air up, on the $x_n = 0$ cross section (dotted border, shaded) of the object. These imaginary opposed forces have the same magnitude, and they allow us to treat the submerged and exposed parts as surrounded by water and air, respectively.

We now have the exposed part surrounded by air acting on the object's surface. Near each surface point, an area $d\sigma$ feels a force $p_0\, d\sigma$ normal to the surface. If we call the outward unit normal $\mathbf{N}$, then the force on that patch is $d\mathbf{F} =$

$p_0\, d\sigma$ (–N). The total force exerted by the air is

$$\mathbf{F}_{\text{air}} = \int_{\text{bd(exposed)}} -p_0 \mathbf{N}\, d\sigma.$$

We do not have theorems for integrals that are vectors, but linearity tells us that the components of $\mathbf{F}_{\text{air}}$ satisfy

$$\mathbf{F}_{\text{air}} \bullet \mathbf{e}_k = \int_{\text{bd(exposed)}} -p_0 \mathbf{e}_k \bullet \mathbf{N}\, d\sigma = \int_{\text{exposed}} \nabla \bullet (-p_0 \mathbf{e}_k) = 0.$$

We conclude that the air applies no force. (Evidently, the decrease of air pressure with altitude, which we ignored, is noticeable to a helium balloon.)

Summing the force elements $d\mathbf{F} = p\, d\sigma(-\mathbf{N})$ applied by the water, we have

$$\mathbf{F}_{\text{water}} = \int_{\text{bd(submerged)}} -p\mathbf{N}\, d\sigma.$$

Hence

$$\begin{aligned}\mathbf{F}_{\text{water}} \bullet \mathbf{e}_k &= \int_{\text{bd(submerged)}} w x_n \mathbf{e}_k \bullet \mathbf{N}\, d\sigma \\ &= \int_{\text{submerged}} \nabla \bullet (w x_n \mathbf{e}_k) = \int_{\text{submerged}} w \frac{\partial x_n}{\partial x_k}.\end{aligned}$$

We conclude that in the horizontal directions, meaning $k \neq n$, $\mathbf{F}_{\text{water}}$ has zero components. Consequently, the force is vertical, with magnitude

$$F_n = \int_{\text{submerged}} w = w\, V(\text{submerged}).$$

In words: The buoyant force is the weight of the displaced water.

We have described ourselves as devotees of analysis. For mathematicians so inclined, it is difficult to think of a greater pleasure than seeing Archimedes' principle as an illustration of Gauss's theorem, or a better place to end the book.

Exercises

1. For the field $\mathbf{F} \equiv x\mathbf{i} + y\mathbf{j} + z\mathbf{k}$ and the cylinder A given by $x^2 + y^2 \leq a^2$, $0 \leq z \leq b$:

   (a) Calculate $\nabla \bullet \mathbf{F}$ and $\int_A \nabla \bullet \mathbf{F}$.

   (b) Evaluate $\int_{\text{bd}(A)} \mathbf{F} \bullet d\boldsymbol{\sigma}$ (outward). Is this supposed to match (a)?

2. Repeat Exercise 1 with the region A given by $(5 - x^2 - y^2)^{1/2} \leq z \leq 5 - x^2 - y^2$.

3. For the gravitational field $\mathbf{F} \equiv -k\frac{\mathbf{x}}{\|\mathbf{x}\|^3}$ in $\mathbf{R}^n$ and the ball $B \equiv B(\mathbf{O}, a)$:

   (a) Calculate $\nabla \bullet \mathbf{F}$ and $\int_B \nabla \bullet \mathbf{F}$.

   (b) Evaluate $\int_{\mathrm{bd}(B)} \mathbf{F} \bullet d\sigma$ (outward). Is this supposed to match (a)?

4. Prove that Gauss's theorem holds for a hollow ball: Assume that $\mathbf{F}$ is C^1 on $B(\mathbf{O}, 3)$, and $A \equiv B(\mathbf{O}, 2) - N(\mathbf{O}, 1)$; then the flux of $\mathbf{F}$ outward from A is $\int_A \nabla \bullet \mathbf{F}$.

5. Show that a curl in $\mathbf{R}^3$ has zero flux: Let A be a simple region in the domain of field $\mathbf{F}$; then

$$\int_{\mathrm{bd}(A)} (\nabla \times \mathbf{F}) \bullet d\sigma = 0.$$

6. We have introduced the operators ∇, $\nabla\times$, and $\nabla\bullet$. For each of the following combinations, decide whether it makes sense in $\mathbf{R}^3$. If practical, write a simplified expression for any legal one. Here f is a scalar function and $\mathbf{F}$ a field.

   (a) $\nabla(\nabla f)$.

   (b) $\nabla \times (\nabla f)$.

   (c) $\nabla \bullet (\nabla f)$.

   (d) $\nabla(\nabla \times \mathbf{F})$.

   (e) $\nabla \times (\nabla \times \mathbf{F})$.

   (f) $\nabla \bullet (\nabla \times \mathbf{F})$.

   (g) $\nabla(\nabla \bullet \mathbf{F})$.

   (h) $\nabla \times (\nabla \bullet \mathbf{F})$.

   (i) $\nabla \bullet (\nabla \bullet \mathbf{F})$.

7. Show that:

   (a) $\nabla \bullet (\mathbf{F} \times \mathbf{G}) = (\nabla \times \mathbf{F}) \bullet \mathbf{G} - \mathbf{F} \bullet (\nabla \times \mathbf{G})$.

   (b) $\nabla \bullet (f\mathbf{G}) = \nabla f \bullet \mathbf{G} + f(\nabla \bullet \mathbf{G})$.

Solutions to Exercises

"For emergency use only."

That is how Buck begins his list of solutions. Ross comments, in the same spirit, that students will simply "cheat themselves" if they turn to this section before working hard to develop their own solutions. We trust that, at this level, our students are mature enough to agree.

The answers here are intended, like the text, to teach. They are therefore sprinkled with comments, alternative approaches, and questions. Except for leaving such questions to be pondered by the reader, these solutions are complete.

Chapter 1

Section 1.1

1. (a) $G_1(x,y) - G_1(a,b) - [3a^2 \;\; -3b^2]\begin{bmatrix} x-a \\ y-b \end{bmatrix} = x^3 - y^3 - a^3 + b^3 - 3a^2(x-a) + 3b^2(y-b) = (x^2+ax+a^2-3a^2)(x-a) + (y^2+by+b^2-3b^2)(y-b)$. Recalling that $\frac{|x-a|}{\sqrt{(x-a)^2+(y-b)^2}}$ and $\frac{|y-b|}{\sqrt{(x-a)^2+(y-b)^2}}$ are bounded by 1, we have

$$\frac{|\text{difference}|}{\sqrt{(x-a)^2+(y-b)^2}} \le |x^2+ax+a^2-3a^2| + |y^2+by+b^2-3b^2| \to 0$$

as $(x,y) \to (a,b)$.

(b) $G_2(x, y) - G_2(a, b) - [b \;\; a]\begin{bmatrix} x-a \\ y-b \end{bmatrix} = xy - ab - b(x-a) - a(y-b) = (x-a)(y-b)$. On division, this becomes $(x-a)[(y-b)/\sqrt{(x-a)^2+(y-b)^2}]$. The bracketed factor has absolute value ≤ 1, and $(x-a) \to 0$ as $(x, y) \to (a, b)$.

2. $\mathbf{F}'(a, b) = \begin{bmatrix} 2(a+b) & 2(a+b) \\ 2(a-b) & 2(b-a) \end{bmatrix}$. First, by the mean value theorem, $(x+y)^2-(a+b)^2 = 2t(x+y-a-b)$ for some t between $x+y$ and $a+b$. From $(x+y)^2-(a+b)^2 = 2t(x-a)+2t(y-b) \approx [2(a+b) \;\; 2(a+b)]\begin{bmatrix} x-a \\ y-b \end{bmatrix}$, we guess $F_1{}' = [2(a+b) \;\; 2(a+b)]$. Confirmation comes from

$$\begin{aligned} F_1(x, y) - F_1(a, b) - [2(a+b) \;\; 2(a+b)]\begin{bmatrix} x-a \\ y-b \end{bmatrix} \\ = (2t-2a-2b)(x-a+y-b), \end{aligned}$$

because

$$\frac{|(2t-2a-2b)(x-a+y-b)|}{\sqrt{(x-a)^2+(y-b)^2}} \leq 4|t-a-b| \to 0$$

as $(x, y) \to (a, b)$. By similar algebra, $F_2{}' = [2(a-b) \;\; -2(a-b)]$. Then apply Theorem 1.

3. $\mathbf{H}'(a, b) = \begin{bmatrix} \cos a & \sin b \\ 2e^{2a} & 3e^{3b} \end{bmatrix}$. Use the techniques of Answer 2:

$$\begin{aligned} \sin x - \cos y - (\sin a - \cos b) &= \cos t\,(x-a) + \sin s\,(y-b) \\ &\approx [\cos a \;\; \sin b]\begin{bmatrix} x-a \\ y-b \end{bmatrix}, \end{aligned}$$

because $|\cos t - \cos a| + |\sin s - \sin b| \to 0$; and analogous constructions for the exponentials.

4. Assume $\mathbf{b} \neq \mathbf{O}$. Apply the mean value theorem to

$$\begin{aligned} \sqrt{t}\colon \|\mathbf{x}\| - \|\mathbf{b}\| &= \sqrt{(x_1{}^2+\cdots+x_n{}^2)} - \sqrt{(b_1{}^2+\cdots+b_n{}^2)} \\ &= 0.5t^{-1/2}[(x_1-b_1)(x_1+b_1)+\cdots+(x_n-b_n)(x_n+b_n)] \end{aligned}$$

for a value t between $x_1{}^2+\cdots+x_n{}^2$ and $b_1{}^2+\cdots+b_n{}^2$ The value is no problem if we confine $\mathbf{x}$ to $N\left(\mathbf{b}, \frac{\|\mathbf{b}\|}{2}\right)$. Then

$$\|\mathbf{x}\| - \|\mathbf{b}\| \approx 0.5t^{-1/2}[2b_1 \cdots 2b_n][x_1-b_1 \cdots x_n-b_n]^t$$

suggests that $f'(x) = \frac{[b_1 \cdots b_n]}{\sqrt{b_1{}^2+\cdots+b_n{}^2}}$. We verify as in Answers 2 and 3.

5. $(\alpha\mathbf{F}+\beta\mathbf{G})(\mathbf{x})-(\alpha\mathbf{F}+\beta\mathbf{G})(\mathbf{b})-(\alpha\mathbf{F}'+\beta\mathbf{G}')(\mathbf{b})\langle\mathbf{x}-\mathbf{b}\rangle = \alpha\big(\mathbf{F}(\mathbf{x})-\mathbf{F}(\mathbf{b})-\mathbf{F}'(\mathbf{b})\langle\mathbf{x}-\mathbf{b}\rangle\big)+\beta\big(\mathbf{G}(\mathbf{x})-\mathbf{G}(\mathbf{b})-\mathbf{G}'(\mathbf{b})\langle\mathbf{x}-\mathbf{b}\rangle\big) = \mathbf{o}(\mathbf{x}-\mathbf{b})+\mathbf{o}(\mathbf{x}-\mathbf{b})$.

6. $\mathbf{H}' = \mathbf{A}$, because $\mathbf{H}(\mathbf{x})-\mathbf{H}(\mathbf{c}) = \mathbf{A}\langle\mathbf{x}\rangle+\mathbf{b}-(\mathbf{A}\langle\mathbf{c}\rangle+\mathbf{b}) = \mathbf{A}\langle\mathbf{x}-\mathbf{c}\rangle$, which certainly says that $\mathbf{A}\langle\mathbf{x}-\mathbf{c}\rangle$ approximates $\mathbf{H}(\mathbf{x})-\mathbf{H}(\mathbf{c})$ to within $\mathbf{o}(\mathbf{x}-\mathbf{c})$.

Section 1.2

1. (a) $(x, y) \to (0, 0) \Rightarrow x \to 0, y \to 0 \Rightarrow (xy)^{1/3} \to 0$.
   (b) $f \equiv 0$ along the x-axis. Hence $\frac{\partial f}{\partial x}(0,0) \equiv \lim_{t\to 0} \frac{f(t,0)-f(0,0)}{t} = 0$; similarly with y.
   (c) $f(t,t)-f(0,0)-[0\ \ 0][t\ \ t]^t = (tt)^{1/3} = t^{2/3}$, which is much larger than $\|(t,t)\| = \sqrt{2}|t|$; $\frac{t^{2/3}}{t} \to \infty$ as $t \to 0^+$.

2. (a) Away from the origin, the partials are given by the quotient rule. At $(0, 0)$, both partials are 0, because $h \equiv 0$ along the axes.
   (b) $h(x,y)-h(0,0)-[0\ \ 0][x\ \ y]^t = \frac{xy}{x^2+y^2}$. Along the line $y = x$, this quantity is not small. Thus, h is not even continuous at $(0, 0)$.

3. Away from the y-axis, $\frac{\partial G}{\partial x} = 2x\ \sin\left(\frac{1}{x}\right) - \cos\left(\frac{1}{x}\right)$ and $\frac{\partial G}{\partial y} = 0$ are continuous. By Theorem 2, G is differentiable there. For a point $(0, b)$ on the y-axis,

$$\frac{\partial G}{\partial x} = \lim_{t\to 0}\frac{G(t,b)-G(0,b)}{t} = \lim_{t\to 0} t\sin\left(\frac{1}{t}\right) = 0,$$

answering (b). But $\lim_{x\to 0}\frac{\partial G}{\partial x}(x,b) = \lim_{x\to 0} -\cos\left(\frac{1}{x}\right)$ does not exist; that answers (c). On the other hand,

$$|G(x,y)-G(0,b)-[0\ \ 0][x\ \ y-b]^t| = \left|x^2\sin\left(\frac{1}{x}\right)\right| \le |x|^2,$$

which is small compared to $\|(x, y-b)\|$ as $(x, y) \to (0, b)$. Therefore, G is also differentiable on the y-axis.

4. (a) Differentiable everywhere; its partials are continuous.

$$\mathbf{f}'(x,y) = \begin{bmatrix} y & x \\ \exp(x+y) & \exp(x+y) \\ \cos x\cos y & -\sin x\sin y \end{bmatrix}.$$

   (b) Differentiable everywhere. The formula

$$\frac{\partial g}{\partial x} = 3x^2y^3\sin\left(\frac{1}{xy}\right) - xy^2\cos\left(\frac{1}{xy}\right)$$

shows that $\frac{\partial g}{\partial x}$ is continuous away from the axes and $\to 0$ as you get close to any point on the axes. At $(0, b)$,

$$\frac{\partial g}{\partial x} = \lim_{t\to 0} \frac{t^3 b^3 \sin(1/tb)}{t} = 0.$$

At $(a, 0)$, $\frac{\partial g}{\partial x} = 0$ because $g \equiv 0$ along the x-axis. Hence $\frac{\partial g}{\partial x}$ is always continuous; analogously for y.

$$g' = \left[3x^2y^3 \sin\left(\frac{1}{xy}\right) - xy^2 \cos\left(\frac{1}{xy}\right) \right.$$
$$\left. 3x^3y^2 \sin\left(\frac{1}{xy}\right) - x^2y \cos\left(\frac{1}{xy}\right)\right]$$

or $[0\ \ 0]$.

(c) Differentiable everywhere, even though the partials are discontinuous along the axes; justifications and derivative matrix mimic Exercise 3.

(d) Differentiable off the axes, plus at the origin. Off the axes, $\frac{\partial G}{\partial x} = \frac{2}{3}x^{-1/3}y^{2/3}$ is continuous; same with $\frac{\partial G}{\partial y}$. At $(0, 0)$, $\frac{\partial G}{\partial x} = \frac{\partial G}{\partial y} = 0$, and

$$\frac{\left[G(x, y) - G(0, 0) - [0\,0]\begin{bmatrix} x \\ y \end{bmatrix}\right]}{\sqrt{x^2 + y^2}}$$
$$= \left(\frac{x}{\sqrt{x^2+y^2}}\right)^{2/3} \left(\frac{y}{\sqrt{x^2+y^2}}\right)^{2/3} (x^2 + y^2)^{1/6} \to 0$$

as $(x, y) \to (0, 0)$. Anywhere else, G has an undefined partial: At $(a, 0)$,

$$\frac{\partial G}{\partial y} = \lim_{t\to 0} \frac{G(a, t) - G(a, 0)}{t} = \lim_{t\to 0} a^{2/3}t^{-1/3} = \infty;$$

by Theorem 1, G is undifferentiable. $G' = \left[\frac{2}{3}x^{-1/3}y^{2/3}\ \frac{2}{3}x^{2/3}y^{-1/3}\right]$ off the axes, $[0\,0]$ at the origin.

(e) Differentiable off the axes only, with $H' = \left[\frac{1}{3}x^{-2/3}y^{1/3}\ \frac{1}{3}x^{1/3}y^{-2/3}\right]$. Away from the axes, both partials are continuous. At $(a, 0)$,

$$\frac{\partial H}{\partial y} = \lim_{t\to 0} \frac{H(a, t) - H(a, 0)}{t} = \lim_{t\to 0} a^{1/3}t^{-2/3} = \infty;$$

by Theorem 1, H is undifferentiable. At $(0, 0)$, $\frac{\partial H}{\partial x} = \frac{\partial H}{\partial y} = 0$, but

$$\left|\frac{H(x, x) - H(0, 0) - [0\,0][x\ x]^t}{\sqrt{2}x}\right| = x^{2/3}/\sqrt{2}|x| \to \infty,$$

making H undifferentiable.

5. Since $\mathbf{f}: \mathbf{R} \to \mathbf{R}^n$, $\mathbf{f}'$ has to be $n \times 1$. Because

$$\mathbf{f}(t) - \mathbf{f}(b) = (t-b)\mathbf{v} = [\mathbf{v}_1 \ \cdots \ \mathbf{v}_n]^t (t-b),$$

we have $\mathbf{f}' = \langle \mathbf{v} \rangle$.

Section 1.3

1. (a)

$$\frac{\partial g}{\partial(x,y)} = \frac{\partial g}{\partial(u,v)}\frac{\partial(u,v)}{\partial(x,y)} = [2u \ 2v]\begin{bmatrix} y\cos xy & x\cos xy \\ -y\sin xy & -x\sin xy \end{bmatrix}$$
$$= [2uy\cos xy - 2vy\sin xy \ \ 2ux\cos xy - 2vx\sin xy] = [0 \ \ 0].$$

(b) $g = (\sin xy)^2 + (\cos xy)^2 = 1$. Hence $\frac{\partial g}{\partial(x,y)} = \mathbf{O}$.

2. Decreasing. Distance $r = \left(x^2 + y^2\right)^{1/2}$ has

$$\frac{dr}{dt} = \frac{\partial r}{\partial x}\frac{dx}{dt} + \frac{\partial r}{\partial y}\frac{dy}{dt} = \frac{x}{r}(-10\sin t) + \frac{y}{r}(8\cos t).$$

When $x = 5$, $\cos t = 0.5$, so $\sin t = \pm\frac{\sqrt{3}}{2}$. Quadrant I forces $y = 4\sqrt{3}$, and $r\frac{\partial r}{\partial t} = 5(-5\sqrt{3}) + 4\sqrt{3}(4) < 0$.

3.

$$\frac{\partial f}{\partial(\rho,\theta,\phi)} = \frac{\partial f}{\partial(x,y,z)}\frac{\partial(x,y,z)}{\partial(\rho,\theta,\phi)} = \begin{bmatrix} \frac{\partial f}{\partial x} & \frac{\partial f}{\partial y} & \frac{\partial f}{\partial z} \end{bmatrix}$$
$$\times \begin{bmatrix} \cos\theta\sin\phi & \rho\sin\theta\sin\phi & \rho\cos\theta\cos\phi \\ \sin\theta\sin\phi & \rho\cos\theta\sin\phi & \rho\sin\theta\cos\phi \\ \cos\phi & 0 & -\rho\sin\phi \end{bmatrix}$$
$$= \left[\cos\theta\sin\phi\left(\frac{\partial f}{\partial x}\right) + \sin\theta\sin\phi\left(\frac{\partial f}{\partial y}\right) + \cos\phi\left(\frac{\partial f}{\partial z}\right) \ \ \text{etc.}\right].$$

4. ML. By Exercise 1.1:6, the derivative of a first-degree function is the associated matrix.

5. Rule: If f and g are differentiable at $\mathbf{b}$ and $g(t) \neq 0$, then $h(\mathbf{x}) \equiv \frac{f(\mathbf{x})}{g(\mathbf{x})}$ is differentiable at $\mathbf{b}$, and

$$h'(\mathbf{b}) = \frac{g(\mathbf{b})f'(\mathbf{b}) - f(\mathbf{b})g'(\mathbf{b})}{g(\mathbf{b})^2}.$$

Proof: $H(u,v) \equiv \frac{u}{v}$ has $\frac{\partial H}{\partial u} = \frac{1}{v}$ and $\frac{\partial H}{\partial v} = -\frac{u}{v^2}$ wherever $v \neq 0$. Set $u \equiv f(\mathbf{x})$, $v \equiv g(\mathbf{x})$. Then $h(\mathbf{x}) = H(f(\mathbf{x}), g(\mathbf{x}))$ is a differentiable composite at $\mathbf{b}$, with

$$\frac{\partial h}{\partial \mathbf{x}} = \frac{\partial H}{\partial u}\frac{\partial u}{\partial \mathbf{x}} + \frac{\partial H}{\partial v}\frac{\partial v}{\partial \mathbf{x}} = \left(\frac{1}{g(\mathbf{b})}\right) f'(\mathbf{b}) + \left(-\frac{f(\mathbf{b})}{g(\mathbf{b})^2}\right) g'(\mathbf{b}).$$

6. $\mathbf{u} \bullet \mathbf{v} = \sum u_j v_j$. By the product rule,

$$\frac{\partial(\mathbf{u} \bullet \mathbf{v})}{\partial \mathbf{x}} = \sum \left(u_j \frac{\partial v_j}{\partial \mathbf{x}} + v_j \frac{\partial u_j}{\partial \mathbf{x}} \right) = \sum u_j \frac{\partial v_j}{\partial \mathbf{x}} + \sum v_j \frac{\partial u_j}{\partial \mathbf{x}}$$

$$= \begin{bmatrix} u_1 & \cdots & u_m \end{bmatrix} \begin{bmatrix} \frac{\partial v_1}{\partial \mathbf{x}} \\ \ddots \\ \frac{\partial v_m}{\partial \mathbf{x}} \end{bmatrix} + \begin{bmatrix} v_1 & \cdots & v_m \end{bmatrix} \begin{bmatrix} \frac{\partial u_1}{\partial \mathbf{x}} \\ \ddots \\ \frac{\partial u_m}{\partial \mathbf{x}} \end{bmatrix}$$

$$= \operatorname{row}(\mathbf{u}) \frac{\partial \mathbf{v}}{\partial \mathbf{x}} + \operatorname{row}(\mathbf{v}) \frac{\partial \mathbf{u}}{\partial \mathbf{x}}.$$

7. (a) Assume that $r = \sqrt{x^2 + y^2} > 0$.
$\Rightarrow$ If H is differentiable at r, then $G(x, y) = H(r)$ is a differentiable composite.
$\Leftarrow$ Assume that G is differentiable at (x, y). We have

$$\frac{H(r+s) - H(r)}{s}$$
$$= \frac{G\left(x + \frac{sx}{r}, y + \frac{sy}{r}\right) - G(x, y) - G'(x, y)\left[\frac{sx}{r}\ \frac{sy}{r}\right]^t}{s} + \frac{G'(x, y)\left[\frac{sx}{r}\ \frac{sy}{r}\right]^t}{s}.$$

As $s \to 0$, the first fraction on the right vanishes, and the second is fixed at $G'(x, y)\left[\frac{x}{r}\ \frac{y}{r}\right]^t$. Hence H is differentiable at r.

(b) $\Rightarrow$ If G is differentiable at $(0, 0)$, then

$$\frac{H(s) - H(0)}{s} = \frac{G(s, 0) - G(0, 0)}{s} \to \frac{\partial G}{\partial x}(0, 0)$$

as $s \to 0$. Hence H is differentiable at 0. Also

$$\frac{H(s) - H(0)}{s} = \frac{G(-s, 0) - G(0, 0)}{s} \to -\frac{\partial G}{\partial x}(0, 0).$$

Of necessity, $H'(0) = \frac{\partial G}{\partial x}(0, 0) = 0$; same with $\frac{\partial G}{\partial y}$.
$\Leftarrow$ If H is differentiable at 0, then

$$\frac{G(x, y) - G(0, 0) - [0\ 0][x\ y]^t}{\|(x, y)\|} = \frac{H(r) - H(0)}{r} \to H'(0)$$

as $(x, y) \to \mathbf{O}$. Hence if $H'(0) = 0$, then G is differentiable at $\mathbf{O}$ (and $G'(0, 0) = \mathbf{O}$).

(c) The equality is trivial at the origin. Away, wherever G is differentiable, $G(x, y) = H(r)$ gives

$$\frac{\partial G}{\partial x} = \frac{dH}{dr} \frac{\partial r}{\partial x} = \frac{H'(r)x}{r}$$

and $\frac{\partial G}{\partial y} = \frac{H'(r)y}{r}$. Clearly $y \frac{\partial G}{\partial x} = x \frac{\partial G}{\partial y}$.

8. $\mathbf{L}\langle\mathbf{x}\rangle = \left[\sum_{k=1}^{n} a_{1k}x_k \ \sum_{k=1}^{n} a_{mk}x_k\right]^t$, so

$$\begin{aligned}\|\mathbf{L}\langle\mathbf{x}\rangle\|^2 &= \left(\sum_{k=1}^{n} a_{1k}x_k\right)^2 + \cdots + \left(\sum_{k=1}^{n} a_{mk}x_k\right)^2\\ &\le \left(\sum_{k=1}^{n} a_{1k}{}^2\right)\left(\sum_{k=1}^{n} x_k{}^2\right) + \cdots + \left(\sum_{k=1}^{n} a_{mk}{}^2\right)\left(\sum_{k=1}^{n} x_k{}^2\right)\\ (\text{Cauchy}) &= \left(\sum_{j=1}^{m}\sum_{k=1}^{n} a_{mk}{}^2\right)\left(\sum_{k=1}^{n} x_k{}^2\right);\end{aligned}$$

take square roots.

Section 1.4

1. (a)

$$\begin{aligned}\frac{\partial^2 f}{\partial x \partial y} &= \frac{\partial}{\partial y}(2xy\exp(x^2y)) = 2x\exp(x^2y) + 2xyx^2\exp(x^2y),\\ \frac{\partial^2 f}{\partial y \partial x} &= \frac{\partial}{\partial x}(x^2\exp(x^2y)) = 2x\exp(x^2y) + x^2 2xy\exp(x^2y).\end{aligned}$$

(b) $f'' = \exp(x^2y)\Big[\left[2y + 4x^2y^2 \ \ 2x + 2x^3y\right] \ \left[2x + 2x^3y \ \ x^4\right]\Big]$

2. $\mathbf{g}' = \begin{bmatrix} 2x & -2y \\ y & x \end{bmatrix}$, $\mathbf{g}'' = \left[\begin{bmatrix} 2 & 0 \\ 0 & 1 \end{bmatrix} \begin{bmatrix} 0 & -2 \\ 1 & 0 \end{bmatrix}\right]$.

3. f''' is straightforward and long; $\mathbf{g}'''$ is clearly $\mathbf{O}$, but requires sixteen 0's.

4. By the formula for f'' found in Example 1,

$$\begin{aligned}&f''(x, y) - f''(a, b)\\ &= \left[[2y - 2b \ \ 2x - 2a - 2y + 2b] \ [2x - 2a - 2y + 2b \ \ -2x + 2a]\right]\\ &= \left[\left[[0\ 2]\begin{bmatrix} x-a \\ y-b \end{bmatrix} \ [2\ -2]\begin{bmatrix} x-a \\ y-b \end{bmatrix}\right] \ \left[[2\ -2]\begin{bmatrix} x-a \\ y-b \end{bmatrix} \ [-2\ 0]\begin{bmatrix} x-a \\ y-b \end{bmatrix}\right]\right]\\ &= \Big[\left[[0\ 2] \ [2\ -2]\right] \ \left[[2\ -2] \ [-2\ 0]\right]\Big]\langle(x - a, y - b)\rangle\end{aligned}$$

5. No. You would have $\frac{\partial^2 h}{\partial x \partial y} = \exp\left(x^2\right)$, $\frac{\partial^2 h}{\partial y \partial x} = \sin^2 y$, continuous but not symmetric.

6. There would exist $\mathbf{x}_1$ and $\mathbf{y}_1 \in N(\mathbf{b}, 1)$ with $\mathbf{g}(\mathbf{x}_1) = \mathbf{f}(\mathbf{y}_1)$, $\mathbf{x}_2$ and $\mathbf{y}_2 \in N\left(\mathbf{b}, \frac{1}{2}\right)$ with $\mathbf{g}(\mathbf{x}_2) = \mathbf{f}(\mathbf{y}_2), \ldots$. Then $\mathbf{x}_i \to \mathbf{b}$ and $\mathbf{y}_i \to \mathbf{b}$, so by continuity

$$\mathbf{g}(\mathbf{b}) = \mathbf{g}(\lim \mathbf{x}_i) = \lim \mathbf{g}(\mathbf{x}_i) = \lim \mathbf{f}(\mathbf{y}_i) = \mathbf{f}(\lim \mathbf{y}_i) = \mathbf{f}(\mathbf{b}).$$

Chapter 2

Section 2.1

1. (a) Directional $= \nabla \mathbf{f} \bullet \mathbf{u} = (2x, -2y) \bullet (\cos\theta, \sin\theta) = 10\cos\theta - 6\sin\theta$.

   (b) The directional is 0 in the directions with $10\cos\theta - 6\sin\theta = 0$, or $\tan\theta = \frac{10}{6}$. The derivative is found implicitly: $2x - 2yy' = 0$ leads to $y' = \frac{x}{y} = \frac{5}{3}$; the tangent is in the zero-growth direction.

2. (a) $\nabla g(x, y) = (2x, 8y), = (2a, 8b)$ at the point given.

   (b) Implicitly, $2x + 8yy' = 0$, $y' = -\frac{x}{4y}$. Hence the tangent at (a, b) lies in the direction of vector $(4b, -a)$. In that direction,

$$\partial_{\mathbf{u}} g = \nabla g \bullet \mathbf{u} = (2a, 8b) \bullet \frac{(4b, -a)}{\|(4b, -a)\|} = 0.$$

3. (a) Excluding the origin, $\frac{\partial h}{\partial x} = 0.5\left(x^2 + y^2\right)^{-1/2} 2x = \frac{x}{r}$. Similarly, $\frac{\partial h}{\partial y} = \frac{y}{r}$, and $\|\nabla h\|^2 = \left(\frac{x}{r}\right)^2 + \left(\frac{y}{r}\right)^2 = 1$.

   (b) Since h is distance from the origin, it increases fastest if you walk outward along the line from you to the origin, in which case it increases 1 m for every meter you walk.

4. The unit vector in the direction of negative x_j is

$$-\mathbf{e}_j \equiv (0, \ldots, 0, -1, 0, \ldots, 0).$$

By Theorem 2(a), the derivative in that direction is

$$\left(\frac{\partial F}{\partial x_1}, \ldots, \frac{\partial F}{\partial x_n}\right) \bullet (-\mathbf{e}_j) = -\frac{\partial F}{\partial x_j}.$$

5. First, $\nabla G(\mathbf{O}) = (0, 1)$, because $G \equiv 0$ along the x-axis, $\equiv y$ along the y-axis. Therefore,

$$G(x, y) - G(\mathbf{O}) - \nabla G(\mathbf{O})\left[x \;\; y\right]' = G(x, y) - y.$$

Along the line $y = x$ in quadrant I,

$$\frac{G(x, y) - y}{\sqrt{(x^2 + y^2)}} = \frac{y/2 - y}{\sqrt{2}y} = -2^{-3/2}$$

does not approach 0.

6. Where the partials exist, the one-variable product rule gives

$$\frac{\partial (fg)}{\partial x_j} = f\frac{\partial g}{\partial x_j} + g\frac{\partial f}{\partial x_j}.$$

Hence

$$\left(\frac{\partial(fg)}{\partial x_1},\ldots,\frac{\partial(fg)}{\partial x_n}\right)=f\left(\frac{\partial g}{\partial x_1},\ldots,\frac{\partial g}{\partial x_n}\right)+g\left(\frac{\partial f}{\partial x_1},\ldots,\frac{\partial f}{\partial x_n}\right).$$

7. (a) By Exercise 6, $\nabla(ff)=f\nabla f+f\nabla f=2f\nabla f$.

   (b) Write $u=f^2$. Then $\frac{\partial u}{\partial x_j}=\frac{du}{df}\frac{\partial f}{\partial x_j}=2f\frac{\partial f}{\partial x_j}$ for each j. Consequently,

$$\left(\frac{\partial u}{\partial x_1},\ldots,\frac{\partial u}{\partial x_n}\right)=\left(2f\frac{\partial f}{\partial x_1},\ldots,2f\frac{\partial f}{\partial x_n}\right)=2f\nabla f.$$

   (c) The fastest increase for f^2 is in the direction of ∇f if f is positive, opposite direction if f is negative. Thus,

$$\nabla([xy]^2)(2xy^2,2x^2y)=2xy(y,z)=2f\nabla f;$$

   at $(5,3)$, $\nabla(f^2)=30\nabla f$; at $(-2,2)$, $\nabla(f^2)=-8\nabla f$. It makes sense: When $f<0$, increasing f makes $|f|$ decrease. Note also that if $f=0$, then f^2 is stationary; its rate of change is zero in every direction.

8. The directional of f along $\mathbf{v}$ is $\partial_{\mathbf{v}}f=\nabla f\bullet\mathbf{v}=v_1\frac{\partial f}{\partial x_1}+\cdots+v_n\frac{\partial f}{\partial x_n}$. The derivative of this quantity is

$$\left[v_1\frac{\partial^2 f}{\partial x_1\partial x_1}+\cdots+v_n\frac{\partial^2 f}{\partial x_n\partial x_1}\quad\cdots\quad v_1\frac{\partial^2 f}{\partial x_1\partial x_n}+\cdots+v_n\frac{\partial^2 f}{\partial x_n\partial x_n}\right]$$

$$=\left[\left[\frac{\partial^2 f}{\partial x_1\partial x_1}+\cdots+\frac{\partial^2 f}{\partial x_n\partial x_1}\right]\begin{bmatrix}v_1\\ \vdots\\ v_n\end{bmatrix}\right.$$

$$\left.\cdots\left[\frac{\partial^2 f}{\partial x_1\partial x_n}+\cdots+\frac{\partial^2 f}{\partial x_n\partial x_n}\right]\begin{bmatrix}v_1\\ \vdots\\ v_n\end{bmatrix}\right]=f''(\mathbf{b})\langle\mathbf{v}\rangle.$$

Then by Theorem 1, the directional of $\partial_{\mathbf{v}}f$ along $\mathbf{u}$ is $[f''(\mathbf{b})\langle\mathbf{v}\rangle]\langle\mathbf{u}\rangle$.

Section 2.2

1. (a) $f(\mathbf{b})-f(\mathbf{a})=5-5=0$, $\nabla f=\left(\frac{x}{r},\frac{y}{r}\right)$, $\mathbf{b}-\mathbf{a}=(0,8)$. Then $0=\frac{8y}{r}$ at $y=0$, point $(3,0)$.

   (b) $f(\mathbf{b})-f(\mathbf{a})=e^3-e^1$, $\nabla f=\left(e^{x+y},e^{x+y}\right)$, $\mathbf{b}-\mathbf{a}=(1,1)$. The equation $e^3-e^1=2e^{x+y}$ is not sufficient. But the segment is given by $y=x-1$. The simultaneous solution of $x+y=\ln\frac{e^3-e^1}{2}$ and $x-y=1$ is $x=0.5+\ln\left[\frac{(e^3-e^1)}{2}\right]$, $y=-0.5+\ln\frac{e^3-e^1}{2}$.

2. $F(x, y) \equiv \frac{x}{|x|}$ has zero derivative for $x > 0$ and for $x < 0$, but has two different values.

3. (a) $f(x, y) \equiv \sqrt{(x-1)^2 + y^2} + \sqrt{(x+1)^2 + y^2}$ is undifferentiable at just $(\pm 1, 0)$.

   (b) $f(1, 0) - f(-1, 0) = 0$,

$$\nabla f = \left(\frac{x-1}{\sqrt{(x-1)^2 + y^2}} + \frac{x+1}{\sqrt{(x+1)^2 + y^2}}, \frac{y}{\sqrt{(x-1)^2 + y^2}} + \frac{y}{\sqrt{(x+1)^2 + y^2}} \right).$$

   The latter is $\mathbf{O}$ at the origin, so $\mathbf{c} \equiv (0, 0)$ is one such point.

4. By Theorem 1.1:1, $\mathbf{F}'$ is a matrix with rows $F_1', \ldots, F_n'$. If $\mathbf{F}' = \mathbf{O}$, then each $F_j' = \mathbf{O}$. By Theorem 3, each F_j is constant.

5. Assume $\mathbf{h}'' = \mathbf{O}$. By Exercise 4, $\mathbf{h}' =$ constant $=$ (matrix) $\mathbf{M}$. Write $\mathbf{H}(\mathbf{x}) \equiv \mathbf{h}(\mathbf{x}) - \mathbf{M}\langle\mathbf{x}\rangle$. Then $\mathbf{H}'(\mathbf{x}) = \mathbf{h}'(\mathbf{x}) - (\mathbf{M}\langle\mathbf{x}\rangle)'$ (linearity, Exercise 1.1:5) $= \mathbf{h}'(\mathbf{x}) - \mathbf{M}$ (Exercise 1.1:6) $= \mathbf{O}$. Therefore $\mathbf{H}(\mathbf{x}) =$ constant $=$ (vector) $\mathbf{b}$, and $\mathbf{h}(\mathbf{x}) = \mathbf{M}\langle\mathbf{x}\rangle + \mathbf{b}$.

6. (a) From $\|\mathbf{g}(\mathbf{x}) - \mathbf{g}(\mathbf{y})\| \le M\|\mathbf{x} - \mathbf{y}\|^r$, we conclude that $\|\mathbf{g}(\mathbf{x}) - \mathbf{g}(\mathbf{y})\| < \varepsilon$ as long as $\|\mathbf{x} - \mathbf{y}\| < \delta \equiv \left(\frac{\varepsilon}{M}\right)^{1/r}$.

   (b) If $\|\mathbf{g}(\mathbf{x}) - \mathbf{g}(\mathbf{b})\| \le M\|\mathbf{x} - \mathbf{b}\|^{1+\varepsilon}$, then

$$\|\mathbf{g}(\mathbf{x}) - \mathbf{g}(\mathbf{b}) - \mathbf{O}\langle\mathbf{x} - \mathbf{b}\rangle\| \le M\|\mathbf{x} - \mathbf{b}\|^{\varepsilon}\|\mathbf{x} - \mathbf{b}\| = \mathbf{o}(\mathbf{x} - \mathbf{b}).$$

   Hence $\mathbf{g}$ is differentiable, with $\mathbf{g}' = \mathbf{O}$; necessarily $\mathbf{g}$ is constant.

   (c) Assume $\|\mathbf{g}'\| \le M$. We apply the mean value theorem separately to the components of $\mathbf{g}$. We get

$$\begin{aligned} \|\mathbf{g}(x) - \mathbf{g}(\mathbf{y})\|^2 &= [g_1(\mathbf{x}) - g_1(\mathbf{y})]^2 + \cdots + [g_n(\mathbf{x}) - g_n(\mathbf{y})]^2 \\ &= [g_1'(\mathbf{c}_1)\langle\mathbf{x} - \mathbf{y}\rangle]^2 + \cdots + [g_n'(\mathbf{c}_n)\langle\mathbf{x} - \mathbf{y}\rangle]^2 \\ &\le \|g_1(\mathbf{c}_1)\|^2\|\mathbf{x} - \mathbf{y}\|^2 + \cdots + \|g_n(\mathbf{c}_n)\|^2\|\mathbf{x} - \mathbf{y}\|^2 \end{aligned}$$

   (operator inequality) $\le nM^2\|\mathbf{x} - \mathbf{y}\|^2$.

   (d) By properties of continuous functions, there is a neighborhood of $\mathbf{b}$ in which $\mathbf{g}'(\mathbf{x})$ is bounded. Apply part (c) there.

7. (a) any constant.

   (b) any discontinuous function.

(c) $f(t) \equiv \sqrt{|t|}$. If s and t are on the same side of zero, say $0 \le t < s$, then $f(s) - f(t) = \sqrt{s} - \sqrt{t} \le \sqrt{s-t}$, because

$$\left(\sqrt{s} - \sqrt{t}\right)^2 = s - 2\sqrt{st} + t \le s - 2\sqrt{tt} + t = s - t,$$

and similarly if $s < t \le 0$. If they are on opposite sides, $t < 0 < s$, then

$$|f(s) - f(t)| = |\sqrt{s} - \sqrt{-t}| \le \sqrt{|s - -t|} \le \sqrt{s-t},$$

because $|s + t| \le |s| + |t| = s - t$. This proves that f is Lipschitz of degree $\frac{1}{2}$. It is not Lipschitz of degree $\frac{1}{2} + \varepsilon$, because

$$\frac{f(s) - f(0)}{s^{1/2+\varepsilon}} = s^{-\varepsilon} \to \infty \qquad \text{as } s \to 0.$$

(d) $g(x) \equiv |x|$ is Lipschitz, because the slope of every secant is between -1 and 1: $\frac{|g(x)-g(y)|}{|x-y|} \le 1$. But $g'(0)$ is undefined.

(e) $h(x) \equiv \frac{1}{1-x}$. A Lipschitz function on a bounded domain has to be bounded (Justify!), and h is not. But for any $b \in (-1, 1)$, h' is (continuous and therefore) bounded on $\left[\frac{-1+b}{2}, \frac{1+b}{2}\right]$, so h is Lipschitz on this latter interval.

(f) Part (c) suggests that t^ε is Lipschitz of degree ε and no higher; you should verify that principle. What we need is an arbitrarily small power of t. This points to $t^{1/t}$, but $t^{1/t}$ does not work. At infinity, $\ln t$ is smaller than any power of t, so we try $f(t) \equiv \frac{1}{\ln|t/2|}$. Since $f \to 0$ as $t \to 0$, we set $f(0) = 0$. By L'Hospital's rule,

$$\frac{f(t) - f(0)}{t^\varepsilon} = \frac{t^{-\varepsilon}}{\ln t - \ln 2} \approx \frac{-\varepsilon t^{-\varepsilon-1}}{t^{-1}} \approx -\infty$$

as $t \to 0^+$, so f is not Lipschitz of any degree.

8. Set $f(x, y) \equiv 0$ if $y < 0$, $\equiv y^2$ if $0 \le y \le 2$ and $x > 0$, $\equiv -y^2$ if $0 \le y \le 2$ and $x < 0$. The domain has the points off the nonnegative y-axis with $y \le 2$. $\frac{\partial f}{\partial x}$ is always 0; $\frac{\partial f}{\partial y}$ is 0 on or below the x-axis, $\pm 2y$ above; both partials are continuous, and $\|f'\| \le 4$. But $f(\varepsilon, 1) - f(-\varepsilon, 1) = 1 - (-1) = 2$, even though the points are close together.

(b) Set $g(x) \equiv x^{3/2} \sin\left(\frac{1}{x}\right)$ on $(-1, 1)$, $g(0) \equiv 0$. Then g' is defined on the interval and continuous except at 0, but unbounded. Specifically, $g'(x)$ has arbitrarily large values near $x = 0$. Thus, given M, there will exist intervals $[s, t]$ throughout which $g'(x) > M$, forcing

$$|g(t) - g(s)| = g'(t^*)|t - s| > M[t - s].$$

Hence g is not Lipschitz in any neighborhood of $x = 0$.

9. Assume that $N(\mathbf{b}, r)$ and $N(\mathbf{c}, s)$ have $\mathbf{d}$ is common. Then $\|\mathbf{b}-\mathbf{c}\| \le \|\mathbf{b}-\mathbf{d}\| + \|\mathbf{d}-\mathbf{c}\| < r+s$, as the hint says. If $\mathbf{x} = (1-t)\mathbf{b} + t\mathbf{c}$ is on the segment, then

$$\|\mathbf{x}-\mathbf{b}\| + \|\mathbf{x}-\mathbf{c}\| = t\|\mathbf{c}-\mathbf{b}\| + (1-t)\|\mathbf{b}-\mathbf{c}\| = \|\mathbf{b}-\mathbf{c}\| < r+s.$$

Of necessity, either $\|\mathbf{x}-\mathbf{b}\| < r$, making $\mathbf{x} \in N(\mathbf{b}, r)$, or $\|\mathbf{x}-\mathbf{c}\| < s$, making $\mathbf{x} \in N(\mathbf{c}, s)$; either way, $\mathbf{x}$ is in their union.

10. (a) Trivial; they are continuous functions of z.

(b) Write $F(x, y) \equiv \int_0^1 f(x, y, z)\,dz$. By the mean value theorem,

$$\frac{F(x+t, y) - F(x, y)}{t} - \int_0^1 \frac{\partial f}{\partial x}(x, y, z)\,dz$$
$$= \int_0^1 \left[\frac{\partial f}{\partial x}(x+t^*, y, z) - \frac{\partial f}{\partial x}(x, y, z)\right] dz.$$

Since $\frac{\partial f}{\partial x}$ is uniformly continuous on the cube,

$$t \to 0 \Rightarrow t^* \to 0 \Rightarrow \frac{\partial f}{\partial x}(x+t^*, y, z) \to \frac{\partial f}{\partial x}(x, y, z)$$

uniformly. Hence $\frac{F(x+t,y)-F(x,y)}{t} \to \int_0^1 \frac{\partial f}{\partial x}(x, y, z)\,dz$. This proves that the integral is $\frac{\partial F}{\partial x}$.

(c) $\frac{\partial F}{\partial x}$ is continuous, because

$$\int_0^1 \frac{\partial f}{\partial x}(x+s, y+t, z)\,dz - \int_0^1 \frac{\partial f}{\partial x}(x, y, z)\,dz$$
$$\le \max\left(\frac{\partial f}{\partial x}(x+s, y+t, z) - \frac{\partial f}{\partial x}(x, y, z)\right) \to 0$$

as $(s, t) \to (0, 0)$ by the uniform continuity of $\frac{\partial f}{\partial x}$, and analogously for $\frac{\partial F}{\partial y}$; therefore, F is differentiable.

Section 2.3

1. You should be able to figure out the critical points and their nature by inspection. Then match your conclusions against the prescribed method.

(a) $f'(x, y) = [2x\ 2y]$, always defined, $= \mathbf{O}$ at the origin. Outward directional at (x, y) is

$$\frac{[2x\ 2y][x\ y]^t}{\|(x, y)\|} = 2(x^2+y^2)^{1/2} > 0.$$

Origin is a strict minimum.

(b) $\nabla g = (2xy^2, 2x^2y) = \mathbf{O}$ everywhere on the axes. Near $(a, 0)$, outward directional is given by

$$(2xy^2, 2x^2y) \bullet \frac{(x-a, y)}{([x-a]^2 + y^2)^{1/2}} = \frac{(4x^2y^2 - 2axy^2)}{([x-a]^2 + y^2)^{1/2}} \approx \frac{2a^2y^2}{|y|} \geq 0$$

when $x \approx a$. Hence the x-axis is all loose minima; symmetrically, so is the y-axis.

(c) $\nabla h = \left(\frac{x}{r}, \frac{y}{r}\right)$, never zero, undefined at origin. Outwards are

$$\left(\frac{x}{r}, \frac{y}{r}\right) \bullet \frac{(x, y)}{r} = 1 > 0;$$

strict minimum.

(d) $\nabla F = (4x^3y^3, 3x^4y^2) = \mathbf{O}$ on the axes. Near $(a, 0)$, outwards have numerator

$$4x^3y^3(x-a) + 3x^4y^3 \approx 3a^4y^3.$$

They are positive or zero above the x-axis, ≤ 0 below. Therefore, along any line through $(a, 0)$, there are higher or equal values above the x-axis, lower or equal below. This means that the point is never a strict maximum or strict minimum along a line; it is not a maximum, minimum, or saddle.

Near $(0, b)$, $b \neq 0$, outwards have numerator

$$4x^3y^3x + 3x^4y^2(y-b) \approx 4x^4b^3.$$

This makes them positive to the right and left of the positive y-axis, but zero on the axis; hence the positive y-axis is all loose minima. By similar reasoning, the negative y-axis is all loose maxima.

(e) $\nabla G = (2x, -2y) = \mathbf{O}$ at the origin. Outwards are equal to

$$(2x, -2y) \bullet \frac{(x, y)}{r} = \frac{2\left(x^2 - y^2\right)}{r}.$$

They are positive in the cone $|y| < |x|$, negative in the conjugate cone $|y| > |x|$. Hence the origin is a minimum along the line $y = 0$, a maximum along $x = 0$; origin is a saddle.

(f) $\nabla H = \left(3x^2y^3, 3x^3y^2\right) = \mathbf{O}$ on the axes. Near the origin, outwards are equal to

$$\left(3x^2y^3, 3x^3y^2\right) \bullet \frac{(x, y)}{r} = \frac{6x^3y^3}{r}.$$

They are positive in quadrants I and III, negative in II and IV; origin is a saddle. Near $(a, 0)$, $a \neq 0$, outwards have numerator $3x^2y^3(x - a) + 3x^3y^2y \approx 3a^3y^3$. They are zero along the x-axis, have one sign above the axis, opposite sign below. Hence $(a, 0)$, and likewise $(0, a)$, is not a strict extreme along any line; no minimum, no maximum, no saddle.

(g) By Exercise 1.2:4d, $K'(x)$ is undefined on the axes, except that $K'(0,0) = \mathbf{O}$. Near $(a, 0)$, outwards are equal to

$$\left(\frac{2x^{-1/3}y^{2/3}}{3}, \frac{2x^{2/3}y^{-1/3}}{3}\right) \bullet \frac{(x-a, y)}{\left([x-a]^2 + y^2\right)^{1/2}}$$
$$\approx \frac{2a^{2/3}y^{2/3}}{3\left([x-a]^2 + y^2\right)^{1/2}} \geq 0;$$

point is loose minimum, and likewise $(0, a)$.

2. Start with one variable. What is needed is a function that oscillates between, say, $f = 2x^2$ and $g = 2x^4$ near $x = 0$. That suggests $h(x) \equiv \left(x^2 + x^4\right) - \left(x^2 - x^4\right)\left(\sin \frac{1}{x}\right)$ for $0 < |x| \leq 1$, with $h(0) \equiv 0$. Clearly, $h(x) \geq \left(x^2 + x^4\right) - \left(x^2 - x^4\right) = 2x^4 > 0$, except when $x = 0$, so 0 is a strict minimum. The outward directional toward the right is

$$h'(x) = 2x + 4x^3 - \left(2x - 4x^3\right)\sin\frac{1}{x} + (1 - x^2)\cos\frac{1}{x},$$

which has alternating signs at $x = \frac{1}{\pi}, \frac{1}{2\pi}, \ldots$. (Verify!) We need only set $G(x, y) \equiv h(\sqrt{x^2 + y^2})$.

Section 2.4

1. (a) For f, Hessian $= \begin{bmatrix} 2 & 0 \\ 0 & 2 \end{bmatrix}$; for g, $\begin{bmatrix} 2 & 0 \\ 0 & -2 \end{bmatrix}$.

   (b) For f, $\Delta_1 = 2$, $\Delta_2 = 4$. For g, $\Delta_1 = 2$, $\Delta_2 = -4$.

2. Here $h'(x) = [y \ x]$, so $(0, 0)$ is the only critical point.

$$h'' = [[0 \ \ 1] \ [1 \ \ 0]],$$

so

$$h''(0,0)\langle \mathbf{v}\rangle^2 = \left[[0 \ \ 1]\begin{bmatrix} v_1 \\ v_2 \end{bmatrix} \ \ [1 \ \ 0]\begin{bmatrix} v_1 \\ v_2 \end{bmatrix}\right]\begin{bmatrix} v_1 \\ v_2 \end{bmatrix} = 2v_2v_1$$

is positive for $\mathbf{v} = (1, 1)$ and negative for $\mathbf{v} = (-1, 1)$. Theorem 2 says that $(0, 0)$ is a saddle.

3. (a)

$$F' = [\cos x \sin y \ \ \sin x \cos y],$$
$$F'' = \Big[\big[-\sin x \sin y \ \ \cos x \cos y\big]\big[\cos x \cos y \ \ -\sin x \sin y\big]\Big].$$

Two kinds of critical points: where x and y are both odd multiples of $\frac{\pi}{2}$; and where x and y are both even multiples of $\frac{\pi}{2}$ = multiples of π.

(b) At the first kind, $\sin x$ and $\sin y$ are ± 1, so their product is as big or small as possible; we expect extremes. At the others, $\sin x = \sin y = 0$; the changing signs should produce saddles.

For the first kind, let k be odd. At the places

$$\left(\frac{k\pi}{2}, \frac{k\pi}{2}\right), \left(\frac{k\pi}{2}, \frac{[k \pm 4]\pi}{2}\right), \ldots,$$

$F'' = [[-1\ 0]\ [0\ -1]]$, so $F''(0, 0)\langle \mathbf{v} \rangle^2 = -v_1^2 - v_2^2 < 0$. Therefore, these places are maxima. Between those are the places

$$\left(\frac{k\pi}{2}, \frac{[k \pm 2]\pi}{2}\right), \left(\frac{k\pi}{2}, \frac{[k \pm 6]\pi}{2}\right), \ldots.$$

At those,

$$F''(0, 0)\langle \mathbf{v} \rangle^2 = [[1\ 0]\ [0\ 1]]\, \langle (v_1, v_2) \rangle^2 = v_1^2 + v_2^2 > 0;$$

they are minima.

At the second kind of critical point, $F'' = \big[[0\ \pm 1]\ [\pm 1\ 0]\big]$ (independent signs), so $F''(0, 0)\langle \mathbf{v} \rangle^2 = \pm 2 v_1 v_2$. For either sign, the product is positive for some $\mathbf{v}$, negative for some; all saddles.

4. (a) $f(x, y) \equiv x^4 + y^4$; clearly strict minimum.

(b) $F(x, y) \equiv x^2 y^2$; has minimal value everywhere on the two axes.

(c) $g(x, y) \equiv x^3 y^3$ has minimum along the line $y = x$, maximum along $y = -x$.

(d) $G(x, y) \equiv x^3 + y^3$. Along every line through the origin, the values on one side are the negatives of those on the other. Hence no line offers a strict maximum or minimum.

5. (a) $h(x, y) \equiv x^2 + y^4$. Clearly, $\mathbf{O}$ is a strict minimum.

$$h''(0, 0) = [[2\ 0]\ [0\ 0]],$$

so $h''(0, 0)\langle \mathbf{v} \rangle^2 = 2v_1^2 \geq 0$ always, with $h''(0, 0)\langle (1, 0) \rangle^2 = 2$, $h''(0, 0)\langle (0, 1) \rangle^2 = 0$.

(b) $H(x, y) \equiv x^2$; loose minimum, everything else like (a).

(c) $k(x, y) \equiv x^2 - y^4$; minimum along the x-axis, maximum along the y-axis, all else like (a).

(d) $K(x, y) \equiv x^2 + y^3$. Along the y-axis, K has positive values above the origin, negative below. Hence $(0, 0)$ is not an extreme. Along $y = mx$, $K(x, y) = x^2(1 + m^3 x)$ always has a strict minimum at $(0, 0)$. Therefore no line gives a strict maximum, and $(0, 0)$ is not a saddle.

6. (a) Let $\mathbf{b} \in N(\mathbf{a}, \delta)$. Write $\mathbf{x}(t) \equiv [1-t]\mathbf{a} + t\mathbf{b}$ and $g(t) \equiv f(\mathbf{x}(t))$, $|t| < \frac{\delta}{\|\mathbf{b}-\mathbf{a}\|}$. In view of the hint, if suffices to show that g has $k+1$ derivatives in its domain, with

$$\frac{d^j g}{dt^j} = f^{(j)}(\mathbf{x}(t))\langle \mathbf{b}-\mathbf{a}\rangle^j.$$

The proof is a straightforward induction (compare the proof of Theorem 2.4:1):

$$\begin{aligned}
f^{(j+1)}(\mathbf{x}(t))\langle \mathbf{b}-\mathbf{a}\rangle^{j+1} &\equiv \left[f^{(j+1)}(\mathbf{x}(t))\langle \mathbf{b}-\mathbf{a}\rangle\right]\langle \mathbf{b}-\mathbf{a}\rangle^j \\
&= \left[f^{(j+1)}(\mathbf{x}(t))\mathbf{x}'(t)\right]\langle \mathbf{b}-\mathbf{a}\rangle^j \\
&= \frac{d[f^{(j)}(\mathbf{x}(t))]}{dt}\langle \mathbf{b}-\mathbf{a}\rangle^j \qquad \text{(chain rule)} \\
&= \frac{d[f^{(j)}(\mathbf{x}(t))\langle \mathbf{b}-\mathbf{a}\rangle^j]}{dt} \qquad \text{(operator algebra).}
\end{aligned}$$

(b) At $(0,0)$, $F = xe^y = 0$, $F' = \left[e^y \; xe^y\right] = \left[1 \; 0\right]$,

$$\begin{aligned}
F'' &= \left[[0 \; e^y] \; [e^y \; xe^y]\right] = \left[[0 \; 1] \; [1 \; 0]\right], \\
F''' &= \left[\left[[0 \; 0] \; [0 \; e^y]\right]\left[[0 \; e^y] \; [e^y \; xe^y]\right]\right] \\
&= \left[\left[[0 \; 0] \; [0 \; 1]\right] \; \left[[0 \; 1] \; [1 \; 0]\right]\right].
\end{aligned}$$

The polynomial is $0 + F'\langle(x,y)\rangle + F''\frac{\langle(x,y)\rangle^2}{2!} + F'''\frac{\langle(x,y)\rangle^3}{3!} = x + \frac{xy+yx}{2!} + \frac{xy^2+y[xy+yx]}{3!} = x\left(1 + y + \frac{y^2}{2!}\right)$.

(c) It is easy to establish the following pattern in part (b): for $\mathbf{c} \equiv (s,t)$,

$$f^{(k+1)}(\mathbf{c}) = \left[\left[\cdots [0 \; 0] \; \cdots \; [0 \; e^t]\right] \; f^{(k)}(\mathbf{c})\right].$$

Hence

$$\begin{aligned}
&f^{(k+1)}(\mathbf{c})\langle(x,y)\rangle^{k+1} \\
&= \left[\left[\cdots [0\;0] \cdots [0 \; e^t]\right]\langle(x,y)\rangle^k \; f^{(k)}(\mathbf{c})\langle(x,y)\rangle^k\right]\langle(x,y)\rangle \\
&= \left[e^t y^k \quad f^{(k)}(\mathbf{c})\langle(x,y)\rangle^k\right]\begin{bmatrix} x \\ y \end{bmatrix} \\
&= xe^t y^k + yf^{(k)}(\mathbf{c})\langle(x,y)\rangle^k.
\end{aligned}$$

Recursively, it follows that

$$f^{(k+1)}(\mathbf{c})\frac{\langle(x,y)\rangle^{k+1}}{(k+1)!} = \frac{(k+1)e^t xy^k + se^t y^{k+1}}{(k+1)!}.$$

Its absolute value is at most $e^t|x|\,|y|^k\frac{k+1+|y|}{(k+1)!}$, which for fixed (x, y) tends to 0 as $k \to \infty$.

(d) In the expression after "it follows" in part (c), set $s = t = 0$. We obtain

$$F(x, y) = 0 + \frac{1x}{1!} + \frac{2xy}{2!} + \frac{3xy^2}{3!} + \cdots = x\left(1 + y + \frac{y^2}{2!} + \cdots\right),$$

which is x times the series for e^y.

(e) Assume that f is twice differentiable near $\mathbf{b}$. With $f'(\mathbf{b}) = \mathbf{O}$, Taylor's theorem gives $f(\mathbf{x}) - f(\mathbf{b}) = \frac{f''(\mathbf{c})(\mathbf{x}-\mathbf{b})^2}{2!}$ for $\mathbf{x}$ in a neighborhood N, $\mathbf{c}$ corresponding to $\mathbf{x}$. Assume further that f'' is positive definite at $\mathbf{b}$. We have seen that there is a positive ε such that $f''(\mathbf{b})\langle\mathbf{v}\rangle^2 \geq 2\varepsilon$ for every unit vector $\mathbf{v}$. Finally, assume that f'' is continuous at $\mathbf{b}$. There is a neighborhood P of $\mathbf{b}$ in which $\|f''(\mathbf{y}) - f''(\mathbf{x})\| < \varepsilon$. Then for $\mathbf{x} \in N \cap P$,

$$\begin{aligned} f''(\mathbf{c})\langle\mathbf{v}\rangle^2 &= f''(\mathbf{b})\langle\mathbf{v}\rangle^2 - [f''(\mathbf{b}) - f''(\mathbf{c})]\langle\mathbf{v}\rangle^2 \\ &\geq 2\varepsilon - \|f''(\mathbf{b}) - f''(\mathbf{c})\|\,\|\mathbf{v}\|^2 > \varepsilon > 0. \end{aligned}$$

It follows that $f(\mathbf{x}) > f(\mathbf{b})$.

Section 2.5

1. (a) Anywhere but $(1, 2)$. $\frac{\partial(x-y^2+4y)}{\partial y} = 4 - 2y = 0$ along the line $y = 2$. Hence, away from $(1, 2)$, Theorem 1 guarantees a solution. Near that point, the equation is not functional: $x - y^2 + 4y = 5 \Leftrightarrow y = 2 \pm \sqrt{x-1}$, which guarantees two solutions for $x > 1$.

   (b) $\frac{dy}{dx} = \frac{-\partial F/\partial x}{\partial F/\partial y} = \frac{1}{4-2y}$.

   c) It is always possible: $x = y^2 - 4y + 5$. Either directly or by Theorem 2,

$$\frac{dx}{dy} = \frac{-\partial F/\partial y}{\partial F/\partial x} = 2y - 4.$$

2. $\frac{\partial e^{xy}}{\partial x} = ye^{xy} = 0$ on the x-axis. Hence you can solve for x on the rest of the graph, meaning the y-axis minus the origin. Obviously the solution is $x = 0$. Similarly, you can solve for y along the x-axis minus the origin. At the origin, you cannot solve for either variable; $x = 0$ gives multiple y's, and vice versa.

   (b) all the points in (a).

3. (a) Clearly, you can always solve: $y = \left(e^{-x} - x^3\right)^{1/3}$.

(b) That function is undifferentiable at the point where $e^{-x} - x^3 = 0$.

(c) Theorem 1:

$$\frac{-\partial f/\partial x}{\partial f/\partial y} = -\frac{e^x 3x^2 + e^x(x^3 + y^3)}{3y^2 e^x} = -\frac{3x^2 e^x + 1}{3y^2 e^x},$$

provided $y \neq 0$. Explicitly,

$$y' = \frac{1}{3}(e^{-x} - x^3)^{-2/3}(-e^{-x} - 3x^2) = \frac{1}{3y^2}\,\frac{-3x^2 e^x - 1}{e^x},$$

provided $y \neq 0$. They match.

4. (a) Write $F \equiv x^2 + y^2 - z^2$. You can solve $F = 0$ for x wherever $\frac{\partial F}{\partial x} = 2x \neq 0$.

(b) $\frac{\partial x}{\partial(y,z)} = \frac{-\partial F/\partial(y,z)}{\partial F/\partial x} = \frac{-[2y\ -2z]}{2x} = \left[-\frac{y}{x}\ \ \frac{z}{x}\right]$.

(c) No. Near any point where $x = 0$, the points $(z, 0, z)$ and $(-z, 0, z)$ both solve the equation; the relation is not functional with respect to x.

5. Let $F(x, y) \equiv g(x) - y$. Clearly, $\frac{\partial F}{\partial x} = g'(x)$, $\frac{\partial F}{\partial y} = -1$. We always have

$$\begin{aligned} F(x, y) - F(s, t) - \left[g'(s)\ \ -1\right]\begin{bmatrix} x - s \\ y - t \end{bmatrix} &= g(x) - y - g(s) + t - g'(s)(x - s) + y - t \\ &= \left[\frac{g(x) - g(s)}{x - s} - g'(s)\right](x - s), \end{aligned}$$

or zero if $x = s$. If g is differentiable at $x = s$, then this expression is $o(\|(x, y) - (s, t)\|)$. Hence $F(x, y)$ is a differentiable function of two variables at (s, t).

Now assume that g is differentiable near $x = b$. Then F is differentiable near $\big(b, g(b)\big)$. If g' is continuous at b, then F' is continuous at $\big(b, g(b)\big)$. Finally, $\frac{\partial F}{\partial y} \neq 0$ there. By Theorem 1, $F = 0$ can be solved for x: $g(x) - y = 0 \Leftrightarrow F(x, y) = 0 \Leftrightarrow x = h(y)$ in some box $b - \delta \leq x \leq b + \delta$, $g(b) - \varepsilon \leq y \leq g(b) + \varepsilon$; and

$$h'(y) = \frac{-\partial F/\partial(y)}{\partial F/\partial x} = \frac{1}{g'(x)}.$$

Section 2.6

1. (a) Vector description:

$$\mathbf{t} = (x', y', z') = (-2\sin\theta, 2\cos\theta, e^\theta) = \left(-\sqrt{3}, 1, e^{\pi/3}\right);$$

then the line is $\left(1, \sqrt{3}, e^{\pi/3}\right) + \langle\langle(-\sqrt{3}, 1, e^{\pi/3})\rangle\rangle$. Parametric equations: $\frac{-(x-1)}{\sqrt{3}} = \frac{y-\sqrt{3}}{1} = \frac{z-e^{\pi/3}}{e^{\pi/3}}$.

(b) $\mathbf{t} = \left(\frac{dx}{dx}, \frac{dy}{dx}\right) = (1, 2x) = (1, 4)$; the line is $(2, 4) + \langle\langle(1, 4)\rangle\rangle$; the equations are $\frac{x-2}{1} = \frac{y-4}{4}$, or $y - 4 = 4(x - 2)$.

2. (a) Write $F(x, y, z) \equiv x^2 + y^2 - z$. Then $\nabla F = (2x, 2y, -1)$ is never $\mathbf{O}$, so the surface is smooth. The tangent plane is given by (Example 5)

$$\begin{aligned} 0 &= (\mathbf{x} - \mathbf{b}) \bullet \nabla F(b) = (x - a, y - b, z - c) \bullet (2a, 2b, -1) \\ &= 2ax + 2by - z - c, \end{aligned}$$

because $c = a^2 + b^2$.

(b) $G(x, y, z) \equiv x^2 + y^2$ has $\nabla G = (2x, 2y, 0)$. The gradient is never zero on the surface, because $\|\nabla G\|^2 = 4x^2 + 4y^2 = 16$; the graph is smooth. The plane is given by

$$0 = (x - a, y - b, z - c) \bullet (2a, 2b, 0) = 2a(x - a) + 2b(y - b),$$

or $ax + by = 4$.

(c) $H(x, y, z) \equiv x^2 + y^2 + z^2$ has $\|\nabla H\|^2 = \|(2x, 2y, 2z)\|^2 = 4r^2 \neq 0$, smooth. The tangent plane equation is

$$\begin{aligned} 0 &= (x - a, y - b, z - c) \bullet (2a, 2b, 2c) \\ &= 2ax + 2by + 2cz - 2(a^2 + b^2 + c^2), \end{aligned}$$

or $ax + by + cz = r^2$.

(d) The given parametrization is undifferentiable where $x = y$, but it is equivalent to $z^3 = x - y$. The latter is smooth, because $\nabla(x - y - z^3) = (1, -1, -3z^2)$. The tangent plane has

$$\begin{aligned} 0 &= (x - a) - (y - b) - 3c^2(z - c) = x - y - 3c^2 z - a + b + 3c^3 \\ &= x - y - 3(a - b)^{2/3} z + 2(a - b). \end{aligned}$$

The plane is vertical wherever $a = b$.

3. The given vector has $y = 2x$, so we try the image of the line $y = 2x \subseteq xy$-plane. Thus, $x = x$, $y = 2x$, $z = x^2 + y^2 = 5x^2$. It crosses the origin, with tangent vector

$$\left(\frac{dx}{dx}, \frac{dy}{dx}, \frac{dz}{dx}\right) = (1, 2, 0).$$

4. Solve for y: $y = x + z^2 - 1$. By Example 4, $\left(1, \frac{\partial y}{\partial x}, 0\right) = (1, 1, 0)$ and $\left(0, \frac{\partial y}{\partial z}, 1\right) = (0, 2c, 1)$ make a basis. Reconcile the geometry: $z^2 = 1 - (x - y)$ is a parabolic cylinder ruled by lines parallel to its spine, which is the line $x = y$, $z = 1$; therefore, the direction of $\mathbf{i} + \mathbf{j}$ is always tangent to the cylinder.

5. Write $G(x, y) \equiv F(x) - y$. By hypothesis, $\nabla G = (F'(x), -1)$ is continuous, and it is never $\mathbf{O}$. By Theorem 3, $\nabla G(a, F(a))$ is normal to the graph of $G = 0$. Hence the tangent is given by

$$0 = (x - a, y - F(a)) \bullet (F'(a), -1),$$

or $y - F(a) = F'(a)(x - a)$.

6. (a) By the definition of curve, $\mathbf{g}$ is differentiable. Therefore,

$$\frac{\mathbf{g}(s) - \mathbf{g}(t)}{\|\mathbf{g}(s) - \mathbf{g}(t)\|} = \frac{\mathbf{g}'(s^*)(s - t)}{\|\mathbf{g}'(s^*)\|\,|s - t|},$$

assuming that the division is legal. If $\mathbf{g}$ is smooth at $\mathbf{g}(t)$, then by definition $\mathbf{g}'$ is continuous and nonzero at t, so $\mathbf{g}'$ *stays* nonzero nearby, making the division legal. Set $\mathbf{u} \equiv \frac{\mathbf{g}'(s^*)}{\|\mathbf{g}'(s^*)\|}$. Then

$$\frac{\mathbf{g}'(s^*)(s - t)}{\|\mathbf{g}'(s)\|\,|s - t|} \to \mathbf{u}$$

as $s \to t^+$, $\to -\mathbf{u}$ as $s \to t^-$, and those are the conclusions sought.

(b) Let $\mathbf{g}(s) = \left(x(s), x(s)^{2/3}\right)$ and $(0, 0) = \mathbf{g}(t)$. By the one-to-one assumption, $x(t^+)$ and $x(t^-)$ are of opposite signs, say $x(t^+) > 0 > x(t^-)$. Then for $s \to t^+$, we have $\mathbf{g}(s) - \mathbf{g}(t) = x(s)^{2/3}(x(s)^{1/3}, 1)$, whose limiting direction is $(0, 1)$. For $s \to t^-$, the forward secant is $\mathbf{g}(t) - \mathbf{g}(s) = -x(s)^{2/3}(x(s)^{1/3}, 1)$. Since the scalar $x(s)^{2/3}$ remains positive, this secant lines up along $-(0, 1)$.

(c) Argue as in (b) with $\mathbf{g}(s) = (x(s), |x(s)|)$. The secant on the t^+ side is $(x(s), |x(s)|) = x(s)(1, 1)$; it is constantly along the line of $(1, 1)$. On the t^- side, $|x(s)| = -x(s)$, and the secant is $(x(s), |x(s)|) = x(s)(1, -1)$, different line.

7. The dot products fit the following pattern:

$$\nabla f \bullet \left(1, 0, \dots, 0, \frac{\partial x_n}{\partial x_1}\right) = \left(\frac{\partial f}{\partial x_1}, \dots, \frac{\partial f}{\partial x_n}\right) \bullet \left(1, 0, \dots, 0, \frac{-\partial f/\partial x_1}{\partial f/\partial x_n}\right)$$

$$= \frac{\partial f}{\partial x_1} - \frac{\partial f}{\partial x_1} = 0.$$

Chapter 3

Section 3.1

1. The equation $x = \sqrt{1+x^2}$ implies $x^2 = 1 + x^2$, impossible.

2. (a) One sample sequence:

$$\begin{aligned}\cos 100 &= -0.1736\ldots;\\ \cos(Ans.) &= -0.9999954073\ldots;\\ \cos(Ans.) &= -0.9998476966\ldots;\\ \cos(Ans.) &= -0.9998477415\ldots;\\ \cos(Ans.) &= -0.9998477415\ldots;\end{aligned}$$

fixed.

(b) $\cos 0° = 1 > 0$, while $\cos 1° = 0.9\ldots < 1$, so $\cos x° = x$ occurs between 0 and 1. There is a different fixed point because $\cos x°$ is not $\cos x$: $\cos x° = \cos\left(\frac{\pi x}{180}\right)$.

(c) From most starts, the iterations get out of $[-1, 1]$. For example, $\cos^{-1} 0.1 \approx 1.47$; similarly, $\cos^{-1} 0.9 = 0.45\ldots$, $\cos^{-1}(Ans.) \approx 1.1$. Once they exit, the result is *ERROR*. From a start near 0.739 RADIAN, the iterations slowly move apart. The reason is the mean value theorem:

$$\frac{|\cos^{-1}(Ans._2) - \cos^{-1}(Ans._1)||}{|Ans._2 - Ans._1|} = \frac{1}{\sqrt{1-t^2}} > 1.$$

(d) The RADIAN answer converges slowly to zero. The absolute value has to drop, because $|\sin x| < |x|$ for $x \neq 0$. But the drop is slow, because

$$\frac{|\sin(Ans.) - 0|}{|Ans. - 0|} = |\cos t| \to 1$$

as $Ans.$ descends to 0.

3. (a) For this function, $G(x) - 7 = 3 - \frac{21}{x} = \frac{3(x-7)}{x}$. Therefore,

$$\frac{x_{i+1} - 7}{x_i - 7} = \frac{G(x_i) - 7}{x_i - 7} = \frac{3(x_i - 7)/x_i}{x_i - 7} = \frac{3}{x_i}.$$

This says that x_{i+1} is on the same side of 7 as x_i is, and at a distance from 7 reduced by a factor $\frac{3}{x_i}$, no more than the fixed $\frac{3}{x_0} < 1$.

(b) A negative x_0 gives $x_1 = G(x_0) = 10 - \frac{21}{x_0} > 10$. If $0 < x_0 < 2.1$, then $x_1 = 10 - \frac{21}{x_0} < 0$, $x_2 = 10 - \frac{21}{x_1} > 10$. Either way, we are back in (a).

(c) G is one-to-one, with $G^{-1}(y) = \frac{21}{10-y}$. If you set x_0 at any of 0, $G^{-1}(0) = 2.1$, $G^{-1}(2.1) \approx 2.66, \ldots$, then you eventually iterate G down to 0. These numbers increase toward 3, because

$$\frac{3 - G^{-1}(y)}{3 - y} = \frac{3(3-y)/(10-y)}{3-y} = \frac{3}{10-y} < \frac{3}{7}.$$

(d) If $0 < x < 3$, then $3 - G(x) = -7 + \frac{21}{x} = \frac{7(3-x)}{x} > 2(3-x)$. In words, the distance from 3 more than doubles with each iteration. If x is on list (c), then the iterates reach zero; if not, then they pass zero, then immediately go over 10.

4. (a) Yes: $\|\mathbf{f}(\mathbf{x}) - \mathbf{f}(\mathbf{y})\| \le \|\mathbf{x} - \mathbf{y}\|$ implies that $\mathbf{x} \to \mathbf{y} \Rightarrow \mathbf{f}(\mathbf{x}) \to \mathbf{f}(\mathbf{y})$.

(b) No: $f(\mathbf{x}) \equiv \frac{\|\mathbf{x}\|}{2}$ is a contraction, because

$$|f(\mathbf{x}) - f(\mathbf{y})| = \frac{|\|\mathbf{x}\| - \|\mathbf{y}\||}{2} \le \frac{\|\mathbf{x} - \mathbf{y}\|}{2},$$

but f is not differentiable at $\mathbf{x} = \mathbf{O}$.

(c) With elementary tools, we can prove at least this: $\|\mathbf{f}'\| \le \sqrt{m}K$. Let $\mathbf{u}$ be a unit vector. From

$$\mathbf{f}'(\mathbf{x})\langle\mathbf{u}\rangle = \lim_{t \to 0}\left[\frac{\mathbf{f}(\mathbf{x} + t\mathbf{u}) - \mathbf{f}(\mathbf{x}) - \mathbf{f}'(\mathbf{x})\langle t\mathbf{u}\rangle}{t} + \frac{\mathbf{f}'(\mathbf{x})\langle t\mathbf{u}\rangle}{t}\right],$$

we deduce that $\|\mathbf{f}'(\mathbf{x})\langle\mathbf{u}\rangle\| = K$. Setting $\mathbf{u} \equiv \mathbf{e}_j$, we get $K \ge \mathbf{f}'(\mathbf{x})\langle\mathbf{e}_j\rangle$ $= \|$column j of $\mathbf{f}'(\mathbf{x})\|$, so that

$$\|\mathbf{f}'(\mathbf{x})\|^2 = \|\text{column } 1 \text{ of } f'(\mathbf{x})\|^2 + \cdots + \|\text{column } m \text{ of } \mathbf{f}'(x)\|^2 \le mK^2.$$

5. If $\mathbf{x}$ and $\mathbf{y}$ are both fixed points, then

$$\|\mathbf{x} - \mathbf{y}\| = \|\mathbf{f}(\mathbf{x}) - \mathbf{f}(\mathbf{y})\| \le K\|\mathbf{x} - \mathbf{y}\| \Rightarrow (1 - K)\|\mathbf{x} - \mathbf{y}\| \le 0.$$

Since $K < 1$, necessarily $\|\mathbf{x} - \mathbf{y}\| = 0$.

6. (a) $f_1(x) = 1 + 0$, $f_2(x) = 1 + \int_0^x 1\,dt = 1 + x$, $f_3(x) = 1 + \int_0^x (1 + t)\,dt = 1 + x + \frac{x^2}{2}, \ldots, f_k(x) = 1 + x + \cdots + \frac{x^{k-1}}{(k-1)!}$.

(b) The limit is clearly e^x. That function solves the initial value problem.

(c) The first five iterates are

$$g_1 = 1 + (\sin x), \quad g_2 = 1 + x + (1 - \cos x),$$

$$g_3 = 1 + x + \frac{x^2}{2} + (x - \sin x),$$

$$g_4 = 1 + x + \frac{x^2}{2} + \frac{x^3}{(2)(3)} + \left(\frac{x^2}{2} - 1 + \cos x\right),$$

$$g_5 = 1 + x + \frac{x^2}{2} + \frac{x^3}{3!} + \frac{x^4}{4!} + \left(\frac{x^3}{3!} - x + \sin x\right).$$

In these, the polynomial tends to e^x. The part in parentheses looks like either

$$\pm\left(\sin x - \left[x - \frac{x^3}{3!} + \frac{x^5}{5!} - \dots\right]\right)$$

or $\pm\left(\cos x - \left[1 - \frac{x^2}{2!} + \frac{x^4}{4!} - \dots\right]\right)$, both of which tend to 0.

Section 3.2

1. The calculation is tedious but straightforward. The result is $\det(\mathbf{h}') = 2$.

2. (a) $\frac{\partial(x,y,z)}{\partial(\rho,\theta,\phi)} = \begin{bmatrix} \cos\theta\sin\phi & -\rho\sin\theta\sin\phi & \rho\cos\theta\cos\phi \\ \sin\theta\sin\phi & \rho\cos\theta\sin\phi & \rho\sin\theta\cos\phi \\ \cos\phi & 0 & -\rho\sin\phi \end{bmatrix}$. See Answer 1.3:3.

   (b) It is straightforward to calculate the cofactors and determinant, so we will give a different view. Observe that

$$\frac{\partial(x,y,z)}{\partial(\rho,\theta,\phi)} \begin{bmatrix} 1 & 0 & 0 \\ 0 & \frac{1}{\rho}\sin\phi & 0 \\ 0 & 0 & \frac{1}{\rho} \end{bmatrix} = \begin{bmatrix} \cos\theta\sin\phi & -\sin\theta & \cos\theta\cos\phi \\ \sin\theta\sin\phi & \cos\theta & \sin\theta\cos\phi \\ \cos\phi & 0 & -\sin\phi \end{bmatrix}.$$

   The last matrix is **orthogonal**: Its rows are orthonormal, as are its columns. For such a matrix, the inverse is the transpose. Our equation says "Jacobian times diagonal = orthogonal." It follows that J^{-1} = diagonal times transpose = transpose with rows 2 and 3 multiplied by $\frac{1}{\rho\sin\phi}$, $\frac{1}{\rho}$, respectively. Thus,

$$J^{-1} = \begin{bmatrix} \cos\theta\sin\phi & \sin\theta\sin\phi & \cos\phi \\ \frac{-\sin\theta}{\rho\sin\phi} & \frac{\cos\theta}{\rho\sin\phi} & 0 \\ \frac{\cos\theta\cos\phi}{\rho} & \frac{\sin\theta\cos\phi}{\rho} & \frac{-\sin\phi}{\rho} \end{bmatrix},$$

   defined as long as $\rho\sin\phi \neq 0$; you have to stay off the z-axis.

   (c) For example,

$$\frac{\partial F}{\partial x} = \frac{\partial f}{\partial(\rho,\theta,\phi)}\frac{\partial(\rho,\theta,\phi)}{\partial x} = \begin{bmatrix} \frac{\partial f}{\partial\rho} & \frac{\partial f}{\partial\theta} & \frac{\partial f}{\partial\phi} \end{bmatrix} \begin{bmatrix} \cos\theta\sin\phi \\ \frac{-\sin\theta}{\rho\sin\phi} \\ \frac{\cos\theta\cos\phi}{\rho} \end{bmatrix}$$

$$= \frac{\partial f}{\partial\rho}\cos\theta\sin\phi - \frac{\partial f}{\partial\theta}\frac{\sin\theta}{\rho\sin\phi} + \frac{\partial f}{\partial\phi}\frac{\cos\theta\cos\phi}{\rho}.$$

3. (a) $\mathbf{g}(x,y,z) \equiv (x^3, y^3, z^3)$ is one-to-one, even though $\frac{\partial\mathbf{g}}{\partial(x,y,z)} = \mathbf{O}$ at $(0,0,0)$.

(b) Under the hypothesis, $\mathbf{f}^{-1}$ exists. Let $\mathbf{x}_t \equiv \mathbf{b} + t\mathbf{e}_1$, $\mathbf{y}_t \equiv \mathbf{f}(\mathbf{x}_t)$. Since $\mathbf{f}$ is continuous at $\mathbf{b}$, $\mathbf{y}_t \to \mathbf{f}(\mathbf{b})$ as $t \to 0$. We have

$$\left\| \frac{\mathbf{f}^{-1}(\mathbf{y}_t) - \mathbf{f}^{-1}(\mathbf{f}(\mathbf{b}))}{\|\mathbf{y}_t - \mathbf{f}(\mathbf{b})\|} \right\| = \left(\frac{\|\mathbf{f}(\mathbf{x}_t) - \mathbf{f}(\mathbf{b})\|}{\|\mathbf{x}_t - \mathbf{b}\|} \right)^{-1}.$$

(Why are all those differences nonzero?) Since the fraction in parentheses approaches $\mathbf{f}'(\mathbf{b})\langle \mathbf{e}_1 \rangle = 0$, the other fraction tends to ∞. Therefore, there cannot be an operator $\mathbf{L}$ with $\mathbf{f}^{-1}(\mathbf{y}_t) - \mathbf{f}^{-1}(\mathbf{f}(\mathbf{b})) \approx \mathbf{L}\langle \mathbf{y}_t - \mathbf{f}(\mathbf{b}) \rangle$; $\mathbf{f}^{-1}$ cannot be differentiable at $\mathbf{f}(\mathbf{b})$.

Section 3.3

1. The counterexample, as always, uses cubics: $x - y - u^3 + v^3 = 0 = x + y - u^3 - v^3$ iff $u = x^{1/3}$ and $v = y^{1/3}$, even though $\frac{\partial(f,g)}{\partial(u,v)} = \mathbf{O}$ at $(x, y, u, v) = \mathbf{O}$.

2. (a) $\begin{vmatrix} a & b & c \\ e & f & g \\ i & j & k \end{vmatrix} \neq 0$; d, h, l are immaterial.

   (b) Set $d = h = l = 0$ and all the coefficients $\equiv 1$.

3. (a) Write $f \equiv x^2 - y^2 - u$, $g \equiv x^2 + y^2 - v$. Then $\frac{\partial(f,g)}{\partial(x,y)} = 8xy$. Away from the axes, Theorem 1 applies and guarantees a solution. We always have $u + v = 2x^2$. Hence near any point on the y-axis, $x = \pm\left(\frac{u}{2} + \frac{v}{2}\right)^{1/2}$ gives distinct solutions in any neighborhood, and the relation is not functional in x; analogously for y near the x-axis.

   (b) The points $v =$ constant $c > 0$ constitute a circle $x^2 + y^2 = c$. The points $u =$ constant $b > 0$ make up a hyperbola $x^2 - y^2 = b$. Sketch such a circle and hyperbola. Along the latter, distance from the origin increases as you move away from the vertex. Therefore, if the two curves meet in a quadrant, then they are transversal. If instead they meet at the x-axis, then they are tangent; that is the bad situation. The same argument applies if $b < 0$, in which case the tangencies occur along the y-axis. If $b = 0$, then the hyperbola becomes its asymptotes, and the intersections are automatically transversal along $y = \pm x$.

   For $v = 0$, we no longer have a curve. Finally, $v < 0$ is not allowed.

4. The plane–cone intersection is a parabola given by $(x + 4)^2 = x^2 + y^2$, or $y^2 = 8x + 16$, along with $z = x + 4$. Where $y = 8$, we have $x = 6$. Hence the tangent vector is

$$\left(1, \frac{dy}{dx}, \frac{dz}{dx}\right) = \left(1, \frac{8}{2y}, 1\right) = (1, 0.5, 1).$$

The equation is $x - 6 = 2(y - 8) = z - 10$. Alternatively,

$$\begin{aligned} 0 &= (x-6, y-8, z-10) \bullet \nabla \left(x^2 + y^2 - z^2\right) \\ &= 12(x-6) + 16(y-8) - 20(z-10) \end{aligned}$$

plus $0 = (x-6, y-8, z-10) \bullet \nabla(x+4-z) = x - 6 - z + 10$, a system that reduces to $z =$ arbitrary, $y = 3 + \frac{z}{2}$, $x = z - 4$.

5. If $\left(\frac{\partial F_1}{\partial x_1}, \ldots, \frac{\partial F_1}{\partial x_n}\right), \ldots, \left(\frac{\partial F_k}{\partial x_1}, \ldots, \frac{\partial F_k}{\partial x_n}\right)$ are independent, then $\frac{\partial(F_1,\ldots,F_k)}{\partial(x_1,\ldots,x_n)}$ has k independent rows, the maximum possible. Hence it has k independent columns, say

$$\frac{\partial(F_1, \ldots, F_k)}{\partial x_1}, \ldots, \frac{\partial(F_1, \ldots, F_k)}{\partial x_k}.$$

The implicit function theorem guarantees that you can solve the system for

$$x_1 = g_1(x_{k+1}, \ldots, x_n), \ldots, x_k = g_k(x_{k+1}, \ldots, x_n).$$

This means that the system is equivalent to

$$\mathbf{x} = \mathbf{g}(x_{k+1}, \ldots, x_n) \equiv (g_1, \ldots, g_k, x_{k+1}, \ldots, x_n),$$

an $(n-k)$-surface.

6. By the argument in Theorem 2(a), we can solve

$$x_1 = g_1(t_1, \ldots, t_{n-k}), \ldots, x_n = g_n(t_1, \ldots, t_{n-k})$$

for n variables, meaning $t_1, \ldots, t_{n-k}$ and (say) $x_1, \ldots, x_k$. Then

$$\begin{aligned} x_1 &= g_1(\mathbf{t}) = g_1\big(\mathbf{h}(x_{k+1}, \ldots, x_n)\big), \ldots, \\ x_k &= g_k(\mathbf{t}) = g_k\big(\mathbf{h}(x_{k+1}, \ldots, x_n)\big) \end{aligned}$$

is the intersection of the k hypersurfaces

$$0 = f_j(\mathbf{x}) \equiv x_j - g_j\big(\mathbf{h}(x_{k+1}, \ldots, x_n)\big), \quad j = 1, \ldots, k.$$

Section 3.4

1. $\nabla f = (y, x) = \lambda \nabla(x + 2y - 5) = (\lambda, 2\lambda)$ becomes $y = \lambda$, $x = 5 - 2y = 2\lambda$, leading to $y = \frac{5}{4}$, $x = \frac{5}{2}$, $f = \frac{25}{8}$. At that point, the line $x + 2y = 5$ has slope $-\frac{1}{2}$, the hyperbola $xy = \frac{25}{8}$ has slope $\frac{-25}{8x^2} = \frac{-25}{50}$, so they are tangent.

2. To minimize $f \equiv \text{distance}^2 = (x-a)^2 + (y-b)^2$, set $\nabla f = (2x - 2a, 2y - 2b) = \lambda(c, d)$. The equation implies $x - a = \frac{\lambda c}{2}$, $y - b = \frac{\lambda d}{2}$, and $f = \frac{\lambda^2(c^2+d^2)}{4}$. (Notice that $(x-a, y-b)$ is in the line of (c, d), perpendicular to the original line.) Also, $\lambda c^2 + \lambda d^2 = 2(x-a)c + 2(y-b)d = 2e - 2ac - 2bd$, so $\lambda = \frac{2(e-ac-bd)}{c^2+d^2}$. Hence the minimal distance is $f = \frac{|e-ac-bd|}{\sqrt{c^2+d^2}}$.

3. $\nabla(x+2y+3z) = \lambda\nabla(x^2+y^2+z^2-1)$ gives $1 = 2\lambda x, 2 = 2\lambda y, 3 = 2\lambda z$. (These say that the radius vector (x, y, z) is perpendicular to the plane.) Divide to get $\frac{y}{x} = 2, \frac{z}{x} = 3$; then $x^2 + (2x)^2 + (3x)^2 = 1$ yields two candidates, $\frac{\pm(1,2,3)}{\sqrt{14}}$. The + gives max, − gives min.

4. From Answer 3.3:4, the intersection satisfies $y^2 = 8x + 16$ and $z = x + 4$. The lowest point there is $(-2, 0, 2)$. If you are at $(0, 0, 2)$ or lower on the z-axis, then it pays to stay in the xz-plane, so the answer should be $(-2, 0, 2)$. We examine:

   (a) $\nabla\left(x^2 + y^2 + (z-2)^2\right) = \lambda\nabla\left(x^2 + y^2 - z^2\right) + \mu\nabla(x+4-z)$ yields $2x = \lambda 2x + \mu$, $2y = \lambda 2y$, $2z - 4 = -\lambda 2z - \mu$. If $y \neq 0$, then we get in succession $\lambda = 1$, $\mu = 0$, $z = 1$, $x = -3$, y undefined. Hence necessarily $y = 0$, which specifies $x = -2$, $z = 2$.

   (b) Change the equations to $2x = \lambda 2x + \mu$, $2y = \lambda 2y$, $2z - 2b = -\lambda 2z - \mu$. As before, $y \neq 0$ implies $\lambda = 1$, $\mu = 0$, $z = \frac{b}{2}$, $x = \frac{b}{2} - 4$, $y = \pm\sqrt{4b - 16}$. Thus, if $b \leq 4$, we are again led to $y = 0$. In words, if $(0, 0, b)$ is on or below the plane of the parabola, then there is no reason to leave the xz-plane. If instead $b > 4$, then you travel to a place other than the vertex.

5. The intersection circle $z = \sqrt{5}$, $x^2 + y^2 = 4$ has vertical tangent planes. At the place $\left(\frac{-2}{\sqrt{10}}, \frac{-6}{\sqrt{10}}, \sqrt{5}\right)$, the normal to the tangent plane is the circle's radius vector $\left(\frac{-2}{\sqrt{10}}, \frac{-6}{\sqrt{10}}, 0\right)$. This vector is also normal to the plane $x+3y = \frac{-20}{\sqrt{10}}$.

6. By Lagrange's method, $f \equiv (x_1 - b_1)^2 + \cdots + (x_n - b_n)^2$ is least when

$$(2x_1 - 2b_1, \ldots, 2x_n - 2b_n) = \lambda_1\nabla f_1(\mathbf{x}) + \cdots + \lambda_k\nabla f_k(\mathbf{x}).$$

   The solution $\mathbf{x} = \mathbf{c}$ therefore satisfies

$$\mathbf{b} - \mathbf{c} = (b_1 - c_1, \ldots, b_n - c_n) = \frac{\lambda_1\nabla f_1(\mathbf{x}) + \cdots + \lambda_k\nabla f_k(x)}{(-2)};$$

   $\mathbf{b} - \mathbf{c}$ is in the orthogonal complement of the tangent plane.

Chapter 4

Section 4.1

1. The "unit cube" $\{\mathbf{x}\colon 0 \leq x_j \leq 1\}$ and its translate to the upper right $\{\mathbf{x}\colon 1 \leq x_j \leq 2\}$ have precisely $(1, 1, \ldots, 1)$ in common, a point on each boundary.

2. (a) $\mathbf{x} \in B \cap C \Leftrightarrow a_j \le x_j \le b_j$ and $c_j \le x_j \le d_j$ for each $j \Leftrightarrow x_j \ge \max\{a_j, c_j\}$ and $x_j \le \min\{b_j, d_j\}$ for each j.

   (b) In view of (a), $B \cap C$ has an interior point $\Leftrightarrow$ there exists $\mathbf{x}$ with $\max\{a_j, c_j\} < x_j < \min\{b_j, d_j\}$ for every $j \Leftrightarrow$ there is $\mathbf{x}$ with both $a_j < x_j < b_j$ and $c_j < x_j < d_j$ for every $j \Leftrightarrow$ there is $\mathbf{x} \in \text{int}(B) \cap \text{int}(C)$.

   (c) Assume that B and C do not overlap and $\mathbf{x} \in B \cap C$. The segment from $\mathbf{x}$ to $\frac{\mathbf{a}+\mathbf{b}}{2}$ is interior to B, except possibly for $\mathbf{x}$. If $\mathbf{x}$ were in int(C), then some neighborhood $N(\mathbf{x}, \delta)$ would be interior to C, so points near but unequal to $\mathbf{x}$ on the segment would be interior to B and C. Hence $\mathbf{x} \in \text{int}(C)$ is impossible. We conclude that $\mathbf{x} \in \text{bd}(C)$, and likewise with B.

   (d) A neighborhood and a line through its center do not overlap, because the line has no interior points, but their intersection consists of points interior to the neighborhood.

   (e) S and T overlap $\Leftrightarrow$ there is $\mathbf{x} \in \text{int}(S) \cap \text{int}(T)$ (definition) $\Leftrightarrow$ there are $N(\mathbf{x}, s) \subseteq S$ and $N(\mathbf{x}, t) \subseteq T \Leftrightarrow$ there is $N(\mathbf{x}, r)$ contained in both S and $T \Leftrightarrow$ there is $\mathbf{x} \in \text{int}(S \cap T)$.

3. In the picture,

$$
\begin{aligned}
V_1 &= (x_1 - x_0)(y_2 - y_0) = (x_1 - x_0)(y_1 - y_0) + (x_1 - x_0)(y_2 - y_1),\\
V_2 &= (x_2 - x_1)(y_2 - y_1),\\
V_3 &= (x_3 - x_1)(y_1 - y_0) = (x_2 - x_1)(y_1 - y_0) + (x_3 - x_2)(y_1 - y_0),\\
V_4 &= (x_3 - x_2)(y_3 - y_1) = (x_3 - x_2)(y_2 - y_1) + (x_3 - x_2)(y_3 - y_2),\\
V_5 &= (x_3 - x_2)(y_4 - y_3),\\
V_6 &= (x_2 - x_0)(y_4 - y_2) = (x_1 - x_0)(y_3 - y_2) + (x_1 - x_0)(y_4 - y_3)\\
&\quad + (x_2 - x_1)(y_3 - y_2) + (x_2 - x_1)(y_4 - y_3).
\end{aligned}
$$

   The twelve right-hand elements sum to $(x_3 - x_0)(y_4 - y_0) = V([\mathbf{a}, \mathbf{b}])$.

4. Suppose $T_k \subseteq S_j$. Since necessarily $T_k \subseteq C$, we have $T_k \subseteq S_j \cap C$. Hence the union of the T_k contained in S_j is a subset of $S_j \cap C$. To prove $S_j \cap C \subseteq$ union, let $\mathbf{x} \in S_j \cap C$. Since S_j and C overlap, Exercise 2b tells us that $S_j \cap C$ is a box. Hence $\mathbf{x} = \lim \mathbf{x}_i$ for some sequence $(\mathbf{x}_i)$ from $\text{int}(S_j \cap C) = \text{int}(S_j) \cap \text{int}(C)$ (Exercise 2e). Those $\mathbf{x}_i$ cannot come from T_k contained in other S_m. They must come from the T_k contained in S_j; that is, $\mathbf{x}$ is a closure point of the union. Since the union is closed, $\mathbf{x} \in$ union, and $S_j \cap C \subseteq$ union.

5. (a) Because B and C overlap, there must exist a neighborhood $N \subseteq B \cap C$. This N must have points interior to some S_j and some T_k. Hence $\mathcal{PR} = \{S_j \cap T_k\colon S_j \text{ and } T_k \text{ overlap}\}$ is nonempty. Any such intersection is a box, by Exercise 2b.

(b) If $S_j \cap T_k$ and $S_l \cap T_m$ overlap, then S_j overlaps S_l, forcing $j = l$, and similarly $k = m$.

(c) Suppose $\mathbf{x} \in B \cap C$. Then $\mathbf{x} \in B = S_1 \cup \cdots \cup S_J$, so that $\mathbf{x} \in$ some S_j. Let $(\mathbf{x}_i) \to \mathbf{x}$ from $\text{int}(S_j)$. An infinity of $\mathbf{x}_i$ must come from a single T_k. Since these points are common to S_j and T_k and are interior to S_j, Exercise 2c implies that S_j and T_k overlap. Also, they are in T_k and converge to $\mathbf{x}$. Since T_k is closed, $\mathbf{x} \in T_k$. We have shown that $\mathbf{x}$ is in some $S_j \cap T_k \in \mathcal{PR}$. Hence $B \cap C \subseteq$ union of the $S_l \cap T_m$. The opposite inclusion is trivial.

The product $\mathcal{PR}$ is the simplest refinement, in that if $\mathcal{Q}$ refines both $\mathcal{P}$ and $\mathcal{R}$, then $\mathcal{Q}$ refines $\mathcal{PR}$: If each Q_l is contained in some S_j and some T_k, then $Q_l \subseteq S_j \cap T_k$.

Section 4.2

1. On any subinterval S_J of any partition $\mathcal{P}$, $\sup f = \inf f = 1$. Therefore, $u(f, \mathcal{P}) = l(f, \mathcal{P}) = V(S_1) + \cdots + V(S_J) = V(B)$; f is integrable, with integral $V(B)$.

2. (a) In the subinterval from (x_{j-1}, y_{k-1}) to (x_j, y_k), $\sup g = \sup x = x_j$. Hence

$$u(f, \mathcal{P}) = \sum_{j=1}^{J} \sum_{k=1}^{K} x_j(x_j - x_{j-1})(y_k - y_{k-1}).$$

Clearly, the k-summation is

$$x_j(x_j - x_{j-1}) \sum_{k=1}^{K} (y_k - y_{k-1}) = 4x_j(x_j - x_{j-1}),$$

and $u(f, \mathcal{P}) = \sum_{1 \le j \le J} 4x_j(x_j - x_{j-1})$.

(b) The answer in (a) is a Riemann sum for the one-variable integral $\int_0^3 4x\,dx = 18$. If $\|\mathcal{P}\|$ approaches 0, then each $x_j - x_{j-1}$ approaches 0, so $u(f, \mathcal{P}) \to 18$.

(c) The box bounded by $x = 0, x = 3, y = 0, y = 4, z = 0, z = 3$, is cut into congruent triangular prisms by the plane $z = g(x, y) = x$. The box has volume $3(4)3 = 36$. Each prism therefore has volume 18.

3. The function is 1 on the squares from $\left(\frac{1}{2}, \frac{1}{2}\right)$ to $(1, 1)$, $\left(\frac{1}{4}, \frac{1}{4}\right)$ to $\left(\frac{1}{2}, \frac{1}{2}\right)$, $\left(\frac{1}{8}, \frac{1}{8}\right)$ to $\left(\frac{1}{4}, \frac{1}{4}\right)$, ... For the lower integral, consider the cross-partition $\mathcal{P}_3$ defined by the first three. Those three are shaded in the figure below, which also shows the $(3+1)^2$ subintervals of $\mathcal{P}_3$. Clearly, $\min h = 1$ on the shaded subintervals, $= 0$ on the others, so $l(h, \mathcal{P}_3) = \frac{1}{2^2} + \frac{1}{4^2} + \frac{1}{8^2}$. We can do the same with the first k squares, producing a partition $\mathcal{P}_k$ with $l(h, \mathcal{P}_k) = \frac{1}{2^2} + \cdots + \frac{1}{(2^k)^2}$. Therefore, $L(h, B) \ge \frac{1}{2^2} + \frac{1}{4^2} + \frac{1}{8^2} + \cdots = \frac{1}{3}$.

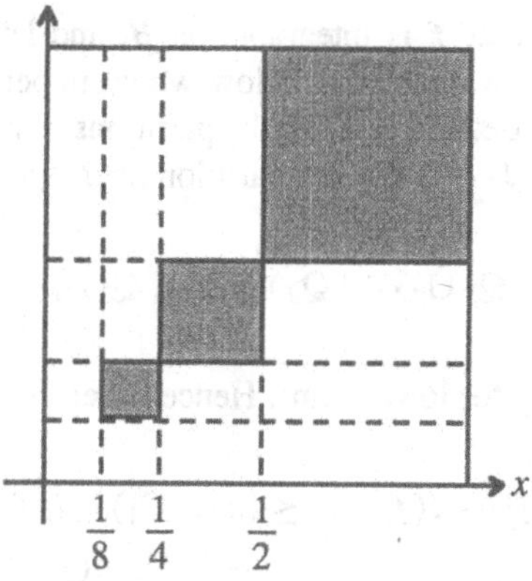

Exercise 3.

For the upper integral, let ε be small. Define $\mathcal{Q}_4$ to be the cross-partition determined by the boxes from $x = y = \frac{(1-\varepsilon)}{2}$ to $x = y = 1$, $x = y = \frac{(1-\varepsilon)}{4}$ to $x = y = \frac{(1+\varepsilon)}{2}$, $x = y = \frac{(1-\varepsilon)}{8}$ to $x = y = \frac{(1+\varepsilon)}{4}$, and $x = y = 0$ to $x = y = \frac{(1+\varepsilon)}{8}$. (These are overlapping boxes that contain the ones where h is nonzero.) On the subintervals of $\mathcal{Q}_4$ not contained in those four, h is identically 0. Hence

$$u(h, \mathcal{Q}_3) < 1\left(\frac{1+\varepsilon}{2}\right)^2 + 1\left(\frac{1+2\varepsilon}{4}\right)^2 + 1\left(\frac{1+2\varepsilon}{8}\right)^2 + 1\left(\frac{1+\varepsilon}{8}\right)^2,$$

the $<$ sign needed because the boxes overlap. Extending the construction to boxes surrounding the first k squares, we build $\mathcal{Q}_k$ with

$$\begin{aligned} u(h, \mathcal{Q}_k) &< 1\left(\frac{1+\varepsilon}{2}\right)^2 + \cdots + 1\left(\frac{1+2\varepsilon}{2^k}\right)^2 + 1\left(\frac{1+\varepsilon}{2^k}\right)^2 \\ &< (1+2\varepsilon)^2\left(\frac{1}{2^2} + \cdots + \frac{1}{2^{2k}} + \frac{1}{2^{2k}}\right) < (1+2\varepsilon)^2\left(\frac{1}{3} + \frac{1}{2^k}\right). \end{aligned}$$

It follows that $U(h, B) \le \frac{(1+2\varepsilon)^2}{3}$. Since this holds for arbitrarily small ε, we have $U(h, B) \le \frac{1}{3}$. Hence h is integrable, and $\int_B h = \frac{1}{3}$.

4. (a) If f and g are integrable, then their integrals are limits of Riemann sums. Then the Riemann sums for $\alpha f + \beta g$ satisfy

$$\sum[\alpha f + \beta g](x_j)V_j = \alpha \sum f(x_j)V_j + \beta \sum g(x_j)V_j \to \alpha \int_B f + \beta \int_B g.$$

This proves both assertions.

(b) If $f \ge g$, then $u(f, \mathcal{P}) \ge u(g, \mathcal{P})$ for every partition, forcing $\int_B f = U(f, B) \ge U(g, B) = \int_B f$.

(c) $\Rightarrow$ Assume that f is integrable on B, and let $\varepsilon > 0$. By Theorem 2(a), there is a fineness δ below which upper sums are within ε of lower sums. Let $\mathcal{Q}_1, \ldots, \mathcal{Q}_J$ be partitions of $S_1, \ldots, S_J$ finer than δ. Clearly, $\mathcal{Q}_1 \cup \cdots \cup \mathcal{Q}_J$ is a partition of B, finer than δ. Also,

$$u(f, \mathcal{Q}_1 \cup \cdots \cup \mathcal{Q}_J) = u(f, \mathcal{Q}_1) + \cdots + u(f, \mathcal{Q}_J),$$

and similarly for lower sums. Hence for each j,

$$\begin{aligned} u(f, \mathcal{Q}_j) - l(f, \mathcal{Q}_j) &\leq [u(f, \mathcal{Q}_1) - l(f, \mathcal{Q}_1)] \\ &\quad + \cdots + [u(f, \mathcal{Q}_J) - l(f, \mathcal{Q}_J)] \\ &= \sum u(f, \mathcal{Q}_j) - \sum l(f, \mathcal{Q}_j) < \varepsilon. \end{aligned}$$

We conclude that f is integrable on each S_j.

$\Leftarrow$ Assume that f is integrable on $S_1, \ldots, S_J$. Given $\varepsilon > 0$, there exist norms $\delta_1, \ldots, \delta_J$ below which partitions $\mathcal{Q}_1, \ldots, \mathcal{Q}_J$ of $S_1, \ldots, S_J$ have upper sums within $\frac{\varepsilon}{J}$ of the corresponding lower sums. Let $\mathcal{R}$ be a partition of B finer than $\min\{\delta_1, \ldots, \delta_J\}$. The partitions $\mathcal{R}\mathcal{Q}_1, \ldots, \mathcal{R}\mathcal{Q}_J$ are refinements of $\mathcal{Q}_1, \ldots, \mathcal{Q}_J$. Let $\mathcal{R}^+ \equiv \mathcal{R}\mathcal{Q}_1 \cup \cdots \cup \mathcal{R}\mathcal{Q}_J$. Then

$$\begin{aligned} u(f, \mathcal{R}^+) - l(f, \mathcal{R}^+) &= \sum u(f, \mathcal{R}\mathcal{Q}_j) - \sum l(f, \mathcal{R}\mathcal{Q}_j) \\ &= \sum [u(f, \mathcal{R}\mathcal{Q}_j) - l(f, \mathcal{R}\mathcal{Q}_j)] \\ &\leq \sum [u(f, \mathcal{Q}_j) - l(f, \mathcal{Q}_j)] \\ &\leq J \frac{\varepsilon}{J} = \varepsilon. \end{aligned}$$

By Theorem 2(a), f is integrable.

For the relation between the integrals, assume that f is integrable. Establish individual partitions $\mathcal{Q}_1, \ldots, \mathcal{Q}_J$ and $\mathcal{Q} \equiv \mathcal{Q}_1 \cup \cdots \cup \mathcal{Q}_J$. These have $u(f, \mathcal{Q}) = u(f, \mathcal{Q}_1) + \cdots + u(f, \mathcal{Q}_J)$. Letting the norms of $\mathcal{Q}_1, \ldots, \mathcal{Q}_J$ tend to 0, we have $u(f, \mathcal{Q}) \to \int_{S_1} f + \cdots + \int_{S_J} f$. That forces $U(f, B) \leq$ sum of integrals. Analogously, we establish $L(f, B) \geq$ sum of integrals, and equality follows.

(d) By extending the boundaries of B, we can create a partition $\{B, S_1, \ldots, S_J\}$ of D. By part (c), $\int_D f = \int_B f + \int_{S_1} f + \cdots + \int_{S_J} f$. The last J terms are nonnegative, by part (b). Hence $\int_D f \geq \int_B f$.

(e) Let $M \equiv \sup f$, $m \equiv \inf f$. Then $mV(B) = m \int_B 1$ (by Exercise 1) $= \int_B m$ (part (a)) $\leq \int_B f$ (part (b)) $\leq \int_B M = MV(B)$.

5. By linearity, $\int(\alpha x+\beta y+\delta z)=\alpha\int x+\beta\int y+\delta\int z$. From Exercise 2, it is clear that $\int_B x$ for a box B has upper sums

$$\begin{aligned}u(x,\mathcal{P}) &= \sum_{j=1}^{J}\sum_{k=1}^{K}\sum_{m=1}^{M} x_j(x_j-x_{j-1})(y_k-y_{k-1})(z_m-z_{m-1})\\ &= (b_2-a_2)(b_3-a_3)\sum_{j=1}^{J} x_j(x_j-x_{j-1})\\ &\to (b_2-a_2)(b_3-a_3)\int_{a_1}^{b_1} x\,dx.\end{aligned}$$

Since the last integral is $\frac{b_1^2}{2}-\frac{a_1^2}{2}=\frac{(b_1-a_1)(b_1+a_1)}{2}$, we see that $\int_B x = V(B)\frac{(b_1+a_1)}{2}$. By symmetry,

$$\begin{aligned}\int(\alpha x+\beta y+\delta z) &= V(B)\frac{(\alpha a_1+\alpha b_1+\beta a_2+\beta b_2+\delta a_3+\delta b_3)}{2}\\ &= V(B)(\alpha,\beta,\delta)\bullet\frac{(\mathbf{a}+\mathbf{b})}{2}.\end{aligned}$$

6. On every subinterval S_j, $\inf F=0$ and $\sup F=1$. Hence every upper sum is $\sum 1V(S_j)=V(B)$, every lower sum is $\sum 0V(S_j)=0$, and $U(F,B)>L(F,B)$.

7. (a) On any subinterval S_j, $|f(\mathbf{x})|-|f(\mathbf{y})|\le|f(\mathbf{x})-f(\mathbf{y})|\le\sup_j f-\inf_j f$ for any $\mathbf{x}$, $\mathbf{y}$. Therefore, $\sup_j|f|-\inf_j|f|\le\sup_j f-\inf_j f$. Summing over the subintervals and separating sups and infs, we obtain

$$u(|f|,\mathcal{P})-l(|f|,\mathcal{P})=u(f,\mathcal{P})-l(f,\mathcal{P}).$$

If f is integrable, then there are $\mathcal{P}$ with $u(f,\mathcal{P})-l(f,\mathcal{P})<\varepsilon$. Hence the same goes for $|f|$, and $|f|$ is integrable.

(b) Let F be Dirichlet's function on the unit square in $\mathbf{R}^2$, and $G\equiv F-\frac{1}{2}$. Then G is unintegrable, because every upper sum is $\frac{1}{2}$, every lower sum is $-\frac{1}{2}$. But $|G|=\frac{1}{2}$ everywhere.

8. Assume that f is continuous on B. Then f has a minimum $f(\mathbf{a})$ and a maximum $f(\mathbf{b})$. By Exercises 1 and 4, $f(\mathbf{a})=\int_B\frac{f(\mathbf{a})}{V(B)}\le\int_B\frac{f(\mathbf{x})}{V(B)}\le\int_B\frac{f(\mathbf{b})}{V(B)}=f(\mathbf{b})$. (Remember that $V(B)$ is required to be positive.) Thus, $\int_B\frac{f(\mathbf{x})}{V(B)}$ is between the extremes of f. By the intermediate value theorem, there exists $\mathbf{c}\in B$ such that $f(\mathbf{c})=\int_B\frac{f(\mathbf{x})}{V(B)}$.

9. By part (d), $l(f,\mathcal{P})\le u(f,\mathcal{R})$ for any $\mathcal{P}$ and $\mathcal{R}$. This says that $l(f,\mathcal{P})$ is a lower bound for $\{u(f,\mathcal{R})\}$, so $l(f,\mathcal{P})$ cannot exceed the infimum of the set: $l(f,\mathcal{P})\le\inf\{u(f,\mathcal{R})\}=U(f,B)$. This last says that $U(f,B)$ is an upper bound for $\{l(f,\mathcal{P})\}$. Therefore, $L(f,B)=\sup\{l(f,\mathcal{P})\}\le U(f,B)$.

Section 4.3

1. It is easy to prove, by induction on the dimension n, that if $\mathbf{x}$ is on one wall of B and $\mathbf{y}$ on a different wall, then the segment $\mathbf{xy}$ is interior to B, except for $\mathbf{x}$ and $\mathbf{y}$. If $B \cap C$ has both $\mathbf{x}$ and $\mathbf{y}$, then it contains $\mathbf{xy}$. That says it has points interior to both B and C; B and C overlap.

2. If $f = 0$ in $\text{int}(B)$, then on every subinterval of a partition $\mathcal{P}$, $\inf f = 0$. Therefore, $l(f, \mathcal{P}) = 0$, and $L(f, B) = 0$. Assuming that $|f| \leq M$ on the walls of B, let $\mathcal{Q}$ be a cross-partition created by a big box C of volume $V(B) - \frac{\varepsilon}{M}$. Then $u(f, \mathcal{Q}) \leq 0V(C) + M \sum V(\text{other subintervals}) = \varepsilon$. We conclude that $U(f, B) = 0$, f is integrable, $\int_B f = 0$.

3. (a) Let B be any box containing $A \equiv \{\mathbf{x}_1, \ldots, \mathbf{x}_m\}$. Clearly, $\text{int}(A)$ is empty, so $v^*(A) = 0$. Let $\mathcal{P}$ partition B into k^n congruent subintervals. A subinterval touches $\text{cl}(A)$ iff it contains some $\mathbf{x}_j$. A given $\mathbf{x}_j$ can lie in only 2^n subintervals (induction on n). Hence $V(A, \mathcal{P}) \leq \frac{2^n V(B)}{k^n}$, true for arbitrary k. We conclude that $V^*(A) = 0$. The set is Archimedean, with $V(A) = 0$.

   (b) It has empty interior, so $v^* = 0$. Also, you can put it in the box

   $$a_1 - \varepsilon \leq x_1 \leq a_1 + \varepsilon,\ a_2 \leq x_2 \leq b_2, \ldots, a_n \leq x_n \leq bn,$$

   whose volume is $2\varepsilon(b_2 - a_2) \cdots (b_n - a_n)$. Let $\varepsilon \to 0$, and conclude that $V^* = 0 = v^*$.

   (c) We may assume that $a, b > 0$. Let k be a positive integer. Draw the grid of lines $x = \frac{ja}{k}$, $y = \frac{jb}{k}$, $0 \leq j \leq k$. They create a partition $\mathcal{P}$ of the box $0 \leq x \leq a$, $0 \leq y \leq b$, covering the set. Each subinterval has volume $\frac{ab}{k^2}$. The lowest row of subintervals has k that meet the closure, no interior ones. The next row up has k meeting the closure, $k - 3$ interior ones; you lose 2 on the right, because the next-to-last one has its upper-right corner on the boundary. In the next row, there are $k - 1$ on the closure, $k - 4$ interior. Continue the pattern to obtain

   $$v(A, \mathcal{P}) = (1 + 2 + \cdots + k - 3)\frac{ab}{k^2} = (k-3)(k-2)\frac{ab}{2k^2},$$

   $$V(A, \mathcal{P}) = (1 + 2 + \cdots + k + k)\frac{ab}{k^2} = k(k+3)\frac{ab}{2k^2}.$$

   Passing to the limit, we conclude that $V^*(A) \leq \frac{ab}{2} \leq v^*(A)$. Therefore, A is Archimedean, with volume $\frac{ab}{2}$.

4. Suppose $\mathbf{b} \in \text{int}(A)$. By definition, some neighborhood $N(\mathbf{b}, \delta)$ is a subset of A. Then the box B from $\left(b_1 - \frac{\delta}{2\sqrt{n}}, \ldots, b_n - \frac{\delta}{2\sqrt{n}}\right)$ to $\left(b_1 + \frac{\delta}{2\sqrt{n}}, \ldots, b_n + \frac{\delta}{2\sqrt{n}}\right)$ is a subset of $N(\mathbf{b}, \delta)$, therefore of A. Any partition $\mathcal{P}$ with B as a subinterval has $v(A, \mathcal{P}) \geq v(B)$, making $v^*(A) \geq v(B) > 0$. By contraposition, $v^*(A) = 0 \Rightarrow \text{int}(A)$ has no members.

5. (a) Unit cube $\cap \mathbf{Q}^n$. It is not Archimedean; its volume is undefined.
   (b) If A is Archimedean, then $V^*(A) = v^*(A)$. If A has empty interior, then $v^*(A) = 0$. If both, then $V(A)$ is defined, equals 0.

6. Assume that A has zero volume and $|f| \le M$. Let B be any box superset of A. There exists a partition of B with $V^*(A, \mathcal{P}) < \frac{\varepsilon}{2M}$. On the subintervals of $\mathcal{P}$ exterior to A, $f\chi_A$ is identically zero. Hence $u(f\chi_A, \mathcal{P}) \le M \sum$ (volumes of closure subintervals) $= MV^*(A, \mathcal{P}) \le \frac{\varepsilon}{2}$, while $l(f\chi_A, \mathcal{P}) \ge -M \sum$ (volumes of closure subintervals) $= -MV^*(A, \mathcal{P}) = \frac{-\varepsilon}{2}$. Therefore $U(f\chi_A, B) \le 0 \le L(f\chi_A, B)$, and the conclusion follows.

7. (a) χ_A, χ_B, and $\chi_{A\cap B}$ all have values 0 and 1, and $\chi_{A\cap B}(\mathbf{x}) = 1 \Leftrightarrow \mathbf{x} \in A$ and $\mathbf{x} \in B \Leftrightarrow \chi_A(\mathbf{x}) = 1$ and $\chi_B(\mathbf{x}) = 1 \Leftrightarrow \chi_A(\mathbf{x})\chi_B(\mathbf{x}) = 1$.
   (b) $A \cup B$ is the disjoint union of $A - B$, $A \cap B$, $B - A$, and the two sides match in each of those sets. The rest of the space is $A^* \cap B^*$, in which both sides are 0.
   (c) If $A \subseteq N(\mathbf{O}, M)$, then $\|\mathbf{x}\| \ge M \Rightarrow \chi_A(\mathbf{x}) = 0$, so $\lim_{\mathbf{x}\to\infty} \chi_A(\mathbf{x}) = 0$. Conversely, if $\lim_{\mathbf{x}\to\infty} \chi_A(\mathbf{x}) = 0$, then there is $N(\mathbf{O}, R)$ outside of which $\chi_A \le \frac{1}{2}$, forcing $\chi_A = 0$. Hence $\|\mathbf{x}\| \ge R \Rightarrow \mathbf{x} \notin A$; that is, $\mathbf{x} \in A \Rightarrow \|\mathbf{x}\| < R$, and A is bounded.

8. (a) Ignore the value $\chi_S(\mathbf{b})$. If every neighborhood of $\mathbf{b}$ has points from both S and S^*, then the limit does not exist. If instead some neighborhood of $\mathbf{b}$ has points from only one of the two, then the limit exists. Hence $\mathbf{b}$ has to be surrounded by S (interior to S or isolated point of S^*) or by S^* (vice versa).
   (b) In view of (a), $\mathbf{b}$ has to belong to the set that surrounds it. Hence $\mathbf{b}$ has to be interior to S or interior to S^* = exterior to S; in fewer words, $\mathbf{b} \notin \mathrm{bd}(S)$.

Section 4.4

1. Unit square $\cap \mathbf{Q}^2$ has the unit square as boundary.

2. By Theorem 2, there is a set of boxes $B_{j1}, \ldots, B_{jk(j)}$ covering D_j with $V(B_{j1}) + \cdots + V(B_{jk(j)}) < \frac{\varepsilon}{J}$. Then $\bigcup_{j=1}^{J} \bigcup_{k=1}^{k(j)} B_{jk}$ is a union of boxes covering $D_1 \cup \cdots \cup D_J$ with volume-sum $< \frac{J\varepsilon}{J} = \varepsilon$. By Theorem 2, $D_1 \cup \cdots \cup D_J$ is meager.

3. If D is meager, then you can cover it with $B_1, \ldots, B_J$ having volume-sum $< \varepsilon$. If $C \subseteq D$, then the same boxes cover C.

4. By induction: Let $(x_1, \ldots, x_n) \in S(\mathbf{O}, a)$, so that $x_1^2 + \cdots + x_n^2 = a^2$. First, $x_1^2 \le a^2$, equivalent to $\left|\frac{x_1}{a}\right| \le 1$. Hence there exists a unique $t_1 \in [0, \pi]$

with $\frac{x_1}{a} = \cos t_1$. Next, $x_1^2 + x_2^2 \le a^2$, or $\left|\frac{x_2}{\sqrt{a^2 - x_1^2}}\right| = \left|\frac{x_2}{a \sin t_1}\right| \le 1$. Hence there exists one $t_2 \in [0, \pi]$ with $\frac{x_2}{a \sin t_1} = \cos t_2$. Next, $\frac{x_3}{\sqrt{a^2 - x_1^2 - x_2^2}} = \frac{x_3}{a \sin t_1 \sin t_2}$ must equal some $\cos t_3$, and so on. Eventually, $x_1^2 + \cdots + x_n^2 = a^2$ forces $|x_n| = a \sin t_1 \cdots \sin t_{n-1}$, and to cover the two possibilities $x_n > 0$ and $x_n < 0$, we extend the domain of t_{n-1} to $[0, 2\pi]$. That parametrization covers the sphere, one-to-one.

Obviously, each $\frac{\partial \mathbf{x}}{\partial t_j}$ is continuous. By induction, we may also show that any two are orthogonal and have lengths $a, a \sin t_1, \ldots, a \sin t_1 \cdots \sin t_{n-2}$. (Look ahead at Example 5.6:3.) Hence they are independent, making the parametrization smooth, as long as $t_1, \ldots, t_{n-2}$ are different from 0 and π.

5. (a) Assume that O_j has some points of O_i. By construction, O_j possesses no points from $\text{bd}(O_i)$. It cannot possess points from $\text{ext}(O_i)$, because then (being connected) it would also have to have points from $\text{bd}(O_i)$. Hence $O_j \subseteq \text{int}(O_i) = O_i$. It cannot be $O_j = O_i$, because even the boundary of O_j has no points from $\text{bd}(O_i)$.

   (b) Suppose $O_{j(1)}, O_{j(2)}, \ldots$ is the subsequence of O_j that intersect O_i. Then O_i intersects $O_{j(1)}$ and $j(1) \le i$. By (a), $O_i \subseteq O_{j(1)}$. This means that $O_{j(2)}, O_{j(3)}, \ldots$ all intersect $O_{j(1)}$. Hence $O_{j(2)}, O_{j(3)}, \ldots$ are all subsets of $O_{j(1)}$, and $O_{j(1)} = O_{j(1)} \cup O_{j(2)} \cup \cdots$. Define $k(i) \equiv j(1)$, and the conclusion follows.

   (c) We cannot predict whether $k(i)$ and $k(m)$ match, or which is larger. However, the following is certain: If $k(i) \ne k(m)$, then $O_{k(i)}$ and $O_{k(m)}$ cannot have anything in common; if they did, then the one with the larger index $K \equiv \max\{k(i), k(m)\}$ would be a proper subset of the other, violating the choice of O_K as the lowest-numbered set intersecting some O_i. Hence $O_{k(i)}$ and

$$O - O_{k(i)} = \bigcup_{k(m) \ne k(i)} O_{k(m)}$$

   are disjoint open sets. That makes $O_{k(i)}$ a maximal connected subset of O.

6. (a) If $\mathbf{x} \in \text{bd}(A \cap B)$, then there is a sequence $(\mathbf{x}_i) \to \mathbf{x}$ from $A \cap B$ and $(\mathbf{y}_i) \to \mathbf{x}$ from outside $A \cap B$. Either a subsequence $(\mathbf{y}_{j(i)})$ comes from outside A, so $\mathbf{x} \in \text{bd}(A)$, or a subsequence $(\mathbf{y}_{k(i)})$ comes from outside B, and $\mathbf{x} \in \text{bd}(B)$. Hence $\mathbf{x} \in \text{bd}(A \cap B) \Rightarrow \mathbf{x} \in \text{bd}(A) \cup \text{bd}(B)$.

   (b) Assume that A and B are Archimedean. Then $\text{bd}(A)$ and $\text{bd}(B)$ are meager. By Exercise 2, their union is meager; by (a), $\text{bd}(A \cap B) \subseteq \text{bd}(A) \cup \text{bd}(B)$; by Exercise 3, $\text{bd}(A \cap B)$ is meager; by Theorem 1, $A \cap B$ is Archimedean.

7. A and B overlap $\Leftrightarrow$ int$(A \cap B)$ is nonempty (Exercise 4.1:2e) $\Leftrightarrow$ $v^*(A \cap B) > 0$ (Exercise 4.3:4 and definition of v^*). Since A and B are Archimedean, so is $A \cap B$ (Exercise 6). Hence $v^*(A \cap B) > 0$ amounts to $V(A \cap B) > 0$.

8. (a) The non-Archimedean set O in Example 2 is the union of Archimedean neighborhoods.

   (b) Unit square – O is not Archimedean, because its boundary is bd(O), which is not meager.

   (c) Again, bd(O) is not meager. However, it has no interior, because it lacks all of $\mathbf{Q}^2$: every $\mathbf{x} \in \mathbf{Q}^2$ is in the interior of O.

9. (a) The given set is a bounded subset of a hyperplane. Such a set is necessarily meager; compare Answer 4.3:3b. Therefore, every bounded function is integrable on it.

   (b) It matches Dirichlet's function on the box from $(-0.5, -0.5, -0.5)$ to $(0.5, 0.5, 0.5)$, is therefore not integrable on that box. By Theorem 5, g cannot be integrable on S.

   (c) This is the restriction of the modified Dirichlet function G, defined on the box from $(-1, -1, -1)$ to $(1, 1, 1)$, to S. By Example 3 and Theorem 5, G is integrable on S.

10. The finite set is meager (Exercise 4.3:3a). Therefore, Theorem 4 guarantees that f is integrable.

11. Assume that f is continuous everywhere and A is Archimedean. Then A is bounded, so cl(A) is closed and bounded. Consequently, f is bounded on cl(A), is therefore bounded and continuous on A, is integrable on A.

12. Assume that every bounded function is integrable on S. Then $g(\mathbf{x}) \equiv 1$ is integrable, so by Theorem 4.3:3(c), S is Archimedean. If S were not meager, then cl(S) would contain some neighborhood, and therefore some box; on that box, Dirichlet's function F would be unintegrable; by Theorem 5, F would be unintegrable on S; contradiction.

13. The Cantor set is meager, because by its construction, it is contained in the union of 2^k intervals of length 3^{-k} each, total length $\left(\frac{2}{3}\right)^k$, for any k.

14. (a) Assume that S is meager. By Theorem 2, there is a class of boxes $C_1, \dots, C_J$ covering S with volume-sum less than $\frac{\varepsilon}{2}$. If finite sequences are allowed, define $(B_i) \equiv (C_1, \dots, C_J)$. If not, set

$$(B_i) \equiv \left(C_1, \dots, C_J, \frac{C_J}{2}, \frac{C_J}{4}, \dots\right),$$

where $\frac{C_J}{k}$ is a box concentric with C_J and $\frac{1}{k}$ as wide. Then

$$\begin{aligned} V(B_1) + V(B_2) + \cdots &= V(C_1) + \cdots + V(C_J) \\ &\quad + V(C_J)\left(\frac{1}{2^n} + \frac{1}{4^n} + \cdots\right) \\ &\leq \frac{\varepsilon}{2} + V(C_J)\left(\frac{1}{2} + \frac{1}{4} + \cdots\right) < \varepsilon. \end{aligned}$$

(b) Unit square $\cap\, \mathbf{Q}^2$. Call it S, and enumerate it as

$$\{(r_1, s_1), (r_2, s_2), \ldots\}.$$

Let B_i be the square centered at (r_i, s_i) with width $2^{-i}\varepsilon^{1/n}$. Clearly, $S \subseteq B_1 \cup B_2 \cup \cdots$, and

$$V(B_1) + V(B_2) + \cdots \leq \sum [2^{-i}\varepsilon^{1/n}]^n \leq \frac{\varepsilon}{2} + \frac{\varepsilon}{4} + \cdots = \varepsilon.$$

That proves that S has zero measure. But S is not Archimedean, so its volume is undefined.

(c) The Cantor set is uncountable and meager (Exercise 13 and references).

(d) Assume that S has positive involume. Then int(S) is nonempty, so S contains a neighborhood, and therefore a box B. Suppose now (B_i) is a sequence of boxes covering S. Then (B_i) also covers B. By the Heine–Borel theorem, some finite subcollection $B_1, \ldots, B_J$ still covers B. By a familiar cross-partition argument, $V(B_1)+\cdots+V(B_J) \geq V(B)$. Hence $V(B_1)+V(B_2)+\cdots \geq V(B)$. We conclude that S cannot have zero measure. By contraposition, if S has zero measure, then it has zero involume.

(e) By (d), A has zero measure $\Rightarrow v^*(A) = 0$. If A is Archimedean, then $V(A) = v^*(A) = 0$.

(f) The words "closed and bounded" should give us a hint. Assume that S is closed, bounded, and of zero measure. The last means that there is a sequence (B_i) of boxes covering S with volume-sum $< \varepsilon$. Since S is compact, the same job can be done by a finite subcollection $B_1, \ldots, B_J$. Then $S \subseteq B_1 \cup \cdots \cup B_J$ with $V(B_1)+\cdots+V(B_J) < \varepsilon$. By Theorem 2, S is meager.

15. (a) Assume $\mathbf{x} \in$ bd(int(S)). Then there are sequences $(\mathbf{x}_i) \to \mathbf{x}$ from int(S) and $(\mathbf{y}_i) \to \mathbf{x}$ from outside int(S). Each $\mathbf{x}_i \in S$. If an infinity of $\mathbf{y}_i$ are in bd(S), then $\mathbf{x}$ is a closure point of bd(S); that forces $\mathbf{x} \in$ bd(S), because boundaries are closed sets. If instead finitely many $\mathbf{y}_i$ are in bd(S), then a subsequence $\mathbf{y}_J, \mathbf{y}_{J+1}, \ldots$ comes from ext(S) $\subseteq S^*$; that makes $\mathbf{x} \in$ bd(S). We have shown that $\mathbf{x} \in$ bd(int(S)) $\Rightarrow \mathbf{x} \in$ bd(S).

(b) Set $S \equiv \left\{(x, y) : x^2 + y^2 \le 1, \text{or } (x, y) \in \mathbf{Q}^2 \text{ and } 1 < x^2 + y^2 \le 2\right\}$. Int($S$) is the open unit disk, whose boundary is the unit circle. The boundary of S is the annulus $1 \le x^2 + y^2 \le 2$.

(c) $\mathbf{b} \in \text{cl}(S) \Leftrightarrow$ every $N(\mathbf{b}, \delta)$ has points from $S \Leftrightarrow$ (every $N(\mathbf{b}, \delta)$ intersects both S and S^*) or (some $N(\mathbf{b}, \delta)$ is all S) $\Leftrightarrow$ ($\mathbf{b} \in \text{bd}(S)$) or ($\mathbf{b} \in \text{int}(S)$).

Chapter 5

Section 5.1

1. Assume that T_k is defined by $a_{k1} \le u_1 \le b_{k1}, \ldots, a_{km} \le u_m \le b_{km}$, S_j by $c_{j1} \le v_1 \le d_{j1}, \ldots, c_{j(n-m)} \le v_{n-m} \le d_{j(n-m)}$.

   (a) The details are straightforward, so we provide an outline: $T_k \times S_j$ is a box, given by the conjunction of all n inequalities; two such boxes do not overlap, because if they did, then two T_k or else two S_j would overlap; and their union is B. That would prove (a).

   (b) The volume of $T_k \times S_j$ is $\left[(b_{k1} - a_{k1}) \cdots (b_{km} - a_{km})\right]\left[(d_{j1} - c_{j1}) \cdots (d_{j(n-m)} - c_{j(n-m)})\right] = V(T_k)V(S_j)$.

   (c) $\left[\text{diag}(T_k \times S_j)\right]^2 = \left[(b_{k1} - a_{k1})^2 + \cdots + (b_{km} - a_{km})^2\right] + \left[(d_{j1} - c_{j1})^2 + \cdots + (d_{j(n-m)} - c_{j(n-m)})^2\right] = \text{diag}(T_k)^2 + \text{diag}(S_j)^2$.

2. (a) Fix a natural number k. For each fraction $\frac{j}{k}$, $0 \le j \le k$, paint the horizontal band $\frac{j}{k} - \frac{1}{k^4} \le y \le \frac{j}{k} + \frac{1}{k^4}$. On those bands, $f(x, y) \le 1$, and the total area is $\left[\frac{(k+1)(k+2)}{2}\right]\frac{2}{k^4}$, contributing at most $\frac{6}{k^2}$ to the upper sum. For the unpainted bands, $f(x, y) < \frac{1}{k}$, and the total area is less than 1, contributing less than $\frac{1}{k}$. Consequently, there are upper sums $= O\left(\frac{1}{k}\right)$, upper integral $= 0 =$ lower integral.

   (b) For fixed $y = \frac{j}{k}$ in lowest terms, $f(x, y) \equiv \frac{1}{k}$ for rational x, $\equiv 0$ otherwise, is Dirichlet's function times $\frac{1}{k}$; it is not integrable.

3. (a) For a fixed y, $g(x, y)$ is constantly 0 or constantly 1.

   (b) If y is rational, then $\phi(y) = \int_0^1 1\,dx = 1$; otherwise, $\phi(y) = \int_0^1 0\,dx = 0$. Thus, ϕ is Dirichlet's function.

4. (a) For a fixed rational y, $h(x, y) = -1$ if $0 \le x \le \frac{1}{2}$, $= 1$ if $\frac{1}{2} < x \le 1$. This is a function with one discontinuity, is integrable. For a fixed irrational y, $h(x, y) = 1$ if $0 \le x \le \frac{1}{2}$, $= -1$ thereafter; likewise integrable.

(b) In (a), for every y, $\int_0^1 h(x, y)\,dx = 0$; ϕ is constant.

(c) In every subinterval of any partition, there are places where $h = \sup h = 1$ and places where $h = \inf h = -1$. Hence the upper integral is 1, the lower integral -1.

5. Assume that f is continuous in B. First, f is integrable on B. Second,

$$\phi_1(x_1) = \int_{(a_2,\dots,a_n)}^{(b_2,\dots,b_n)} f(x_1, \dots, x_n)d(x_2, \dots, x_n)$$

is defined, because $f(x_1, \dots, x_n)$ is a continuous function of $(x_2, \dots, x_n)$ for fixed x_1. Third, ϕ_1 is continuous, and therefore integrable, because

$$\begin{aligned}
&|\phi_1(x_1) - \phi_1(y_1)| \\
&\le \int_{(a_2,\dots,a_n)}^{(b_2,\dots,b_n)} |f(x_1, \dots, x_n) - f(y_1, x_2, \dots, x_n)|d(x_2, \dots, x_n) \\
&\le \max |f(\mathbf{x}) - f(\mathbf{y})|(b_2 - a_2) \cdots (b_n - a_n),
\end{aligned}$$

and $|f(\mathbf{x}) - f(\mathbf{y})|$ is uniformly small. Hence Fubini's theorem applies, to give

$$\int_{[\mathbf{a},\mathbf{b}]} f = \int_{a_1}^{b_1} \left[\int_C f_{x_1}(x_2, \dots, x_n)\right] dx.$$

That proves the reduction formula, beginning a recursive proof.

6. Set $C \equiv \mathbf{Q} \cap [0, 1]$, $A \equiv \{(x, x) \colon x \in C\}$, $g(x) = G(x) \equiv x$. We know that C is not Archimedean. In $\mathbf{R}^2$, A is a subset of a line, is therefore meager.

7. By Fubini,

$$\int e^{x+y+z} = \int_{-1}^1 \int_{-1}^1 \int_{-1}^1 e^x e^y e^z\,dz\,dy\,dx = \left(\int_{-1}^1 e^t dt\right)^3 = \left(e - e^{-1}\right)^3.$$

8. (a) Since the area is 1, the average is

$$1 \int_B (x + y) = \int_0^1 \int_0^1 (x + y)\,dy\,dx = \int_0^1 \left(x + \frac{1}{2}\right) dx = 1.$$

This makes sense: Draw the line $x + y = 1$; for every patch of area to its lower left where $x + y = 1 - \varepsilon$, there is a corresponding one to the upper right with $x + y = 1 + \varepsilon$.

(b) Now the area is $\frac{1}{2}$, so the average is

$$\begin{aligned}
2 \int_A (x + y) &= 2 \int_0^1 \int_0^{1-x} (x + y)\,dy\,dx \\
&= 2 \int_0^1 \left(x[1 - x] + \frac{[1 - x]^2}{2}\right) dx = \frac{2}{3}.
\end{aligned}$$

This may seem surprising, because it is closer to the maximal $x+y=1$ than to the minimal $x+y=0$. But again it makes sense, because there is more area near $x+y=1$ than near the origin.

9. The region is given by $0 \le z \le 3 - x^2 - 3y^2$ for $x^2 + y^2 \le 1$. Clearly, the (x, y) integral is better switched to polar coordinates. Then

$$\begin{aligned} V &= \int_{x^2+y^2\le 1} \int_0^{3-x^2-3y^2} dz\, dy\, dx \\ &= \int_0^{2\pi} \int_0^1 \left(3 - r^2\cos^2\theta - 3r^2\sin\theta\right) r\, dr\, d\theta \\ &= 2\pi\left(\frac{5}{4}\right) = 2\pi. \end{aligned}$$

10. $V = \int 1\, dx\, dy\, dz\, dw\, dv = \int_{-a}^{a} \int_{x^2+y^2+z^2+w^2 \le a^2 - v^2} 1 d(x, y, z, w)\, dv$. The inside integral is the volume $V^{\#}$ of a ball of radius $\sqrt{a^2 - v^2}$ in $\mathbf{R}^4$. By Example 2, we see that $V^{\#} = \frac{\pi^2 (a^2 - v^2)^2}{2}$. Hence

$$V = \int_{-a}^{a} \frac{\pi^2 \left(a^2 - v^2\right)^2}{2}\, dv = \frac{8\pi^2 a^5}{15}.$$

Section 5.2

1. (a) Let $f \equiv 2$, $g \equiv 1$, $A \equiv$ line segment.
   (b) Set $h(x, y) \equiv 1$ if $y = 1$, $\equiv 0$ otherwise, on $A \equiv$ unit square. Then $h \ge 0$,

$$\int_A h = \int_{\text{int}(A)} h = 0, \qquad V(A) = 1.$$

   (c) Same function h on $A^- \equiv \{(x, y)\colon 0 \le x \le 1, \text{ either } 0 \le y \le \frac{1}{2} \text{ or } y = 1\}$. Alternative: $A^+ \equiv$ unit square $\cup$ segment from $(-1, 0)$ to $(0, 0)$, with $H(x, y) \equiv x - |x|$.

2. Assume that $f \ge 0$ is continuous on A. Theorem 1(c) says that $\int_A f \ge 0$. Suppose $f(\mathbf{b}) > 0$ for a place $\mathbf{b} \in \text{int}(A)$. There must exist $N(\mathbf{b}, \delta)$ in which $f(\mathbf{x}) > \frac{f(\mathbf{b})}{2}$ and $N(\mathbf{b}, \varepsilon) \subseteq A$. Let $N \equiv N(\mathbf{b}, \delta) \cap N(\mathbf{b}, \varepsilon)$. Then

$$\int_A f = \int_N f + \int_{A-N} f > \int_N \frac{f(\mathbf{b})}{2} + \int_{A-N} 0 = \frac{f(\mathbf{b}) V(N)}{2} > 0.$$

Thus, the integral is positive, or else there are no interior $\mathbf{b} \in A$ with $f(\mathbf{b}) > 0$. (The equivalence part comes from linearity.)

3. $V(A \cup B) = \int_{A\cup B} 1 = \int_A 1 + \int_B 1$ (Theorem 2(b)) $= V(A) + V(B)$.

4. Assume $S \subseteq T$, both Archimedean. By Theorem 4.4:5(a), f is integrable on T iff it is integrable on both S and $T - S$. Hence if $f \geq 0$ is integrable on T, then f is integrable on S and $T - S$, and $\int_T f = \int_{T-S} f + \int_S f \geq \int_S f$.

5. By Theorem 1, $k \leq f \leq K \Rightarrow V(A) = k \int_A 1 = \int_A k \leq \int_A f \leq \int_A K = KV(A)$.

6. By Cauchy's inequality,

$$\sum_{k=1}^{K} |f(\mathbf{x}_k)| \sqrt{V(S_k)} \sqrt{V(S_k)} \leq \left(\sum_{k=1}^{K} f(\mathbf{x}_k)^2 V(S_k) \right)^{1/2} \left(\sum_{k=1}^{K} V(S_k) \right)^{1/2}.$$

If f is integrable, then so are $|f|$ and f^2, so these Riemann sums tend to

$$\int_B |f| \leq \left(\int_A f^2 \right)^{1/2} (V(A))^{1/2}.$$

The stated inequality follows.

Section 5.3

1. Call the region A. Then $(x, y) = \Phi(u, v) \equiv (au, bv)$ transforms the uv-unit disk onto A. (Clearly one-to-one, with absdet $\Phi' = |ab| > 0$, since we take a and b to be positive.) Hence $\int_A 1 = \int_{\Phi(D)} 1 = \int_D (1 \text{ absdet } \Phi') = ab \text{ area}(D) = \pi ab$.

2. The region $0 \leq x \leq 3$, $0 \leq y \leq \frac{4x}{3}$ is given by $0 \leq \theta \leq \tan^{-1}(\frac{4}{3})$, $0 \leq r \leq \frac{3}{\cos\theta}$. Hence

$$\begin{aligned}
\text{area} &= \int_0^{\tan^{-1}(4/3)} \int_0^{3/\cos\theta} 1r\, dr\, d\theta \\
&= \int_0^{\tan^{-1}(4/3)} \left(\frac{9}{2}\right) \sec^2\theta\, d\theta \\
&= \left(\frac{9}{2}\right)\left(\frac{4}{3}\right) = 6, \\
x - \text{average} &= \frac{1}{6} \int_0^{\tan^{-1}(4/3)} \int_0^{3/\cos\theta} r\cos\theta\, r\, dr\, d\theta \\
&= \int_0^{\tan^{-1}(4/3)} \frac{\cos\theta(3/\cos\theta)^3}{18}\, d\theta = \left(\frac{3}{2}\right)\left(\frac{4}{3}\right) = 2,
\end{aligned}$$

$$y-\text{average} = \frac{1}{6}\int_0^{\tan^{-1}(4/3)}\int_0^{3/\cos\theta} r\sin\theta r\,dr\,d\theta$$
$$= \frac{3}{2}\int_0^{\tan^{-1}(4/3)} \cos^{-3}\theta\sin\theta\,d\theta$$
$$= \frac{3}{4}[\cos^{-2}\theta]_0^{\tan^{-1}4/3} = \left(\frac{3}{4}\right)\left(\frac{16}{9}\right) = \frac{4}{3}.$$

3. The cone is given by $0 \le \phi \le \frac{\pi}{4}, 0 \le \theta \le 2\pi, 0 \le \rho \le \frac{4}{\cos\phi}$. The transformation is singular along the z-axis, $\phi = 0$, but we proceed as in Example 1, as though the substitution hypothesis were met.

   (a) Volume is
   $$\int_0^{2\pi}\int_0^{\pi/4}\int_0^{4/\cos\phi} 1\rho^2\sin\phi\,d\rho\,d\phi\,d\theta$$
   $$= 2\pi\int_0^{\pi/4}\frac{4^3}{3\cos^3\phi}\sin\phi\,d\phi = \frac{64\pi}{3}.$$

   (b) Distance from the origin is ρ, so average distance is
   $$\frac{3}{64\pi}\int_0^{2\pi}\int_0^{\pi/4}\int_0^{4/\cos\phi}\rho\rho^2\sin\phi\,d\rho\,d\phi\,d\theta$$
   $$= 6\int_0^{\pi/4}\cos^{-4}\phi\sin\varphi\,d\phi = 2\left(2\sqrt{2}-1\right).$$

   (c) Symmetry suggests that x-average and y-average are 0. The spherical integral agrees, because $x = \rho(\sin\phi)\cos\theta$ and $y = \rho(\sin\phi)\sin\theta$ make $\int_0^{2\pi}\cos\theta$ and $\int_0^{2\pi}\sin\theta$ factors. The z-average is
   $$\frac{3}{64\pi}\int_0^{2\pi}\int_0^{\pi/4}\int_0^{4/\cos\phi}(\rho\cos\phi)\rho^2\sin\phi\,d\rho\,d\phi\,d\theta$$
   $$= 6\int_0^{\pi/4}\cos^{-3}\phi\sin\varphi\,d\phi = 3.$$

4. The Jacobian $\frac{\partial x}{\partial(r,s,t,u)}$ is $r(r\sin s)(r\sin s\sin t)$ (= product of the lengths of the columns; consult Answer 4.4:4). We want the region $r \le a$. Its volume is
$$\int_0^{\pi}\int_0^{\pi}\int_0^{2\pi}\int_0^{a} 1(r^3\sin^2 s\sin t\,dr\,du\,dt\,ds$$
$$= \frac{2\pi a^4}{4}[-\cos t]_0^{\pi}\left[\frac{s}{2}-\frac{\sin 2s}{4}\right]_0^{\pi} = \frac{\pi^2 a^4}{2}.$$

5. From $\phi'(x) \neq 0$, we conclude (intermediate value theorem) that ϕ' is of one sign. Suppose first $\phi' > 0$. Then ϕ is a strictly increasing function, mapping $[a, b]$ one-to-one onto $[c, d]$. The hypothesis of Theorem 5 is satisfied, so we conclude that

$$\int_{[c,d]} f(u)\,du = \int_{[a,b]} f(\phi(x))|\phi'(x)|\,dx = \int_a^b f(\phi(x))\phi'(x)\,dx.$$

Suppose instead that $\phi' < 0$. Then ϕ maps $[a, b]$ decreasingly onto $[d, c]$. By Theorem 5,

$$\int_{[d,c]} f(u)\,du = \int_{[a,b]} f(\phi(x))|\phi'(x)|\,dx.$$

The definition of $\int_c^d f(u)\,du$ is $-\int_{[d,c]} f(u)\,du$, and $|\phi'| = -\phi'$. Hence we have again

$$\int_c^d f(u)\,du = \int_a^b f(\phi(x))\phi'(x)\,dx;$$

the change of sign required by the absolute-value sign is accomplished by the order on **R**.

6. (a) Theorem 3.2:2 says that Φ maps some open set one-to-one onto some neighborhood $N(\Phi(\mathbf{a}), \varepsilon)$. Theorem 3.2:3 says that Φ^{-1} is differentiable. Any continuous function with a continuous inverse, from one open set to another, will map open sets to open sets: P open $\Rightarrow$ image of $P = \Phi(P) = (\Phi^{-1})^{-1}(P) =$ inverse image of open set, and the last is necessarily open [Guzman, Section 4.4].

   (b) Assume $J(\mathbf{x}) \neq 0$ and $P \subseteq O$ is open. Let $\mathbf{b} \in P$. By Theorem 3.2:2, Φ maps some open set Q invertibly onto some neighborhood $N(\Phi(\mathbf{b}), \varepsilon)$, some $N(\mathbf{b}, \delta) \subseteq Q \cap P$. By (a), $\Phi(N(\mathbf{b}, \delta))$ is an open subset of $\Phi(P)$. Thus, for each $\mathbf{b} \in P$, there exists an open set $\Phi(N(\mathbf{b}, \delta))$ with $\Phi(\mathbf{b}) \in \Phi(N(\mathbf{b}, \delta)) \subseteq \Phi(P)$; this proves that $\Phi(P)$ is open.

   (c) Under these hypotheses, Φ maps the open set int(A) into an open subset of $\Phi(A)$, and $O - \text{cl}(A) = O \cap \text{ext}(A)$ into an open subset of $\Phi(O) - \Phi(A)$. If $\mathbf{x} \in \text{bd}(A)$, then neighborhoods of $\Phi(x)$ are images of open sets surrounding $\mathbf{x}$, open sets that must intersect A and A^*; hence $\Phi(\mathbf{x}) \in \text{bd}(\Phi(A))$. This means that int$(\Phi(A))$ can get images only from int(A), so $\Phi(\text{int}(A))$ has to be all of int$(\Phi(A))$, similarly for the boundary and the remainder.

   (d) Set $A \equiv \left\{(r, \theta)\colon 1.1 \leq r \leq 1.9, \frac{\pi}{180} \leq \theta \leq \frac{269\pi}{180}\right\}$. In the mapping $(r, \theta) \mapsto (r, 2\theta)$, the point at $\left(1.5, \frac{269\pi}{180}\right)$ is on the boundary of A, but its image

$$\left(1.5, \frac{538\pi}{180}\right) = \left(1.5, \frac{178\pi}{180}\right)$$

is interior to the image of A.

7. Assume that $J(\mathbf{x})$ is continuous near $\mathbf{a}$. Then

$$V(\Phi(A)) = \int_{\Phi(A)} 1 = \int_A J(\mathbf{x}) = J(\mathbf{x}^*)V(A)$$

for some $\mathbf{x}^*$ in A. Hence $\frac{V(\Phi(A))}{V(A)} = J(\mathbf{x}^*) \to J(\mathbf{a})$ as long as $A \subseteq N(\mathbf{a}, \delta)$ with $\delta \to 0$.

Section 5.4

1. (a) Suppose $|f(\mathbf{x})| \le K\|\mathbf{x}\|^{-p}$ for $\mathbf{x}$ outside the radius-R ball. Then for any A out there,

$$\left|\int_A f\right| \le \sum_{k=R}^{\infty} \int_{k\le\|\mathbf{x}\|\le k+1} |f| \le \sum_{k=R}^{\infty} Mk^{-p}\big[(k+1)^n - k^n\big]V\big(B(\mathbf{O}, 1)\big).$$

The series converges, because $(k+1)^n - k^n = O\big(k^{n-1}\big)$. Hence $\left\{\int_A f\right\}$ is bounded, and f is integrable on $B(\mathbf{O}, R)^*$.

(b)

$$\int_{\mathbf{R}^2} \left(1 + \sqrt{x^2+y^2}\right)^{-p} = \int_0^{2\pi}\int_0^{\infty} (1+r)^{-p} r\, dr\, d\theta$$
$$= 2\pi\left(\frac{u^{2-p}}{2-p} - \frac{u^{1-p}}{1-p}\right]_1^{\infty} = \frac{2\pi}{(p-2)(p-1)}.$$

(c) Since $(1 + \|\mathbf{x}\|)^{-p} \ge (1 + \|\mathbf{x}\|)^{-n}$, it suffices to do the latter.

$$\int_{\|\mathbf{x}\|\le R} (1+\|\mathbf{x}\|)^{-n} = \sum_{k=1}^{R} \int_{k-1\le\|\mathbf{x}\|\le k} (1+\|\mathbf{x}\|)^{-n}$$
$$\ge \sum_{k=1}^{R} k^{-n}\big[k^n - (k-1)^n\big]V\big(B(\mathbf{O}, 1)\big).$$

Since $(k+1)^n - k^n = nt^{n-1}$ (by the mean value theorem) $\ge n\left(\frac{k}{2}\right)^{n-1}$ beginning with $k = 3$, the series looks like $\sum_{k\ge1} \frac{n}{2^{n-1}k}$, which diverges. Hence the integrals are unbounded.

2. (a) Suppose $A \subseteq$ unit ball is closed and does not include the origin. Then

$$\left|\int_A f\right| \le \sum_{k=1}^{\infty} \int_{1/(k+1)\le\|\mathbf{x}\|\le 1/k} |f|$$
$$\le \sum_{k=1}^{\infty} M(k+1)^q \left[\frac{1}{k^n} - \frac{1}{(k+1)^n}\right] V\big(B(\mathbf{O}, 1)\big).$$

Since $(k+1)^q \le (2k)^q$ and $\frac{1}{k^n} - \frac{1}{(k+1)^n} = -nt^{-n-1}(-1) \le nk^{-n-1}$, the terms in the series are dominated by multiples of k^{q-n-1}. The series converges, and the integrals of f are bounded.

(b) $\int_D (x^2+y^2)^{-q/2} = \int_0^{2\pi}\int_0^1 r^{-q} r\,dr\,d\theta = 2\pi\left[\frac{r^{-q+2}}{-q+2}\right]_0^1 = \frac{2\pi}{2-q}$.

(c) Let D be the disk, V its volume. For any K,

$$\int_D f \ge \sum_{k=1}^{K}\left(\frac{1}{k}\right)^{-q}\left[\frac{1}{k^n} - \frac{1}{(k+1)^n}\right]V.$$

In the series, the terms exceed $k^q n(2k)^{-n-1} \ge$ constant/k; the series diverges.

3. (a) No in both. Between the lines $y = x$ and $y = 2x$, $\frac{xy}{(x^2+y^2)^2} \ge \frac{x^2}{25x^4}$ is too big at $\mathbf{O}$ and at ∞. Thus,

$$\int_{\pi/4}^{\tan^{-1}2}\int_{\delta}^{\varepsilon}\frac{r\cos\theta r\sin\theta}{r^4} r\,dr\,d\theta = K\int_{\delta}^{\varepsilon}\frac{dr}{r}$$

diverges with either $\varepsilon \to \infty$ or $\delta \to 0$.

(b) Yes in D, not D^*. Near the origin, $\frac{\sin xy}{xy} \approx 1$, no problem. Between the hyperbolas $y = \frac{(2k+1/6)\pi}{x}$ and $y = \frac{(2k+5/6)\pi}{x}$, there is infinite area and $\frac{\sin xy}{xy} > \frac{(2k+1)\pi}{2}$.

(c) Yes and yes. It is bounded in the circle, and is $O(r^{-4})$ at infinity.

(d) No in D, yes in D^*. We have $x^4 + y^4 \le (x^2+y^2)^2$. That makes $(x^4+y^4)^{-1} \ge \|(x,y)\|^{-4}$, much too big near $(0,0)$. At infinity,

$$x^4 + y^4 \ge \max\{x^4, y^4\} = \left(\max\{x^2, y^2\}\right)^2 \ge \left(\frac{x^2+y^2}{2}\right)^2,$$

so $(x^4+y^4)^{-1} = O(r^{-4})$.

4.

$$\int_{[0,\pi]\cup\cdots\cup[2k\pi,(2k+1)\pi]}\frac{\sin x}{x} > \frac{1}{\pi}\int_0^{\pi}\sin x\,dx + \cdots + \frac{1}{(2k+1)\pi}\int_0^{\pi}\sin x\,dx$$
$$= \left(1 + \frac{1}{3} + \cdots + \frac{1}{(2k+1)}\right)\frac{2}{\pi},$$

which tends to infinity as $k \to \infty$.

5. The arguments in the proof of Theorem 4.4:3 and Answer 4.4:6a, showing that the boundaries of $S \cup T$, $S - T$, and $S \cap T$ are all subsets of $\text{bd}(S) \cup \text{bd}(T)$, apply to all sets. Also, $\text{bd}(S^*) = \text{bd}(S)$. Suppose S and T are LA. Then $\text{bd}(S)$ and $\text{bd}(T)$ are locally meager, so that

$$[\text{bd}(S) \cup \text{bd}(T)] \cap N = [\text{bd}(S) \cap N] \cup [\text{bd}(T) \cap N]$$

is meager for every neighborhood N. It follows that union, intersection, difference, and complement all have locally meager boundaries.

6. $\Rightarrow$ Assume that A is Archimedean. Then N is a neighborhood $\Rightarrow$ $A \cap N$ is Archimedean (Theorem 4.4:3); this says that A is LA. Boundedness is part of the definition.

   $\Leftarrow$ Assume that S is bounded and LA. Bounded means that $S \subseteq$ some $B(\mathbf{O}, R)$. Because S is LA, $S = S \cap N(\mathbf{O}, 2R)$ is Archimedean.

7. (c) First, $S \cup T$ and $S \cap T$ are LA (Exercise 5). By extension of part (b), f is integrable on $S \cup T$ $\Leftrightarrow$ f is integrable on $S - T$, $S \cap T$, and $T - S$ $\Leftrightarrow$ f is integrable on S and T; and

$$\int_{S \cup T} f = \left(\int_{S-T} f + \int_{S \cap T} f\right) + \left[\int_{T-S} f\right]$$
$$= \left(\int_S f\right) + \left[\int_T f - \int_{S \cap T} f\right].$$

   If S and T do not overlap, then $\text{int}(S \cap T)$ is empty, and (a) says that $\int_{S \cap T} f = 0$.

   (d) Assume $f \geq 0$ on S and $T \subseteq S$. Then for every $A \subseteq T$, $\int_A f \geq 0$ (Theorem 5.2:1). Therefore the inf and sup of $\{\int_A f\}$ are nonnegative, making

$$\int_T f \equiv \sup\left\{\int_A f\right\} + \inf\left\{\int_A f\right\} \geq 0.$$

   Similarly, $\int_{S-T} f \geq 0$. Then by (c), $\int_S f = \int_T f + \int_{S-T} f \geq \int_T f$.

8. On an Archimedean $A \subseteq S$, $\left|\int_A g\right| \leq \int_A |g|$ (Theorem 5.2:1(d)) $\leq \int_A f$ (Theorem 5.2:1(c)) $\leq \int_S f$ (Theorem 5.4:2(d), because $f \geq |g| \geq 0$). Hence $\text{Itg}(g, S)$ is a bounded set.

9. No. By Exercise 2, $\left(x^2 + y^2\right)^{-1/2}$ is integrable on the unit disk, but its square is not.

10. (a) If $f \geq 0$, then $f^+ = f$ and $f^- = 0$, so $f = f^+ - f^- = f^+ + f^- = |f|$. If instead $f < 0$, then $f^+ = 0$, $f^- = -f$, and $f = f^+ - f^-$, $|f| = -f = f^+ + f^-$.

(b) $\Rightarrow$ Assume that f is locally integrable on S, meaning that f is integrable on every closed $A \subseteq S$ – some U. Let A be one such set. Necessarily $|f|$ is integrable on A (Theorem 5.2:1(d)). Hence $f^+ = \frac{|f|+f}{2}$ and $f^- = \frac{|f|-f}{2}$ are integrable on A (Theorem 5.2:1(a)). This proves that f^+ and f^- are locally integrable.

$\Leftarrow$ Assume that f^+ and f^- are locally integrable, meaning that f^+ is integrable on every closed $A \subseteq S$ – some U, f^- is integrable on every closed $A \subseteq S$ – some V. Suppose now that A is a closed subset of $S - (U \cup V)$. Then f^+ and f^- are integrable on A, so that $f = f^+ - f^-$ is integrable on A (Theorem 5.2:1(a)).

(c) f is integrable on $S \Leftrightarrow f$ is integrable on T^+, T^0 (automatic), and $T^- \Leftrightarrow f^+$ is integrable on T^+ and $-f$ is integrable on $T^- \Leftrightarrow f^+$ is integrable on S (because f^+ is automatically integrable on $S - T^+$, where $f^+ = 0$) and f^- is integrable on S (similarly). If true, then

$$\int_S f = \int_{T^+} f + \int_{T^0} f + \int_{T^-} f = \int_{T^+} f + 0 + \int_{T^-} f = \int_S f^+ - \int_S f^-.$$

(d) Dirichlet's function has $T^+ = \mathbf{Q}^n$, not LA.

Section 5.5

1. $\int_0^1 \sqrt{\left(\frac{dx_1}{dt}\right)^2 + \cdots + \left(\frac{dx_n}{dt}\right)^2}\, dt = \sqrt{(b_1 - a_1)^2 + \cdots + (b_n - a_n)^2}$.

2. (a) $\mathbf{g}(t) \equiv \left(2\left|t - \frac{1}{2}\right|, 0\right)$ has two linear pieces of length $= 1$. For an example whose endpoints are $(0, 0)$ and $(1, 0)$, set $\mathbf{h}(t) \equiv (2t, 0)$ if $0 \le t \le \frac{3}{8}$, $\equiv (\frac{3}{2} - 2t, 0)$ if $\frac{3}{8} \le t \le \frac{5}{8}$, $\equiv \left(2[t - \frac{1}{2}], 0\right)$ if $\frac{5}{8} \le t \le 1$; its three pieces have lengths $\frac{3}{4}, \frac{2}{4}, \frac{3}{4}$.

   (b) $\mathbf{g}(t) \equiv \left(t\left|\sin(\frac{\pi}{2t})\right|, 0\right)$ if $0 < t \le 1$, $\mathbf{g}(0) \equiv (0, 0)$. Adapt Example 1(a) for the argument that $\text{len}(\mathbf{g}) = \infty$.

   (c) There must exist a and b with $\mathbf{g}(a) = (0, 0)$, $\mathbf{g}(b) = (1, 0)$. Then the partition
   $$T = \{0, \min\{a, b\}, \max\{a, b\}, 1\}$$
   gives $\text{len}(\mathbf{g}, T) \ge \|\mathbf{g}(a) - \mathbf{g}(b)\| = 1$, forcing $\text{len}(\mathbf{g}) \ge 1$.

3. $\mathbf{g}(t) \equiv (t, h(t))$, $a \le t \le b$, is a smooth parametrization. By Theorem 3(d),
$$\text{len}(\mathbf{g}) = \int_a^b \sqrt{\left(\frac{dx}{dt}\right)^2 + \left(\frac{dy}{dt}\right)^2}\, dt = \int_a^b \sqrt{1 + h'(t)^2}\, dt.$$

4. (a) $x = r\cos\theta = \frac{\cos\theta}{\theta}$, $y = \frac{\sin\theta}{\theta}$ leads to
$$\int_{2\pi}^M \sqrt{\left(\frac{dx}{d\theta}\right)^2 + \left(\frac{dy}{d\theta}\right)^2}\, d\theta = \int_{2\pi}^M \sqrt{1 + \frac{1}{\theta^2}}\, d\theta > M - 2\pi \to \infty;$$

the length is not bounded, and the spiral is unrectifiable.

(b) $x = e^{-\theta}\cos\theta$, $y = e^{-\theta}\sin\theta$ leads to

$$\int_0^M \sqrt{\left(\frac{dx}{d\theta}\right)^2 + \left(\frac{dy}{d\theta}\right)^2}\, d\theta = \int_0^M \sqrt{2}e^{-\theta}\, d\theta \to \sqrt{2}$$

as $M \to \infty$. The spiral is rectifiable.

5. (a) It is simply the fundamental theorem. Write the distance as

$$L(t) \equiv \text{len}(\mathbf{g}) = \int_0^t \sqrt{g_1'(u)^2 + \cdots + g_n'(u)^2}\, du.$$

Then

$$L'(t) = \sqrt{g_1'(t)^2 + \cdots + g_n'(t)^2} = \|(g_1'(t), \ldots, g_n'(t))\| = \left\|\frac{d\mathbf{g}}{dt}\right\|.$$

(b) By (a), $L'(t)^2 = \mathbf{g}'(t) \bullet \mathbf{g}'(t)$. Derivatives of those quantities are $2L'(t)L''(t) = 2\mathbf{g}'(t) \bullet \mathbf{g}''(t)$ (dot-product rule). If speed, meaning L', is constant, then $0 = \mathbf{g}'(t) \bullet \mathbf{g}''(t)$, as desired.

6. (a) Let $\mathbf{g} = (g_1, \ldots, g_n)$. Then for each k and any partition $\{t_0 < \cdots < t_J\}$,

$$\begin{aligned}\sum_{j=1}^J |g_k(t_j) - g_k(t_{j-1})| &\le \sum_{j=1}^J \|\mathbf{g}(t_j) - \mathbf{g}(t_{j-1})\| \\ &\le \sum_{k=1}^N \sum_{j=1}^J |g_k(t_j) - g_k(t_{j-1})|.\end{aligned}$$

This shows that the middle sum is bounded iff each of the individual sums on the left is bounded.

(b) Clearly,

$$\sum_{j=1}^J |h(t_j) - h(t_{j-1})| \le K \sum_{j=1}^J |t_j - t_{j-1}| = K(b-a).$$

If h is differentiable and h' is bounded, then h is Lipschitz, owing to the mean value theorem.

(c) $H'(x) = 3x^{1/2}\sin(1/x)/2 - x^{-1/2}\cos(1/x)$ has values near $\pm\infty$ near $x = 0$. Let (u_i) be the sequence of extremes (critical points) of H, in the intervals $\frac{1}{(i+1)\pi} \le u_i \le \frac{1}{i\pi}$ (at whose ends $H = 0$). Given $T \equiv \{0 = t_0 < \cdots < t_J = 1\}$, assume $u_{k+1} \le t_1 < u_k$, and let $U = T \cup \{u_1, \dots, u_k\}$. Then $\mathrm{Var}(H, U)$, in reverse order, can be broken up into

$$[H(1) - H(u_1)] + [H(u_2) - H(u_1)] + \cdots + |H(u_k) - H(t_1)| \\ + |H(t_1)| \le 2 + \left(2 \sum_{i=1}^{\infty} |H(u_i)|\right) + 1.$$

Since $|H(u_i)| \le u_i^{3/2} \le \left[\frac{1}{i\pi}\right]^{3/2}$, the series converges, and $H \in BV$.

(d) First, some motivation: part (c) shows that $x^{3/2}\sin\left(\frac{1}{x}\right)$ has bounded variation, and clearly what is important there about $\frac{3}{2}$ is that it exceeds 1. We have also seen that $x\sin\left(\frac{1}{x}\right)$ has unbounded variation. The latter does not fit this question, because it is not differentiable at $x = 0$. It appears that we need something intermediate between x^1 and all the powers $x^{1+\varepsilon}$. Such a function is $\frac{x}{\ln x}$. (Compare Answer 2.2:7f.) Accordingly, set $G(x) \equiv \frac{x\sin(1/x)}{\ln(x/2)}$. It is differentiable on $[0, 1]$, because $\frac{G(x)-G(0)}{x} = \frac{\sin(1/x)}{\ln(x/2)} \to 0$ as $x \to 0$. Let

$$T \equiv \left\{0 < \frac{1}{2J\pi} < \frac{1}{(2J-0.5)\pi} < \cdots < \frac{1}{0.5\pi} < 1\right\}.$$

Then

$$\mathrm{Var}(G, T) = \frac{4}{(4J-1)\pi[\ln(4J-1)\pi]} + \frac{4}{(4J-3)\pi[\ln(4J-3)\pi]} \\ + \cdots + \frac{2}{\pi\ln[\pi]} + \left(\frac{\sin 1}{\ln 2} - \frac{2}{\pi\ln[\pi]}\right)$$

is unbounded, because $[\pi\ln\pi]^{-1} + [3\pi\ln 3\pi]^{-1} + \cdots$ diverges. Hence $G \notin BV$.

(e) If f is increasing, then

$$\mathrm{Var}(f, T) = \sum |f(t_{j+1}) - f(t_j)| = \sum [f(t_{j+1}) - f(t_j)] = f(b) - f(a)$$

for every partition. If f is decreasing, then similarly $\mathrm{Var}(f, T) = f(a) - f(b)$.

(f) $\mathrm{sgn}(x) \equiv \frac{|x|}{x}$ if $x \ne 0$, $\equiv 0$ at 0, has variation $= 2$ on $[-1, 1]$.

(g) See part (d).

7. (a) Yes. Set $f(t) \equiv t^2\sin\left(\frac{1}{t}\right)$, $f(0) \equiv 0$. At 0, f has zero derivative; elsewhere, $f'(t) = 2t\sin\left(\frac{1}{t}\right) - \cos\left(\frac{1}{t}\right)$ is bounded. By Exercise 6b, $f \in BV$. By Exercise 6a, $g(t) = (t, f(t))$ is rectifiable.

(b) Yes; $g(t) \equiv t^{3/2}\sin\left(\frac{1}{t}\right)$ is of bounded variation, by Exercise 6c.

8. (a) The argument is identical to that in Theorem 1, except that $\mathbf{h}^{-1}(\mathbf{g})$ is necessarily strictly decreasing. Therefore, $c = \mathbf{h}^{-1}(\mathbf{g}(t_J)) < \cdots < \mathbf{h}^{-1}(\mathbf{g}(t_1)) < \mathbf{h}^{-1}(\mathbf{g}(t_0)) = d$ is the partition, and $u_j \equiv \mathbf{h}^{-1}(\mathbf{g}(t_{J-j}))$, $j = 0, \ldots, J$.

   (b) By the argument in Theorem 1, $\mathbf{h}^{-1}(\mathbf{g})$ is an increasing map of $[a, b)$ to $[c, d)$. Hence if $a < t < b$, then there is a unique $t^* \in (c, d)$ with $\mathbf{h}(t^*) = \mathbf{g}(t)$. Thus, $\mathbf{g}$ on $[a, t]$ and $\mathbf{h}$ on $[c, t^*]$ are one-to-one, with same start, same end. By Theorem 1, $\text{len}(\mathbf{g}[a, t]) = \text{len}(\mathbf{h}[c, t^*])$. Let $t \to b$. Then $t^* = \mathbf{h}^{-1}(\mathbf{g}(t)) \to d$. By Theorem 2(c), $\text{len}(\mathbf{g}[a, t]) \to \text{len}(\mathbf{g})$, $\text{len}(\mathbf{h}[c, t^*]) \to \text{len}(\mathbf{h})$; hence $\text{len}(\mathbf{g}) = \text{len}(\mathbf{h})$.

9. (a)

$$
\begin{aligned}
X &= \frac{1}{\text{len}(C)} \int_C x\, ds \\
&= \frac{1}{2\pi (a^2 + b^2)^{1/2}} \int_0^{2\pi} a \cos t \left(a^2 + b^2\right)^{1/2} dt = 0,
\end{aligned}
$$

which makes sense, given the symmetry. Similarly, $Y = 0$.

$$
Z = \frac{\int_0^{2\pi} bt \left(a^2 + b^2\right)^{1/2} dt}{2\pi \left(a^2 + b^2\right)^{1/2}} = b\pi,
$$

which also makes sense, because half the helix is below $z = b\pi$, half above.

   (b) The mass center should be close to $(a, 0, 1)$, where the helix is densest.

$$
\begin{aligned}
X &= \frac{\int_C x\rho\, ds}{\int_C \rho\, ds} = \frac{\int_0^{2\pi} a \cos t e^{-bt} \left(a^2 + b^2\right)^{1/2} dt}{\int_0^{2\pi} e^{-bt} \left(a^2 + b^2\right)^{1/2} dt} \\
&= a \left[\frac{e^{-bt}}{b^2 + 1} (-b \cos t + \sin t) \right]_0^{2\pi} \Bigg/ \left[\frac{e^{-bt}}{-b} \right]_0^{2\pi} = \frac{b^2 a}{b^2 + 1} \approx a, \\
Y &= \frac{\int_0^{2\pi} a \sin t e^{-bt} \left(a^2 + b^2\right)^{1/2} dt}{\int_0^{2\pi} e^{-bt} \left(a^2 + b^2\right)^{1/2} dt} \\
&= a \left[\frac{e^{-bt}}{b^2 + 1} (-b \sin t - \cos t) \right]_0^{2\pi} \Bigg/ \left[\frac{e^{-bt}}{-b} \right]_0^{2\pi} = \frac{ba}{b^2 + 1}, \\
Z &= \frac{\int_0^{2\pi} bt e^{-bt} \left(a^2 + b^2\right)^{1/2} dt}{\int_0^{2\pi} e^{-bt} \left(a^2 + b^2\right)^{1/2} dt} \\
&= \left[\frac{1}{b} (-bt - 1) e^{-bt} \right]_0^{2\pi} \Bigg/ \left[\frac{e^{-bt}}{-b} \right]_0^{2\pi} = \frac{1 - (2\pi b + 1) e^{-2\pi b}}{1 - e^{-2\pi b}}.
\end{aligned}
$$

10. (a) With $x = g_1 \cos g_2$, $y = g_1 \sin g_2$,

$$\left(\frac{ds}{dt}\right)^2 = \left\| \left(\frac{dx}{dt}, \frac{dy}{dt}\right) \right\|^2$$
$$= \left\| \left(g_1' \cos g_2 - g_1 g_2' \sin g_2, g_1' \sin g_2 - g_1 g_2' \cos g_2\right) \right\|^2$$
$$= (g_1')^2 + (g_1 g_2')^2 .$$

(b) For this form, θ is the parameter, and $g_1(\theta) \equiv g(\theta)$, $g_2(\theta) \equiv \theta$. Hence

$$\left(\frac{ds}{d\theta}\right)^2 = g_1'(\theta)^2 + g_1(\theta)^2 g_2'(\theta)^2 = g'(\theta)^2 + g(\theta)^2.$$

11. The needed arguments are in:

(a) Answer 4.2:4a.
(b) Answer 4.2:4c, with Theorem 3(b).
(c) Answer 4.2:4b.
(d) Answer 4.2:4e.
(e) Theorem 5.2:1(d).
(f) Theorem 5.2:2(d).
(g) Theorem 5.2:3(b).

12. (a) The segment is given by $x = 2t$, $y = t$, $0 \le t \le 1$. Then

$$\int_C x^2 + y^2 \, ds = \int_0^1 5t^2(2^2 + 1^2)^{1/2} \, dt = \frac{5\sqrt{5}}{3},$$
$$\int_C x^2 + y^2 \, dx = \int_0^1 5t^2 2t \, dt = \frac{10}{3},$$
$$\int_C x^2 + y^2 \, dy = \int_0^1 5t^2 1 \, dt = 5.$$

(b) Same segment, opposite sense: $x = 2 - 2t$, $y = 1 - t$, $0 \le t \le 1$. The integrand and ds stay the same, $dx = -2dt$, $dy = -dt$. Hence the x-integral and y-integral change sign.

(c) $x = 2\cos\theta$, $y = 2\sin\theta$, $0 \le \theta \le \pi$. The integrand $x^2 + y^2 = 4$ is constant. Therefore,

$$\int_C x^2 + y^2 \, ds = 4(\text{length}) = 8\pi,$$
$$\int_C x^2 + y^2 \, dx = 4\big(x(\pi) - x(0)\big) = -16,$$
$$\int_C x^2 + y^2 \, dy = 4\big(y(\pi) - y(0)\big) = 0.$$

(d) Same equations as (c), $0 \leq \theta \leq 2\pi$, and again the integrand is constant. Hence

$$\int_C x^2 + y^2\, ds = 4(\text{length}) = 16\pi,$$

the others both zero.

13. (a) is an immediate consequence of operator boundedness and $-|f| \leq f \leq |f|$.
(b) See the x-integrals in Exercises 12b, c.
(c) Set $f(x, y) \equiv y$, and let C be the broken line from $(-1, -1)$ to $(0, 0)$ to $(-2, 1)$. Then

$$\int_C f\, dx = \int_{-1}^{0} x\, dx + \int_0^{-2} \frac{-x}{2}\, dx = \frac{-3}{2},$$

and $\frac{\int_C f\, dx}{x(\text{end}) - x(\text{start})} = \frac{3}{2}$ is not a value of f.

Section 5.6

1. We are looking at $z = x^2 + y^2$, over the disk D given by $x^2 + y^2 \leq 2$. The area is

$$\begin{aligned}\sigma &= \int_D \sqrt{1 + \left(\frac{\partial z}{\partial x}\right)^2 + \left(\frac{\partial z}{\partial y}\right)^2} = \int_D \sqrt{1 + 4x^2 + 4y^2} \\ &= \int_0^{2\pi} \int_0^{\sqrt{2}} \sqrt{1 + 4r^2}\, r\, dr\, d\theta \\ &= 2\pi \left[\frac{(1 + 4r^2)^{3/2}}{12}\right]_0^{\sqrt{2}} = \frac{13\pi}{3}.\end{aligned}$$

2. $\frac{\partial(x,y,z)}{\partial r} = [\cos\theta\ \sin\theta\ 2r]^t$, $\frac{\partial(x,y,z)}{\partial\theta} = [-r\sin\theta\ r\cos\theta\ 0]^t$. They are evidently orthogonal, so area $\left(\frac{\partial f}{\partial r}, \frac{\partial f}{\partial \theta}\right)$ is the product $\left(1 + 4r^2\right)^{1/2} r$ of their lengths. Then

$$\text{area}(\mathbf{f}) = \int_0^{2\pi} \int_0^{\sqrt{2}} \sqrt{1 + 4r^2}\, r\, dr\, d\theta,$$

same as (1).

3. The right triangle from $(0, 0, 0)$ to $(0, 0, z)$ to (x, y, z) has bottom angle

$$\alpha \equiv \cos^{-1} \frac{z}{\sqrt{x^2 + y^2 + z^2}} = \cos^{-1} \frac{z}{a}.$$

Hence $\sec\alpha = \frac{a}{z}$.

(a) $1+\left(\frac{\partial z}{\partial x}\right)^2+\left(\frac{\partial z}{\partial y}\right)^2 = 1+\left(\frac{-x}{z}\right)^2+\left(\frac{-y}{z}\right)^2 = \frac{z^2+x^2+y^2}{z^2} = \frac{a^2}{z^2}$. For $z > 0$, the square root is $\frac{a}{z}$.

(b) $\nabla F = (2x, 2y, 2z)$, $\|\nabla F\|^2 = 4\left(x^2+y^2+z^2\right) = 4a^2$, $\nabla F \bullet \mathbf{k} = 2z$. Hence $\frac{\|\nabla F\|}{|\nabla F \bullet \mathbf{e}_n|} = \frac{2a}{2z}$.

(c) Call the stated angle β. It is determined by

$$\left(\frac{\partial \mathbf{x}}{\partial \phi} \times \frac{\partial \mathbf{x}}{\partial \theta}\right) \bullet \mathbf{k} = \left\|\frac{\partial \mathbf{x}}{\partial \phi} \times \frac{\partial \mathbf{x}}{\partial \theta}\right\| \cos \beta.$$

Reading $\frac{\partial \mathbf{x}}{\partial \phi}$ and $\frac{\partial \mathbf{x}}{\partial \theta}$ from Example 1, we have

$$\left\|\frac{\partial \mathbf{x}}{\partial \phi} \times \frac{\partial \mathbf{x}}{\partial \theta}\right\| = \left\|\begin{vmatrix} \mathbf{i} & a\cos\theta\cos\phi & -a\sin\theta\sin\phi \\ \mathbf{j} & a\sin\theta\cos\phi & a\cos\theta\sin\phi \\ \mathbf{k} & -a\sin\phi & 0 \end{vmatrix}\right\| = a^2 \sin\phi$$

and $\left(\frac{\partial \mathbf{x}}{\partial \phi} \times \frac{\partial \mathbf{x}}{\partial \theta}\right) \bullet \mathbf{k} = a^2 \sin\phi\cos\phi$ (the cofactor of $\mathbf{k}$). Assuming $\sin\phi \neq 0$, meaning off the z-axis, the determining equation gives $\cos\beta = \cos\phi$. Hence $\sec\alpha$ matches $\sec\beta$.

4. (a) This is the area of the part of the plane $x = z$ within the cylinder $x^2 + y^2 \le a^2$. The plane makes a 45° angle with the xy-plane, so $\sec\alpha = \sqrt{2}$. Hence

$$\int \sec\alpha \, dx\, dy = \sqrt{2}(\text{area of circle}) = \pi a^2 \sqrt{2}.$$

(b) $\|\nabla G\| = \|(1, 0, -1)\| = \sqrt{2}$, $|\nabla G \bullet \mathbf{k}| = 1$, so

$$\int \frac{\|\nabla G\|}{|\nabla G \bullet \mathbf{k}|}\, dx\, dy = \pi a^2 \sqrt{2}.$$

(c) $\frac{\partial \mathbf{x}}{\partial r} = \begin{bmatrix} \cos\theta \\ \sin\theta \\ \cos\theta \end{bmatrix}$, $\frac{\partial \mathbf{x}}{\partial \theta} = \begin{bmatrix} -r\sin\theta \\ r\cos\theta \\ -r\sin\theta \end{bmatrix}$, so

$$\int \left\|\frac{\partial \mathbf{x}}{\partial r} \times \frac{\partial \mathbf{x}}{\partial \theta}\right\| dr\, d\theta = \int_0^{2\pi} \int_0^a \|(-r, 0, r)\|\, dr\, d\theta = \pi a^2 \sqrt{2}.$$

5. Our current definition is $\text{area}(\mathbf{g}) = \int_A \sqrt{\det\left[\frac{\partial \mathbf{g}}{\partial t_j} \bullet \frac{\partial \mathbf{g}}{\partial t_k}\right]} dt_1 \cdots dt_n$. By Theorem 5.3:1,

$$\sqrt{\det\left[\frac{\partial \mathbf{g}}{\partial t_j} \bullet \frac{\partial \mathbf{g}}{\partial t_k}\right]} = \operatorname{absdet}\left[\frac{\partial \mathbf{g}}{\partial t_1} \cdots \frac{\partial \mathbf{g}}{\partial t_n}\right] = \text{absolute Jacobian of } \mathbf{g}.$$

Therefore, by Theorem 5.3:5: (a) $\text{area}(\mathbf{g}) = \int_A |\text{Jacobian}| = \text{area}(\mathbf{g}(A))$;

(b)

$$\int_{\mathbf{g}(A)} f\,d\sigma \equiv \int_A f(\mathbf{g}(\mathbf{t}))\sqrt{\det\left[\frac{\partial \mathbf{g}}{\partial t_j} \bullet \frac{\partial \mathbf{g}}{\partial t_k}\right]}$$

$$= \int_A f(\mathbf{g}(\mathbf{t}))\,|\text{Jacobian}| = \int_{\mathbf{g}(A)} f(\mathbf{u})\,d\mathbf{u}.$$

The one-to-one condition is always needed. Applying to Example 3.2:1, Answer 3.2:1 gives Jacobian = 2. Hence

$$\text{area}(\mathbf{h}) \equiv \int_0^{3\pi/2}\int_1^2 2r\,dr\,d\theta = \frac{9\pi}{2}.$$

The range of **h** is the ring $1 \le r \le 2$, whose area is 3π. Thus, compared to the area of the range, area(**h**) is 1.5 times too high, because the parametrization **h** is "one-and-a-half-to-one."

6. The end on the left is a disk of radius $f(a)$, area $= \pi f(a)^2$; similarly, $\pi f(b)^2$ on the right. The surface may be described by $f(x) =$ distance from x-axis $= \sqrt{y^2+z^2}$, leading to $x = x$, $y = f(x)\cos\beta$, $z = f(x)\sin\beta$, for $a \le x \le b, 0 \le \beta \le 2\pi$. Then

$$\frac{\partial \mathbf{x}}{\partial x} \times \frac{\partial \mathbf{x}}{\partial \beta} = \begin{bmatrix} 1 \\ f'(x)\cos\beta \\ f'(x)\sin\beta \end{bmatrix} \times \begin{bmatrix} 0 \\ -f(x)\sin\beta \\ f(x)\cos\beta \end{bmatrix} = \begin{bmatrix} f'(x)f(x) \\ -f(x)\cos\beta \\ -f(x)\sin\beta \end{bmatrix},$$

so

$$\int \text{area}\left(\frac{\partial \mathbf{x}}{\partial x}, \frac{\partial \mathbf{x}}{\partial \beta}\right)$$

$$= \int_0^{2\pi}\int_a^b \sqrt{[f'(x)f(x)]^2 + f(x)^2\cos^2\beta + f(x)^2\sin^2\beta}\,dx\,d\beta$$

$$= \int_a^b 2\pi f(x)\sqrt{1+f'(x)^2}\,dx.$$

This integral matches $\int_a^b 2\pi f(x)\,ds$ along the curve; see Answer 5.5:3.

7. (a) Imagine painting the outside of the sphere with a layer of paint dr thick. Then the volume dV added to the ball is area × thickness $= 4\pi r^2\,dr$. Thus, $\frac{dV}{dr} =$ area.

(b) In $\mathbf{R}^4$, $V = \frac{\pi^2 r^4}{2}$ (Example 5.1:2) and $\frac{dV}{dr} = 2\pi^2 r^3 =$ area (Example 3 here).

(c) In $\mathbf{R}^5$, the volume of the ball is $\frac{8\pi^2 r^5}{15}$ (Answer 5.1:10). This suggests an area of $\frac{8\pi^2 r^4}{3}$. You should check that the usual parametrization (consult Answer 4.4:4) leads to

$$\sigma = \int_0^{\pi}\int_0^{\pi}\int_0^{\pi}\int_0^{2\pi} r^4 \sin^3 s \sin^2 t \sin u \, dv \, du \, dt \, ds$$

$$= 2\pi r^4 \left[-\cos s + \frac{\cos^3 s}{3}\right]_0^{\pi} \left[\frac{t}{2} - \frac{\sin 2t}{4}\right]_0^{\pi} [-\cos u]_0^{\pi} = \frac{8\pi^2 r^4}{3}.$$

8. In defining the cross product, we wrote $\mathbf{v}_2 \times \cdots \times \mathbf{v}_n = C_1\mathbf{e}_1 + \cdots + C_n\mathbf{e}_n$, where the C_j are cofactors of the first column in a formal determinant. Therefore, $\mathbf{v}_1 = v_{11}\mathbf{e}_1 + \cdots + \mathbf{v}_{1n}\mathbf{e}_n$ gives $\mathbf{v}_1 \bullet (\mathbf{v}_2 \times \cdots \times \mathbf{v}_n) = v_{11}C_1 + \cdots + v_{1n}C_n = \det[\mathbf{v}_1 \cdots \mathbf{v}_n]$. The last has, according to Theorem 5.3:2, absolute value matching the volume of the parallelepiped of $\mathbf{v}_1, \ldots, \mathbf{v}_n$.

Chapter 6

Section 6.1

1. Theorem 2 makes clear that the field does no work if the object moves along a circle centered at the origin, because then the field, pointing along the radius, is perpendicular to the path. Further, the symmetry of the field implies that the work from one such circle to another is the same along any radial path. Accordingly, you get the same work over the path from $\mathbf{p}$ radially to the circle of $\mathbf{q}$ (at point $\frac{\|\mathbf{q}\|}{\|\mathbf{p}\|}\mathbf{p}$), then counterclockwise α degrees to $\mathbf{q}$, whether you go directly ($0 \le \alpha < 360°$) or circuitously ($\alpha > 360°$).

2. Go from $\mathbf{p}$ vertically to the x-axis, along the x-axis to the vertical line of $\mathbf{q}$, vertically to $\mathbf{q}$. The work is zero, because $\mathbf{G}$ is perpendicular to the two verticals and $\mathbf{G} \equiv \mathbf{O}$ on the axis, no matter how you actually travel along the three parts.

3. It looks like a whirlpool, circulating counterclockwise with speed increasing directly with distance from the origin.

(a) Use the definition: $H_1 = -y$, $H_2 = x$; $x = a + t(c-a)$, $y = b + t(d-b) \Rightarrow dx = (c-a)dt, dy = (d-b)dt$;

$$\int H_1\,dx + H_2\,dy = \int_0^1 -(b+td-tb)(c-a) + (a+tc-ta)(d-b)\,dt = ad - bc.$$

(b) Use Theorem 2: On the straight part, $\mathbf{H} = 0\mathbf{i} + x\mathbf{j}$, $\mathbf{T} = \mathbf{i}$, work $=$ 0; on the circle, $\mathbf{H} = -y\mathbf{i} + x\mathbf{j}$, $\mathbf{T} =$ unit perpendicular to $(x, y) =$

$\frac{(-y,x)}{(y^2+x^2)^{1/2}} = \frac{(-y,x)}{4}$,

$$\int \mathbf{H} \bullet \mathbf{T}\, ds = \int \frac{(y^2+x^2)}{4} = 4(\text{length}) = 8\pi.$$

(c) As in (b), $\int \mathbf{H} \bullet \mathbf{T}\, ds = 4(\text{circumference}) = 32\pi$.

4. By Theorem 1,

$$\int \mathbf{F} \bullet d\mathbf{s} = \int_a^b \mathbf{F}(\mathbf{g}(t)) \bullet \mathbf{g}'(t)\, dt = \int_a^b \mathbf{F}(\mathbf{g}(t)) \bullet \mathbf{T}\|\mathbf{g}'(t)\|\, dt,$$

and by Theorem 5.5:3(d), the last is $\int_C \mathbf{F} \bullet \mathbf{T}\, ds$.

5. (a) The field is tangential, so the flux should be zero. In detail: The normal is along the radius, $\mathbf{N} = \frac{(x,y)}{\|(x,y)\|} = \frac{(x,y)}{a}$. Hence $\mathbf{H} \bullet \mathbf{N} = \frac{(-yx+xy)}{a} = 0$, and flux $= 0$.

(b) Think of $\mathbf{H}$ as air flow. For the part of the region (shaded in the figure below) near the origin, the air circulates slowly within the region, producing no flux. Further outward, the air enters (negative outward flux) from lower right, leaves (positive flux) at the same rate to upper left. Furthest outward, the same thing happens at higher speed. The net flux is zero.

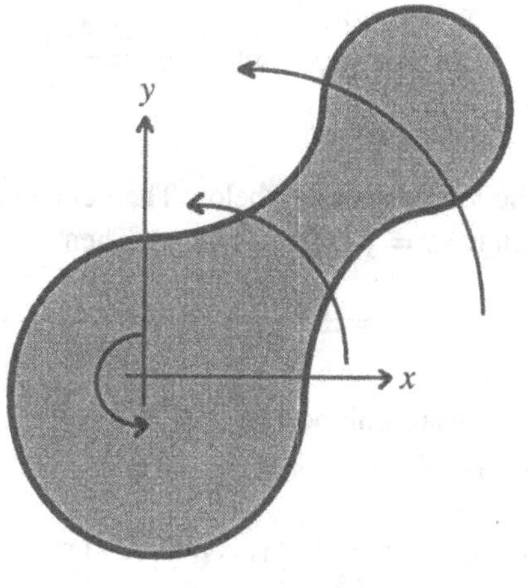

Exercise 5b.

6. (a) The upward normal is along the radius, $\mathbf{N} = \frac{(x,y,z)}{a}$. Hence

$$\int_S \mathbf{V} \bullet \mathbf{N}\, d\sigma = \int_S v\mathbf{k} \bullet \frac{(x, y, z)}{a}\, d\sigma$$
$$= \left(\frac{v}{a}\right) \int_S z\, ds = \frac{v}{a} \int_{x^2+y^2 \le a^2} z\sqrt{1 + \frac{x^2}{z^2} + \frac{y^2}{z^2}}\, dx\, dy$$
$$= \frac{v}{a} \int_{\text{disk}} a\, dx\, dy = \pi a^2 v.$$

(b) We have a uniform upward flow. To escape through the hemisphere, air must enter the upper half of the ball through the disk, to which the flow is perpendicular.

7. The symmetry allows us to find the flux at the face $z = a$, then multiply by 6. For $z = a$, $\mathbf{E} = Kq\frac{(x,y,a)}{(x^2+y^2+a^2)^{3/2}}$, $\mathbf{N} = \mathbf{k}$, $1 + \left(\frac{\partial z}{\partial x}\right)^2 + \left(\frac{\partial z}{\partial y}\right)^2 = 1$. Hence

$$\int \mathbf{E} \bullet \mathbf{N}\, ds = \int_{-a}^{a} \int_{-a}^{a} \frac{Kqa}{\left(x^2 + y^2 + a^2\right)^{3/2}}\, dx\, dy.$$

For this integral, we are best off if we let $0 \le \theta \le \frac{\pi}{4}$, $0 \le r \le \frac{a}{\cos\theta}$, evaluate, and multiply by 8:

$$\int_0^{\pi/4} \int_0^{a/\cos\theta} \frac{Kqa}{\left(r^2 + a^2\right)^{3/2}}\, r\, dr\, d\theta$$
$$= Kqa \int_0^{\pi/4} \left(\frac{1}{a} - \frac{1}{\sqrt{a^2/\cos^2\theta + a^2}}\right) d\theta$$
$$= Kq\left(\frac{\pi}{4} - \int_0^{\pi/4} \frac{\cos\theta}{\sqrt{2 - \sin^2\theta}}\, d\theta\right)$$
$$= Kq\left(\frac{\pi}{4} - \sin^{-1}\frac{\sin(\pi/4)}{\sqrt{2}}\right) = \frac{Kq\pi}{12}.$$

Multiply by 8 and 6 to get $4\pi Kq$.

8. (a) Consult the long discussion below Theorem 5.6:2. Suppose the parametrization is $x_n = f(x_1, \ldots, x_{n-1})$. Then

$$\mathbf{n} = \pm\left(\frac{\partial f}{\partial x_1}, \ldots, \frac{\partial f}{\partial x_{n-1}}, -1\right).$$

Hence the upward unit normal is $\mathbf{N} = \left(-\frac{\partial f}{\partial x_1}, \ldots, -\frac{\partial f}{\partial x_{n-1}}, 1\right) / \|\mathbf{n}\|$, $d\sigma = \|\mathbf{n}\| dx_1 \cdots dx_{n-1}$, and

$$\int_S \mathbf{V} \bullet \mathbf{N}\, ds = \int_A \mathbf{V}(x_1, \ldots x_{n-1}, f(x_1, \ldots, x_{n-1})) \bullet \left(-\frac{\partial f}{\partial x_1}, \ldots, -\frac{\partial f}{\partial x_{n-1}}, 1\right) dx_1 \cdots dx_{n-1}.$$

(b) Suppose the parametrization is $F(\mathbf{x}) = 0$, and $\frac{\partial F}{\partial x_n} = \nabla F \bullet \mathbf{e}_n > 0$. Then F increases in the positive x_n-direction. By (a), the flux to that direction is

$$\begin{aligned}
&\int_A \mathbf{V}(x_1, \ldots, x_{n-1}, x_n(x_1, \ldots, x_{n-1})) \\
&\qquad \bullet \left(-\frac{\partial x_n}{\partial x_1}, \ldots, -\frac{\partial x_n}{\partial x_{n-1}}, 1\right) dx_1 \cdots dx_{n-1} \\
&= \int_A \mathbf{V} \bullet \left(\frac{\partial F/\partial x_1}{\partial F/\partial x_n}, \ldots, \frac{\partial F/\partial x_{n-1}}{\partial F/\partial x_n}, 1\right) dx_1 \cdots dx_{n-1} \\
&= \int_A \mathbf{V} \bullet \frac{\nabla F}{\partial F/\partial x_n} dx_1 \cdots dx_{n-1},
\end{aligned}$$

with $\frac{\partial F}{\partial x_n} = \left|\frac{\partial F}{\partial x_n}\right|$. If $\frac{\partial F}{\partial x_n} < 0$, then F increases toward the negative x_n-direction, and

$$\begin{aligned}
\text{flux} &= \int_A \mathbf{V} \bullet \left(\frac{\partial x_n}{\partial x_1}, \ldots, \frac{\partial x_n}{\partial x_{n-1}}, -1\right) dx_1 \cdots dx_{n-1} \\
&= \int_A \mathbf{V} \bullet \frac{\nabla F}{-\partial F/\partial x_n} dx_1 \cdots dx_{n-1},
\end{aligned}$$

and $-\frac{\partial F}{\partial x_n} = \left|\frac{\partial F}{\partial x_n}\right|$.

9. (a) We know that $L(t) = \int_a^t \|\mathbf{g}'(u)\|\, du$. $L'(t) = \|\mathbf{g}'(t)\|$ is just the fundamental theorem, and $\|\mathbf{g}'(t)\| > 0$ because $\mathbf{g}$ is smooth.

(b) (c) First, (a) shows that L^{-1} exists. Write $t = L^{-1}(s)$. By the inverse function theorem, L^{-1} is differentiable, with $\frac{dL^{-1}(s)}{ds} = \frac{1}{L'(t)}$. Hence $\mathbf{G} = \mathbf{g}(L^{-1})$ is a differentiable composite, with

$$\frac{d\mathbf{G}}{ds} = \left(\frac{d\mathbf{g}}{dt}\right)\left(\frac{dt}{ds}\right) = \frac{\left(\frac{d\mathbf{g}}{dt}\right)}{L'(t)},$$

answering (c). Clearly, $\mathbf{G}'(s) \neq \mathbf{O}$, so $\mathbf{G}$ is smooth. Also, $\mathbf{G}(0) = \mathbf{g}(L^{-1}(0)) = \mathbf{g}(a)$, $\mathbf{G}(\text{len}(C)) = \mathbf{g}(b)$, and the range vector $\mathbf{g}(t)$ is $\mathbf{G}(L(t))$; that takes care of (b).

(d) Length of $\mathbf{G}$ is $\int_0^{\text{len}(C)} \|\mathbf{G}'(s)\|\, ds$. Make the substitution $s = L(t)$. Then

$$\begin{aligned}
\int_0^{\text{len}(C)} \|\mathbf{G}'(s)\|\, ds &= \int_a^b \left\|\frac{d\mathbf{g}/dt}{L'(t)}\right\| L'(t) dt \\
&= \int_a^b \|\mathbf{g}'(t)\|\, dt = \text{len}(C).
\end{aligned}$$

(e) By parts (c) and (a), $\frac{d\mathbf{G}}{ds} = \frac{\mathbf{g}'(t)}{L'(t)} = \frac{\mathbf{g}'(t)}{\|\mathbf{g}'(t)\|}$; the last is **T**.

10. (a) The definition is $K(s) = \|\mathbf{G}''(s)\|$. In terms of t, $\mathbf{G}'(s) = \frac{\mathbf{g}'(t)}{\|\mathbf{g}'(t)\|}$ (Exercise 9e). We calculate the s-derivative of $\mathbf{G}'(s)$ by the chain and quotient rules:

$$\mathbf{G}''(s) \equiv \frac{d\mathbf{G}'(s)}{ds} = \frac{d\mathbf{G}'(s)/dt}{ds/dt}$$
$$= \frac{\|\mathbf{g}'(t)\|\mathbf{g}''(t) - \mathbf{g}'(t)d\|\mathbf{g}'(t)\|/dt}{\|\mathbf{g}'(t)\|^2} \Big/ \|\mathbf{g}'(t)\|.$$

That establishes the desired denominator. From $\|\mathbf{g}'(t)\| = (\mathbf{g}' \bullet \mathbf{g}')^{1/2}$, we get

$$\frac{d\|\mathbf{g}'(t)\|}{dt} = \frac{1}{2}(\mathbf{g}' \bullet \mathbf{g}')^{-1/2} 2(\mathbf{g}' \bullet \mathbf{g}'') = \frac{(\mathbf{g}' \bullet \mathbf{g}'')}{\|\mathbf{g}'\|}.$$

The two-term numerator becomes $\|\mathbf{g}'\|\mathbf{g}'' - \frac{\mathbf{g}'(\mathbf{g}' \bullet \mathbf{g}'')}{\|\mathbf{g}'\|}$, whose dot product with itself works out to

$$\|\mathbf{g}'\|^2\mathbf{g}'' \bullet \mathbf{g}'' - (\mathbf{g}' \bullet \mathbf{g}'')^2 = \|\mathbf{g}'(t)\|^2\|\mathbf{g}''(t)\|^2 - [\mathbf{g}'(t) \bullet \mathbf{g}''(t)]^2.$$

This completes the proof.

(b) Recall that $\|\mathbf{g}' \times \mathbf{g}''\|$ is the area of their parallelogram, so

$$\|\mathbf{g}' \times \mathbf{g}''\| = \text{area}(\mathbf{g}', \mathbf{g}'') = \sqrt{\det \begin{bmatrix} \mathbf{g}' \bullet \mathbf{g}' & \mathbf{g}' \bullet \mathbf{g}'' \\ \mathbf{g}'' \bullet \mathbf{g}' & \mathbf{g}'' \bullet \mathbf{g}'' \end{bmatrix}}$$
$$= \|\mathbf{g}'(t)\|^2\|\mathbf{g}''(t)\|^2 - [\mathbf{g}'(t) \bullet \mathbf{g}''(t)]^2.$$

Fit this into part (a).

(c) On the line, $\mathbf{g}''(t) = \mathbf{O}$, so $K = 0$ by part (a).

(d) Yes. $0 = K(s) = \|\mathbf{G}''(s)\|$ forces $\mathbf{G}'(s) =$ constant row $= [c_1 \cdots c_n]$. Then mean value theorem gives

$$\mathbf{G}(u) - \mathbf{G}(0) = [c_1 \cdots c_n]\langle u - 0\rangle,$$

putting $\mathbf{G}(u)$ on the line through $\mathbf{G}(0)$ parallel to $(c_1, \ldots, c_n)$.

(e) Write

$$s(\theta) = \int_0^\theta \left(a^2 \sin^2 u + a^2 \cos^2 u\right)^{1/2} du = a\theta,$$

so that

$$\mathbf{G}(s) = (x, y) = \left(a \cos\left[\frac{s}{a}\right], a \sin\left[\frac{s}{a}\right]\right).$$

Then $\mathbf{G}''(s) = (-\cos[\frac{s}{a}], -\sin[\frac{s}{a}])/a$, and $\|\mathbf{G}''(s)\| = \frac{1}{a}$. If we switch to cosine and sine of e^t, then we get simply $s(t) = a(e^t - 1)$,

$$(x, y) = \left(a\cos\left[1+\frac{s}{a}\right], a\sin\left[1+\frac{s}{a}\right]\right).$$

Now $\mathbf{G}''(s) = (-\cos[1+\frac{s}{a}], -\sin[1+\frac{s}{a}])/a$, same norm. The curvature is independent of parametrization; it is intrinsic to the curve.

(f) The principle sounds right, but the helix is a counterexample: $s = (a^2+b^2)^{1/2}\,t$ (adapt Example 5.5:4(a)), making

$$(x, y, z) = \left(a\cos\frac{s}{\sqrt{a^2+b^2}}, a\sin\frac{s}{\sqrt{a^2+b^2}}, \frac{bs}{\sqrt{a^2+b^2}}\right),$$

and $\|\mathbf{G}''(s)\| = \frac{a}{a^2+b^2}$; constant curvature, but not a plane curve.

(g) This time, we may as well use part (a) on $\mathbf{g}(x) = (x, f(x))$. We find that $\mathbf{g}' = (1, f')$, $\mathbf{g}'' = (0, f'')$, and

$$K = \frac{\sqrt{(1+f'^2)(f''^2) - (f'f'')^2}}{(1+f'^2)^{3/2}} = \frac{|f''|}{(1+f'^2)^{3/2}}.$$

(h) Here $y' = \frac{-x}{(a^2-x^2)^{1/2}}$, $y'' = \frac{-a^2}{(a^2-x^2)^{3/2}}$. Part (g) gives $K = \frac{1}{a}$, as expected.

(i) We have $\mathbf{T} = \frac{(d\mathbf{g}/dt)}{(ds/dt)} = \frac{\mathbf{g}'}{\|\mathbf{g}'\|}$. Within part (a), we obtained

$$\frac{d\mathbf{T}}{ds} = \frac{d\mathbf{G}'}{ds} = \frac{\|\mathbf{g}'\|\mathbf{g}'' - \mathbf{g}'(\mathbf{g}' \bullet \mathbf{g}'')/\|\mathbf{g}'\|}{\|\mathbf{g}'\|^3} = \frac{\mathbf{g}''}{\|\mathbf{g}'\|^2} - \alpha\mathbf{g}'.$$

Therefore,

$$\mathbf{T} \times \frac{d\mathbf{T}}{ds} = \frac{\mathbf{g}'}{\|\mathbf{g}'\|} \times \left(\frac{\mathbf{g}''}{\|\mathbf{g}'\|^2} - \alpha\mathbf{g}'\right) = \frac{\mathbf{g}' \times \mathbf{g}''}{\|\mathbf{g}'\|^3},$$

so that $\|\mathbf{T} \times \frac{d\mathbf{T}}{ds}\| = \frac{\|\mathbf{g}' \times \mathbf{g}''\|}{\|\mathbf{g}'\|^3}$, which by (b) is K.

We also characterized the norm of a cross product as the area of its parallelogram: $\|\mathbf{T} \times \frac{d\mathbf{T}}{ds}\| = \text{area}\left(\mathbf{T}, \frac{d\mathbf{T}}{ds}\right)$. At the beginning of Section 5.6 we matched that area with $\|\mathbf{T}\|\,\|\frac{d\mathbf{T}}{ds}\|\sin\theta$, where θ is the angle between the vectors. The component of $\frac{d\mathbf{T}}{ds}$ perpendicular to $\mathbf{T}$ is precisely $\|\frac{d\mathbf{T}}{ds}\|\sin\theta$.

Section 6.2

1. On the broken line from $(0, 0)$ to $(a, 0)$ to (a, b),

$$\int \mathbf{G} \bullet \mathbf{T}\,ds = \int_0^a 0\mathbf{i} \bullet \mathbf{i}\,dx + \int_0^b y\mathbf{i} \bullet \mathbf{j}\,dy = 0,$$

and clearly $\mathbf{G} \neq \nabla 0$.

(b) Assume $a > 0$ and $b \geq 0$, and write $c^2 = a^2 + b^2$. Go from $(0, 0)$ straight to $(c, 0)$, then counterclockwise around the radius-c circle to (a, b). Then

$$\int \mathbf{H} \bullet d\mathbf{s} = \int_0^c (0\mathbf{i} + x\mathbf{j}) \bullet \mathbf{i}\, dx + \int_0^{\tan^{-1}(b/a)} (-y, x) \bullet \frac{(-y, x)}{c}\, ds$$

$$= c(\text{length of arc}) = \left(a^2 + b^2\right) \tan^{-1}\left(\frac{b}{a}\right).$$

But

$$\nabla\left([x^2 + y^2] \tan^{-1}\left[\frac{y}{x}\right]\right)$$

$$= \left(-y + 2x \tan^{-1}\left(\frac{y}{x}\right), x + 2y \tan^{-1}\left(\frac{y}{x}\right)\right) \neq \mathbf{H}.$$

2. (a) $G_1 = y$, $G_2 = 0$, and $\frac{\partial G_1}{\partial y} = 1 \neq 0 = \frac{\partial G_2}{\partial x}$.

   (b) $H_1 = -y$, $H_2 = x$, and $\frac{\partial H_1}{\partial y} = -1 \neq 1 = \frac{\partial H_2}{\partial x}$.

3. Yes. Can you tell by inspection: $2xy\mathbf{i} + x^2\mathbf{j} = \nabla\left(x^2 y\right)$? If not, integrate from $(0, 0)$ to $(x, 0)$ to (x, y):

$$\int \mathbf{F} \bullet d\mathbf{s} = \int_0^x x^2 \mathbf{j} \bullet \mathbf{i}\, dt + \int_0^y \left(2xt\mathbf{i} + x^2\mathbf{j}\right) \bullet \mathbf{j}\, dt = x^2 y.$$

4. and 5. Travel from $\mathbf{a} \equiv (a_1, \ldots, a_n)$ radially to the circle of radius $\|\mathbf{b}\|$, then around to $\mathbf{b}$. On the circle, the sense is immaterial, since the field is orthogonal to the tangent. On the radial segment from $\mathbf{a}$ to $\frac{\|\mathbf{b}\|\mathbf{a}}{\|\mathbf{a}\|}$, we have $\mathbf{F} = f(r)\mathbf{x}$, $\mathbf{T} = \frac{\mathbf{x}}{r}$, and $ds = dr$. Hence $\int \mathbf{F} \bullet \mathbf{T}\, ds = \int_{\|\mathbf{a}\|}^{\|\mathbf{b}\|} f(r) r\, dr$. This proves (5), which implies (4).

6. If $\mathbf{F} = f_1(x_1)\mathbf{e}_1 + \cdots + f_n(x_n)\mathbf{e}_n$, then

$$\int_C \mathbf{F} \bullet d\mathbf{s} = \int_C f_1(x_1)dx_1 + \cdots + f_n(x_n)dx_n$$

is, by Theorem 5(b), the sum of PI integrals.

Section 6.3

1. By Green's theorem,

$$\int_C -y\, dx = \int_A \left(\frac{\partial 0}{\partial x} - \frac{\partial(-y)}{\partial y}\right) = \text{area}(A),$$

$$\int_C x\, dy = \int_A 1 = \text{area}(A),$$

$$\int_C x\, dy - y\, dx = \int_A 2 = 2\ \text{area}(A).$$

2. (a) At each point (x, y), $-y\mathbf{i}+x\mathbf{j}$ is normal to the radial segment from the origin. Therefore, $(-y\mathbf{i}+x\mathbf{j})\bullet ds$ is the component of ds perpendicular to that radius. This component determines the rate at which the line from $\mathbf{O}$ to (x, y) sweeps out area as (x, y) moves. In the figure for this exercise, the shaded triangle has area roughly

$$d\sigma^{+} = \frac{1}{2}\|(x, y)\| \left(ds \bullet \frac{-y\mathbf{i}+x\mathbf{j}}{\|-y\mathbf{i}+x\mathbf{j}\|} \right) = \frac{1}{2}(-y\,dx + x\,dy).$$

The integral of this quantity is the area of the region from the origin to the remote part of A minus the area of the region from O to the near part of A. Along this near part, $(-y\mathbf{i}+x\mathbf{j}) \bullet ds$ is negative, and the integration subtracts the area of the part of the shaded triangle located below the letter "C."

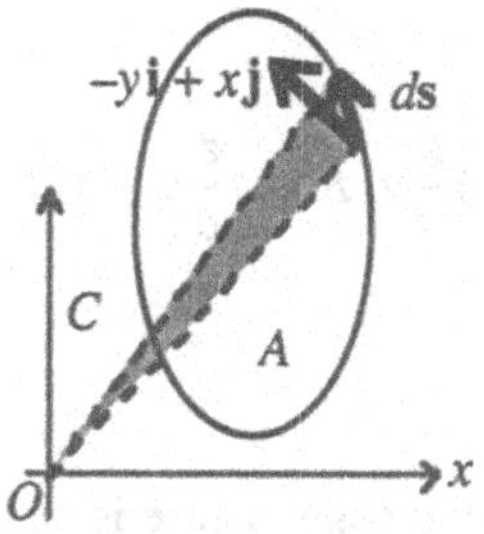

Exercise 2a.

(b) If the curve has $r = f(\theta)$, then it is parametrized by $x = f(\theta)\cos\theta$, $y = f(\theta)\sin\theta, 0 \le \theta \le 2\pi$. The relation becomes

$$\begin{aligned}
\text{area}(A) &= \frac{1}{2}\int_C x\,dy - y\,dx \\
&= \frac{1}{2}\int_0^{2\pi} f(\theta)\cos\theta[f'(\theta)\sin\theta + f(\theta)\cos\theta] \\
&\quad - f(\theta)\sin\theta[f'(\theta)\cos\theta - f(\theta)\sin\theta]d\theta \\
&= \int_0^{2\pi} \frac{f(\theta)^2}{2}\,d\theta\,.
\end{aligned}$$

3. The counterclockwise sense in $\mathbf{R}^2$ around $x^2+y^2 = 4$ goes with the normal in the positive z-direction; we use counterclockwise tangent and upward normal. Along the edge, $\mathbf{F} = -y\mathbf{i} + x\mathbf{j} + 0\mathbf{k}$ is tangent to the circle, so

$$\int_C \mathbf{F} \bullet \mathbf{T}\,ds = \|\mathbf{F}\|\text{ length} = 2(2\pi 2) = 8\pi.$$

On the surface,

$$\nabla \times \mathbf{F} = \begin{vmatrix} \mathbf{i} & \frac{\partial}{\partial x} & -y \\ \mathbf{j} & \frac{\partial}{\partial y} & x \\ \mathbf{k} & \frac{\partial}{\partial z} & e^z \ln(1+z) \end{vmatrix} = 2\mathbf{k}$$

and $d\boldsymbol{\sigma} = \left(-\frac{\partial z}{\partial x}, -\frac{\partial z}{\partial y}, 1\right) dx\, dy$, so

$$\int_S (\nabla \times \mathbf{F}) \bullet d\boldsymbol{\sigma} = \int_A 2dx\, dy = 2(\text{area of disk}) = 8\pi.$$

4. Recall that $\rho \equiv \|\mathbf{x}\|$ has $\frac{\partial \rho}{\partial x_k} = \frac{x_k}{\rho}$. Then

$$\begin{aligned} \nabla \times \mathbf{F} &= \begin{vmatrix} \mathbf{i} & \frac{\partial}{\partial x} & F(\rho)x \\ \mathbf{j} & \frac{\partial}{\partial y} & F(\rho)y \\ \mathbf{k} & \frac{\partial}{\partial z} & F(\rho)z \end{vmatrix} \\ &= \left(F'(\rho)\frac{y}{\rho}z - F'(\rho)\frac{z}{\rho}y, F'(\rho)\frac{z}{\rho}x - F'(\rho)\frac{x}{\rho}z, F'(\rho)\frac{x}{\rho}y - F'(\rho)\frac{y}{\rho}x\right) \\ &= \mathbf{O}, \end{aligned}$$

as expected.

5. Think in terms of $\mathbf{R}^2$; the general case is similar. In the first quadrant, the field will exert a counterclockwise torque on the lower right edge of the pinwheel, and an equally strong, oppositely directed torque on the upper left. The net torque will be zero. The same happens outside the first quadrant, owing to the symmetry.

6. The **i**-component of $\nabla \times (f\mathbf{F})$ is

$$\begin{aligned} \frac{\partial(fF_3)}{\partial y} - \frac{\partial(fF_2)}{\partial z} &= f\frac{\partial F_3}{\partial y} + \frac{\partial f}{\partial y}F_3 - f\frac{\partial F_2}{\partial z} - \frac{\partial f}{\partial z}F_2 \\ &= f\left(\frac{\partial F_3}{\partial y} - \frac{\partial F_2}{\partial z}\right) + \left(\frac{\partial f}{\partial y}F_3 - \frac{\partial f}{\partial z}F_2\right), \end{aligned}$$

which is the **i**-component of $f(\nabla \times \mathbf{F}) + (\nabla f) \times \mathbf{F}$. We may similarly check the other components.

Section 6.4

1. (a) $\nabla \bullet \mathbf{F} = \frac{\partial x}{\partial x} + \frac{\partial y}{\partial y} + \frac{\partial z}{\partial z} = 3$, so $\int_A \nabla \bullet \mathbf{F} = 3V(A) = 3\pi a^2 b$.

   (b) The top of the cylinder is given by $z = b$, so that $\mathbf{N} = \mathbf{k}$,

$$\int_{\text{top}} \mathbf{F} \bullet \mathbf{N}\, d\sigma = \int_{\text{top}} z\, d\sigma = b\left(\pi a^2\right).$$

On the bottom, $\mathbf{N} = -\mathbf{k}$, so $\mathbf{F} \bullet \mathbf{N} = -z = 0$. On the curved part, the normal at (x, y, z) is the radius from the z-axis, $\frac{(x,y,0)}{(x^2+y^2)^{1/2}}$. Hence

$$\int_{\text{side}} \mathbf{F} \bullet \mathbf{N} d\sigma = \int_{\text{side}} (x, y, z) \bullet \frac{(x, y, 0)}{\left(x^2 + y^2\right)^{1/2}} d\sigma$$
$$= a \text{ area(side)} = a\, 2\pi ab.$$

The total flux is $3\pi a^2 b$. The match is predicted by the divergence theorem.

2. The defining inequalities require $x^2 + y^2 \leq 4$; otherwise, the square root is biggest.

(a) Again $\nabla \bullet \mathbf{F} = 3$, so $\int_A \nabla \bullet \mathbf{F} = 3V(A)$. The volume is

$$\int_A 1 = \int_{x^2+y^2\leq 4} \int_{\sqrt{5-x^2-y^2}}^{5-x^2-y^2} dz\, dy\, dx$$
$$= \int_0^{2\pi} \int_0^2 \left(5 - r^2 - \sqrt{5 - r^2}\right) r\, dr\, d\theta = \frac{2\pi\left(19 - 5^{3/2}\right)}{3}.$$

That makes $\int_A \nabla \bullet \mathbf{F} = 2\pi\left(19 - 5^{3/2}\right)$.

(b) The paraboloid $z = 5 - x^2 - y^2$ has $d\boldsymbol{\sigma} = \left(-\frac{\partial z}{\partial x}, -\frac{\partial z}{\partial y}, 1\right) dx\, dy$, giving

$$\int_{\text{top}} \mathbf{F} \bullet d\boldsymbol{\sigma} = \int_{x^2+y^2\leq 4} (x, y, z) \bullet (2x, 2y, 1)\, dx\, dy$$
$$= \int_0^{2\pi} \int_0^2 \left(r^2 + 5\right) r\, dr\, d\theta = 28\pi.$$

The hemisphere has $d\boldsymbol{\sigma} = \left(\frac{\partial z}{\partial x}, \frac{\partial z}{\partial y}, -1\right) dx\, dy$ (downward normal), so

$$\int_{\text{bottom}} \mathbf{F} \bullet d\boldsymbol{\sigma} = \int_{x^2+y^2\leq 4} (x, y, z) \bullet \left(-\frac{x}{z}, -\frac{y}{z}, -1\right) dx\, dy$$
$$= \int_{x^2+y^2\leq 4} \frac{-x^2 - y^2 - z^2}{z} dx\, dy$$
$$= \int_0^{2\pi} \int_0^2 -\frac{5}{\sqrt{5 - r^2}} r\, dr\, d\theta = 10\pi\left(1 - \sqrt{5}\right).$$

Total flux is $38\pi - 10\pi\sqrt{5}$.

We are dealing with a simple region; the match is as expected.

3. (a) We have $\nabla \bullet \mathbf{F} = \sum_{j=1}^{n} \frac{\partial(-kx_j/r^3)}{\partial x_j} = \frac{-k}{r^6}\sum_{j=1}^{n}\left(r^3 - \frac{x_j 3r^2 x_j}{r}\right) = \frac{k(3-n)}{r^3}$, $r > 0$. If $n = 2$, then $\nabla \bullet \mathbf{F} = \frac{k}{r^3}$ is not integrable. If $n = 3$, then $\nabla \bullet \mathbf{F} = 0$ except at one point, so that $\int_B \nabla \bullet \mathbf{F} = 0$. Finally, if $n \geq 4$, we may integrate based on the differential $dV =$ area(r-sphere) dr (see Answer 5.6:7) $= r^{n-1}$area(1-sphere) dr. Thus,

$$\int_B \nabla \bullet \mathbf{F} = \int_0^a \frac{k(3-n)}{r^3} r^{n-1} \text{area}(S_1)dr = -ka^{n-3}\text{area}(S_1).$$

(b) On the sphere, $d\boldsymbol{\sigma} = \frac{\mathbf{x}}{a}\, d\sigma$, so

$$\begin{aligned}\int_{\text{bd}(B)} \mathbf{F} \bullet d\boldsymbol{\sigma} &= \int_{\text{bd}(B)} -\frac{k\mathbf{x}}{a^3} \bullet \frac{\mathbf{x}}{a}\, d\sigma \\ &= -\frac{k}{a^2}\text{ area(sphere)} = -ka^{n-3}\text{area}(S_1).\end{aligned}$$

Here we had no reason to expect a match, because B is not a subset of the domain of $\mathbf{F}$. They do match, however, for $n \geq 4$. We leave it to the reader to consider why.

4. The boundary of A is the 2-sphere $S(\mathbf{O}, 2)$, with normals pointing away from the origin, together with the 1-sphere $S(\mathbf{O}, 1)$, normals pointing toward the origin. Write B for the unit ball, oriented in the standard way, with normals pointing outward. We have

$$\int_{\text{bd}(A)} \mathbf{F} \bullet d\boldsymbol{\sigma} + \int_{\text{bd}(B)} \mathbf{F} \bullet d\boldsymbol{\sigma} = \int_{S(\mathbf{O},2)} \mathbf{F} \bullet d\boldsymbol{\sigma},$$

because the flux over the inner border of A is the negative of the flux out of B. Next,

$$\int_{S(\mathbf{O},2)} \mathbf{F} \bullet d\boldsymbol{\sigma} = \int_{B(\mathbf{O},2)} \nabla \bullet \mathbf{F}$$

by Gauss's theorem, and $\int_{B(\mathbf{O},2)} \nabla \bullet \mathbf{F} = \int_A \nabla \bullet \mathbf{F} + \int_B \nabla \bullet \mathbf{F}$ by set additivity. Finally, $\int_B \nabla \bullet \mathbf{F} = \int_{\text{bd}(B)} \mathbf{F} \bullet d\boldsymbol{\sigma}$ (Gauss). Consequently, $\int_{\text{bd}(A)} \mathbf{F} \bullet d\boldsymbol{\sigma} = \int_A \nabla \bullet \mathbf{F}$.

5. Straightforward: By Gauss's theorem, $\int_{\text{bd}(A)} (\nabla \times \mathbf{F}) \bullet d\boldsymbol{\sigma} = \int_A \nabla \bullet (\nabla \times \mathbf{F})$, and

$$\begin{aligned}\nabla \bullet (\nabla \times \mathbf{F}) &= \frac{\partial\left(\frac{\partial F_3}{\partial y} - \frac{\partial F_2}{\partial z}\right)}{\partial x} + \frac{\partial\left(\frac{\partial F_1}{\partial z} - \frac{\partial F_3}{\partial x}\right)}{\partial y} + \frac{\partial\left(\frac{\partial F_2}{\partial x} - \frac{\partial F_1}{\partial y}\right)}{\partial z} \\ &= \frac{\partial^2 F_3}{\partial y \partial x} - \frac{\partial^2 F_2}{\partial z \partial x} + \frac{\partial^2 F_1}{\partial z \partial y} - \frac{\partial^2 F_3}{\partial x \partial y} + \frac{\partial^2 F_2}{\partial x \partial z} - \frac{\partial^2 F_1}{\partial y \partial z} = 0\end{aligned}$$

as long as the mixed partials are symmetric.

Tricky: Draw the "equator" $z = \frac{H(x,y)+h(x,y)}{2}$, for (x, y) in the boundary of the plane region that defines A. Going eastward, this curve is the boundary of the upper half U of the surface bd(A), so by Stokes' theorem,

$$\int_U (\nabla \times \mathbf{F}) \bullet d\sigma = \int_{\text{equator east}} \mathbf{F} \bullet d\mathbf{s}.$$

Going westward, the same curve is the edge of the lower half L of bd(A), so that

$$\int_L (\nabla \times \mathbf{F}) \bullet d\sigma = \int_{\text{equator west}} \mathbf{F} \bullet d\mathbf{s}.$$

Hence

$$\int_{\text{bd}(A)} (\nabla \times \mathbf{F}) \bullet d\sigma = \int_{\text{east}} + \int_{\text{west}} = 0.$$

6. (a) Undefined; you may not apply ∇ to a vector.

   (b) Defined and equals $\mathbf{O}$. We can calculate directly. Alternatively, in a convex open set, use Theorems 6.2:2 and 6.3:6.

   (c) Defined;

$$\nabla \bullet (\nabla f) = \frac{\partial(\partial f/\partial x)}{\partial x} + \frac{\partial(\partial f/\partial y)}{\partial y} + \frac{\partial(\partial f/\partial z)}{\partial z}$$
$$= \frac{\partial^2 f}{\partial x^2} + \frac{\partial^2 f}{\partial y^2} + \frac{\partial^2 f}{\partial z^2}.$$

   This is an important construction in mathematical physics, beginning with one type of flow: heat. [See Kline, p. 672 among others.]

   (d) Undefined, just as (a).

   (e) Defined. It is inconvenient to work with, but we may show that its $\mathbf{i}$ component is

$$\frac{\partial^2 F_2}{\partial x \partial y} - \frac{\partial^2 F_1}{\partial y^2} - \frac{\partial^2 F_1}{\partial z^2} + \frac{\partial^2 F_3}{\partial x \partial z} = \frac{\partial}{\partial x}\left(\frac{\partial F_1}{\partial x} + \frac{\partial F_2}{\partial y} + \frac{\partial F_3}{\partial z}\right) - \left(\frac{\partial^2}{\partial x^2} + \frac{\partial^2}{\partial y^2} + \frac{\partial^2}{\partial z^2}\right) F_1.$$

   Doing likewise with the $\mathbf{j}$ and $\mathbf{k}$ components, we have the symbols $\nabla \times (\nabla \times \mathbf{F}) = \nabla(\nabla \bullet \mathbf{F}) - (\nabla \bullet \nabla)\mathbf{F}$.

   (f) Defined, $= 0$ by Exercise 5.

   (g) Defined, no special form or significance.

   (h) Undefined; you may not apply $\nabla \times$ to a scalar.

   (i) Undefined; you may not apply $\nabla \bullet$ to a scalar.

7. (a)

$$\begin{aligned}
&\nabla \bullet (F_2G_3 - F_3G_2, F_3G_1 - F_1G_3, F_1G_2 - F_2G_1) \\
&= \frac{\partial F_2}{\partial x}G_3 + F_2\frac{\partial G_3}{\partial x} - \frac{\partial F_3}{\partial x}G_2 - F_3\frac{\partial G_2}{\partial x} \\
&+ \frac{\partial F_3}{\partial y}G_1 + F_3\frac{\partial G_1}{\partial y} - \frac{\partial F_1}{\partial y}G_3 \\
&- F_1\frac{\partial G_3}{\partial y} + \frac{\partial F_1}{\partial z}G_2 + F_1\frac{\partial G_2}{\partial z} - \frac{\partial F_2}{\partial z}G_1 - F_2\frac{\partial G_1}{\partial z} \\
&= G_1\left(\frac{\partial F_3}{\partial y} - \frac{\partial F_2}{\partial z}\right) + G_2\left(\frac{\partial F_1}{\partial z} - \frac{\partial F_3}{\partial x}\right) + G_3\left(\frac{\partial F_2}{\partial x} - \frac{\partial F_1}{\partial y}\right) \\
&- F_1\left(\frac{\partial G_3}{\partial y} - \frac{\partial G_2}{\partial z}\right) - F_2\left(\frac{\partial G_1}{\partial z} - \frac{\partial G_3}{\partial x}\right) - F_3\left(\frac{\partial G_2}{\partial x} - \frac{\partial G_1}{\partial y}\right) \\
&= (\nabla \times \mathbf{F}) \bullet \mathbf{G} - \mathbf{F} \bullet (\nabla \times \mathbf{G}).
\end{aligned}$$

(b)

$$\begin{aligned}
\nabla \bullet (f\mathbf{G}) &= \frac{\partial fG_1}{\partial x} + \frac{\partial fG_2}{\partial y} + \frac{\partial fG_3}{\partial z} \\
&= \frac{f\partial G_1}{\partial x} + \frac{f\partial G_2}{\partial y} + \frac{f\partial G_3}{\partial z} + G_1\frac{\partial f}{\partial x} + G_2\frac{\partial f}{\partial y} + G_3\frac{\partial f}{\partial z} \\
&= f(\nabla \bullet \mathbf{G}) + \mathbf{G} \bullet \nabla f.
\end{aligned}$$

References

Alan F. Beardon, *Limits: A New Approach to Real Analysis*, Springer-Verlag, New York, 1997.

This interesting book is a development of one-variable analysis founded entirely on the notion of limit. Its Chapter 6 gives strictly analytic definitions for the exponential and trigonometric functions. The context there is functions of a complex variable, where we have no experience. Still, the proof we cited—Theorem 6.3.1 on page 90—uses from complex variables only the formulas for sine and cosine of a sum. Those formulas are derivable from just rearrangement of real Taylor series.

R.C. Buck, *Advanced Calculus*, 2nd edition, McGraw Hill, New York 1965.

Originally published in 1956, "Buck" (no further identification was needed) is a classic, and the reader is urged to look there for a breadth of coverage that we could not hope to match.

Alberto Guzman, *Continuous Functions of Vector Variables*, Birkhäuser Boston, 2002.

This is a favorite of ours. The material most needed from it is from Chapters 4–5, on properties of continuous functions and the topology of Euclidean space.

Morris Kline, *Mathematical Thought from Ancient to Modern Times*, Oxford University Press, New York, 1972.

Kline's book is a monumental achievement among histories of the sciences and mathematics. Our focus on calculus restricts our interest in history to only Europeans during (mostly) the nineteenth century. However, Kline is a wonderful source of information about the development of mathematical ideas worldwide and over thousands of years.

David C. Lay, *Linear Algebra and Its Applications*, 2nd edition, Addison Wesley Longman, Reading MA, 1997.

Lay is a good source for material we have assumed from elementary linear algebra.

L. Mirsky, *An Introduction to Linear Algebra*, Clarendon Press, Oxford, 1995, reprinted by Dover, Mineola, NY, 1990.

E.R. Peck, *Electricity and Magnetism*, McGraw Hill, New York, 1953.

This is a sentimental favorite; the author was taught from it by a wonderful teacher named Fred Rose.

Kenneth A. Ross, *Elementary Analysis: The Theory of Calculus*, Springer-Verlag, New York, 1980.

This is the best introduction to advanced calculus we know. The subtitle suggests that Ross's mission is an axiomatization of elementary calculus. The book carries out that mission with an admirable combination of mathematical rigor and attention to pedagogy. It has all the material we require with respect to functions of one variable, together with (Section 13) much of the topology of Euclidean space.

Index